AF613903

Physics and Applications of Non-Crystalline Semiconductors in Optoelectronics

NATO ASI Series

Advanced Science Institutes Series

A Series presenting the results of activities sponsored by the NATO Science Committee, which aims at the dissemination of advanced scientific and technological knowledge, with a view to strengthening links between scientific communities.

The Series is published by an international board of publishers in conjunction with the NATO Scientific Affairs Division

A Life Sciences **B Physics**	Plenum Publishing Corporation London and New York
C Mathematical and Physical Sciences **D Behavioural and Social Sciences** **E Applied Sciences**	Kluwer Academic Publishers Dordrecht, Boston and London
F Computer and Systems Sciences **G Ecological Sciences** **H Cell Biology** **I Global Environmental Change**	Springer-Verlag Berlin, Heidelberg, New York, London, Paris and Tokyo

PARTNERSHIP SUB-SERIES

1. Disarmament Technologies	Kluwer Academic Publishers
2. Environment	Springer-Verlag / Kluwer Academic Publishers
3. High Technology	Kluwer Academic Publishers
4. Science and Technology Policy	Kluwer Academic Publishers
5. Computer Networking	Kluwer Academic Publishers

The Partnership Sub-Series incorporates activities undertaken in collaboration with NATO's Cooperation Partners, the countries of the CIS and Central and Eastern Europe, in Priority Areas of concern to those countries.

NATO-PCO-DATA BASE

The electronic index to the NATO ASI Series provides full bibliographical references (with keywords and/or abstracts) to more than 50000 contributions from international scientists published in all sections of the NATO ASI Series.
Access to the NATO-PCO-DATA BASE is possible in two ways:

– via online FILE 128 (NATO-PCO-DATA BASE) hosted by ESRIN, Via Galileo Galilei, I-00044 Frascati, Italy.

– via CD-ROM "NATO-PCO-DATA BASE" with user-friendly retrieval software in English, French and German (© WTV GmbH and DATAWARE Technologies Inc. 1989).

The CD-ROM can be ordered through any member of the Board of Publishers or through NATO-PCO, Overijse, Belgium.

3. High Technology – Vol. 36

Physics and Applications of Non-Crystalline Semiconductors in Optoelectronics

edited by

Andrei Andriesh

Moldavian Academy of Sciences,
Centre of Optoelectronics,
Kishinau, Republic of Moldova

and

Mario Bertolotti

Department of Energetics,
University of Rome "La Sapienza",
Rome, Italy

Springer-Science+Business Media, B.V.

Proceedings of the NATO Advanced Research Workshop on
Physics and Applications of Non-Crystalline Semiconductors in Optoelectronics
Kishinau, Moldova
October 1996

A C.I.P. Catalogue record for this book is available from the Library of Congress.

DOI 10.1007/978-94-011-5496-3

Printed on acid-free paper

CONTENTS

PREFACE

The Workshop on Physics and Application of Non-crystalline Semiconductors in Optoelectronics was held from 15 to 17 October 1996 in Chisinau, republic of Moldova and was devoted to the problems of non-crystalline semiconducting materials. The reports covered two main topics: theoretical basis of physics of non -crystalline materials and experimental results.

In the framework of these major topics there were treated many subjects, concerning the physics of non-crystalline semiconductors and their specific application:

- optical properties of non-crystalline semiconductors;
- doping of glassy semiconductors and photoinduced effects in chalcogenide glasses and their application for practical purposes;
- methods for investigation of the structure in non-crystalline semiconductors
- new glassy materials for IR trasmittance and optoelectronics.

Reports and communications were presented on various aspects of the theory, new physical principles, studies of the atomic structure, search and development of optoelectronics devices. Special attention was paid to the actual subject of photoinduced transformations and its applications.

Experimental investigations covered a rather wide spectrum of materials and physical phenomena. As a novel item it is worth to mention the study of nonlinear optical effects in amorphous semiconducting films. The third order optical non-linearities, fast photoinduced optical absorption and refraction, acusto-optic effects recently discovered in non-crystalline semiconductors could potentially be utilised for optical signal processing. The important problems of photoinduced structural transformations and related phenomena, which are very attractive and actual both from the scientific and practical points of view, received much attention in discussions at the conference.

Considerable attention was also given to technical applications of the photoinduced phenomena in optoelectronics, applications and devices development based on non crystalline semiconductors with the focus on optoelectronics. Optical and photoelectrical information storage media, high resolution lithography, optical diffraction elements, optoelectronic parts and circuits can be based on the photoinduced structural transformation phenomena.

In conclusion the workshop showed that non-crystalline semiconductors represent an important class of semiconductors, which have large potentialities for optolectronics and the most significative advances were discussed. The present book contains the proceedings of the conference.

Before concluding on behalf of all participants and guests of the Workshop, of all Chisinau organizers, we would like to express our deep and sincere gratitude to NATO Scientific Affairs Division for financial support without which the Workshop in Chisinau would have never taken place.

We are sincere grateful to Dr. J.M.Cadioiu, Dr. Jose Rausell-Colom, Dr. Alain Jubier and Dr. James Bombace for encouragement and valuable advice, for contributing so much to the success of the Workshop.

We highly appreciate all the support and assistance offered by Co-sponsoring Organizations: the Government of Moldova Republic, Chisinau Municipality, the Company "Moldtelecom", Air Moldova International Agency, SPIE- the International Society for Optical Engineering, University di Roma "La Sapienza", the Academy of Sciences of the Moldova Republic.

We gratefully acknowledge the support from the International Advisory Committee: Professor A.Owen, Professor S. Dembovsky, Professor T.Necsoiu, Professor J.Lucas, Professor P.Nagels, Professor Linke, Professor H.Fritzsche and Professor V.Vlad.

A special note of thanks is addressed to the members of Local Committee for all their permanent efforts in the period of the preparation and functioning of the Workshop: M.Iovu, I.Culeac, E.Hancevskaia, V.Bivol, S.Sutov, A.Popescu. Many thanks to the staff of the Center of the Optoelectronics of the Institute of Applied Physics for providing clerical and technical assistance.

We also appreciate the invaluable help by V.Chumash, E.Fazio, G.Liakhou and Mrs S.Serafini in the preparation of the manuscript and Mrs Annelies Kersbergen from Kluwer Academic Publishers for her help.

The Editors

A.A.Andriesh M.Bertolotti

MEASUREMENTS OF THIRD-ORDER NONLINEARITIES IN AMORPHOUS MATERIALS

M. Bertolotti, E. Fazio, G. Liakhou, R. LiVoti, F. Michelotti, S. Paoloni, F. Senesi, C. Sibilia

Dipartimento di Energetica dell'Università degli Studi "La Sapienza" and Gruppo Nazionale di Elettronica Quantistica e Plasmi del C.N.R. and Istituto di Fisica della Materia ITALY

Abstract

A description is given of some experimental results obtained using the photothermal deflection, the z-scanning and the pump-probe techniques for measuring the nonlinear part of the refractive index and of the absorption of several amorphous materials.

1. Introduction

It is well appreciated nowdays that amorphous materials present interesting optical nonlinearities which may be very useful for the design of a number of devices.

As it is well known, electric polarisation in a dielectric material in the presence of a strong electric field can be given as a series expansion as follows:

$$\mathbf{P}=\varepsilon_0 \chi \mathbf{E} + \varepsilon_0 \chi^{(2)}:\mathbf{EE} + \varepsilon_0 \chi^{(3)}\vdots\mathbf{EEE} + \ldots\ldots \qquad (1)$$

where the first term is the usual linear one, and the successive terms describe second, third-order nonlinearities etc... Symmetry considerations show at once that the second order nonlinear term that is responsible for example of second harmonic production, field rectification, parametric processes, etc. is absent in centrosymmetric materials. Its presence is effective only in media lacking of a symmetry center. It is therefore absent in amorphous materials except in special cases where the symmetry is broken (for example at interfaces). The important term in amorphous materials is therefore the third-order nonlinearity which in general is responsible for third-harmonic production, parametric processes, stimulated Raman, Brillouin, Rayleigh scattering and, in the case in which all fields have the same frequency and interact in such a way to give a polarisation at this frequency, is responsible of a refractive index change that is proportional to the light intensity or

$$n=n_0+n_2 I \qquad (2)$$

A. Andriesh and M. Bertolotti (eds.),
Physics and Applications of Non-Crystalline Semiconductors in Optoelectronics, 1–16.

where n_0 is the ordinary linear refractive index and n_2 is the nonlinear contribution which arises from a nonlinear dielectric constant that in general is complex. One may have accordingly both a nonlinear refractive index as described by eq. (2) and a nonlinear change in absorption.

A material in which eq. (2) is fulfilled is called "Kerr-like". In real life usually the refractive index saturates by increasing the light intensity and eq.(2) is valid only for the first part of the curve. We will limit our discussion to the case in which the material behaves "Kerr-like".

Phase changes of a wave traveling in the medium induced by the nonlinear change of refractive index can be detected by several beam-distortion methods. In some cases these techniques are also able to measure absorption variations.

The optical nonlinearities producing these effects have of many origins. In resonant conditions, when absorption is present, the predominant effect may be a thermal change of complex refractive index (real and imaginary part). Superimposed to this contribution there are effects connected to resonant excitation of carriers or nonresonant distortion of electronic orbits.

Nonresonant third-order nonlinearities have very short response times but are usually weak.

Phase changes are obtained also in some second-order processes usually indicated as "cascaded" or $\chi^{(2)}:\chi^{(2)}$ processes, and also higher order processes are effective. In particular, whenever a saturation of the refractive index as a function of intensity occurs, it means that higher order terms enter into play. The cascading effect has the advantage of being a nonresonant effect, therefore having very fast response time without thermal side-effects. It needs however the lack of an inversion symmetry of the material and is therefore not effective in amorphous materials.

In the following, measurements of nonlinear refractive index and absorption based on the distortion of the beam will be presented focusing on some special cases based on the use of the photodeflection, the self-diffraction (z-scanning) and pump-probe methods.

2. The photodeflection method used to detect refractive index nonlinearities.

The photodeflection method enables to measure a refractive index gradient by the deflection it produces on a probe beam. In most cases the methods for the measurement of this deflection are relatively slow, so that the technique is particularly well suited to measure the refractive index nonlinearities of thermal origin [1-2]. A pump laser beam is focused at the sample surface and chopped at a frequency ν. A probe beam obtained from a different wavelength laser propagates through the sample, and its deflection angle is measured with a position detector (s. Fig. 1).

Very small deflection angles of the order 10^{-3}-10^{-6} rad can be measured correspondig to temperature increases less than 1 K. In these conditions, the thermal constants of the material do not change and we may write

$$n = n_0 + n_2 I, \qquad (3)$$

where I is the pump light intensity, and

$$n_2 = (dn/dT)\, f(\rho c, \alpha), \qquad (4)$$

where c is the heat capacity per unit volume, ρ is the density, α is the absorption coefficient of the material and dn/dT and f can be obtained by a fit of the photodeflection signal as a function of the chopper frequency.

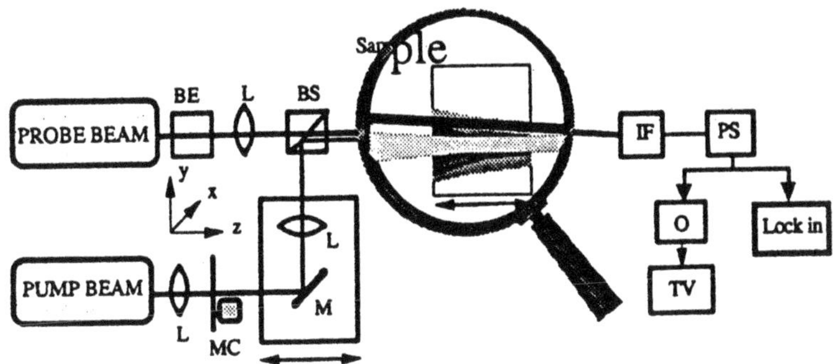

Fig. 1 Photothermal collinear configuration. Experimental set-up: L are lenses, M a mirror, O the oscilloscope, IF an interference filter, TV a digital TV set, BE a beam expander, MC the mechanical chopper, PS the position sensor, and BS is a beam splitter.

An example of the obtained data, in which the amplitude of the deflection signal is plotted as a function of the chopper frequency, is shown in Fig. 2. Fitting the experimental curves with the theory using the optothermal coefficient as a fitting parameter gives dn/dT values comparable with the ones obtained with other methods.

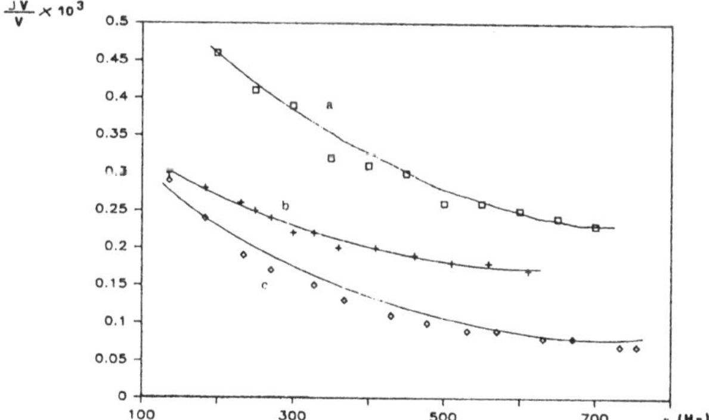

Fig. 2 Photothermal signal vs. frequency. Experimental points and theoretical curves for (squares) CS-2611 colour filter, with a dn/dT=3.6.10^{-7} K^{-1}, (crosses) CS-3681 colour filter, with a dn/dT=2.4.10^{-7} K^{-1}, (diamonds) sodium disilicate metallic (Mn) glass, with a dn/dT=8.8. 10^{-7} K^{-1}, from ref. [1].

The method can be used also in pulsed regime [3]. In this case, it is possible to modify the set-up introducing a time delay between the pump and probe beams which allows to measure very fast changes of the refractive index [4]. When operating in this way the electronic contribution to changes in refractive index can easily be detected.

In some cases it is possible to have slow processes of electronic origin. Two cases are considered here. In chalcogenide glasses one may have a nonlinear variation of the refractive index, produced by a weak He-Ne laser beam. The refractive index variation needs long time exposure (hours) to be created and lives for a long time (few days) after the laser is switched off. The effect is due to self-organization of electrons and holes under weak pumping with bandgap energy beams. The ordered state is long living and gives contribution to the refractive index variation.

The main idea of the electron-hole self-organization model is the following. Under light pumping with band-gap energy, free carriers are generated. The probability of an electron or a hole trapping in some region of an amorphous semiconductor depends on the spatial distribution of already trapped particles. In fact, a static electric field is produced by a trapped electron or hole. If in some unoccupied trap this is too strong, this trap is spoiled, and an electron (hole) can not be trapped in this place. The recombination rate of an electron (hole) also depends on the relative positions of holes (electrons). The rate of recombination is exponentially decreasing with the relative distance. As a result, a state with strong electron-electron, electron-hole, and hole-hole correlations appears under light pumping.

Only a strongly correlated (ordered) state is long living and survives after switching the laser off. In spite of Coulomb repulsion electron and hole domains appear at weak pumping. The extra-carriers in the ordered state give contribution to the refractive index.

The model has been tested by computer simulations which have confirmed the self-organization and the formation of domains [5].

Applying this model, one can see that a probe beam produces by itself a channel in which a change of refractive index that is negative at the beam's center occurs in As_2S_5 and in the far field a ring structure develops, as is well known in the analogous situations of focusing or defocusing. The effect can be observed inserting a diaphragm in front of the photodiode which detects the probe beam [5]. The intensity changes as the ring structure develops, are shown in Fig. 3 in c.w. regime, and pulsed conditions (insert in the figure) for As_2S_5.

Other experimental evidence of ordering can be obtained in the experiment shown in Fig. 4 [6]. The probe beam is made to travel in the As_2S_5 sample volume at a variable distance z from the pumped surface. Fig. 5 shows the time behavior of the deflection signal $\delta V/V_0$, given as the change in voltage of the position sensor over its continuous bias, for several distances z. The pump beam is switched on at t=0, and the switching off time is marked with the symbol ■ on the curves. Except for the first curve (1, made at z=0.5mm), the other curves (made for larger values of z), near t=0, show at first a negative signal, which subsequently changes its sign. For the chosen polarization of the position sensor, this means that the deflection is first directed away from the sample surface, and then towards the surface. After switching off the pump laser, the deflection at first continues to increase with a greater slope, and then starts to decrease. The deflection signal at z=0.5mm varies slowly according to heating and cooling of the sample. It is due to an increase of refractive index of thermal origin ($dn/dT > 0$ in chalcogenide). For $z > 1$mm, the initial negative part is faster than the thermal one. An analysis of this experiment shows that the negative contribution (and the positive contribution when the laser is switched off) can be interpreted by taking into account the channel of varied refractive index produced by the probe beam [6].

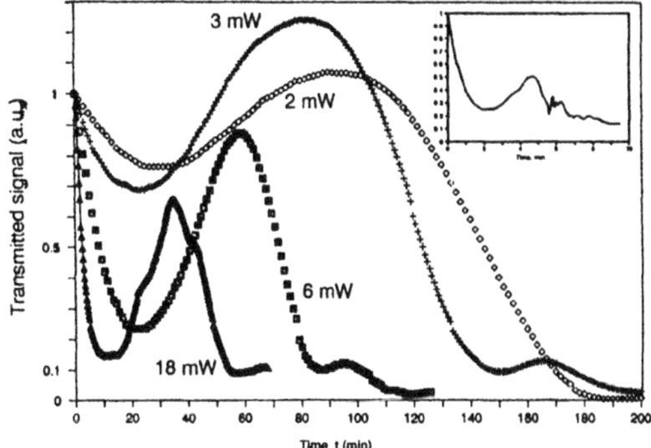

Fig. 3 Transmitted He-Ne signal through a sample of As_2S_5 observed through a diaphragm for different input powers as a function of time. In the inset the signal when pulsed pumping is used is shown.

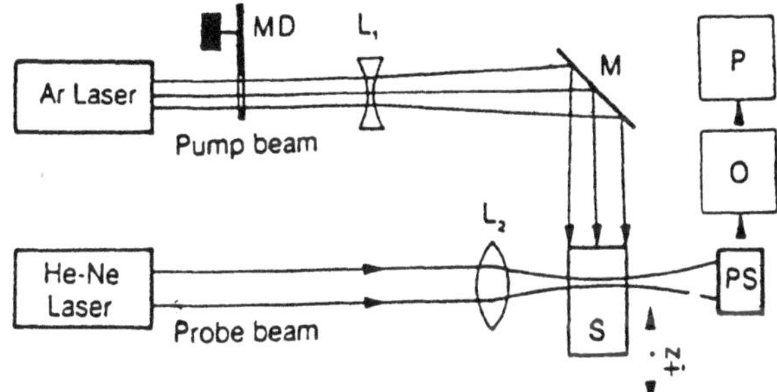

Fig. 4 Experimental set-up. MD is mechanical chopper, L_1 and L_2 are lens, M is a mirror, S is the sample, PS the position sensor, o an oscilloscope, an P the plotter.

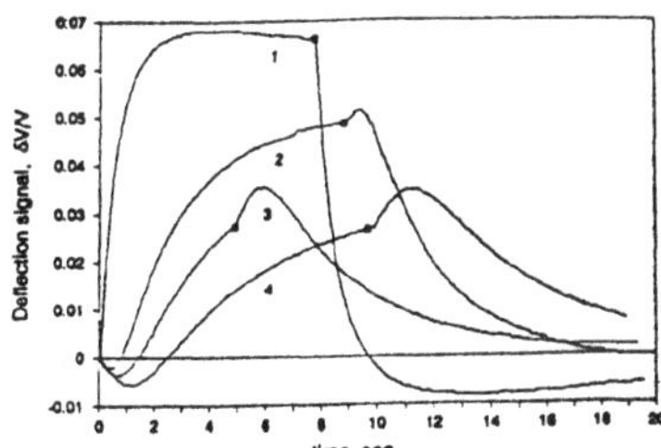

Fig. 5 Time dependence of the probe beam deflection as a function of the distance z of the propagating beam from the illuminated sample surface. (1) z=0.5mm, (2) z=1mm, (3) z=1.25mm, (4) z=1.5mm.

The computer modelling of the self-organization process allows also to study its dynamics. Fourier transform of the electron (hole) density time fluctuations (and, hence, of the refractive index variations) indicated that a $1/\omega$ noise is present (for $\omega \to 0$) in the ordered state (s. Fig. 6), while disordered electron-hole systems reveal white noise. The Fourier transform of Fig. 3 is shown in Fig. 7, for the curves at 3 and 6 mW, and shows $1/\omega$ noise for $\omega \to 0$, confirming the ordered state model [7].

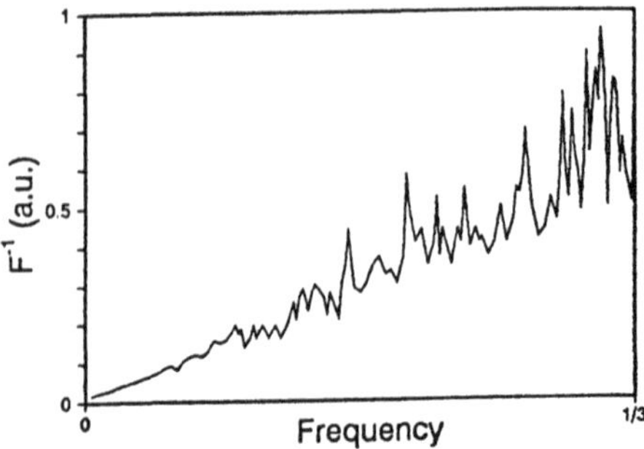

Fig. 6 Computer simulation of the Fourier transform of the time concentration of carriers.

In semiconductor doped glasses, carriers produced by the light beam can be trapped at interface states between the semiconductor microcrystals and the glass matrix [8-13], producing a change of refractive index which relaxes very slowly with the decay time of trapped carriers (milliseconds).

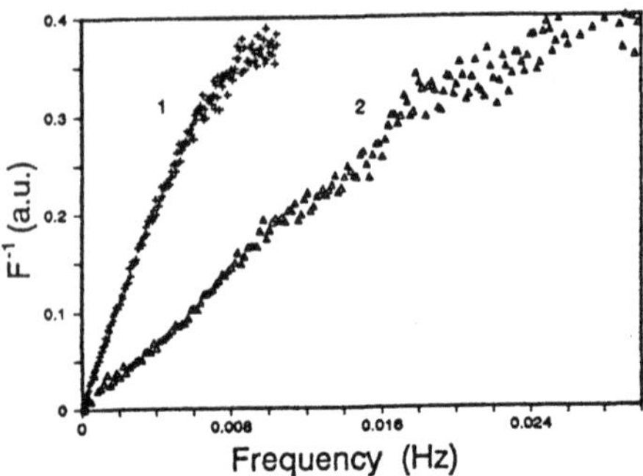

Fig. 7 The Fourier transform of curves from Fig. 3. (1) is for the 3mW curve, and (2) is for the 6 mW curve.

The photodeflection method is very well suited, also in this case, to measure the resulting refractive index changes. Fig. 8 shows the chopped pump signal and the probe beam deflection change. An initial small deviation of the probe beam is observed which is of opposite sign with respect to the subsequent thermal deflection and is of electronic origin produced by the pump beam photoexcited electrons trapped at the semiconductor crystallites in the glass [14].

Also changes in absorption can be studied. Fig. 9 shows the change in absorption of the probe beam when a train of Ar pump beam pulses is switched on. Absorption first drops down very rapidly, due to an electronic bleaching, and then slowly increases due to heating of the material, which increases absorption. As time progresses, and more and more pulses are sent on the material, the initial bleaching decreases because electrons are trapped at the microcrystals boundaries.

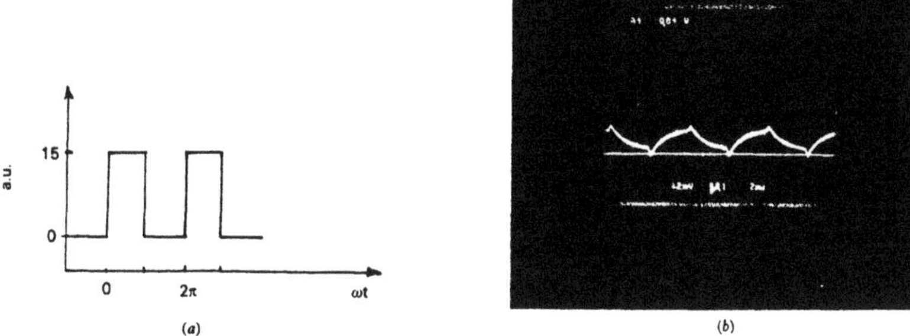

Fig. 8 (a) chopper signal, (b) the probe beam deflection signal during short time exposure as a function of time.

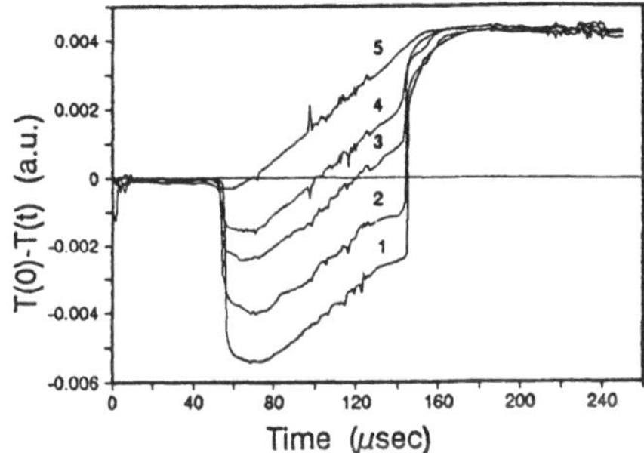

Fig. 9 Change in absorption of a probe beam when a train of Ar pump beam pulses is switched on at time 60 μs and off at 140 μs for a semiconductor doped glass. Curve 1 is obtained at time 0 when only one pulse was applied. Curve 2 after 1.5 minuts, curve 3 after 6.5, curve 4 after 15 min, and curve 5 after 140 min. Repetition rate of pulses was 20 pulses per second.

3. The z-scanning technique

The so called z-scan technique is very suitable to measure both refractive and absorbitive nonlinearities in any transparent material.

In this technique, the transmission of a focused laser beam through a finite aperture in the far field is measured as a function of the displacement (z) of a nonlinear sample along the propagation direction with respect to the focal plane [15] (see Fig. 10). The results can be understood by considering the sample as a thin lens whose focal length is determined by the nonlinear change of refractive index produced by the intensity on the sample itself.

For a positive change of refractive index (focusing nonlinearity, $n_2>0$) a curve of transmission as a function of z shows first a valley and then a peak (s. Fig. 11), while for negative n_2 one has first a peak and then a valley. If the aperture is fully open, then the transmission as a function of z allows to determine changes in absorption as a function of intensity (which is a function of z).

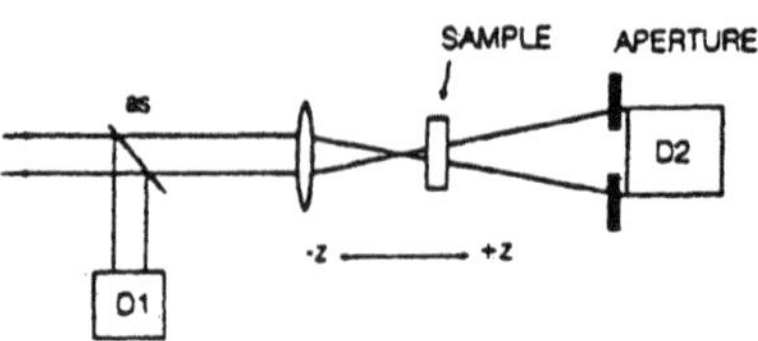

Fig. 10 Simple z-scanning experimental apparatus in which the transmittance ratio D_2/D_1 is recorded as a function of the sample position z. BS is beam splitter.

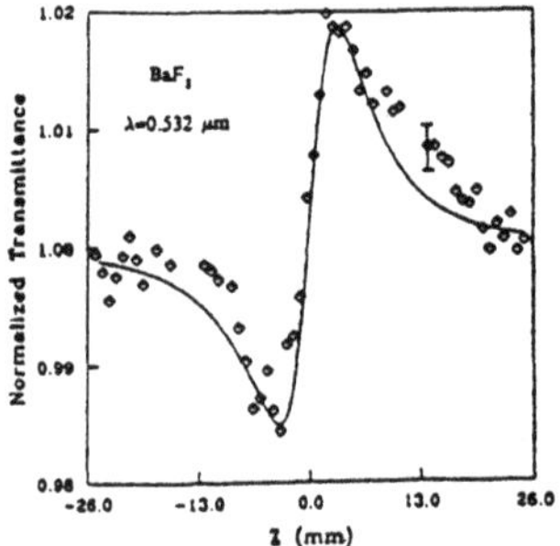

Fig. 11 Measured z-scan of a 2.5 mm thick BaF_2 sample using 27 ps pulses, indicating the self-focusing due to the electronic Kerr effect (from [16]).

In the case of pulsed excitation it is possible, by varying the pulse duration and the repetition rate, to put into evidence the competition between thermal and electronic effects. This is particularly evident, for example, in the case of semiconductor doped glasses in which, at suitable wavelengts, the thermal and the electronic contribution to the refractive index have opposite sign [17]. The contribution from heating of the sample can be decreased by decreasing the repetition rate of the pulses. Fig. 12 shows z-scan plots obtained for two different repetition rates of the pump. Simple inspection shows that for high repetition rate the z-scan shows a thermal change of the refractive index (positive n_2), while a reduction of the repetition rate puts into evidence the change of sign of n_2, now of electronic origin, that at the particular wavelength used (578 nm) is of opposite sign with respect to the thermal contribution.

In the case of chalcogenide glasses and amorphous silicon, peculiar effects are

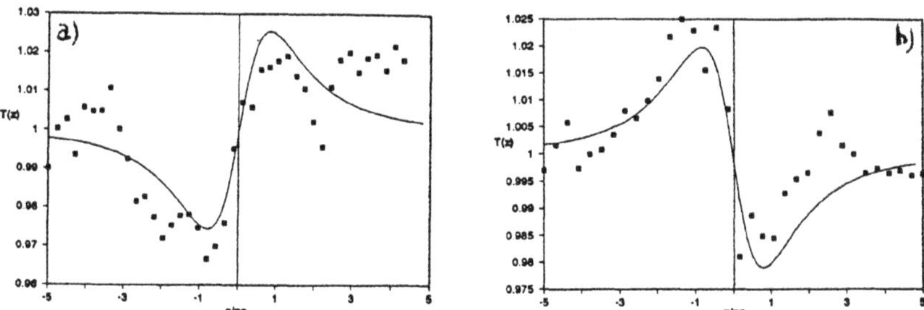

Fig. 12 Normalized transmission for a Shott yellow filter at=578nm, for different repetition rates of 4ps pulses: (a) 9.5 MHz, (b) 148 KHz.

present due to photostructuring of the materials which give rise to changes of absorption and refractive index, depending on input intensity and illumination time, as shown in Fig. 13 [18-19]. Here the normalized transmission is plotted as a function of the irradiation time, for different light intensities through a As_2S_3 chalcogenide sample. These changes can produce, during z-scanning, a permanent lens in the material which can be detected and must be taken into account in order to obtain the real nonlinear response of the material.

In Fig. 14 the z-scan transmittance signal for an As_2S_3 sample in cw detection (λ=514.5 nm) at low power is shown, produced by the thin lens created by the photostructuring effect. It is evident that the curve simulates quite well a negative nonlinearity (diamond). The continuous line superposed with the signal obtained with partially closed aperture has been calculated as the effect of a positive lens of 35μm diameter and focal lenght 40mm. It fits most of the signal points that are therefore produced by the effect of thin lens rather than by a dynamic change of refractive index. After eliminating the background signal due to the lens induced by the photostructural changes one can obtain a true z-scan signal raising the power, as shown in Fig. 15. From the data one obtains a maximum change of the refractrive index at the focal position equal to $\Delta n=8.10^{-3}$ [20]. Similar results have been obtained in the case of a-Si:H. In this case, the results are again the sum of the nonlinear change of refractive index and the effect of the photostructural change. A four level system can be considered for a-Si:H (s. Fig. 16) [20]. When photons of the right energy irradiate the sample, electrons from the valence band (E_1) are excited to the conduction band (E_3) thereafter relaxing via a multiple-trapping mechanism to the ground state or to the dangling bond level (E_2).

Concentration of metastable defects (dangling bonds) is described by the self-limiting bond breaking model [21], where defects are created by direct electron-hole non radiative recombination events and thermal annealing is also taken into account with an activation energy E_A which is the height of the barrier separating the metastable configurations.

In this model the absorption coefficient is given by

$$\alpha=\sigma_1(N_1- N_3)-\sigma_2(N_2- N_4) \quad (5)$$

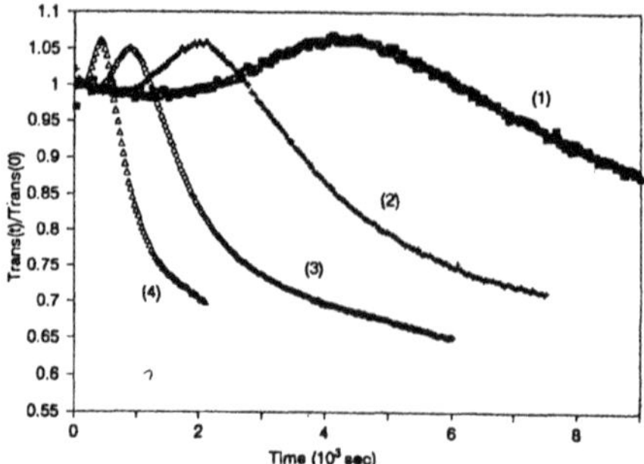

Fig. 13 Kinetics of the relative transmittance of Ar laser radiation for (1) I=1.7.10^{-3} W/cm^2, (2) I=3.5.10^{-3} W/cm^2, (3) I=7.10^{-3} W/cm^2; (4) I=1.5.10^{-2} W/cm^2. The sample is As_2S_3 chalcogenide glass.

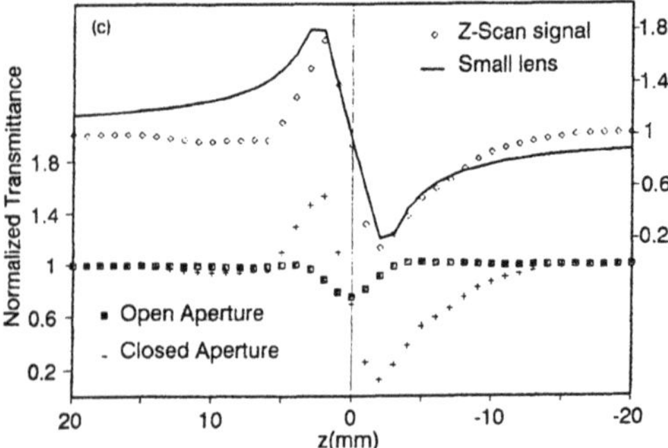

fig. 14 Effect of a thin lens.

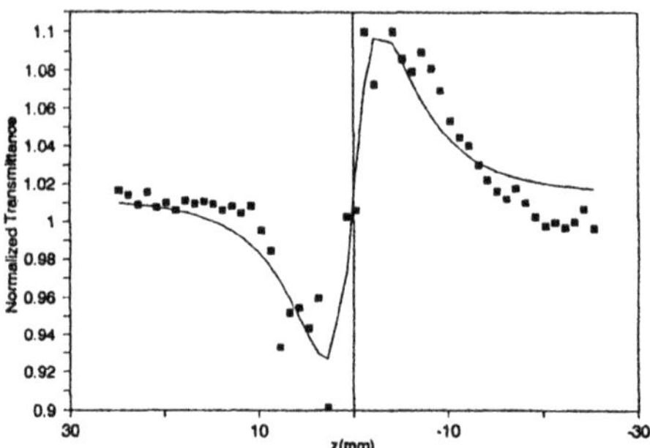

Fig. 15 Effective z-scan signal.

where N_1 are the population densities on the i-th level, and σ_1,σ_2 are the absorption cross-sections for the transition 1-3 and 2-4, respectively.

This expression can be used to calculate the z-scan transmittance curve for an open aperture, calculating the rise of temperature during illumination as

$$\Delta T(z)=\sqrt{\frac{2}{\pi}}\frac{P_m\alpha}{K\omega(z)} \tag{6}$$

which is valid under steady state illumination, where K is the thermal conductivity, P_m is the average input power, and $\omega(z)$ is the beam waist at position z.

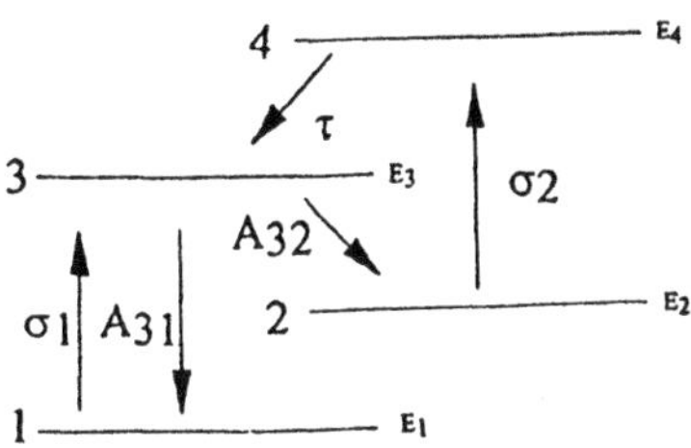

Fig. 16 Four level scheme for a-Si:H.

In Fig. 17 the experimental z-scan signal for open aperture (crosses) and the corresponding simulation (continuous curve), in which the values of the parameters are those reported in the literature, are shown. Good agreement exist between experiment and theory, except in the minimum of the simulated curve where the approximation for the temperature fails.

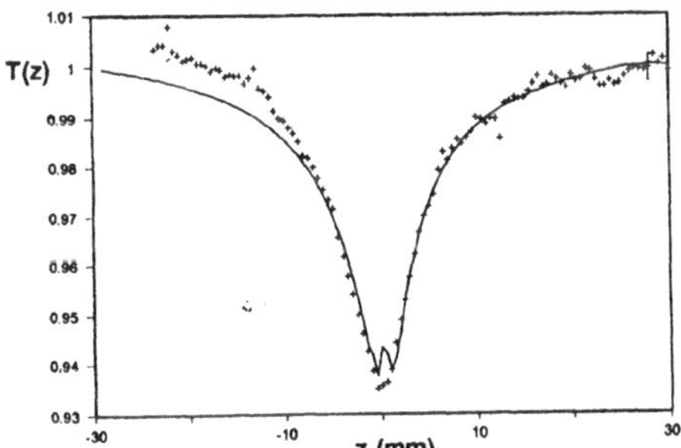

Fig. 17 Z-scan transmittance curves with full open aperture of an intrinsic a-Si:H sample for an intensity of 50 MW/cm^2.

In Fig. 18 three consecutive scans are shown for different input powers on the same area of the sample. In curve 2 (made at high energy) a balance between the production of light induced dangling bonds and annealing is taking place during the measurement itself. The smooth peak is due to the incomplete compensation of the two phenomena near the focal plane, where annealing is dominating.

The changes of refractive index are obtained by measurements with a partially open aperture, and after correcting for sample inhomogeneities, one obtains a z-scan signal as shown in Fig. 19.

From the curve, using standard theory [15-16] one obtains a very large maximum change of refractive index $\Delta n \cong -0.07$. This large negative value of the nonlinearity is due to the high concentration of electron-hole pairs that can be easily photogenerated in the amorphous silicon.

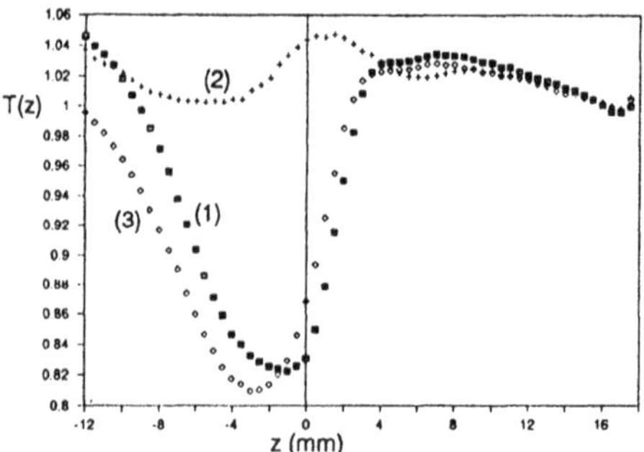

Fig. 18 Z-scan transmittances with full open aperture for an a-Si:H sample for several input intensities on the same site: (1) $5MW/cm^2$, (2) $15MW/cm^2$, (3) $5.8MW/cm^2$.

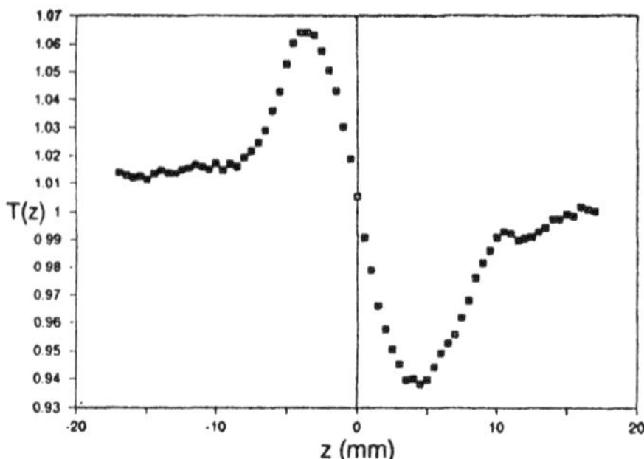

Fig. 19 Z-scan transmittance with partially closed aperture of an a-Si:H sample for an intensity $0.3MW/cm^2$ at 590nm with 9.5MHz repetition rate.

4. Pump-probe technique

The ultrafast dynamics of the carriers out of equilibrium can be investigated using the pump-probe technique. The experimental set-up is shown in Fig. 20 [22]. A 100 fs dye laser beam at λ=620nm is divided into two part, the former with high intensity and the latter with much lower one. Both these sub-beams are focussed on the sample with a suitable time delay in order to perform a stroboscopic measurement of the absorption modification induced by the strong light beam on the second. To probe the absorption changes at energies different from the pumping, before focusing on the sample the lower beam is broadened in spectrum by means of self-phase modulation in a water cell. Following this procedure a continuous spectrum of frequency could be used to test the material. In the following only few of them will be considered, in order to point out the material behavior in direct absorption resonance or out of it.

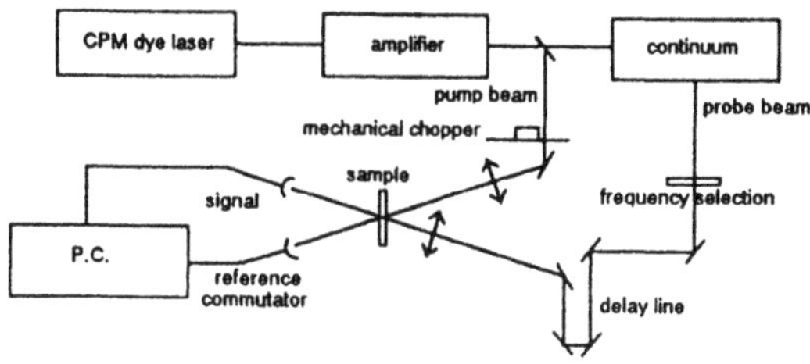

fig. 20 Experimental set-up femtosecond pump-probe spectroscopy.

Using different samples of As_2S_3 (Eg=2.4 eV) and As_2Se_3 (Eg=1.84 eV) one may study the three situations in which:

1 - pump (2eV) and probe (2.1 eV) are both absorbed (As_2Se_3 sample);

2 - pump (2eV) only is absorbed (probe at 1.8 eV) (As_2Se_3 sample);

3 - pump (2eV) and probe (1.8 eV) are both not absorbed (As_2S_3 sample).

An example of the obtained results is shown in Fig. 21 which gives the change of absorption induced by the pump pulse as function of the time delay of the arrival of the probe beam. In particular in this case the pump beam was strongly absorbed by the As_2Se_3 sample, while the probe was just below the gap (second case). The curve is composed by two relaxation parts, a fast one of the order of few picosecond followed by a longer one of several hundreds of picosecond.

The response of the material can be described by a four-level system model, shown in fig. 22, to account for the valence band, the conduction band, the shallow traps and the deep traps.

From the experimental results we characterize the complete dynamics of the excitation-relaxation processes: in particular we pointed out that interband relaxation of free carriers is usually really fast, occurring in few hundreds of femtoseconds, towards the bottom of the conduction band. The levels at the bottom of the conduction band are streakly connected with shallow traps created by a short-range loosing of symmetry of the lattice. In particular the bottom of the conduction band could be seen as an ensemble of levels confined around each elementary cell which are separated one each other by deep potential walls. As soon as free carriers arrive to these confined levels, they could relax towards the valence band through radiative channels but usually go to saturate dangling bonds or other deep traps. This last relaxation occurs in several tens of picoseconds. As soon as this regime is reached, a metastable state is created which requires a further light excitation to be overcome. This process is semi-permanent, known as darkening.

It is interesting to note that the excitation of carriers in amorphous materials can occur both by one- or two-photon processes. In this case however the two-photon process may require transitions through both virtual and real states. In fact at variance with the crystalline case, during an absorption process in amorphous and glassy semiconductors the momentum is not conserved because the electron momentum is not defined. This leads to the possibility of having strong two- or three-step absorption and also free-carrier absorption.

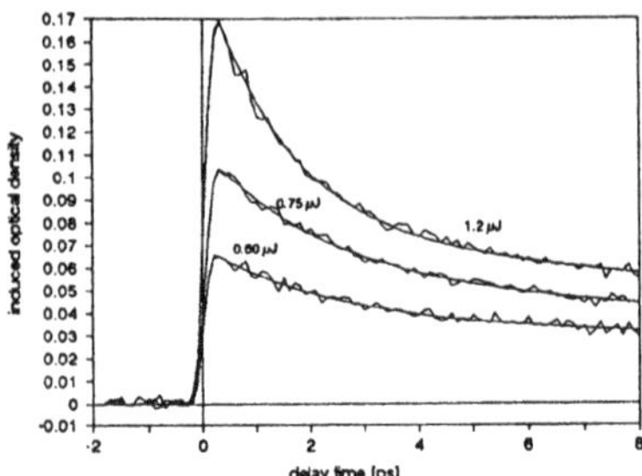

fig. 21 Experimental results of pump probe measurement on As_2Se_3 samples obtained by using a pump absorbed and probe not absorbed.

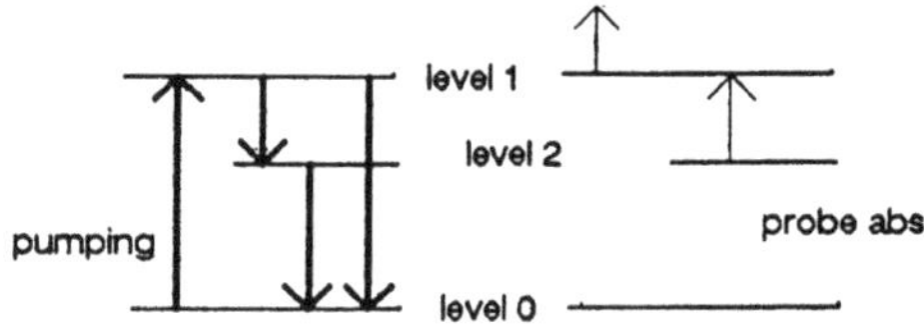

fig. 22 Scheme of the four-level system used to describe the material response

5. Conclusions

The methods based on self-distorsion of an optical beam are very effective to measure optical nonlinearities of very different origin. Thermal changes of refractive index and absorption can be easily measured and their competition with "electronic" nonlinearities can be seen in some cases.

On the other hand situations can be considered in which the two contributions, thermal and electronic, can be easily separated. Permanenet effects due to photostructuring of materials are also important for some materials and can give rise to spourious signals, as e.g. in the case of z-scanning, which must carefully be considered when extracting numbers. On the other hand pump and probe techniques using ultrashort pulses are well suited to study the dynamics of nonlinearities.

References

1. Bertolotti M., Ferrari A., Gnappi G., Montenero A., Sibilia C., Suber G., (1988) The thermal nonlinearity of semiconductor doped glasses, *J. Phys. D. : Appl. Phys.* **21**, S17-S19.
2. Bertolotti M., Ferrari A., Sibilia C., Suber G., Apostol D., Jani P., (1988) Photothermal deflection technique for measuring thermal non-linearities in semiconductor glasses, *Appl. Opt.* **27**, 1811-1813.
3. Bertolotti M., Fazio E., Ferrari A., Liakhou G., Gnappi G., Montenero A., Rossi M., Sibilia C., Zimin L.G., (1990) Thermal nonlinearities of semiconductor doped glasses in the near-ir region, *Material Science and Engineering* **B5**, 143-145.
4. Albrecht H.S., Heist P., Kleinschmidt J., Lap D.V., (1993) Ultrafast beam-deflection method and its application for measuring the transient refractive index of materials, *Appl. Phys.* **B57**, 193-198.
5. Antonyuk B.P., Kiselev S.A., Bertolotti M., Sibilia C., Liakhou G. L., Andriesh A. M., (1993) Creation of long-living ordered electron-hole structures in a disordered system: slow nonlinear response and formation of a beam channel by weak pumping, *Phys. Rev. B* **47**, 10186-10192.
6. Antonyuk B.P., Bertolotti M., Kiselev S.A., Liakhou G. L., Popescu A. A., Sibilia C., (1994) A photothermal experiment interpreted as evidence of electron-hole self-ordering in As-S chalcogenide glasses, *Appl. Phys.* **B59**, 639-643.
7. Antonyuk B.P., Musichenko S. F., Bertolotti M., Li Voti R., Sibilia C., Liakhou G. L., (1995) Electro-hole ordering and 1/ω noise in amorphous semiconductors, *Phys. Lett.* **A201**, 77-80.
8. Zheludev N., Ruddock I. S., Illingrowrth R., (1987) Generation and amplification of subharmonics semiconductor-doped glass by a modulated argon-ion laser, *J. Mod. Opt.* **34**, 1257-1262.
9. Remillard J. T., Steel D. G., (1988) Narrow nonlinear-optical resonances in CdSSe-doped glass, *Opt. Lett.* **13**, 30-32.
10. Tomita M., Matsumoto T., Jatsuoka M., in Ultrafast Phenomena 1988, p. 340.
11. Remillard J. T., Wang H., Webb M. D., Steel D. G., (1989) Optical phase-conjugation and nonlinear optical bandpass filter characteristics in CdSSe microcristalline-doped glass, *IEEE J. Quant. Electr.* **QE25**, *408-413.*
12. Tomita M., Matsumoto T., Jatsuoka M., (1989) Nonlinear dynamical relaxation processes in semiconductor-doped glasses at liquid nitrogen temperature, *JOSA B* **6**, 165-170.
13. Burkitbaev S., Bertolotti M., Fazio E., Ferrari A., Liakhou G., Sibilia C., (1992) Luminescence kinetics of semiconductor doped glasses in the long time region, *J. Appl. Phys.* **71**, 942-945.

14. Antonyuk B.P., Bertolotti M., Fazio E., Liakhou G., Sibilia C., (1992) Refractive index changes in CdS_xSe_{1-x} doped glasses through the photodeflection method. Evidence for an electronic contribution together with reversible and irreversible effects, *J. of Mod. Opt.* **39**, 1965-1976.

15. Sheik-Bahae M., Said A. A., Van Stryland E. W., (1989) High sensitivity single-beam n_2-measurements, *Opt. Lett.* **14**, 955-957.

16. Sheik-Bahae M., Said A. A., Tai-Hull W., Hagan D.J., Van Stryland E. W.,Sensitive measurement of optical nonlinearities by using a single beam, (1991) *IEEE J. Quant. Electr.* **QE-26**, 760-765.

17. Bertolotti M., Liakhou G., Michelotti F., Senesi F., Sibilia C., (1992) A beam distortion (Z-Scan) Technique applied to the measuremets of nonlinear in CdS_xSe_{1-x} doped glasses, *Pure Appl. Opt.* **1**, 145-156.

18. Michelotti F., Fazio E., Senesi F., Bertolotti M., Chumash V., Andriesh A., (1993) Nonlinearity and photostructural changes in glassy AsS thin films, *Opt. Comm.* **101**, 74-78.

19. Michelotti F., Bertolotti M., Chumash V., Andriesh A., (1992) Chalcogenide glass thin films: Z-scan measurements of refractive index changes, *SPIE* **1773,** 423-432.

20. Paoloni S., Senesi F., Michelotti F., Fazio E., Bertolotti M., (1994) Optical nonlinearities and defect generation in a-Si:H thin films, *J. Non-crystalline Solids* **176**, 247-252.

21. Stutzmann M., Jackson W.B., Tsai C.C., (1985) Light induced metastable defects in hydrogenided amorphous silicon: a systematic study, *Phys. Rev. B* **32**, 23-47.

22. Fazio E., Hulin D., Chumash V., Michelotti F., Andriesh A.M., Bertolotti M., (1994) On-off resonance femtosecond non-linear absorption of chalcogenide glassy films, *J. Non-Crystalline Solids* **168**, 213-222.

PHOTOINDUCED PHENOMENA IN CHALCOGENIDE GLASS SEMICONDUCTORS

A.M.ANDRIESH
Centre of Optoelectronics
Institute of Applied Physics
Academy of Sciences of Moldova
Chishinau, 2028 Moldova

1. Introduction

The interaction of radiation with vitreous materials leads to irreversible and reversible changes of atomic and electronic structures. These changes in their term manifest themselves in the experiment through changes of their mechanical, thermal, optical, photoelectrical and other characteristics. Under the radiation the structure of vitreous materials changes, new defects appear, film crystallization or amorphization take place, phase transition or transition from one unstable state to another unstable state are observed.

It is evident that such changes radically influence the optical and photoelectrical properties of the material: the refraction index, reflection and absorption of light, as well as photoelectrical parameters. The most sensitive characteristic is the kinetics of the mentioned parameters which under certain conditions can be non-linear. Especially effective is the influence of laser radiation which permits to create high energy densities in a small volume of material and to use effectively the coherence of radiation, as well as the radiation of several sources simultaneously.

The above mentioned convincingly demonstrates that the photoinduced phenomena permit to study the physics of non-crystalline materials under different aspects, to investigate physical phenomena unexplored before. Not of less importance is the appearance of new possibilities for application of vitreous materials as optical and photoelectrical recording media, fibers and planar waveguides, active and passive elements for optoelectronics, including non-linear elements and sensors of different physical quantities.

This paper covers a review of works dedicated to the research of various photoinduced electronic phenomena in thin films, waveguids and fiber samples of chalcogenide glassy semiconductors. Many of these phenomena are studied at the Centre of Optoelectronics of the Institute of Applied Physics of the Academy of Sciences of Moldova.

2. Photoinduced conductivity

Even during first investigations of interaction of radiation with vitreous materials there was found out that after stopping the illumination of samples their conductivity keeps an increased value during a long period of time. As an example we can mention investigation of the photoconductivity in vitreous samples $Tl_2SeAs_2Te_3$ at the temperature of liquid nitrogen /1/.

A. Andriesh and M. Bertolotti (eds.),
Physics and Applications of Non-Crystalline Semiconductors in Optoelectronics, 17–30.

As shown in the Fig.1 during the first cycle of illumination photoinductivity reaches a stationary value during a long period of time. In the course of the next cycles of illumination the stationary value of photoconductivity is established rather fast. At the same time after stopping the illumination a residual photoconductivity or, to be more concrete, an increased value of conductivity was observed, which could be called photoinduced conductivity as it appears as the result of the preliminary optical excitation of samples. This phenomenon was later noted by many authors. The photoinduced conductivity can be reduced to zero either by heating of samples up to higher temperatures, i.e., in the regime of manifestation of thermostimulated current /2/, or by infrared illumination, i.e., in the regime of photoinduced photoconduction /3/.

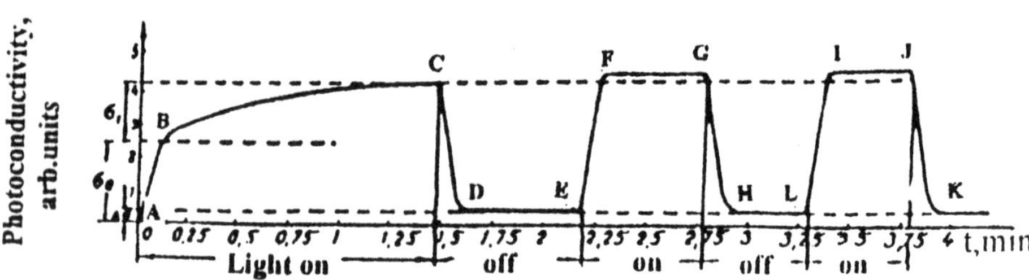

Fig. 1. Decay of the relative photoconductivity in $Tl_2Se{\cdot}As_2Te_3$ samples at T=77 K. Illumination - 500 lux.

Returning back to the above mentioned curves of photoconductivity kinetics, described in one of the first work on chalcogenide glassy semiconductors /1/, one can note the presence of strong trapping of non-equilibrium carriers, which at the temperature of experiment are not exposed to the thermal transitions in the conduction band. Supposing that in this case the conductivity has a monopolarity character which is confirmed by the investigations of termo-emf and of charge transport phenomena, then the rate equation, which describes the behavior of concentration of non-equilibrium holes in the valence band is as follows:

$$\frac{dp}{dt} = \beta kI - \frac{p}{\tau_p} - \gamma_0 p(M - m) + \gamma_p m P_V M \qquad (1)$$

where p – is the concentration of non-equilibrium holes in the valence band;

βkI – the number of carriers of light generation per unit time;

τ – the life-time of holes;

γ_p – the trapping coefficient for holes;

M – the concentration of all hole traps;

m – the concentration of all holes, captured by the traps;

P_V – effective density of states in valence band, normalized to an effective level of traps M.

In the case of rather low temperature the thermal transitions of carriers from the traps into the valence band is hardly probable, that's why the last term of the equation (1) can be neglected and the kinetics of photoconduction will be determined by the third term, i.e. by trapping of non-equilibrium carriers until filling up of the hole traps.

This is the cause of the delay in the achievement of the stationary state and of the absence of the delay in the next cycles of illumination. Such a photoconductive behavior is not characteristic under higher temperature when it is impossible to neglect the thermal transitions from the traps.

As other experiments on transport phenomena in glassy semiconductors showed, illumination of glass samples results not only in changing of the concentration of non-equilibrium free carriers by their capturing in localized states, but also in changing of carrier drift mobility; i.e. we can speak about photoinduced mobility.

A number of papers, for example /4/, have investigated photoinduced change of xerographic parameters of chalcogenide glassy semiconductor films. After irradiation of $As_{50}Se_{50}$ films the initial potential charge of the film decreases by 2-3 times, the time of dark discharge - by 4-6 times, the photosensitivity - by 6-8 times. All the parameters recovers after annealing under the temperature of 180 °C. Many authors connect these changes with the appearance of new defects, the nature of which was described rather complete in many papers, for example, monograph /5/.

At the same time, as it was shown by authors /6/, such processes can also be explained in terms of electron processes only.

3. Photoinduced absorption

Photoinduced absorption in vitreous materials is related to many effects: photostructural transformations, photocrystallization, amorphization, partial fusion, and metal diffusion, etc. These phenomena are largely described in the literature, that is why they will not be discussed in this report.

In this section the reversible photoinduced phenomena, which are conditioned mainly by the interaction of radiation with charge carriers or charge carriers and phonons, will be discussed.

The investigations were carried out on thin films and fiber samples.

3.1. Photoinduced effects in chalcogenide waveguides

One of the very important aspect of the integrated optics is the possibility to obtain of very high light intensities in small cross-sections. As a result the power of optical devices sharply decreases, and non-linear or non-stationary phenomena are realized at small levels of light power, the efficiency of photodetectors and acoustooptical devices increases. For example, in As_2S_3 waveguides non-stationary optical processes can be observed already at light power levels at about 10^{-3} Wt /7/. It was demonstrated, that the reduction of output intensity of light is the result of increasing of optical absorption as well as of the change in the index of refraction

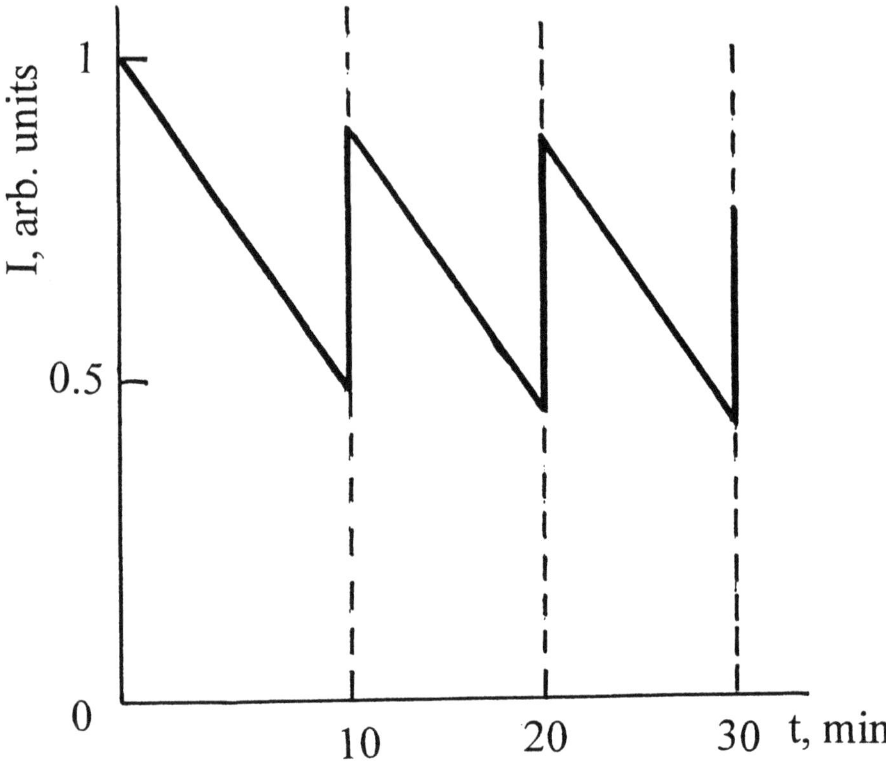

Fig. 2. Decreasing of the output intensity of waveguide under the illumination with the constant input intensity of light ($\lambda = 0.63$ μm, I = 10^3 W/cm^2).

which influences first of all the conditions of light coupling input as well as the process of scattering along the track (Fig 2).

As it was determined, the parameters of waveguide can be restored by reduction of the input light intensity or by keeping the waveguide in the dark.

The changing rate og the waveguide parameters may be significantly increased by its additional exposure to short wave length radiation (Fig 3). When the radiation is switched on optical losses increase rather quickly and afterwards optical transmission change more slowly. When the radiation is switched off, one can see a similar behavior: "the quick" initial section alters with a slow one.

It should be noted that even at rather low intensity of light it is possible to change the value of the waveguide index of refraction, and in this way to change the conditions of waveguide mode propagation. The influence of light radiation on the waveguide light propagation is manifested stronger on the high order modes, which are more sensitive to the boundary conditions where the index of refraction is changed to a

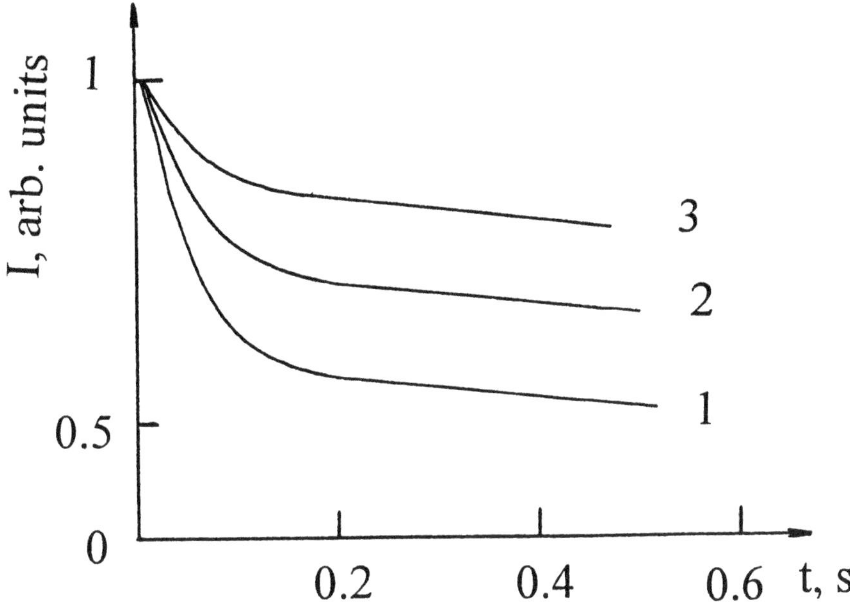

Fig. 3. Decreasing of the waveguide output intensity at different intensities of the input radiation I (%): 1 - 100; 2 - 30; 3 - 10; $\lambda = 0.63$ μm .

higher degree.

The possibility to control the refractive index valueis very important because, in reality, all parameters of planar waveguides are determined first of all by the index of refraction. Its value can be controlled by light as well as by x-radiation, or electron beam. Though the changes of the index of refraction, as a rule, are not relatively high (in As_2S_3 films it changes within 2.447-2.569 /8/) they are sufficient for changing the parameters of optical waveguides, i.e. the constant of modes propagation, mode distribution, etc. This effect can be used to create different elements: periodical and non-periodical structures, strip line waveguides, planar optical lenses, optical filters, input-output devices for computers, and other passive or active elements for integrated optics /9/.

On the bases of photoinduced changes of the index of refraction in the As_2S_3 glass the authors of /10/ have shown the possibility of making Bragg reflectors.

It should be noted, that as a result of irradiation of As_2S_3 waveguide by He-Ne laser a drop in transmittance appears on the wavelength of 0,633 μm, conditioned by the reflection increasing on this wavelength. On the authors opinion such Bragg reflectors can be created in other spectral intervals, too.

It seems that for the explanation of the radiation induced change of the refractive index and of the photoinduced absorption a rather reliable explanation, connected with the formation of new defects, and with structural transformations, was found.

However, as the authors of paper /6/ showed, the explanation of the change of index of refraction and photodarkening under the action of bandgap light can be explained also by the model of long-living electron-holes structures. This model will be commented further.

3.2. Photoinduced absorption in fiber samples of chalcogenide glassy semiconductors

Investigation of interaction of light radiation with fiber samples of chalcogenide glassy semiconductors led to similar results, received earlier in planar waveguides. The intensity of probing light at the fiber output (under the constant probing light intensity at the fiber input) decreases in time and this decrease is stronger when the intensity of the input radiation is higher /9/.

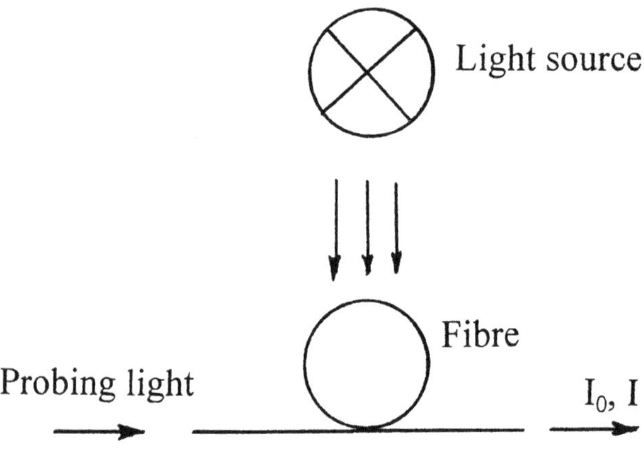

Fig. 4. Illustration of experimental set-up.

More fruitful are the experiments when a low intensity light beam from the midgap was used as a probe light and the fiber was illuminated from lateral surface with bandgap light /11-12/. (Fig.4). The time dependence of photoabsorption in As_2S_3 fibers is given in Fig.5. Under certain conditions (low probe energy and high levels of illumination) one can observe a non-monotonous kinetics (Fig.6). When changing the wavelength of the probe light the photoinduced absorption drastically changes. The spectral distribution $\Delta\alpha = f(h\nu)$ is presented in the Fig.7.

In order to explain the photoinduced absorption in chalcogenide glass fibers a model with carrier redistribution on localized states was proposed /Fig.8/. The electrons and holes generated by light are trapped in quasi-continuous distribution localized states. The probing light with shorter energy $h\nu \angle E_g$ excites the trapped carriers in the band of localized states, leading to an additional optical absorption.

The solutions of corresponding kinetics equations show that $\Delta\alpha$ will increase with the decrease of temperature and with increasing of photon energy of the probing

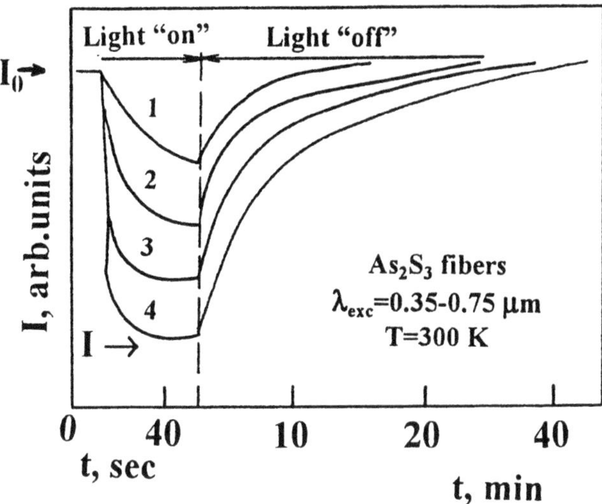

Fig. 5. The time dependence of photoabsorption in As_2S_3 fibers for $\lambda_p = 1.15$ μm. The intensity of exciting light $P_{exc} = P_o$ (1) ; $8P_o$ (2) ; $20P_o$ (3) ; and $700\ P_o$ (4).

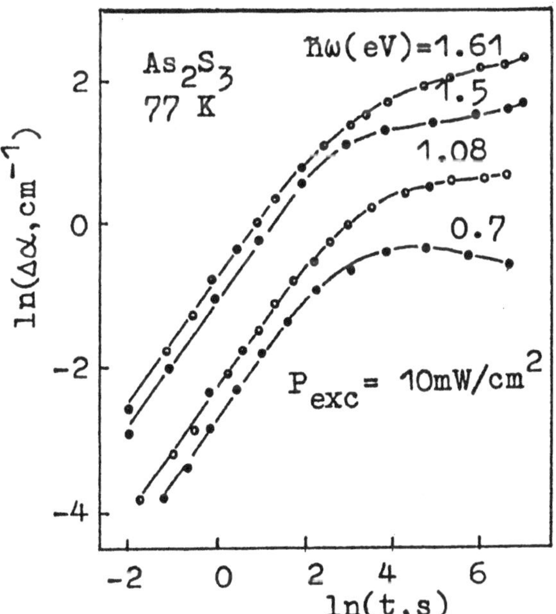

Fig. 6. Photoabsorption kinetics in As_2S_3 fibers at 77 K after switching on the Ar-Laser excitation $\lambda_{exc} = 0.46\text{-}0.52$ μm .

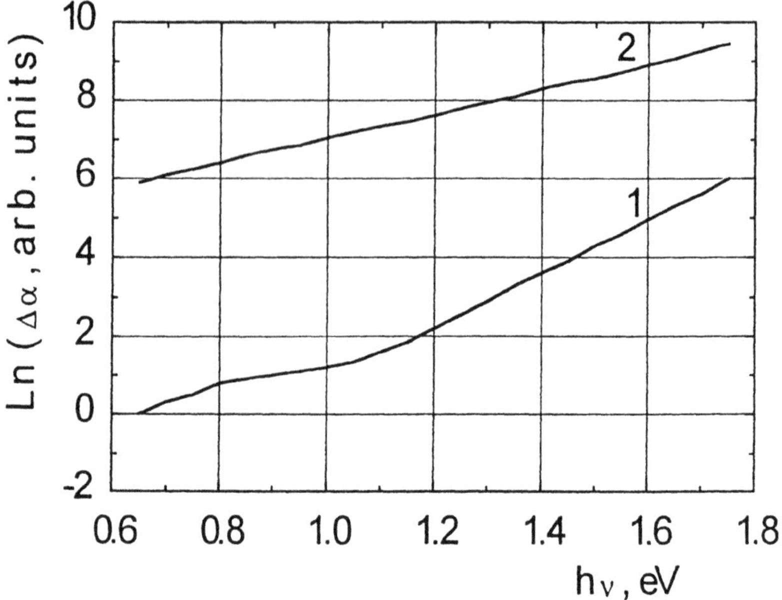

Fig. 7. Spectral distribution of photoabsorption steady-state coefficient $\Delta\alpha$ in As_2S_3 fibers at T = 300 K - (1) and T = 77 K - (2), after irradiation with Ar-laser light (λ_{exc} = 0.46-0.52 μm). The excitation intensity P_{exc} = 10 mW/cm^2.

light /11/. The latter dependence has an exponential character. As it can be seen from Fig.7 the exponential dependence $\Delta\alpha = f(h\nu)$ stretches in a rather large interval of energy (1.0-1.9 eV). At the same time, in the region 0.6-1.0 eV we can see features which, perhaps, are determined by the distribution of the localized states in the gap, which by the way, were established by many independent experiments /5/.

The analysis of the kinetics of photoinduced absorption, at different temperatures, give the possibility to determine the parameter of the exponential distribution of localized states, which for As_2S_3 fibers was found to be about 0.16 eV, which does not much differ from the value of this parameter, measured by other methods.

It is very interesting to study the effect of combined excitation of charge carriers with energy $h\nu \geq E_g$ and $h\nu \angle E_g$. As the experiments showed, subsequent illumination of the samples with light $h\nu \angle E_g$ after illumination of samples with light $h\nu \geq E_g$ can result both in bleaching (Fig. 9) of previously induced absorption, as well as in enhancement of induced absorption (Fig. 10). The rate

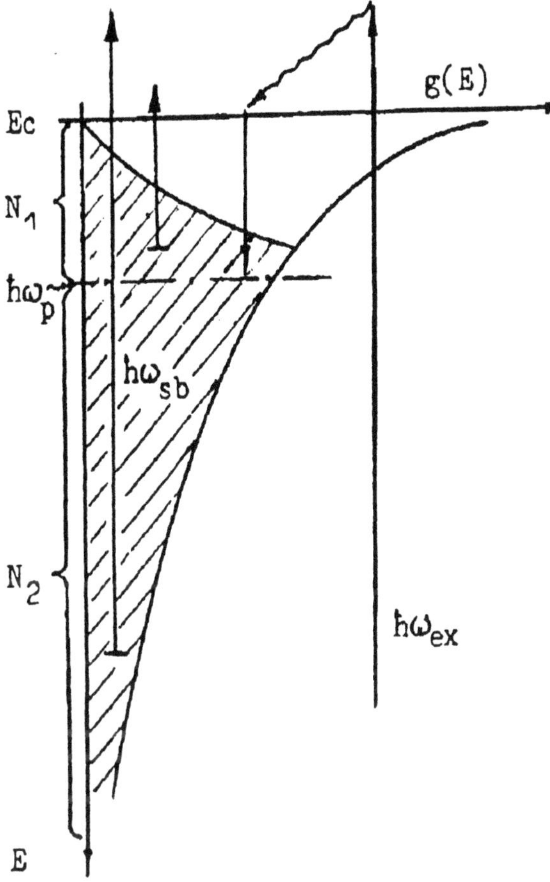

Fig. 8. The steady-state excess carrier distribution in the exponential tail of localized states.

of the effects depends upon the level of population of localized states /12/.

The results of these experiments can be explained by the same model of multi-trapping in the quasi-continuous distribution of the localized states. However, it should be noted the presence of optical transitions with the participation of localized states in the case when together with optical transitions we can observe thermal transitions and when thermal transitions are very small, i.e. in cases of deep and shallow traps.

Infrared irradiation $h\nu \angle E_g$ leads to optical transitions which results in depopulation of the shallow traps and will manifest itself through increasing of the rate of recovering of the initial fiber transparency (Fig.8). Simultaneously there is present the process of carrier excitation from deep traps with their following trapping in shallow and deep traps. However, the last process can be observed only in the case

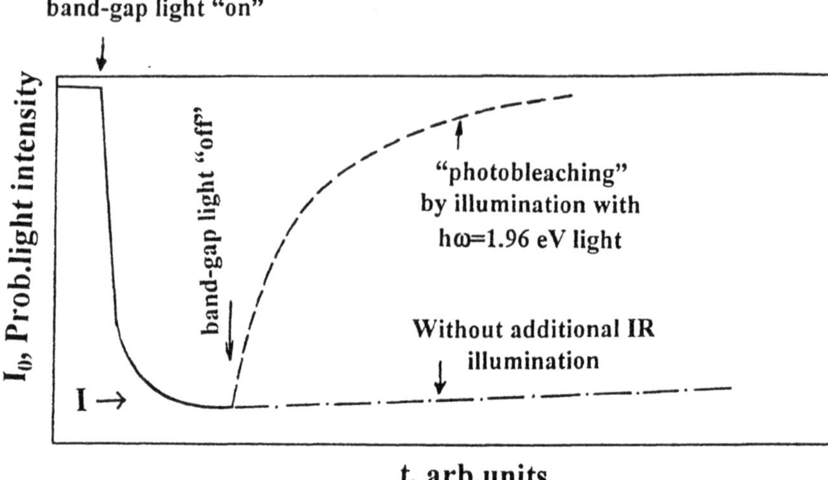

Fig. 9. Time-dependence of the photoabsorption at T = 77 K.

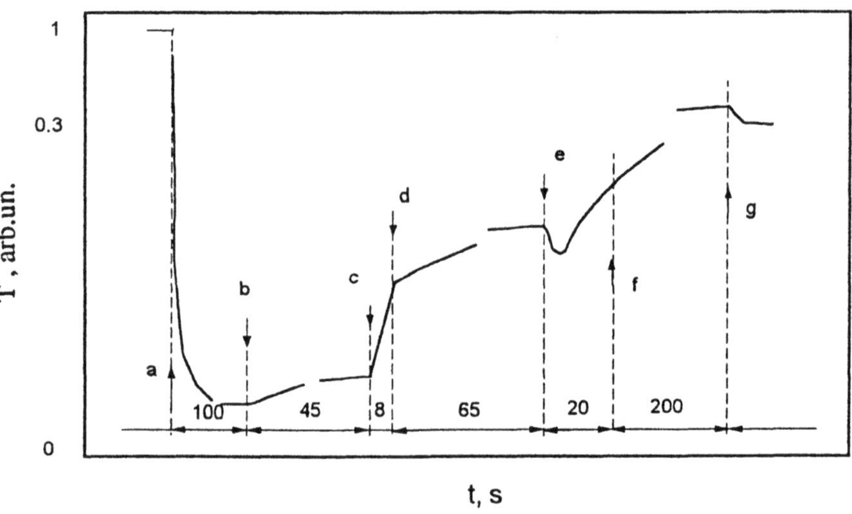

Fig. 10. The change of the photoabsorption of probe light with photon energy $h\nu_p$ = 1.5 eV in As_2S_3 fibers upon sub-band-gap illumination:
(a) - band-gap light $h\nu_{exc}$=2.41 eV is switched on;
(b) - $h\nu_{exc}$=2.41 eV is switched off;
(c, e, g) - sub-band-gap light $h\nu_{sb}$=1.96 eV is switched on;
(d, f) - $h\nu_{sb}$ = 1.96 eV is switched off.

when the concentration of trapped carriers in shallow traps will be comparable to the concentration of captured carriers due to optical transitions with the participation of deep levels. That is why the effect of additional photoinduced absorption under the action of infrared radiation will be seen only after corresponding depopulation of shallow traps. This process has been simulated in the framework of kinetic equations and a reasonable consent between the theory and experiment has been found out /12/.

The above mentioned phenomena have a great interest in connection with understanding of physics of noncrystalline semiconductors as well as in connection with development of various sensors, attenuators and other elements of optoelectronics.

4. Photoinduced nonlinearity

The experiments mentioned above are carried out at low levels of excitation. When the intensity of the laser beam is increased above a certain threshold the nonlinear propagation of light can appear.

It was shown that in the strong field of laser pumping the optical hysteresis of the light transmission can be observed in thin films of different calcogenide glasses (Fig.11) and even in a-Si:H thin films /13-15/.

A typical change of time profile of the laser pulses is shown in Fig.12 for microsecond laser pulse.

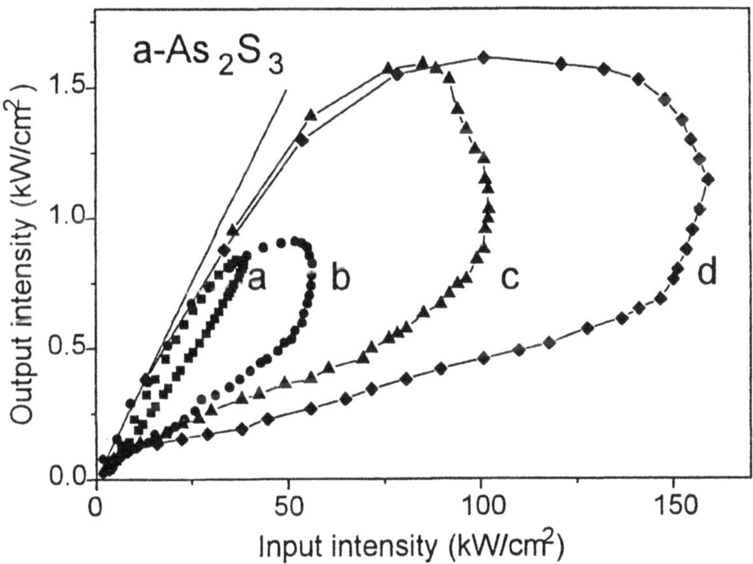

Fig. 11. Hysteresis dependencies of output light intensity versus input one. Incident pulse peak intensity (I_o) : 40 (a), 55 (b), 105 (c), and 160 kW/cm2 (d). Film thickness - 2,6 μm, hν = 2.58 eV.

The similar picture was observed for nanosecond and picosecond region. Such a change of the laser pulse profile leads to a hysteresis-like dependence of the output light intensity on the corresponding value of the input light (Fig.11). The area surrounded by the optical histeresis loop increases when cooling the samples or when increasing the input light intensity. These peculiarities were observed (in the region of high absorption of samples) when the input energy density of the excitation pulses were of the order or more than 1 - 10 mJ/cm^2.

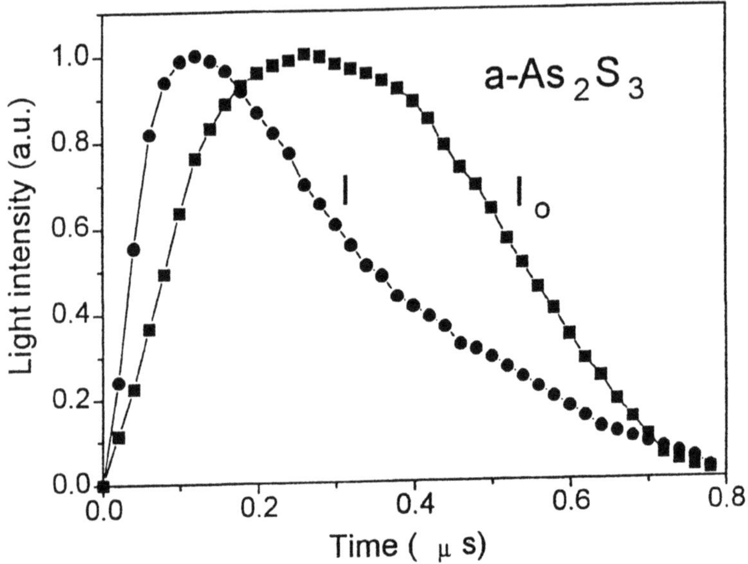

Fig. 12. Oscillograms of incident (I_o) and transmitted (I) pulses. Film thickness - 26 μm, hν = 2.58 eV.

The authors explain the peculiarities of nonlinear light propagation in chalcogenide glasses by appearance of nonequilibrium phonons generated as a result of nonequilibrium carrier relaxation [15]. Interaction of light with generated nonequilibrium localized vibratinal states and phonons leads to an increase of optical absorption coefficient.

5. Conclusions

First of all it is necessary to note that all experiments described above have many common peculiarities. All of them take place in the process or after the light excitation. Hence in all cases nonequilibrium carriers or nonequilibrium vibrational states [15] are generated.

The kinetics of the linear or nonstationary phenomena are controlled by capture of the carriers by localized states. It is seen from the process of delay of establishing of the steady-state photoconductivity, from the appearance of a new channel of absorption, from the appearance of the change of refractive index. After secession of the excitation light, all characteristics of these phenomena (photoconductivity, transmittance etc.) return to their initial values, and the process of restoration to the initial state has a short time constant following by the next process with permanent increasing of the instant time constant of the process. The longer is the period of time after excitation the slower is the process of restoration. In the case of low temperatures one can not even observe any changes as in the case of frozen conductivity.

Therefore it would be reasonable to find a common mechanism for explanation of these phenomena. While one compares the mentioned peculiarities with peculiarities of the model described in the paper /6/ one could find that qualitatively the model can explain the given above results. The model of long-living ordered electron-hole structures takes into account strong correlation between carriers in the process of population and depopulation of trapped states in the system with random potential due to strong disorder /6/.

According to this model a short-distance pairs recombine rapidly and structure with large electron-hole distance survive, leading to ordered structures. This causes the long living conductivity, long living absorption, long living change of refractive index, etc. after switching the light off.

References.

1. Andriesh, A.M. (1968). Influence of the trapping levels on the decay of photoconductivity in vitreous $Tl_2Se{\cdot}As_2Te_3$. In the book: Isledovania po poluprovodnicam, Kishinev, 105-108.
2. Andriesh A.M., Kolomiets B.T. (1963). The localized levels in vitreous $Tl_2Se{\cdot}As_2Te_3$. Solid State Phys., (rus.) **5**, 1461-1465. Andriesh A.M., Kolomiets B.T. (1964). Thermostimulated current in vitreous $Tl_2Se{\cdot}As_2Te_3$. Izvestia AN SSSR, seria fizicescaia, 28, 1291-1292 (rus); Gubanov A.I., Mazets T.E. (1964). Investigation of d.c. conductivity in vitreous semiconductors like As_2Te_3. Izvestia AN SSSR, seria fizicescaia, **28**, 1276-1278.
3. Kolomiets B.T., Mamontova T.N., Stepanov G.I. (1965). The fluctuation levels in vitreous semiconductor $Tl_2Se{\cdot}As_2Te_3$. Fiz.Tverd. Tela **9**, 27-32.
4. Kasimova A.G., Koiava O.V., Lantratova S.S., Lyubin V.M., Tihomirov. V.C. (1989) Photostimulated process in chalcogenide glasses: new results and possibilities for applications in optoelectronics. Abstracts of the workshop: "Application of chacogenide vitreous semiconductors in optoelectronics. Kishinev, 36-37.
5. Popescu M., Andries A., Ciumas V., Iovu M., Sutov S., Tiuleanu D. (1996) Fizica sticlelor calcogenice. Ed. Stiintifica, Bucuresti & Ed. Stiinta, Chisinau, 487 pag.
6. Antonyuk B.P., Kisilev S.A., Bertolotti, Sibilia C., Leakhou G., Andriesh

A.M. (1993). Creation of long-living ordered electrone-hole structures in a disordered system: slow non-liniar response and formation of a beam chanel by weak pumping. Phys.Rev.B., **47**, 10186-10192.

7. Andriesh A.M., Bolshakov O.V., Zhitar V.V., Popescu A.A. (1987) Optical effects in thin film waveguides of glassy arsenic sulphide. Journ. of Non-Cryst.Solids, **90**, 565-568.
8. Keneman S.A. (1974) Thin Solid Films , **21**(2), 281
9. Andriesh a.M., Ponomari V.V., Smirnov V.L., Mironos A.V. (1986) Application of chalcogenide glassy in integrated and fiber optics. Kvantovaia electronica., **13**, 1093-1117.
10. Shirmine K., Hisakuni H and Tanaka K. (1994). Photoinduced Bragg reflector in As_2S_3 glass, Appl.Phys.Lett.,**64** (14), 1771.
11. Andriesh A.M., Culeac I.P., Loghin V.M. (1992) Photoinduced changes of optical absorption in chalcogenide glass fibres, Pure Appl. Opt. **1**, 91-92.
12. Andriesh A.M., Enachi N.A., Culeac I.P., Copaci T.N., Binchevici V.A.(1995). The mechanism of enhancement of photoinduced absorption in As_2S_3 glass at sub-bandgap illumination. Journ. of Non-Cryst.Sol. **189,** 147-153.
13. Andriesh A.M., Enaki N.A., Cojocaru I.A., Ostafeichuk N.D., Cerbari P.G., Chumash V.N. (1988). Optical hysteresis in amorphous semiconductor. Soviet Pisma JTF, **14,** 1985-1989.
14. Andriesh A.M., Bogdan O.I., Enaki N.A., Cojocaru I.A., Chumash V.N. (1992). Optical hysteresis and nonlinear absorption at short laser pulses in chalcogenide glasses. Izvestia RAN, ser.phys. **56,** 92-109.
15. Chumash V., Cojocaru I., Fazio E., Michelotti F., Bertolotti M. (1996). Nonliniar propagation of strong laser pulses in chalcogenide glass films. Progress in Optics. **36,** 1-47.

ELECTRONIC MICRO-FABRICATION OF CHALCOGENIDE GLASS

KEIJI TANAKA
Department of Applied Physics, Faculty of Engineering,
Hokkaido University, Sapporo 060, Japan

Micrometer-scale deformations can be produced in chalcogenide glasses only by exposure to laser or electron beams. The deformations are assumed to be induced through electronic (athermal) processes, while the details depend upon materials and exposure beams.

1. Introduction

It is well-known that chalcogenide glasses exhibit a variety of photoinduced phenomena such as the reversible photodarkening and the Ag-photodoping [1-4]. These phenomena are induced through atomic structural changes, which provide modifications in mechanical, optical and chemical properties. Then, utilizing the chemical change in combination with some developing processes such as wet or dry etching, we can produce macroscopic geometrical patterns. Such a technique has been applied to ultrahigh resolution photolithography [1, 2, 5] and so forth.

On the other hand, Tanaka *et al.* have recently demonstrated that light or electron-beam exposures can directly manipulate chalcogenide glasses [6-10]. That is, we can shape and deform chalcogenide glasses such as (Ag)-As(Ge)-S(Se) through purely electronic means, without using any developing procedures. Such phenomena seem to be quite new to materials science, and accordingly underlying mechanisms and applications may be of great interests. In details, depending upon the kind of exposure beams and chalcogenide glasses employed, some different phenomena appear, which may be referred to as photoinduced fluidity (photoinduced glass transition) [8], giant photoexpansion [6], electron-beam-induced deformation [10], and electron-beam-induced chemical modification [9]. These phenomena are promising to produce micro-fabricated chalcogenide glasses which can be utilized as fiber filters, micro-lenses [7], optical gratings [10] and so forth. Although these phenomena have been just uncovered recently and the detailed features have not yet been fully

A. Andriesh and M. Bertolotti (eds.),
Physics and Applications of Non-Crystalline Semiconductors in Optoelectronics, 31–44.

understood, it may be timely to review these deformation processes from a unified point-of-view.

2. Photoinduced Fluidity

An optically-processed As_2S_3 glass is shown in Fig. 1 [8]. As_2S_3 was prepared through vacuum evaporation, and then annealed at the glass-transition temperature (~ 450 K), a process commonly employed for stabilizing glass structures [1]. Then, a free flake was obtained by peeling off from a substrate. Next, one end of the sample was pasted to a glass slide (the white edge in Fig. 1), and the other end was bent with a stick. The bending was confirmed to be elastic, because the curved sample recovered when the stress was removed. However, when the curved As_2S_3 flake was illuminated locally by a loosely-focused light beam (the spot diameter of 50 - 100 μm) from a He-Ne laser (633 nm, 10 mW) for 2 hours, a permanent deformation occurred, as shown in Fig. 1. It seemed that the illuminated part became fluidal and the sample was bent segmentally. In a similar way, we can modify also the shape of As_2S_3 optical fibers [8].

In marked contrast, for oxide glasses, some trials inducing similar phenomena have been unsuccessful. Deformations appear only under intense illumination.

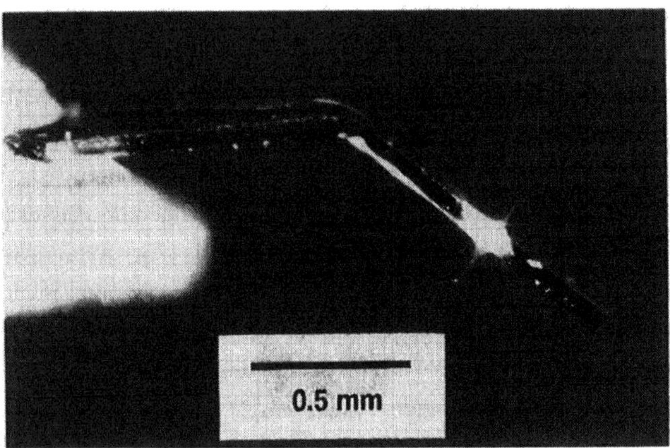

Figure 1. An optically-processed As_2S_3 film. The sample is 50 μm in thickness, $\sim$ 0.2 mm in depth, and $\sim$ 2 mm in length.

It has been demonstrated that the phenomenon observed in As_2S_3 becomes more conspicuous if illumination is provided at lower temperatures [8]. This fact manifests that the photoinduced fluidity occurs through athermal processes. If the temperature rise induced by light illumination were responsible, the phenomenon would become more prominent at higher temperatures. Note that the glass-transition temperature of As_2S_3 is ~ 450 K [1], and an estimated temperature rise is less than 0.1 K [8].

An interesting feature which should be underlined here is the small photon energy of illumination. That is, the photon energy 2.0 eV (= 633 nm) of illumination is substantially smaller than the Tauc optical bandgap of ~ 2.4 eV in As_2S_3 [1]. In this sense, 2.0 eV photons can be regarded as sub-bandgap light, or more precisely Urbach-tail light, since at this photon energy As_2S_3 exhibits the so-called Urbach tail [1].

Why can the Urbach-tail light give rise to the photoinduced change ? Koseki and Odajima have demonstrated that the photoinduced fluidity (photoinduced stress relaxation) is observed in amorphous Se when subjected to bandgap illumination [11]. Accordingly, we may be interested in the role of bandgap and sub-bandgap photons in chalcogenide glasses.

This problem still remains unresolved [12], while the following consideration gives a new insight into the role of photoexcited electrons and holes. A photoconductive measurement of As_2S_3 glass using the constant photocurrent method shows that the photoresponse induced by 2.0 eV light is smaller by $10^{-2} - 10^{-3}$ than that by 2.4 eV light [13]. That is, the number of photoexcited carriers by the Urbach-tail light is considerably smaller. However, the ratio merely reflects the absorption coefficients. If the numbers of generated carriers normalized to an *absorbed* photon are evaluated, no appreciable difference exists between 2.0 and 2.4 eV photons. This fact implies that, since holes are responsible for photocurrents [1], only free holes are needed for the photoinduced change, irrespective as to whether electrons being trapped (for Urbach-tail excitation) or free (for bandgap excitation). We must admit here, however, that our understanding of the Tauc optical bandgap and the Urbach tail in chalcogenide glasses is still unsatisfactory [1].

Microscopically, some electro-atomic processes follow the photoexcitation of holes, and then slipping of molecular clusters will occur, which appears as the photoinduced fluidity. However, since the molecular clusters are entangled and networked [1], scission of covalent bonds may also be required. Why such atomic changes can be induced by free holes and trapped electrons is largely speculative at present [11, 14].

Electron-beam irradiation may also cause the fluidity enhancement, while such a phenomenon has not yet been demonstrated directly.

3. Giant Photoexpansion

If As_2S_3 glass is illuminated by tightly-focused He-Ne laser light under unstressed condition, macroscopic surface expansion appears as shown in Fig. 2. The sample is a 50 μm-thick annealed As_2S_3 film, which has been illuminated for 1000 s with focused (using a microscope objective lens of $\times 10$) light from a He-Ne laser of 10 mW. We see in the figure a convex deformation at the position, where the laser beam has impinged. The lateral scale of the deformation approximates the Gaussian beam profile of the focused laser spot. On the other hand, the height ΔL is ~ 1 μm. In this case (when $L \leq 100\mu$m), a similar expansion appears also on the rear surface. Accordingly, the fractional thickness change $\Delta L/L$, where L is the sample thickness, reaches 4 %. This magnitude is greater by an order than that (~ 0.4 %) observed in the conventional photoexpansion phenomenon in As_2S_3 [16].

The giant photoexpansion has appeared also in As_2Se_3 and GeS_2 when excited by 1.6 and 2.7 eV light, respectively. These observations suggest that the phenomenon is inherent to chalcogenide glasses subjected to Urbach-tail illumination.

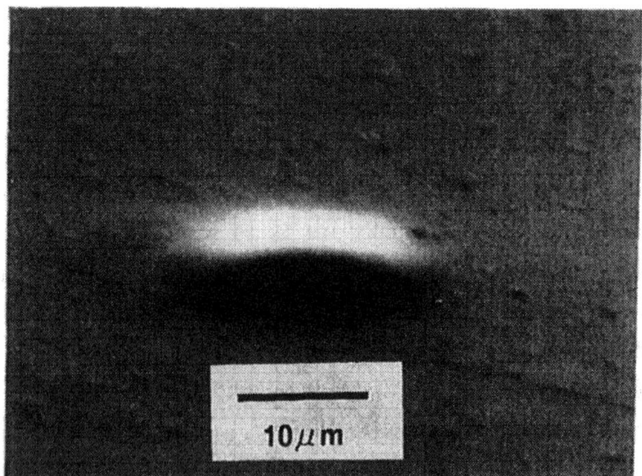

Figure 2. A giant expansion appearing on a surface of a 50 μm-thick As_2S_3 film. The photograph is taken using a scanning electron microscope.

As shown in Fig. 3, the expansion becomes greater if the sample is illuminated at lower temperatures. This temperature dependence clearly manifests that the giant photoexpansion is induced with athermal processes. In the figure, it may also be surprising that the maximal expansion observed amounts to 20 μm. (This seems to correspond $\Delta L/L = 5$ %, since $L = 400$ μm. However, when $L \geq 100$ μm, L is determined by the self-focused depth of light ($\simeq$ 100 - 200 μm) [6, 15], and accordingly $\Delta L/L \simeq$ 10 - 20 %.)

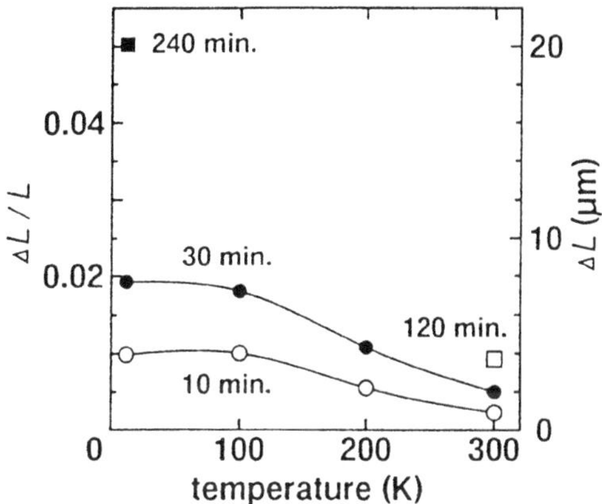

Figure 3. Giant photoexpansion in a 0.4 mm-thick As_2S_3 as a function of temperature, at which the sample is illuminated. The exposure time is indicated. The light source is a 25 mW He-Ne laser, which is focused by a $\times 5$ objective lens. The expansion is measured at room temperature.

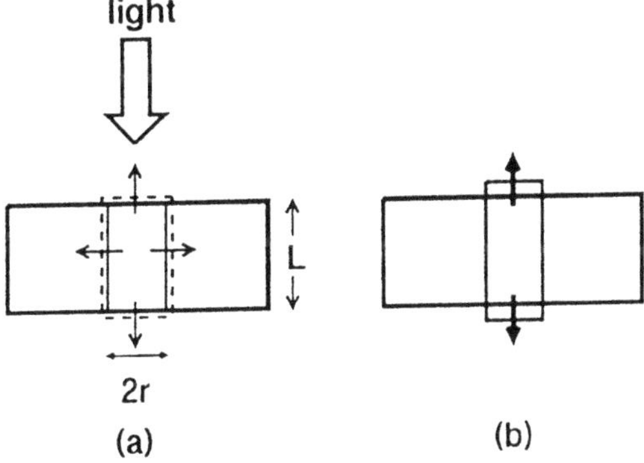

Figure 4. A model for the giant photoexpansion (cross-sectional view).

As illustrated in Fig. 4, we can assume that the giant photoexpansion is induced through a combination of the conventional photoexpansion [6] and the photoinduced fluidity (discussed in 2.). That is, the irradiated volume with a thickness of L and a diameter of $2r$ is apt to expand with a ratio α ($\simeq 0.4$ % at room temperature) of the conventional photoexpansion [16, 17]. However, the lateral expansion is practically impossible due to the existence of unilluminated region, and accordingly the strain components will flow to the vertical direction through the photoinduced fluidity. In this case, the apparent expansion $\Delta L/L$ becomes $\Delta L/L = \alpha(1 + L/r)$. That is, the conventional photoexpansion is seemingly amplified by L/r [6]. This model is consistent with the observations that the giant photoexpansion is prominent only when $L \gg r$.

Microscopic mechanism of the conventional photoexpansion giving rise to α is still speculative, while it may be related to an increase in structural randomness of amorphous networks. It is known that a glass is less dense than the corresponding crystal by $\sim$ 10 % [1]. It is also known that the density of As_2S_3 can be modified by $\sim$ 1 % through some treatments such as annealing [1]. Accordingly, illumination may also be capable to increase the specific volume by $\leq$ 1 %, which is comparable to the magnitude observed in the (giant) photoexpansion.

4. Electron-beam-induced Deformation

Electron-beam irradiation gives a different effect upon As_2S_3 from that induced by light illumination. Figure 5 shows an AFM image of an As_2S_3 film, which has been exposed to a line-scanned electron beam of 20 keV. We see that the irradiated region expands while the periphery is grooved. Closer inspection has shown that the volumes of the expanded and the grooved region are comparable, which implies that the electron-beam exposure induces a flow of As_2S_3 toward the irradiated region from the periphery.

In details, the deformation changes with exposure as follows: The height of the deformation monotonically increases with the exposure time t (1 - 200 s/line), the acceleration voltage V (10 - 30 kV), and the absorption current I (1 - 20 nA). The groove depth always approximates 1/2 - 1/3 of the height. In contrast, the lateral dimension does not change appreciably with t and I, while it increases with V: $\sim$ 3 μm at 10 kV and $\sim$ 6 μm at 30 kV.

Here, we should mark the following features. First, the deformation can be produced only by scanned or pulsed electron beams. If a point is continuously irradiated by a focused electron-beam spot, deformation has hardly appeared. Otherwise, if some area is raster-scanned, only the edge is deformed. Second, prominent deformations appear in thicker samples than $\sim 5\ \mu m$. In thinner films, deformation becomes smaller. Third, deformations disappear with flood exposure. For instance, the linear deformation produced by a line-scanned exposure of t = 60 s, V = 20 kV and I = 2 nA (shown in Fig. 5) was completely erased by a raster-scanned exposure of t = 30 min, V = 20 kV and I = 10 nA. Deformations become reduced also with annealing at 450 - 500 K, which is nearly equal to the glass-transition temperature [1]. Fourth, surface deformations induced under a fixed exposure condition (V = 20 kV, I = 10 nA, t = 60 s) have been the greatest in As_2S_3, intermediate in As_2Se_3 and GeS_2, and it has not been detected in As_2Te_3. No deformation was observed in crystalline As_2S_3 (orpiment).

These observations suggest that the surface deformation is induced through an athermal mechanism. If a temperature rise caused by electron-beam irradiation were responsible, spatially-fixed irradiation should provide a more prominent effect. In addition, if a thermal process governed the deformation, it would be greater in $As_2S(Se,Te)_3$ than in GeS_2, since the thermal effect becomes appreciable at around the glass-transition temperature, which is substantially lower ($\sim$ 450 K) in $As_2S(Se,Te)_3$ than that (760 K) in GeS_2 [1].

Figure 5. An atomic-force-microscope image of a 6 μm-thick As_2S_3 film on Si-wafer which has been exposed to an electron beam accelerated at 20 kV for 60 s/line. For the three lines from the left-hand to the right-hand side, the absorption currents are 10, 5 and 2 nA.

At present, electrostatic force and electroinduced fluidity seem to be responsible for the surface deformation. We first note that the range of electrons accelerated at 10 - 30 kV is $\sim 5\ \mu m$ [18] in the materials of interest, which is comparable to the width of the deformation. It is also known that electrons with $V = 10 - 30$ kV can induce electronic excitation, while the knock-on process can mostly be neglected [19]. Accordingly, for pulsed spot irradiation, we can envisage a hemispherical region with a diameter of $\sim 5\ \mu m$, where incident electrons are scattered (Fig. 6(a)). In the region, a number of electrons and holes are excited by the incident electrons: the number being equal to the order of $qV/E_g \simeq 10^4$ [19], where $E_g(\simeq 2$ eV [1]) is the bandgap energy of the material. However, since As_2S_3 is nearly insulating [1], incident electrons make the region negatively charged. These electrons are probably trapped at localized sites, giving rise to Coulomb repulsion among the sites, which forces to expand the hemispherical region. In addition, we have seen in 2. the photoinduced fluidity in As_2S_3, which implies that electron-beam exposure can also enhance the fluidity through generating a number of free carriers. Accordingly, as illustrated in Fig. 6(b), the free surface expands with accompanying peripheral depression, which is necessary to maintain the material density. Meanwhile electrons and holes will undergo dielectric relaxation and/or recombination, since holes are mobile in the materials of interest [1]. Under pulsed (or scanned) irradiation, this process will be repeated, giving rise to a prominent deformation.

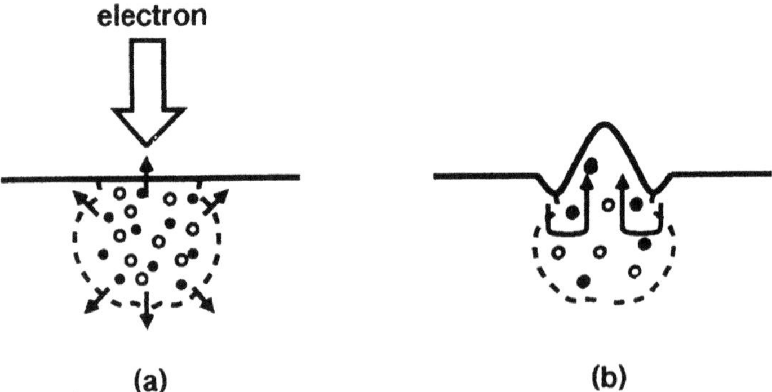

Figure 6. A model of the electron-beam-induced deformation (cross-sectional view). Solid and open circles represent electrons and holes, and arrows show the material flow.

5. Electron-beam-induced Chemical Modification

If we expose Ag-As-S glass to electron beams, only expansions appear. No depressions accompany in this case as shown in Fig. 7. The sample is an $Ag_{25}As_{25}S_{50}$ glass, which has been irradiated by a line-scanned electron beam accelerated at 10 kV. Absorption current was 0.5 nA, and exposure time was 10 s/line. We see that the irradiated region expands. A marked difference from that induced in As_2S_3 (described in 4.) is that the deformation in Ag-As-S can be induced also by spatially-fixed exposures.

Related features are summerized as follows: First, this phenomenon is observed in Ag-As-S(Se) glasses with the Ag content of 15 - 30 at.%. The phenomenon has not been detected in Cu-As-Se glasses. It is also mentioned that the geometrical change has hardly appeared in crystalline Ag-As(Ge)-S samples. Second, the deformation is prominent when $V \leq 10$ kV. Electron beams accelerated at $V = 20$ - 30 kV produce greater blurred deformations, and spatial resolution becomes worse. The beam appears too energetic. Third, similarly to that in 4, the relief pattern can be reduced with thermal annealing. However, a flood exposure of electron beams gives blurred deformation.

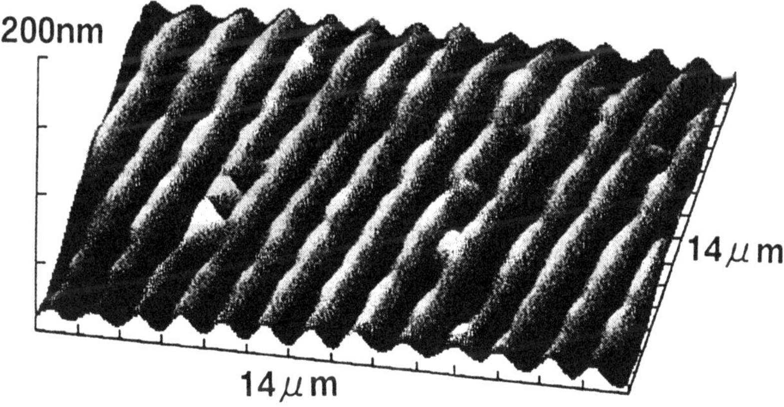

Figure 7. An atomic-force-microscope image of a grating pattern fabricated on an $Ag_{25}As_{25}S_{50}$ glass surface. The grating is written by laterally shifting a line-scanned electron beam.

As illustrated in Fig. 8, the volume expansion induced by electron beams is assumed to be caused by the accumulation of Ag^+ ions into an irradiated region. Here, we should first note that Ag-As(Ge)-S(Se) glasses with the Ag content higher than 10 at.% are the ion-conducting amorphous semiconductor, in which the electric conductivity being governed by Ag^+ ions and holes being more mobile than electrons [9, 20, 21]. When such Ag-As(Ge)-S(Se) glasses are exposed to electron beams, Ag^+ ions can migrate toward the irradiated region from surroundings, and in consequence, the Ag content at the irradiated region dramatically increases ($\sim$ 60 at.%), which has actually been demonstrated through compositional studies [9]. Therefore, the Ag-accumulated region will expand.

It might be assumed that volume contraction would occur at the region where Ag^+ ions are depleted, while such contractions have not been detected. It is plausible that the glass network consisting of covalent As-S bonds is fairly rigid, and accordingly Ag^+ ions can solely migrate in the As-S network without leaving any traces. Only Ag^+-accumulated regions will expand.

Note that the chemical modification can also be induced by light illumination [21], while no geometrical changes have been detected. We may assume that the non-existence of geometrical changes is due to a small changes ($\leq$ 5 at.%) in the Ag content upon light illumination.

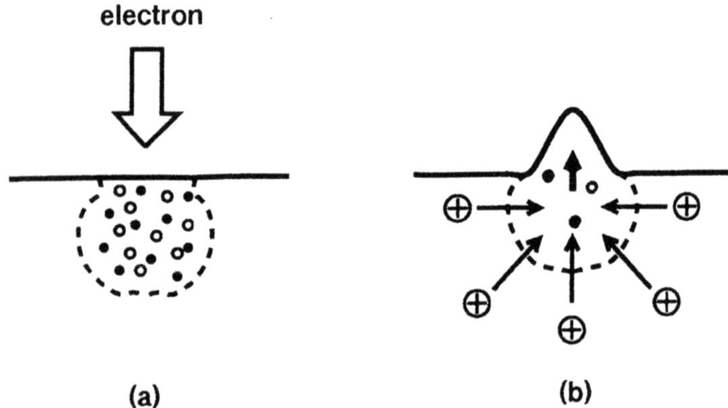

Figure 8. A model for the electron-beam-induced chemical modification (cross-sectional view). Solid, open, and crossed circles indicate, respectively, electrons, holes, and Ag^+ ions.

6. Excitation of Chalcogenide Glasses by Photon and Electron

It seems to be valuable to compare the effects induced by light and electron beams. These beams can easily be focused, which is important to prepare micro-patterns. Ion beams may also be focused, while these inevitably give compositional changes. Another common feature to electron and photon beams is that these can induce electronic excitation, which seems to cause the fluidity in As_2S_3. In contrast, the details are substantially different as summarized in TABLE 1.

TABLE 1. A comparison of photon and electron excitations. For super-bandgap illumination, see [22].

excitation	penetration depth [μm]	quantum efficiency	dominant effect
super-bandgap light (40 eV)	1	10^{-2}	structural
bandgap light	1	≤ 1	structural
sub-bandgap light	10^4 (10^2)	≤ 1	structural
e-beam (10 kV)	5	10^4	electrical

First, penetration characteristics are largely different. Bandgap and sub-bandgap (Urbach-tail) light can penetrate $\sim$ 1 μm and $\sim$ 1 mm for all the materials of interest [1]. Accordingly, bandgap illumination can excite only thin specimens, while sub-bandgap light can excite bulk glasses. (In detail, as described in 3., sub-bandgap illumination can practically modify As_2S_3 to a depth of $\sim$ 0.1 mm due to the self-focusing effect [15].) As demonstrated in 2. and 3., the longer penetration depth of sub-bandgap illumination is essentially important to induce the prominent deformations. In contrast, the penetration depth of electron beams depends upon two factors; the depth increasing with the acceleration voltage and decreasing with the density of irradiated materials. Typically, for As_2S_3 and 10 kV electrons, it is $\sim$ 5 μm. 10 kV electron beams can be focused into a spot with a diameter of 0.1 μm, while the penetration depth, which is approximately equal to the scattering range, seems to govern the size of deformations, as demonstrated in 4. and 5.

Second, the quantum efficiency, that is the number of excited carriers per an absorbed photon/electron, is different. For bandgap and sub bandgap illumination this is approximately unity. Needlessly, for sub-bandgap light the ratio of the absorbed to the incident photon number is much less than unity, and accordingly the external quantum efficiency appears to be very small. For electron beams, the quantum efficiency is approximately given by qV/E_g ($\simeq 10^4$ for $V = 10$ kV) as described previously.

Finally, since the electron has the negative charge, the irradiated region in a material charges negatively, if the material is insulating and the dielectric relaxation time is long. As we see in 4. and 5., this charge effect appears to govern the electron-beam effects. Electron beams may also induce structural effects, which are actually observed in ionic crystals and so forth [23], while the effect seems to be less appreciable in the present case. In contrast, photons can modify amorphous networks only through electronic excitation.

At present, the phenomena described in this article are observed only in chalcogenide glasses, and the reason may be of interest. Why are the phenomena unique to glasses ? It can be assumed straightforwardly that these dramatic changes cannot occur in crystalline materials, since *continuous* changes in structure and composition are very difficult. Because, the crystal possesses in principle a thermodynamically-equilibrium structure. In addition, in a polycrystalline material, the domain structure tends to kinetically suppress atomic motions. In a glass, which is thermally quasi-equilibrium, no distinct domains exist in principle, so that it can be deformed smoothly.

In addition, among a variety of glasses, the chalcogenide glass is unique in a respect that it behaves as a *soft glassy semiconductor*. The softness is due to the two-fold-coordinated chalcogen atoms, which are susceptible to exhibit electro-atomic responses. The glass exemplifies a flexible electron-lattice coupling system. In contrast, oxide glasses (and amorphous tetrahedral semiconductors) are more rigid due to three-dimensional network structures.

The author would like to thank N. Yoshida and N. Toyosawa for their assistence. The present work was partially supported by grants from Ministry of Education, Suhara Memorial Foundation, Casio Science Foundation, Showa Electric Wire and Cable Co. Ltd, and Hitachi Microcomputer Engineering Ltd.

References

[1] Elliott, S.R. (1991) Chalcogenide Glasses, in J. Zarzycki (ed.), *Materials Science and Technology* Vol. 9, VCH, Weinheim, pp. 375-454.

[2] Kolobov, A.V. and Elliott, S.R. (1991) Photodoping of amorphous chalcogenides by metals, *Adv. Phys.* **40**, 625-684.

[3] Shimakawa, K., Kolobov, A., and Elliott S.R., (1995) Photoinduced effects and metastability in amorphous semiconductors and insulators, *Adv. Phys.* **44**, 475-588.

[4] Tanaka, K., (1996) Photoinduced processes in chalcogenide glasses, *Current Opinion in Solid State & Mater. Sci.* **1**, 567-571.

[5] Saito, K., Utsugi, Y., and Yoshikawa, A., (1988) X-ray lithography with a Ag-Se/Ge-Se inorganic resist using synchrotron radiation, *J. Appl. Phys.* **63**, 565-567.

[6] Hisakuni, H. and Tanaka, K., (1994) Giant photoexpansion in As_2S_3 glass, *Appl. Phys. Lett.* **65**, 2925-2927.

[7] Hisakuni, H and Tanaka, K., (1995) Optical fabrication of microlenses in chalcogenide glasses, *Opt. Lett.* **20**, 958-960.

[8] Hisakuni, H and Tanaka, K., (1995) Optical microfabrication of chalcogenide glasses, *Science* **270**, 974-975.

[9] Yoshida, N. and Tanaka, K., Ag migration in Ag-As-S glasses induced by electron-beam irradiation, *J. Non-Cryst. Solids* (in press).

[10] Tanaka, K. *et. al* (Unpublished).

[11] Koseki, H., Odajima, A., (1982) Photo-induced stress relaxation in amorphous Se films, *Jpn. J. Appl. Phys.* **21**, 424-428.

[12] Tanaka, K., Hisakuni, H., (1996) Photoinduced phenomena in As_2S_3 glass under sub-bandgap excitation, *J. Non-Cryst. Solids* **198-200**, 714-718.

[13] Tanaka, K., Nakayama, S., and Toyosawa, N., The constant-photocurrent method applied to chalcogenide materials, *Philos. Mag. Lett.* (in press).

[14] Fritzsche, H., (1996) Photo-induced fluidity of chalcogenide glasses, *Solid State Commun.* **99**, 153-155.

[15] Hisakuni, H., Tanaka, K., (1994) Photo-induced persistent self-focusing in As_2S_3 glass, *Solid State Commun.* **90**, 483-486.

[16] Hamanaka, H., Tanaka, K., and Iizima, S., (1977) Reversible photostructural change in melt-quenched As_2S_3 glass, *Solid State Commun.* **23**, 63-65.

[17] Tanaka, K., (1980) Reversible photostructural change: mechanisms, properties and applications, *J. Non-Cryst. Solids* **35&36**, 1023-1034.

[18] Nishihara, H., Handa, Y., Suhara, T., and Kojima, J., (1978) Scanning-electron-microscope-written gratings in chalcogenide films for optical integrated circuits *Appl. Opt.* **17**, 2342-2346.

[19] Robbins, D.J., (1980) On predicting the maximum efficiency of phosphor systems excited by ionizing radiation *J. Electrochem. Soc.* **127**, 2694-2702.

[20] Tanaka, K., Itoh, M., Yoshida, N., and Ohto, M., (1995) Photoelectric properties of Ag-As-S glasses *J. Appl. Phys.* **78**, 3895-3901.

[21] Yoshida, N., Tanaka, K., (1995) Photoinduced Ag migration in Ag-As-S glasses, *J. Appl. Phys.* **78**, 1745-1750.

[22] Hayashi, K., Kato, D., and Shimakawa, K., (1996) Photoinduced effects in amorphous chalcogenide films by vacuum ultra-violet light, *J. Non-Cryst. Solids* **198-200**, 696-699.

[23] Itoh, N., Tanimura, K., (1986) Radiation effects on ionic solids, *Radiation Effects* **98**, 268-287.

METAL-PHOTODISSOLUTION IN AMORPHOUS SEMICONDUCTORS OF THE As-S SYSTEM

G. DALE, A.E. OWEN and P.J.S. EWEN
Department of Electrical Engineering
University of Edinburgh
The King's Buildings
Edinburgh EH9 3JL
Scotland, UK.

1. Introduction

It was just 30 years ago that Kostyshin et al. [1] first reported the phenomenon which is now generally known as metal-photodissolution in amorphous (or glassy) chalcogenides and the literature, up to about 1990-91, has been comprehensively reviewed by Kolobov and Elliott [2]. Since the initial paper by Kostyshin et al. there have been numerous experimental studies and several mechanisms have been proposed but there is still no generally accepted model or interpretation. The problem is in unifying all of the experimental data within a general model. There are several reasons for this difficulty. One is the large number of metal/chalcogenide systems investigated - different compositions of the chalcogenide glass and possibly also of the metallic source may give rise to different experimental characteristics. Most studies of the process involve thin film bilayers with the metal source either on top of or below the chalcogenide layer. In this case the metal-photodissolution may be influenced by the thickness of the respective film, both absolutely and in terms of a ratio (most experiments have used a thin metal layer, ≈ 20nm, on top of a chalcogenide layer which is several times thicker). In addition, the results with such bilayers may be significantly affected by the preparation method (e.g. thermal evaporation or sputtering etc) as well as the conditions used during film preparation. Some studies, on the other hand, relate to "bulk" materials which may cause further variations in the experimental results. This can be either a bulk metallic source with a thin film chalcogenide coating [e.g. 3], or it can involve the preparation of the chalcogenide glass in bulk form which is subsequently coated with a thin film of metal [e.g. 4]. A further and very important complication arises in devising a simple and direct method of investigating the metal-photodissolution process. For example, a method commonly used is to measure the resistivity of the thin metal layer source, as a function of time, as it dissolves into the chalcogenide. This technique requires an accurate calibration of the metal sheet resistance which is a strongly non-linear function of its thickness, particularly when the metal is very thin (e.g. ≤ 30nm).

A. Andriesh and M. Bertolotti (eds.),
Physics and Applications of Non-Crystalline Semiconductors in Optoelectronics, 45–60.

In a relatively short paper it is not possible to cover all aspects of the subject. We will concentrate therefore on some of our own recent work on a coherent set of experiments aimed primarily at measuring the kinetics of metal-photodissolution in evaporated films of the As-S system, by optical means. Wherever possible it will be instructive to relate the **photo**dissolution process with straightforward **thermal** dissolution (or diffusion). While it is known that several group I and II metals may be made to photodissolve in chalcogenide glasses [2], much the greatest amount of work reported in the literature has been done with Ag and, to a lesser extent, Cu. We will be concerned entirely with those two group Ib metals. The experimental data will be considered, briefly, in the context of relevant material from the literature.

2. Thermal and Photoinduced Dissolution of Ag and Cu into Evaporated As-S Films

2.1 OPTICAL MONITORING

Metal/chalcogenide bilayer systems were studied and three basic methods used to investigate the kinetics of metal dissolution:

(i) Measuring the optical transmission through the bilayer.

(ii) Monitoring the reflectivity of the metal layer during dissolution.
Either of these two methods can be used *in situ* or, alternatively, by taking measurements at appropriate intervals between cycles of heating/exposure. In general, it was found that wavelengths in the near-IR (900-2500nm) were preferred because they provide better discrimination between the presence or absence of undissolved metal which is highly reflecting (and absorbing) in that spectral region.

(iii) The method of Firth et al. [5] in which near-normal reflection from the chalcogenide side is measured to follow the dissolution process by recording the periodic oscillations resulting from the interference between two reflected beams: one from the chalcogenide/air interface and the other from the interface between the metal-free chalcogenide and the growing reaction product underneath.

The two simpler methods, (i) and (ii), have the disadvantage that they yield only an end-point of the metal-dissolution process, i.e. when the transmission is saturated at a maximum or the reflectance at a minimum. This means that a rate-law for the process of metal-dissolution cannot be determined directly from the experimental data but must be assumed, generally as either a square-root or linear time dependence for the growth of the reaction-product layer. The actual time dependence of photodissolution can follow more complicated relationships and it may divide into several phases. For the thermally stimulated effect, however, a dependence on $\sqrt{t}$ for the penetration of Ag ions seems reasonable because the process is expected to be diffusion-limited.

When these two optical methods have been used in the present work the rate-law has been assumed to follow a linear or square-root time dependence for the photodissolution and thermally stimulated dissolution processes respectively. This is what Plocharski et al. [6] observed when studying the illuminated and the dark phenomena using the method of continuous resistivity monitoring. Zekak [7] also observed a linear rate-law for photodissolution when using a similar white-light source to that which was used in the present work.

The interference technique, method (iii), has the advantage that it gives an indication on how far the reaction has proceeded, rather than just the end-point. Information relating to intermediate stages is obtained from the oscillations which have a periodicity of (λ/2n), where λ is the wavelength of the visible probe-beam and n is the refractive index of the metal-free As-S layer [5]. IR wavelengths cannot be used in this method because they are weakly absorbing in both the metal-rich and metal-free glass layers and therefore do not offer the required discrimination to monitor consumption of the As-S layer. The progression of the reaction-product layer growth (i.e. As-S layer consumption) can then be shown by plotting thickness increments corresponding to (λ/4n) for each successive maxima and minima in the reflectivity curve.

The method is illustrated by modelling the reflectivity of the metal-chalcogenide system during evolution of the intermediate reaction-product phase, as shown in figure 1. The system has been treated as a multilayer stack of thin films with the Ag, Ag-As-S and As-S being considered as homogeneous films deposited on a transparent glass substrate, in that order. This model is slightly different from the experimental arrangement actually used where Ag is on top and the incident beam passes through the transparent substrate. That should have little effect on the principle because the substrate is very much thicker than the coherence length of the monochromatic beam used. Each of the layers in the model is characterised by a refractive index, *n*, and extinction coefficient, *k*. The values used in the present model and given in figure 1 have been taken from the literature [5]. Pedrotti et al. [8] have given a logically consistent method for modelling a multilayer stack of films by representing each individual layer as a single transfer matrix so that the total system can be presented by a transfer matrix resulting from the successive multiplication of all the individual matrices. From this overall transfer matrix the reflection and transmission coefficients can be calculated. This matrix approach was used to generate the curve in figure 1, which gives a computer simulation of the process of metal-dissolution. The initial system (t = 0) consists of two layers, the thickness of the reaction-product layer being taken as zero before any metal-dissolution has occurred. The intermediate layer is then assumed to grow progressively and simultaneously in both directions, with the Ag layer consumption occurring at one-third the rate of the As-S layer consumption. The form of this curve (with constant T) would be unchanged if time was plotted on the abscissa provided that the Ag-As-S layer growth occurred at a linear rate.

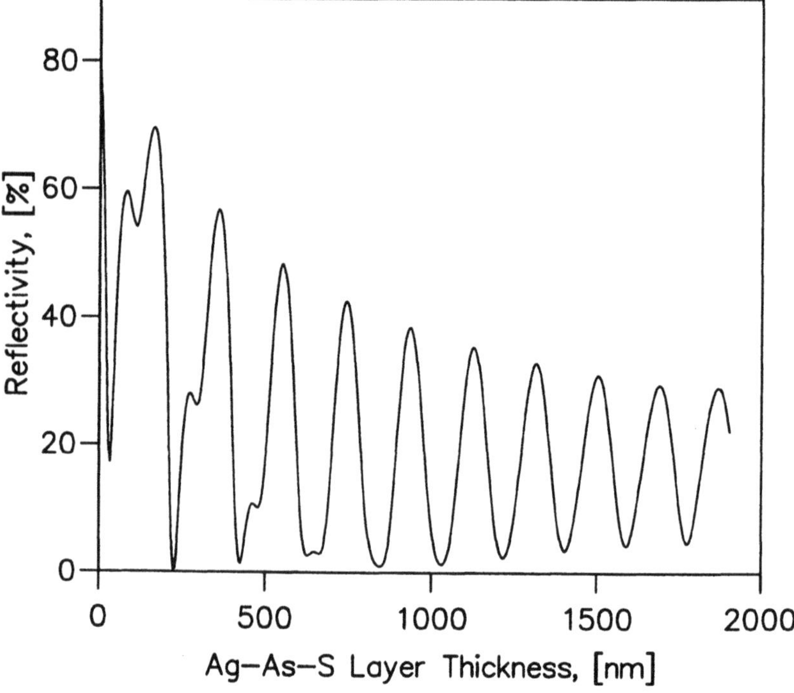

Figure 1: simulation of a 400nm Ag layer dissolving into an $As_{40}S_{60}$ layer. Refractive indices for the $As_{40}S_{60}$, Ag-As-S reaction-product layer and Ag layer are respectively assumed as 2.27, 3.3 and 0.5. The corresponding extinction coefficients are assumed as 0.0, -0.1 and -3 respectively. The Ag layer is taken to be on a transparent substrate with refractive index 1.52. It is also assumed that the Ag-rich/Ag-free glass interface propagates through the Ag-free glass region at a rate three times the rate of Ag layer consumption.

The oscillations from the model appear to be near-sinusoidal, except during the initial period where the curve is more complicated. This is because the Ag-As-S intermediate layer is relatively thin at this stage, such that the total absorption determined by the absorption-thickness product (αd), is insufficient to eliminate interference from the Ag/Ag-As-S interface.

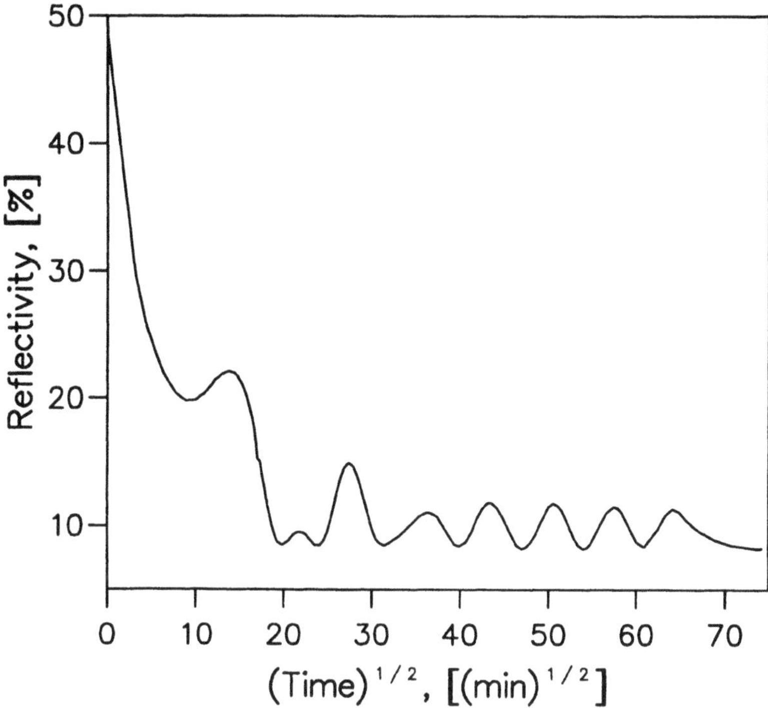

Figure 2: reflectivity curve obtained from near-normal incidence ($\lambda = 633$nm) on a Ag/$As_{33}S_{67}$ bilayer during thermally stimulated dissolution of the 475nm Ag layer into the $As_{33}S_{67}$ glass film.

An example of a corresponding reflectivity curve obtained experimentally from thermally stimulated dissolution in a Ag/$As_{33}S_{67}$ bilayer system is given in figure 2. Note that in this case the data is plotted against $\sqrt{t}$ on the abscissa, rather than the layer thickness as in figure 1. The initial high reflectivity falls off rapidly as Ag diffusion proceeds and an intermediate reaction-product layer is formed. As in the case of the model in the previous figure, some of the initial oscillations appear anomalous, perhaps also due to the relatively thin intermediate layer at the early stages. As the reaction proceeds, however, the oscillations become more regular and near-sinusoidal in character. The presence of these oscillations is an indication that the intermediate layer is growing as a distinct material phase providing a sharp change in the optical constants with depth.

The observation of these oscillations in the present work is important because the reflectivity method has not, to the author's knowledge, previously been applied to the thermal process where the profile of the Ag concentration might be expected to follow a continuous Fickian distribution rather than the well-known step-like characteristic associated with photodissolution. In raising the issue it should also be pointed out that the origin of the observed metal-dissolution may not be exclusively thermal, but may also be stimulated by the incident beam used to monitor the oscillations in reflectivity, albeit a very weak beam with below band-gap wavelength. The method has, however, recently being applied to the case of thermally enhanced **photo**dissolution by Wagner et al. [9]. The existence of oscillations indicates that the reaction-product is propagating as a distinct material phase which may be explained by a step-change in the concentration dependence of the diffusion coefficient. The low amplitude of the experimentally observed oscillations is however a deviation from the idealised model presented in figure 1. This may be interpreted as an indication that the interface between the Ag-rich and Ag-free glass phases is not step-like but more gradual. Unfortunately, it is not possible to relate the absolute magnitude of the oscillations to the sharpness of the step; a low magnitude could just as easily result from factors related to the experimental set-up. On the other hand, a relative change in the amplitude of the oscillations during the experiment may well indicate a broadening of the "diffusion" front as the reaction proceeds.

2.2 TEMPERATURE DEPENDENCE

2.2.1 *Thermal and Photodissolution of "Thin" Ag Films*

The simple optical transmission method ((i) in para. 2.1, above) was used to monitor the progress of thermally stimulated metal dissolution in $Ag/As_{33}S_{67}$ bilayers with film thicknesses of 95 and 650nm respectively. (It was not possible to use the preferred bilayer reflectively technique, (iii), because with such thin Ag layers the periodic oscillations could not be resolved). The samples were heated *in situ* in an IR spectrophotometer and in this case, therefore, there was also some exposure to weak illumination (<0.1 Wcm^{-2}) from a broad-band IR source. A rather narrow temperature range, 98.5 - 117°C, was investigated, partly because of the practical difficulty of observing thermal dissolution at low temperatures where the diffusion times become very long.

The recovery process for the optical transmission of the heated bilayers is given in figure 3. The average rate of change of optical transmission is different for each curve, which correspond to different temperatures, the most rapid change occurring for the greatest temperature. The diffusion coefficient of Ag obviously increases with increasing temperature, as normally expected. Using the assumptions in the previous section, the diffusion coefficients for Ag in the $As_{33}S_{67}$ glass film for the stated temperature range have been calculated as between 1.19 - 8.89 × $10^{-14}cm^2sec^{-1}$. A similar calculation from the data of Mizuno et al. [10] gives $D = 1.6 \times 10^{-14}cm^2sec^{-1}$ for the thermal diffusion of a 30nm Ag layer into bulk $As_{40}S_{60}$ at 120°C. This compares with $D = 8.89 \times 10^{-14}cm^2sec^{-1}$ at 117°C from figure 7.6, i.e. the diffusion coefficient in the present data is approximately 5.5 times that observed by Mizuno et al. at similar temperatures. The greater diffusion coefficient found in the present work is what would be expected from the differences in composition of the As-S host materials.

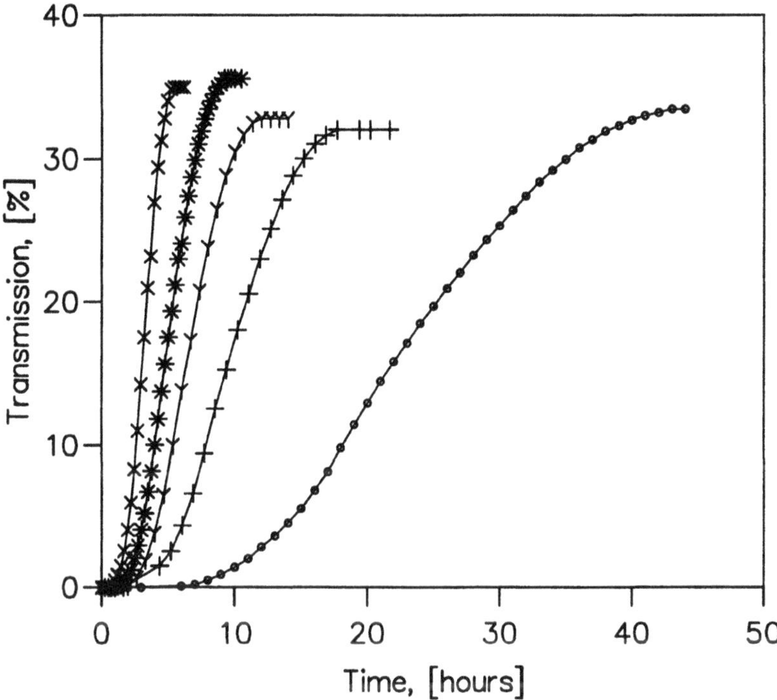

Figure 3: optical transmission of $Ag/As_{33}S_{67}$ bilayers (95/650nm) during thermally stimulated metal-dissolution at T = 98.5, 104.7, 107.8, 110.9, and 117°C for the curves from right to left respectively. The calculated diffusion coefficients are D = 1.19, 2.78, 4.63, 5.53 and 8.89 $\times 10^{-14} cm^2 sec^{-1}$ respectively.

For example, Tanaka [11] found that higher temperatures were required to effect similar levels of Ag diffusion into bulk $As_{40}S_{60}$ glass compared with bulk $As_{33}S_{67}$ glass. It has also been found that the photodissolution rate is a maximum for the composition with approximately 67 at.% of sulphur [12]. A similar set of data has been obtained for the same $Ag/As_{33}S_{67}$ bilayer system with respective film thicknesses of 120 and 650nm. The thermally stimulated metal-dissolution in these samples was studied by annealing the bilayers face-down on a hot-plate surface **in the dark** using the broader temperature range of 90-150°C. In this case the end-point of the process, i.e. the point of Ag exhaustion, was determined by the metal-side reflectivity method, i.e. (ii) in section 2.1.

The Arrhenius plots of the diffusion coefficients for these samples together with those obtained using the optical transmission method are shown in figure 4. For the samples annealed on the hot-plate, the calculated diffusion coefficients show good linearity on the Arrhenius plot with a calculated activation energy of E_A = 1.4eV. This is comparable with the figure of 1.55eV obtained by Plocharski et al. [6] for the thermal diffusion of Ag (d = 17-20nm) into $As_{40}S_{60}$ films in the temperature range 120-175°C.

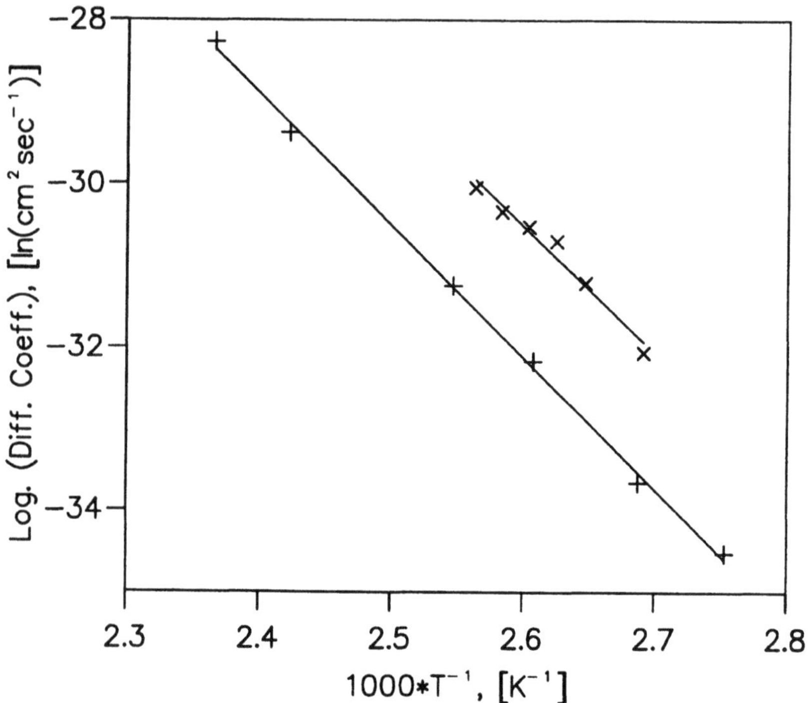

Figure 4: Arrhenius plots for the thermal diffusion of Ag into $As_{33}S_{67}$ layers. X's represent samples heated in temperature cell which had some broad-band illumination from the IR spectrophotometer source (E_A = 1.33eV). +'s represent samples heated on a hot-plate in the dark (E_A = 1.41eV).

The calculated diffusion coefficients for the samples heated within the IR spectrophotometer also show approximate linearity on the Arrhenius plot. The activation energy E_A = 1.33eV, which is marginally less than for the samples heated on the hot-plate, although the temperature range is more restricted. In addition, the diffusion coefficients for these samples appear significantly larger. These two observations are consistent with weak light from the broad-band spectrophotometer source increasing the rate of Ag-dissolution for the samples heated *in situ*.

The effect of thermally enhanced **photo**dissolution was also studied by using the hot-plate approach described earlier. The bilayer samples were taken from the same evaporations as those used for the dark process just discussed. A 50W quartz-tungsten-halogen (QTH) source was used to provide white-light illumination in the spectral region for which it is known that photodissolution is sensitive and the sample irradiance was measured at ~3 mWcm^{-2}. The rate of metal-dissolution was defined as the ratio of the original Ag thickness to the time required for exhaustion of the Ag layer, as indicated by the end-point of the Ag-side reflectivity. Thus the rate-law for the process was assumed to be linear, which need not be the case, although that is was observed by Zekak [7] who also used a white-light source.

An Arrhenius plot for the process is given in figure 5, the photodissolution taking place in the temperature range 30-140°C. The activation energy for the photodissolution process calculated from this data is 0.2eV, i.e. E_A has been reduced by around one order of magnitude by the illumination. It compares with values of E_A of ~0.24eV and 0.18eV reported by Zekak [7] and Plocharski et al. [6] (respectively) for Ag photodissolution into $As_{40}S_{60}$ evaporated films.

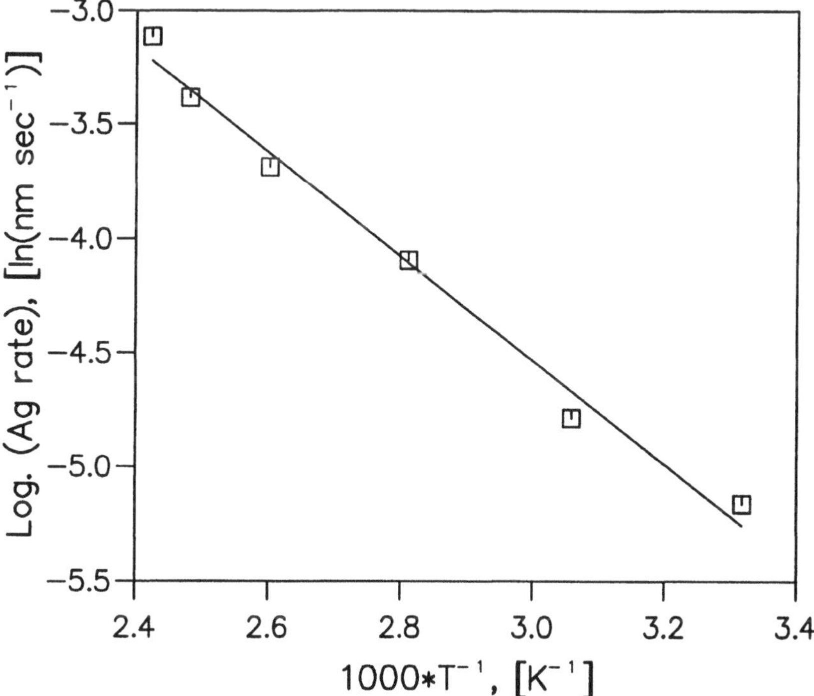

Figure 5: Arrhenius plot for photodissolution of 150nm Ag layer into 650nm $As_{33}S_{67}$ layer stimulated by a QTH source; $E_A = 0.2$eV.

Wagner et al. [9] considered the $Ag/As_{33}S_{67}$ system and identified two stages for the process based on chemical kinetics modelling. The first stage followed an exponential rate-law, whilst the second was linear. The two stages had E_A = 0.1 and 0.15eV respectively.

2.2.2 *Thermally Stimulated Dissolution of "Thick" Ag Layers*

To study the thermally stimulated dissolution of thicker Ag films into evaporated $As_{33}S_{67}$ films the interferometric reflectivity method was used. This is the more complicated but preferred technique for studying the dissolution of thicker metal layers because it provides information about the thickness of the reaction-product at intermediate stages of the process. The samples consisted of $Ag/As_{33}S_{67}$ evaporated bilayers with respective film thicknesses of 475 and 1730nm. The bilayers were heated to 112.5, 122.5, 132.5, 162.5°C and probed with monochromatic red light (λ = 633nm, ~2 $mWcm^{-2}$) from within the spectrophotometer. Data showing the movement of the diffusion front of the A-As-S layer (estimated as 4/3 times the thickness of the $As_{33}S_{67}$ layer which is actually measured by the method) are plotted in figures 6 and 7. Figure 6 is for T = 162.5°C and straight-line fits have been drawn through the points (corresponding to the oscillation peaks and troughs) plotted against the square-root of time.

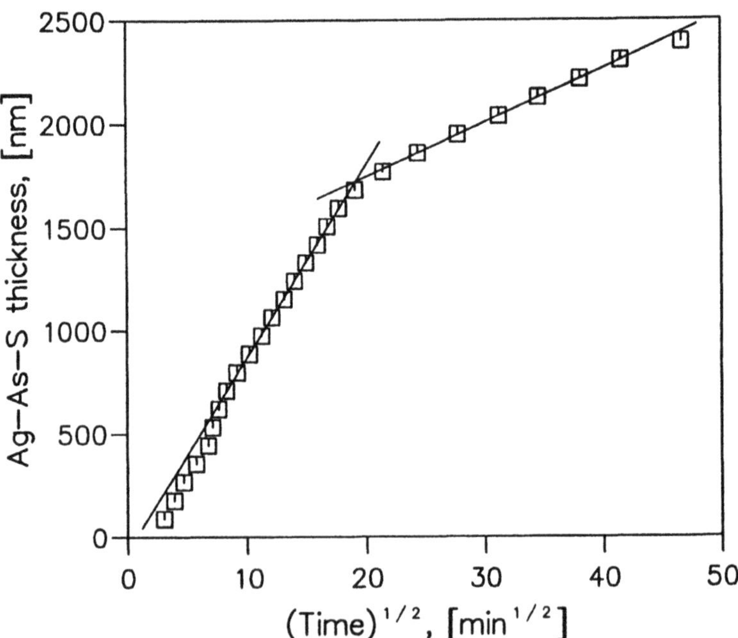

Figure 6: growth of intermediate Ag-As-S layer in $Ag/As_{33}S_{67}$ (475/1730nm) monitored by the oscillations obtained from reflectance interferometry when the sample is heated to 162.5°C.

Thus, the diffusion front propagates through the $As_{33}S_{67}$ film following a clear square-root time dependence, the diffusion coefficient being obtainable from the gradient of the straight line fit. After a period of time a second stage of the diffusion process begins, apparently with a significantly reduced diffusion coefficient. The transition between these two stages might at first be thought to be related to the exhaustion of the Ag layer source, but visual examination of the samples indicated that there was still significant Ag left on the surface after this transition.

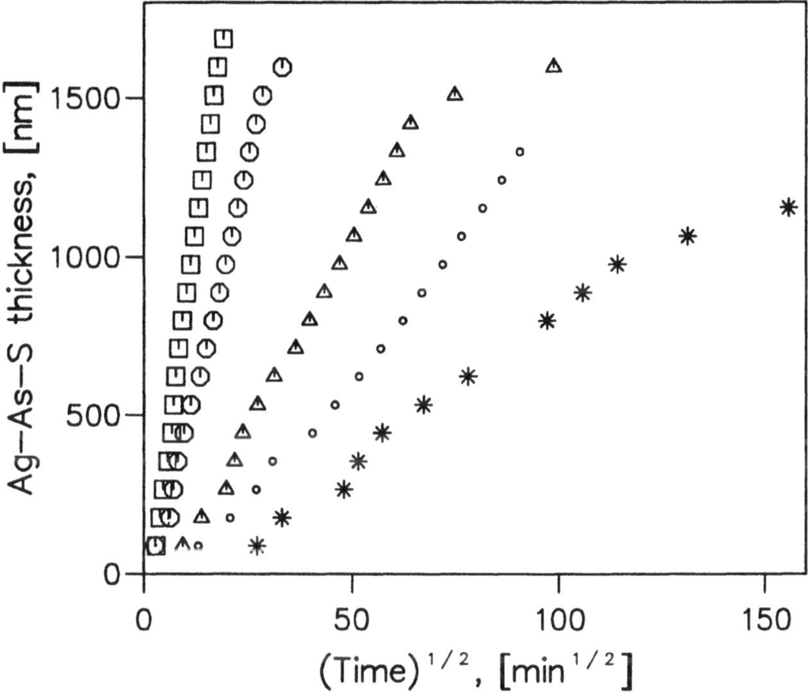

Figure 7: thermally stimulated dissolution of "thick" Ag layers (d = 475nm) into evaporated $As_{33}S_{67}$ glass films monitored by the oscillations in optical reflectivity at 162.5, 152.2, 132.5, 122.5 and 112.5°C reading consecutively from the steepest gradient downwards.

This type of two-stage behaviour has been observed in **photo**dissolution experiments by other researchers [9] who have also concluded that it is not connected with exhaustion of the Ag. During the second stage the observed oscillations have a very small amplitude, which might suggest a spreading of the Ag-As-S diffusion front into a less sharp, more gradual interface. A sharp interface propagating towards the upper $As_{33}S_{67}$ surface should result in a clear end-point in the process, evident from the optical reflectivity curves. Such end-points were not observed in the present examples, which is consistent with a continuous profile in Ag concentration at the later stages of the process. The final thickness of the reaction product in figure 6 also appears to exceed the total (Ag + $As_{33}S_{67}$) for the actual sample (d = 2205nm) but that could be accounted for by the known volume expansion of the reaction product [13]. An Arrhenius plot for the diffusion coefficients derived from the slopes of the lines in figure 7 gives an activation energy E_A = 1.3eV which is close to the 1.4eV found for the thermal process using the hot-plate (para. 2.2.1, figure 4). The slightly smaller E_A may be due to some weak optical stimulation in the present case, though that would be expected to be small for the very low irradiance of the red probe beam.

2.2.3 *Thermally Stimulated Dissolution of Cu Layers*

Although Kostyshin et al. [1] identified Cu as a suitable metal source for photodissolution in their original paper, there has been much less work on Cu/As-S in comparison with the more commonly studied Ag/As-S systems. It is known however that the reaction between Cu and the As-S film can occur spontaneously, i.e. without illumination at room temperature and this led Zekak [7] and Ewen et al [12] to investigate Cu/Ag alloys with the aim of increasing the photodissolution rate in an controllable manner.

We have therefore measured the thermally stimulated dissolution of Cu films with thickness 150nm into 1100nm thick $As_{40}S_{60}$ layers in the temperature range 87.5 - 137.5°C, with the same optical reflectivity method used for the thicker Ag films described in the previous paragraph. As with the case of thermally stimulated dissolution of thick Ag layers, the presence of oscillations in the reflectivity curve indicates that the Cu-As-S layer grows as a relatively distinct region. From plots of the peaks and troughs of the oscillations against $\sqrt{t}$, diffusion coefficients have been calculated and the results, plotted in Arrhenius form, are shown in figure 8 from which an activation energy E_A = 1.06eV is obtained for the thermal dissolutions of Cu in $As_{40}S_{60}$. The point at 111°C, designated by an asterisk (*) in figure 8 was obtained by the simple optical transmission technique ((i) in para. 3.1) assuming a $\sqrt{t}$ time dependence.

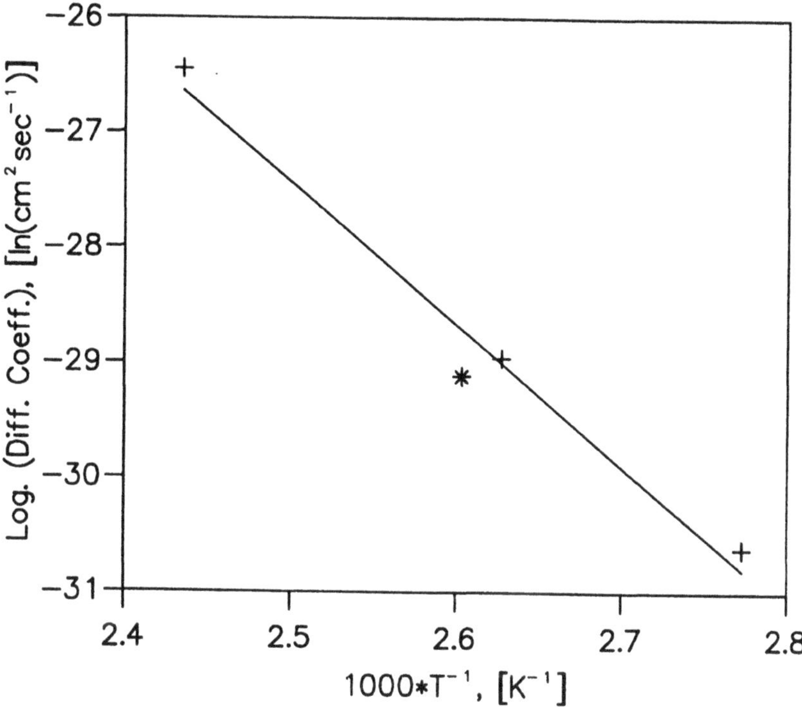

Figure 8: Arrhenius plot for thermally stimulated dissolution of Cu; (*) represents a diffusion coefficient obtained from simple optical transmission measurements.

3. Discussion

Figure 9 is a simplified phase diagram for the Ag-As-S system showing, in particular, the small region of homogeneous glass formation near the middle of the triangle. That small glass forming region is roughly centred on the composition $AgAsS_2$, which is also known in two polymorphic crystalline mineral forms: smithite and trechmanite [14]. The experiments described in section 2 were concerned mostly with the combination $Ag/As_{33}S_{67}$ for the good reason that the tie-line joining the two components passes through the central glass forming region and through the composition $AgAsS_2$. It is known that the rate of Ag photodissolution in the As-S system is a maximum at around the $As_{33}S_{67}$ composition [12]. Moreover, there is evidence that provided some of the Ag source remains undissolved, with $As_{33}S_{67}$ as the starting point, the final Ag concentration is a minimum (compared with As-S compositions on either side) [15], and that the final reaction product is a homogeneous glass of composition $AgAsS_2$ (i.e. 25 atomic % Ag) [16, 17].

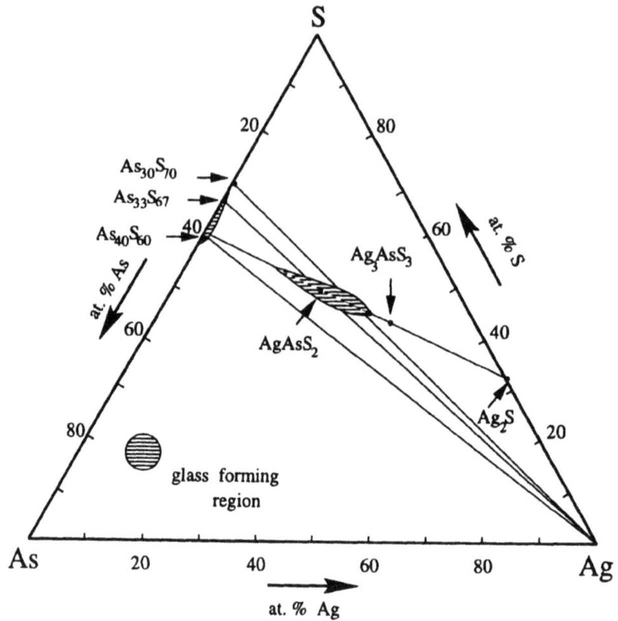

Figure 9: Simplified phase diagram of the Ag-As-S system showing the regions of glass formation and the location of important known crystalline compounds.

The most significant point from the results of the kinetic experiments on the $Ag/As_{33}S_{67}$ combination is to be found in a comparison of the activation energies for the two processes: thermally stimulated and photo-dissolution of Ag, the former having much the larger activation energy, i.e.-

thermal dissolution, $E_A \approx 1.4eV$
photodissolution, $E_A \approx 0.2eV$.

The important comparison to be made is with activation energies for silver ion (Ag^+) conduction (E_σ) and the self-diffusion of silver (E_{SD}) in the same or similar compositions in the absence of illumination. It has been known for some time that silver chalcogenide glasses are mixed electronic/ionic conductors and that if the silver content is increased beyond a few atomic percent, ionic (Ag^+) (dark) conduction rapidly becomes dominant [18, 19]. Hajto et al. [20] have measured the d.c. conductivity (and thermopower) of bulk Ag-As-S glasses, including the $AgAsS_2$ composition. Kawamoto and Nishida have studied both d.c. conductivity and silver self-diffusion in similar bulk glasses [18]. The activation energies, from both sets of authors, are:

d.c. conductivity, $E_\sigma \approx 0.6eV$ [20]
$E_\sigma \approx 0.55eV$ [18]

and

Ag self-diffusion, $E_{SD} \approx 0.55eV$ [18]

Plocharski et al. report a rather lower activation energy for d.c. conduction, $E_\sigma \approx 0.34eV$, in a glass with a somewhat greater Ag content (34 atomic %) [6]. Kawamoto and Nishida point out that their d.c. conductivity and self-diffusion data are consistent with the Nernst-Einstein equation, with a correlation coefficient in the region 0.4 - 0.5, implying that silver self-diffusion in the $AgAsS_2$ glass involves the same transport mechanism as d.c. conduction, i.e. the migration of silver ions (Ag^+). Relating this to the present experiments and, in particular, the much larger activation energy for thermal dissolution ($E_A \approx 1.4eV$), the conclusion must be that whatever else the rate limiting mechanism is in the **thermally** stimulated dissolution of silver it **is not** the diffusion of silver ions. The comparatively large energy involved suggests that it is a reaction controlled process, in which case it is more likely to be a reaction at the reaction product (notionally $AgAsS_2$)/$As_{33}S_{67}$ interface rather than at the Ag/product interface. Presumably the reaction product immediately adjacent to the metallic silver is saturated with its "equilibrium" concentration of Ag.

The crucial question is: what is it about the kinetic barrier to the reaction between the silver-chalcogenide ($AgAsS_2$, or something close to it) and the unreacted chalcogenide ($As_{33}S_{67}$ in the present case) which is so sensitive to light that on illumination its activation energy is reduced by almost an order of magnitude (~1.4eV -> 0.2eV)? That question remains unresolved.

4. References

1. Kostyshin M.T., Mikhailovskaya E.V. and Romanenko P.F. (1966). Photographic sensitivity effect in thin semiconducting films on metal substrates, *Sov. Phys. Solid State*, **8**, 451-452.
2. Kolobov A.V. and Elliott S.R. (1991). Photodoping of amorphous chalcogenides by metals, *Adv. in Phys.*, **40**, 625-684.
3. Ewen P.J.S., Taylor W.T., Firth A.P. and Owen A.E. (1983). A Raman study of photochemical reactions in amorphous As_2S_3 on polycrystalline Ag_2S substrates, *Phil. Mag.*, **B48**, L15-L21.
4. Holmquist G.A. and Pask J.A. (1979). Reaction and diffusion in silver-arsenic chalcogenide glass systems, *J. Am. Ceram. Soc.*, **62**, 183-188.
5. Firth A.P., Ewen P.J.S. and Owen A.E. (1985). An optical reflectivity study of Ag photodissolution into As-S films, *J. Non-cryst. Sol.*, **77-78**, 1153-1156.

6. Plocharski J., Przyluski J. and Teodorczyk M. (1987). Ionic conductivity of Ag photodoped As_2S_3 glass, *J. Non-cryst. Sol.*, **93**, 303-310.

7. Zekak A. (1993). *The Optical Characterisation and Kinetics of Ag Photodissolution in Amorphous As-S Films*, Ph.D. Thesis, University of Edinburgh, Scotland, UK.

8. Pedrotti F.L. and Pedrotti L.S. (1987). *Introduction to Optics*, Chapter 22, Prentice Hall Inc.

9. Wagner T., Vlcek M., Smrcka V., Ewen P.J.S. and Owen A.E. (1993). Kinetics and reaction products of the photo-induced solid state chemical reaction between silver and amorphous $As_{33}S_{67}$ layers, *J. Non-cryst. Sol.*, **164-166**, 1255-1258.

10. Mizuno H., Tanaka K. and Kikuchi M. (1973), Photo- and thermal-diffusion of metals into As_2S_3 glass, *Sol. St. Commun.*, **12**, 999-1001.

11. Tanaka K. (1994). Photodoping of Ag in Ag-As-S glasses, *J. Non-cryst. Sol.*, **170**, 27-31.

12. Ewen P.J.S., Zakery A., Firth A.P. and Owen A.E. (1988). Optical monitoring of photodissolution kinetics in amorphous As-S films, *Phil. Mag.*, **B57**, 1-12.

13. Lee J., Ogawa T., Kudo H. and Arai T. (1994). Volume expansion and Ag doping amounts in amorphous As_2S_3, *Jpn. J. Appl. Phys.*, **33**, 5865-5869.

14. Roland G.W. (1970). Phase relations below 575°C in the system Ag-As-S, *Econ. Geol.*, **65**, 241-252.

15. Petrova S., Simidchieva P. and Buroff A. (1984). *Proceedings of the International Conference on Amorphous Semiconductors*, Eds: Farhi-Vateva E. and Buroff A., Bulgarian Acad. Sci., Sofia, p256.

16. Firth A.P., Ewen P.J.S. and Owen A.E. (1982). Structural changes in amorphous arsenic sulphide films on photodoping with silver, studied by Raman spectroscopy in *Structure of Non-Crystalline Materials*, Eds: Gaskell P.H., Parker J.M. and Davis E.A., Taylor and Francis, 286-293.

17. Steel A.T., Greaves G.N., Firth A.P. and Owen A.E. (1989). Photodissolution of silver in arsenic sulphide films, *J. Non-cryst. Sol.*, **107**, 155-162.

18. Kawamoto Y. and Nishida M. (1977). Silver diffusion in $As_2S_3-Ag_2S$, GeS_2-GeS-Ag_2S and $P_2S_5-Ag_2S$ glasses, *Phys. Chem. Glasses*, **18**, 19-23.

19. Kitao M., Ishikawa T and Yamada S. (1986). Estimation of ionic conduction in glassy As_2Se_3: Ag, *J. Non-cryst. Sol.*, **79**, 205-207.

20. Hajto E., Belford R.E., Ewen P.J.S. and Owen A.E. (1991). Electrical properties of silver doped As-S glasses, *J. Non-cryst. Sol.*, **137-138**, 1039-1042.

Amorphous superlattices of chalcogenides

E. VATEVA
G. Nadjakov Institute of Solid State Physics, Bulg. Acad. Sci.
72 Tzarigradsko Chaussee Blvd., 1784 Sofia, Bulgaria

1. Introduction

After the first work on superlattices by Esaki and Tsu [1] in 1970 the investigations on low-dimensionality hetero- and homostructures increase progressively. The molecular beam epitaxy opened possibilities for the growth of semiconductor atomic layer upon atomic layer. This and other techniques have been used to grow ultrathin, well-controlled systems. The various structures may be classified by means of their band diagrams. Single and multiple quantum wells, single and double barrier tunneling structures, incoherent multilayer tunneling structures and superlattices, are the main ones [2]. It can be mentioned that in the structures defined as multiple quantum wells the barrier thickness may be large enough to prevent tunneling. In the superlattice structure disorder and scattering must be low enough to allow the coherent superlattice band states to be built up and to prevent destruction of the phase coherence between the tunneling states by disordered interface fluctuations. Nevertheless, usually all the mentioned multilayers are refered as superlattices. Since the mid-Seventies, bandgap engineered multilayer structures with desired properties and new devices based on them have been prepared from III-V materials.

Abeles and Tiedje [3] are among the first who have shown in 1983 that the range of materials from which superlattices can be fabricated can be extended to hydrogenated amorphous semiconductors. The amorphous multilayers (a-ML) or superlattices (SL) are synthetic semiconductor materials which can be divided into two groups: (i) compositional ML composed of thin barrier and well layers of two types of materials in alternating order and (ii) doping modulated ML composed of one and the same material but differently doped. The a-ML as counterparts of the crystalline structures may also be divided in multiple quantum wells, incoherent ML tunneling structures and superlattices. The a-ML have some advantages, the main of which is that the problems of lattice matching of the different materials do not exist. A greater choice of material combinations is possible and simpler deposition techniques can be used. With bandgap engineering new properties can be achieved based both on classical effects, connected mainly with the influence of the interlayer barriers, and on quantum-size effect (QSE) when the sublayers are sufficiently thin and the period (or the well layer width) are comparable with the de Broglie wave lengths of the electrons and holes. It is under discussion up to now whether or not the QSE exists, because in amorphous semiconductors the coherence length of the wave functions is smaller than that in crystalline semiconductors. Finally, new

A. Andriesh and M. Bertolotti (eds.),
Physics and Applications of Non-Crystalline Semiconductors in Optoelectronics, 61–75.

application can be expected, for example in solar cells, in thin film transistors, in photodetectors and xerographic photoreceptors.

The a-ML are usually prepared on the basis of a-Si:H or a-Ge:H, which are 4-fold coordinated materials. The chalcogenide a-ML are a new type of a-ML which we began to study intensively since 1989 [4-7]. Little work has been done before on amorphous chalcogenide ML [8-10] and some authors maintain that such structures could not be prepared with a period less than 10 nm. We have supposed that using suitable deposition methods high quality SL on the basis of Se can be obtained with thinner, smooth and parallel sublayers. The advantages of the chalcogenides are associated with their lower coordination. The a-Se is composed of trans-polymer chains and cis-ring molecules. Covalent bonds exist between nearest-neighbor atoms and van der Waals forces - between chains and rings . The coordination is nearly 2-fold which ensures a greater flexibility than in the layers of 4-fold coordinated materials. The greater flexibility has been confirmed by Raman scattering investigations (as pointed out below). So a-Se as well as its alloys are suitable for thin sublayers with sharp interfaces and the observation of QSE can be expected.

Using step-by-step deposition methods we have composed novel amorphous chalcogenide ML of pp^+pp^+ type, Se/Se-Te, and pnpn type Se/CdSe and Se-Te/CdSe. The first results are summarized by Vateva and Nesheva [11]. Here new results will be taken into account when presenting and discussing the preparation conditions, structural peculiarities and stability, electrical and optical properties, including phenomena, which could be related to QSE and possibilities for applications.

2. Preparation conditions

Two step-by-step deposition methods are chosen - laser sputtering and thermal vacuum evaporation. Previous experiments on SL based on a-Si:H have shown that step-by-step deposition leads to smoother layers than by continuos deposition [12]. Such a result may be expected with greater probability for Se-based a-ML taking into account that in a-Se and its alloys there exist slow relaxations to the stable state even at 300 K.

The laser sputtering was used predominantly for preparing Se/Se-Te ML (although Se/Bi-Ge-Se and Se/CdSe ML have also been obtained). The deposition was realized in a vacuum chamber (Fig. 1) with a pulsed Nd laser ($\leq$30 J in $\sim$2 ms duration pulse) from two alternatively changed targets from pressed glassy or crystalline powder. Usually Corning 7059 glass substrates held at room temperature have been used. The layer thicknesses were measured in situ with quartz thickness monitor (MIKI-FFV). The pause between the laser pulses was $\sim$15 s. The layer thickness change after each pulse was kept nearly the same. This method has two important advantages: it insures an effectively higher vacuum during the deposition step than in the vacuum chamber which affects on the properties of some layers and also insures nearly the same composition in the layers as in the target which is difficult to be reached for the Se-Te alloys by evaporation.

The step-by-step thermal vacuum evaporation was elaborated mainly by

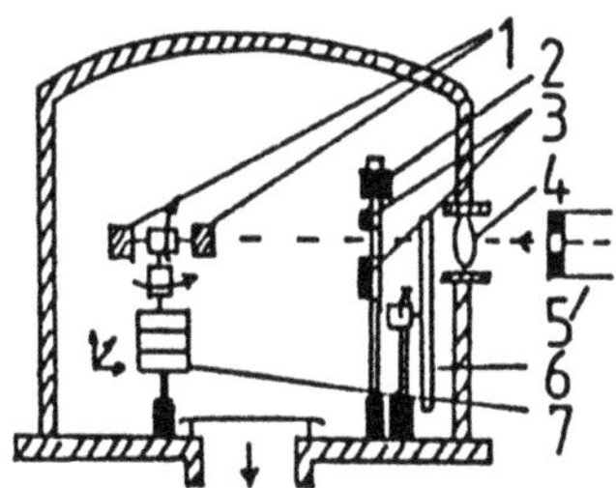

Figure 1. The vacuum chamber arrangement used for ML preparation by laser sputtering: 1) targets, 2) quartz head, 3) substrates, 4) lens, 5) Nd laser, 6) rotating glass disk, 7) x-y-z table [13].

Nesheva et al [14]. Se/CdSe, CdSe/Se-Te as well as Se/Se-Te ML were grown in a vacuum chamber (with a diameter of ~0.55 m) at ~10^{-4} Pa by means of alternative evaporation from two Ta sources mounted as far as possible apart from each other. Arrangement of Jena glass cylinders installed above the sources ensures evaporation in quasi-closed volume. The quartz thickness and deposition rate monitor heads were fixed above the cylinders. The substrates were placed on a rotating (0.13 Hz) disk with ratio of the time above the sources to the entire turn period about 1/13. The substrates were stopped at the greatest possible distance from the sources at the moment the evaporation source was changed.

The materials used in the two methods of preparation were pure Se (Fluka 99.999), pure Se and Te (Fluka 99.99) alloyed by usual melt-quenching techniques, and CdSe (Merck, superpure).

Planar or sandwich type electrodes were used for parallel or perpendicular conductivity measurements, respectively. Planar carbon electrodes, deposited after preliminary scratching of the structures, were established to be the most suitable ohmic contacts. For sandwich type arrangements top Al or carbon, and bottom Al, c-Si or ITO electrodes were used [6,15].

3. Structural peculiarities and stability

Different kinds of structural characterization of the chalcogenide ML are of interest - the presence of periodic structures and the structural stability.

To prove directly the periodic structure and the interface sharpness of our amorphous chalcogenide multilayers X-ray diffraction investigation have been made using automated "D 500 Siemens" X-ray diffractometer [16] on Se/CdSe and Se-Te/CdSe ML. Results on small-angle X-ray diffraction (SAXRD) measurements are shown on Fig. 2 for structures of up to 20 periods of Se/CdSe with repeat distance d = 9.2 nm, Se/CdSe with d = 5.7 nm and Se-Te/CdSe with d = 7 nm.

The repeat distance d is determined by Bragg's formula $2d\sin\theta_N = N\lambda$, where λ is the wavelength of the X-ray monochromatic irradiation (0.154 nm) and N is the diffraction order. The value of the full width at half-maximum (FWHM) in our last ML [17] reached 0.04^0 (in 2θ scale) even for second order peaks, which is close to values obtained for superlattices based on a-Si:H with

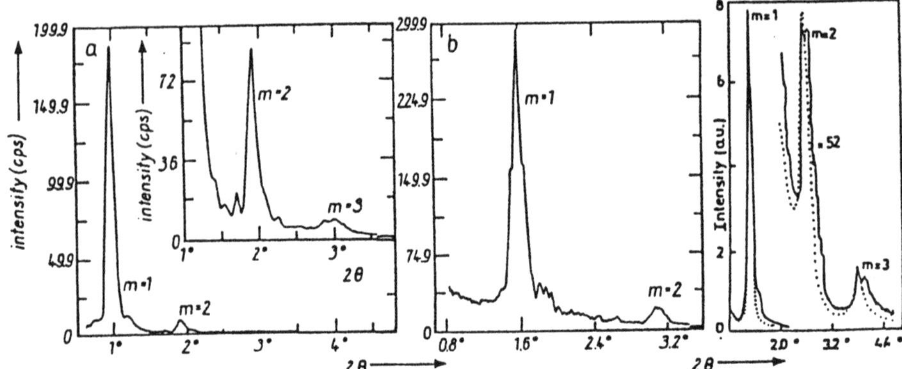

Figure 2. X-ray diffraction patterns vs the double scattering angle for: (a), (b) Se/CdSe [16] and (c) Se-Te/CdSe [14] multilayers with periods 9.2, 5.7 and 7 nm, respectively.

atomically abrupt interfaces [3]. This opens the possibilities for observation of quantum-size effects at small thicknesses of the sublayers in the a-ML of chalcogenides.

Here it could be noted that while the Se and Se-Te layers, deposited by the described methods, are amorphous, the single layers of CdSe are polycrystalline. The X-ray diffraction pattern in the range where peaks typical for the crystalline structure can be expected displayed a broadened peak at $2\theta \approx 25.5^0$ [18]. But sufficiently thin CdSe layers in the ML are amorphous. Indeed, the measurements at high diffraction angles of ML with CdSe sublayers with thickness ~11 nm reveal only broad distribution of the diffracted intensity [17].

The SAXRD measurements have shown that good ML on the basis of chalcogenides can be produced using sublayers with width ~2.5 nm. The thermodynamic (DSC) investigations in the first work on chalcogenide a-ML [8] have shown that in Se/As_2Se_3 the two phases of Se and As_2Se_3 are separated even at 1 nm sublayers width, but there is no information about the quality of the ML with this sublayer width.

The property dependences of the ML on the deposition conditions were also of interest. We have studied the influence of the deposition rate on vacuum thermal evaporated ML. The Se/CdSe structures considered above were deposited at a rate of 0.5 nm/s. At deposition rates of the order of 2 to 3 nm/s the peaks are strongly smeared even when the sublayers are not too thin. The analysis of the dark decay of the surface potential of electrophotographic Se-structures overcoated with Se/CdSe ML have shown [19] that when ML are evaporated with 3 nm/s deposition rate the transport is considerably affected by different kinds of traps not present when the ML are deposited at lower rate. In addition, higher Urbach energy values have been measured from Nesheva et al [20] in Se/CdSe ML obtained with higher deposition rates, related to a high degree of disorder. Probably, the method of deposition may be the cause for the presence of 2 to 3 nm wide rough interfaces in the Se/CdSe structures with a period ~20 nm reported in [10], as well as in the deposition of chalcogenide structures by other authors.

The intensity of the diffraction patterns for Se-Te/CdSe ML on Fig. 2c are simulated by Ionov and Nesheva [14] by means of a computational model for a ML with d = 7 nm. A good fit with the experimental data have been reached supposing an interface roughness about 0.6 nm.

We have studied [21] the problem of the interface sharpness of Se/CdSe ML in details, as well as their thermal stability.

An approach similar to that proposed by Santos et al [22] has been applied for the first time in X-ray measurements in order to determine the effective interface thickness, d_i, of chalcogenide multilayers. It is based on the intensity dependence of the first X-ray diffraction peak of a series of ML on their period d:

$$I \sim \sin^2(\pi d_w/d)\exp(-2\pi\sigma/d)^2$$

Here d_w is the well layer width and σ is the effective half-width of the interfaces ($d_i = 2\sigma$). σ is defined as $\sigma^2 = \sigma_m{}^2 + \sigma_l{}^2 + \sigma_r{}^2$. It takes into account material intermixing σ_m, interface roughness σ_l, and thickness fluctuations σ_r. From Fig. 3, curve a, values for the effective interface thickness of about 1.6 nm have been obtained in as deposited multilayers. After annealing at 363 K (curve b) d_i is around 1.3 nm, although one can expect some deterioration because of interdiffusion.

Another approach similar to that used by Hattori et al [23] has been employed in order to study in detail the dependences of σ on the annealing temperature and the annealing time at a constant temperature. In the small-angle approximation the diffraction intensity, I_N, of each peak (with peak number N = 1,2,3...) of one and the same ML follows the relation

$$I_N \sim \sin^2(N\pi d_w/d)N^{-3}\exp(-(2\pi N\sigma/d)^2) = F_N\exp(-(2\pi\sigma/d)^2N^2).$$

From this equation σ can be obtained from the slope of the linear dependence of the normalized intensity I_N/F_N on N^2 of only one ML. Investigations of ML with not too large periodicity, d < 10 nm, have been done after a short time annealing. An increase of the peak intensities as well as a decrease of the FWHM has been established when the annealing temperature, T_a, was $\leq$ 363 K. As can be seen from Fig. 4, an improvement of the interface sharpness could also be found from the decrease of the slope of the curves when T_a increases up to 363 K. If the annealing is carried out at T_a > 363 K the peak parameters deteriorate.

At temperatures close to the optimal T_a annealing for different time, t_a, was performed. At T_a = 353 K an interface improvement is established with increasing t_a which shows that in the investigated ML d_i decreases from 1.6 to 0.8 nm.

The results imply that two different thermally dependent processes occur in the ML during annealing. When the peak parameters improve a process of phase separation at interfaces may dominate in the material diffusion. This change leads to a negative interdiffusion coefficient. The second process may be an

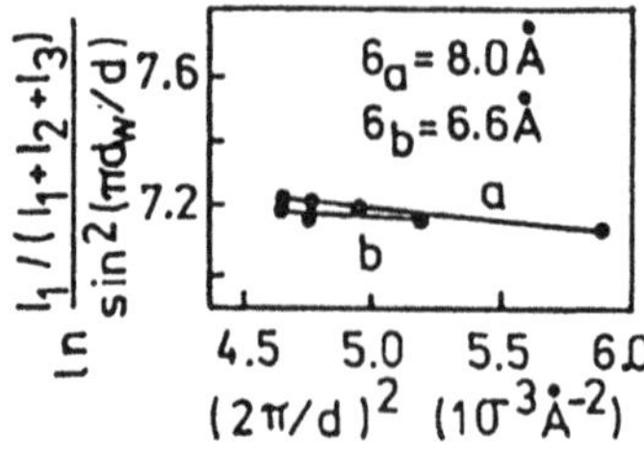

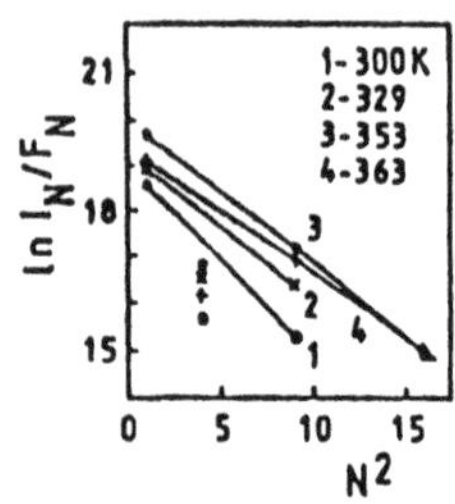

Figure 3. Normalized intensity of the first X-ray diffraction peak vs the inverse ML period, d, for a series of Se/CdSe ML (a) in as-deposited state and (b) after 30 min annealing at 360 K in Ar [21].
Figure 4. Normalized intensity , I_N/F_N, vs N^2 of the peaks of a ML of 20CdSe(43 Å)/20Se(48 Å) in as-deposited state (1) and after 30 min annealing at different temperatures (2-4) [21].

intermixing of the material of the different layers. It leads to a positive interdiffusion coefficient D. At $T_a \approx 380$ K a value of $D \approx 2.10^{-18}$ cm^2/s has been calculated, which shows that the ML with d < 10 nm investigated here show a good thermal stability at short time annealing up to $T_a \approx 380$ K. It could be emphasized that these ML retain their super-structure many years at normal room temperature conditions.

The thermal stability of Se/CdSe ML with large peroidicity (d ≈ 22 nm) has also been investigated, but at long time annealing [17]. A sequential annealing at T_a between 323 and 428 K in steps of ≅ 10 K has been carried out. After every thermal treatment the sample was measured by X-ray diffraction in the small and the high angle region. The multulayer coherence is rapidly vanishing for annealing above 340 K. After annealing at 360 K a crystalline Se phase appears and above 390 K a CdSe crystallization starts. At 390 K the amorphous background vanishes. The recrystallization processes facilitate the boundaries blurring of the Se/CdSe ML with d = 22.4 nm. But up to 330 K the ML with large periodicity are stable even at long time annealing which is interesting for practical uses.

As shown, the stability in the Se/CdSe ML depends on the periodicity and respectively on the sublayer width d_s (the widths of the two sublayers in the examined ML are nearly the same) and different behavior can be seen at d_s values lower or higher than 5 nm. This has been confirmed with Raman studies of the structural stability of Se/CdSe ML [24].

The higher frequency (vibrational) Raman spectra of ML of up to 20 periods and various sublayer widths (3.5, 5, 6.5 and 10 nm) have been measured. Three Raman bands have been observed in this spectral region, Fig. 5. The band at 209 cm^{-1} is connected with LO phonons of CdSe while the bands at 256 cm^{-1} and 237 cm^{-1} are typical for a-Se. Using the last interpretation of these bands [25], we accept that the 256 cm^{-1} band is due to chains (trans-coupling) as well as to local segments with ring configurations (cis-coupling), typical for the a-Se state. (For the cis-coupling two A_1 symmetry modes are characteristic, at 113 cm^{-1} and 256 cm^{-1}, but in our ML there is no clear evidence of the 113 cm^{-1} peak). Due to inter-chain interactions this frequency goes down to 237 cm^{-1} in trigonal Se or in small region in a-Se sublayer with higher ordering similar to the trigonal-like clusters proposed in [26]. In the framework of this model it is easy to understand the changes of the bands with the d_s decrease and after annealing of the ML with different sublayer thicknesses, obtained from spectra similar to those shown in Fig. 5 and summarized in the inset.

At d_s < 5 nm the 256 cm^{-1} band increases and the 237 cm^{-1} shoulder decreases. This is interpreted in terms of increasing the disorder in a-Se sublayers with decreasing the sublayer thickness. Simultaneously 209 cm^{-1} increases which is connected with an increase in the number of Cd-Se vs Se-Se bonds.

The short time (30 min) annealing at 360 K leads to: (i) an ordering in both CdSe and Se sublayers with d_s < 5 nm, as the band at 209 cm^{-1} increases and the shoulder at 237 cm^{-1} increases, which is connected with an interface smoothening and (ii) a decrease in ordering in Se sublayer at d_s > 5 nm, as the band at 237 cm^{-1} decreases. The last effect is explained with the prevalence of the great difference in thermal expansion coefficients of Se and CdSe over the effects of the interface smoothening.

The two opposite effects of annealing on the sublayer disorder for the extreme cases of sublayer thickness have been also seen in [27] where structural studies have been connected with investigations on the disorder characterized by the Urbach energy E_U determined by the constant photocurrent method (CPM). A decrease in E_U accompanied by improvement of the interface quality after annealing for short times of the ML with d_s < 5 nm at $T_a \leq 360$ K has been found. An E_U increase accompanied by a deterioration of the interfaces has been obtained after annealing at T_a > 360 K. But in ML with $d_s \approx 10$ nm a monotonous increase in E_U has been observed. Phase separation, Cd atomic diffusion into the Se sublayers and structural disorder changes in CdSe sublayers have been considered as possible reasons for these changes.

Not only the influence of the thermal annealing but also of the laser beam illumination on the ML structure has been investigated with Raman scattering measurements under gradually increasing laser power.

Fig. 6 presents the spectra of an a-ML of Se/CdSe with d_s = 3.5 nm taken at various laser light intensities. It can be seen: (i) an increase of the 237 cm^{-1} peak that is well expressed over a threshold laser power of about 480 mW/mm^2, which can be connected with a crystallization to trigonal form in the Se sublayers; (ii) a decrease of the 256 cm^{-1} peak and its "red" shift to 249 cm^{-1}, connected with a formation of α-monoclinic regions and a superposition of the modes of α-monoclinic Se; (iii) an increase of the 209 cm^{-1} peak and its shift to 205 cm^{-1}, which can be explained with the appearance of nano-crystallites in the CdSe sublayers.

Measurements at lower temperatures (25 K) shift the threshold to 2500 mW/mm^2 (to 5 times the highest power density used at 300 K). This shows that the photoinduced changes are primarily thermal effect.

It sould be noted that for thick sublayers ($d_s \approx 10$ nm) the threshold laser power for crystallization is 320 mW/mm^2, which manifests an increase of the crystallization temperature with decreasing the sublayer thickness. All this confirms the higher stability of the chalcogenide a-ML with lower sublayer thickness.

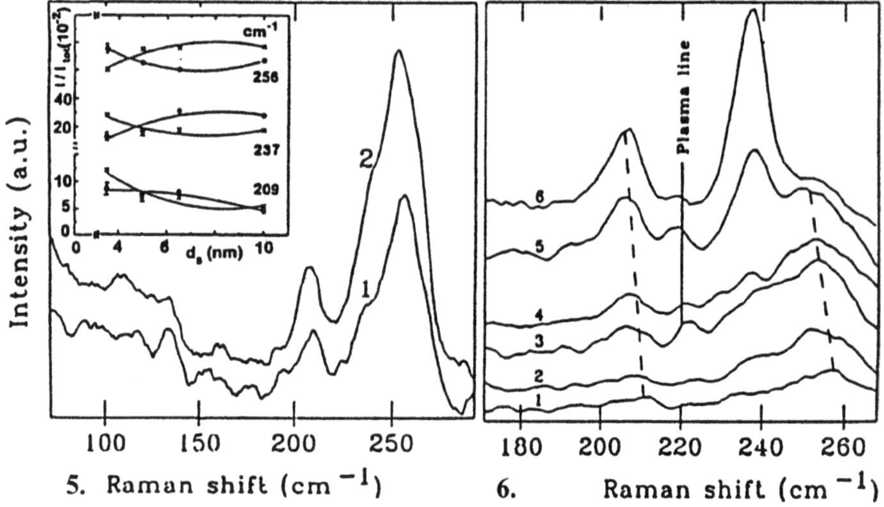

Figure 5. Raman spectra of as-deposited (1) and annealed (2) Se/CdSe ML with 18 periods and 3.5 nm sublayer thickness, measured at 80 mW/mm^2 power density (vertically displaced for clarity). The inset shows the thickness dependence of the normalized intensity of the Raman bands of as-deposited (o) and annealed (×) ML [24].

Figure 6. Raman spectra of Se/CdSe ML with d_s = 3.5 nm measured at a power density of 80 (1), 240 (2), 320 (3), 400 (4), 480 (5) and 560 (6) mW/mm^2 [24].

In addition it can be mentioned, that in the Raman investigations of the Se sublayers in the chalcogenide ML [28] lower values have been determined for the deviations $\Delta\theta$ of the bond angle at the interfaces and the distortion energy stored in these deviations compared to this of a-Si:H based SL. This result, based on the proportionality between peak width an $\Delta\theta$, was related to the higher flexibility of the a-Se structure.

4. Electrical and optical properties

The interesting properties of the a-ML are related to the quantum confinement effects, which produce quantized levels or sub-bands in the extended states of the well layers, as well as to the classical phenomena based on the presence of barriers.

Electrical and optical properties have been investigated and the energy band structure of the studied ML by us were found. The schematic band structure of Se/Se-Te, Se/CdSe and Se-Te/CdSe are shown in Fig. 7. The Fermi-level position and the optical band gap, E_g^0, of the single layers were determined from the temperature dependence of the dark conductivity and from transmission measurements (using Swanepoel's method), respectively.

Measuring the electrical parameters of the chalcogenide a-ML, we have established [6] a strong dielectric polarization, which causes different voltage-linked metastability effects. Studying in detail the polarization and depolarization process in the ML it has been established that the decay of the depolarization

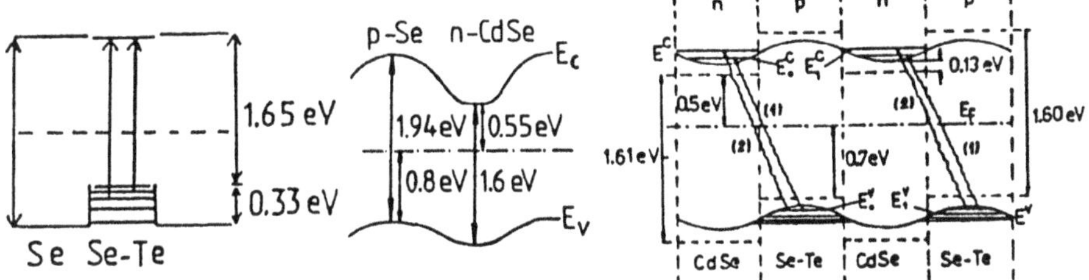

Figure 7. Band diagrams of the Se/$Se_{85}Te_{15}$ [5] and Se/CdSe [20] and Se-Te/CdSe [36] ML.

current follows an extended exponential law

$$I = I_0 \exp(ct^{\beta})$$

where c is a constant depending on the dispersion parameter β ($0<\beta<1$). This law typical for a number of dispersive processes in disordered materials has not been employed for polarization processes so far. The process of electron polarization which we suppose in our ML is connected with creating heterocharges (with a sign opposite to that of the electrodes), usually captured in traps, and homocharges compensating them, which may be injected from the electrodes. It has been established that the value of the collected homocharge in the ML is higher: (i) when the applied fields are higher and they are applied for a longer time and (ii) when there are more interfaces in the ML and when the sublayers are thinner. At high enough voltages the homocharge accumulated near the electrodes diminishes the effective interelectrode distance so that the sample could switch through S-type negative differential resistance to high conductivity state. An example is given in Fig. 8.

Also other metastability effects have been observed such as low and high conductivity states measured in the temperature dependence of the conductivity after annealing the samples and subsequent cooling with and without high voltage applied [4,5], as well as effects similar to persistent conductivity and the Steabler-Wronski effect. These phenomena are similar to phenomena observed in a-Si:H, but up to now it is not clear if they are of the same nature or not. In their comparison and explanation the differences in the dielectric constants must be taken into account.

The polarization must be taken into account in the electrical measurements also when studying the dependences which are interesting for the presence of QSE.

There are different kinds of features in chalcogenide a-ML which may be connected with quantum confinement of the carriers in the quantum wells. But for some similar features in a-ML based on a-Si:H alternative explanations have been proposed which are not on the basis of QSE. Here one can mention: (i) the decrease in the parallel conductivity, σ_{II}, when the well layer width in Se/Se-Te ML decreases down to 2 nm [7]. This decrease may be due to the increase in the distance between the Fermi-level and possible quantum levels existing in the well layers. However, it may also be connected with a mobility decrease. (ii) The

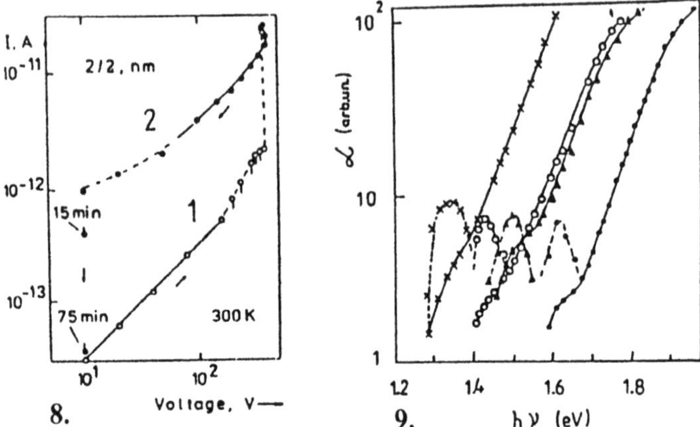

Figure 8. Dark current dependence on the increasing (1) and decreasing (2) voltage in Se/Se-Te ML with 28 periods and 2 nm thick sublayers [6].

Figure 9. α (hν) dependence of ML with different d_{CdSe} thicknesses: 29 nm, 10 nm, 5 nm and 2.5 nm. The respective dashed curves have been obtained by substracting the α value which obey the exponential law from the total α value [20].

appearance of current bumps in the I-V characteristics in direction perpendicular to the layers direction in Se/Se-Te structures with 9 barriers with d_{Se} < 10 nm [6]. Regions with negative conductivity are well pronounced at 77 K which may be taken as evidence for resonant tunneling. It is interesting to note that the bubbles appear at voltages expected for a square quantum well model, and the effective mass of holes in the Se-Te alloy determined on this base is in good coincidence with the known data. Alternative explanations have been discussed for current bumps observed in some double barrier structures on the base of a-Si:H [29]. They are based mainly on the pessimismus with regard to the existence of resonant tunneling because of the low values calculated for the mean free path of the carriers which can lead to loss of phase coherence of the wave function. But some of the assumptions in these calculations (the use of a thermal velocity instead of the velocity resulting from the energy of the first subband) are under question. Also some theoretical predictions [30] lead to the conclusion that QSE would be present even if the mean free path is an order of magnitude smaller than expected. So, new experiments in chalcogenide a-ML on this problem are of current interest. (iii) The "blue shift", i.e. the increase, of the optical band gap, E_g^0, with decreasing well thickness. The major objection against relating it to QSE in ML on the basis of a:Si:H are based on the use of the Tauc-plot (under the assumption of constant refractive index). A careful evaluation of the various factors leading to errors in E_g^0 determination has been carried out for the chalcogenide a-ML [31-33]. In Se/Se-Te ML the measured E_g^0 increase (using Cody plot) is not too high. Although its value is near to the expected shift of the first quantum level in different potential well models (≤0.07 eV at $d_{Se\text{-}Te}$ = 2 nm) it is not a well expressed proof for the QSE [31]. But in the case of Se/CdSe ML the Cody plot is not appropriate. A good coincidence between the higher shifts of the experimentally determined Tauc E_g^0 values and the E_g^0 values calculated using the Pöschl-Teller potential well has been obtained [32]. As the reasons leading to an artificial E_g^0 increase are taken in mind one can conclude that the QSE seems to be the most possible cause for the observed "blue shift" of E_g^0 in the Se/CdSe ML. It is possible that the different results on the "blue shift" obtained by many authors studying a-ML have been affected in different degree by multilayer optics [34], or by different changes in sublayer refractive index [35] or effective mass of carriers with decreasing well thickness, not taken into account so far.

It can be pointed that in the chalcogenide a-ML there are some phenomena which have been connected with QSE and for which there are not any other explanations so far.

Studying the band and subband absorption, α, in Se/CdSe ML Nesheva et al [20] have observed a "shoulder" in the lowest energy region of the $\alpha(h\nu)$ dependence, connected with indirect transitions from the Se valence band to the CdSe conduction band. It has been found that the energy of the shoulder transition depends on the CdSe sublayer thickness (Fig. 9). The energy variation (between 1.35 and 1.6 eV) has been explained assuming quantization of the electron states in the conduction band of CdSe sublayers. The assumption is checked by a comparison between the experimentally determined (with CPM) and the calculated indirect transition energies using the Pöschl-Teller potential well model. So a new evidence is given for the QSE in Se/CdSe ML.

Another persuasive proof for existing QSE in Se-Te/CdSe [36] and Se/Se-Te [33] is the optically induced stepwise behavior observed in their transmission spectra. On scanning the samples from higher to lower frequences steps appears leading to bleaching or darkening depending on the duration and intensity of a preliminary illumination. For the appearance of a step leading to bleaching it is necessary to create a high concentration of nonequillibrium carriers populating the states of the lowest subband in the wells. If the absorption begins from the second subband in a well, on scanning the sample the photon energy becomes less efficient for filling the levels with holes and at a certain moment a transition to bleaching may take place after which the absorption starts from the first quantum subband.

The energies of the transitions between the subbands have been estimated for different Se/Se-Te ML. The experimental shifts of E_g^0 obtained from the Cody plots, from the energy range where the absorption in Se is more important than in Se-Te, are close to the calculated energy differences between the quantum levels in the square potential well model. On analysing the absorption in the range of lower energies one can use the plot of Hattori et al [23] according to which the absorption coefficient for transitions between subband states in a-ML followes the relationship

$$\alpha/h\nu \sim \sum_{n}(h\nu - E_n)U(h\nu - E_n).$$

Here E_n denotes energy separation between the n-th subband edges in the conduction and valence band and U is a step function ($\alpha/h\nu$ is used instead of $\alpha h\nu$, since Tauc's suggestions are replaced by Cody's ones). This equation indicates that the quantization effect leads to changes in the slope of this dependence at each transition energy E_n. The derivative spectra will exhibit a stair-case form with steps at E_n.

The absorption in the low energy range in structures with two different well layer width is shown in Fig. 10 and Fig. 11. Changes in the slopes at the evaluated transition energies E_n have been observed, as well as a staircase derivate spectrum. So, these results confirm the idea that in the chalcogenide a-ML the QSE affects the absorption.

At last it can be pointed that in another kind of optical phenomena and in other chalcogenide a-ML proofs for QSE have been given recently from Hamanaka et al [37]. In As_2Se_3/As_2S_8 ML a blue shift of the excitation spectra of the photoluminescence compared with that of the single a-As_2Se_3 has been observed, as well as a lowering of this shift when the barrier layer thickness is

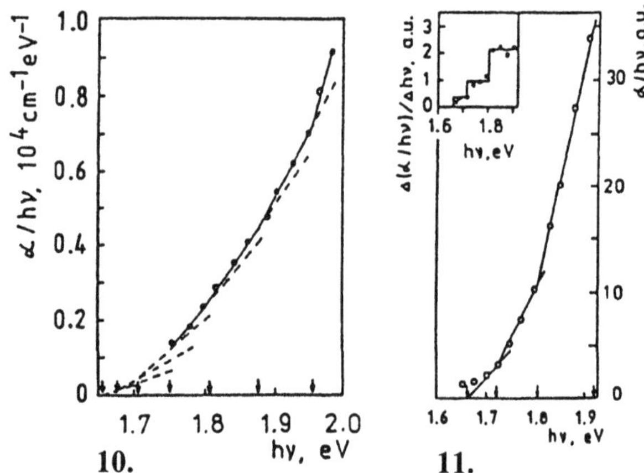

Figure 10. α/hν spectra for Se/Se-Te SL with Se-Te well layer thickness of about 3.5 nm; the arrows indicate the transition energy position E_N (N=1-7) calculated for well layer thickness of 3.6 nm. Dashed lines are calculated according the equation of Hattori et al (α is obtained from the transmission using Swanepoel's procedure).

Figure 11. α/hν spectra for Se/Se-Te SL with well layer thickness of about 2 nm; the arrows indicate the transition energy position E_N (N = 1-4) calculated for well layer thickness 2.1 nm. The inset shows the derivative α/hν with respect to hν (α is obtained by the CPM) [33].

changed so that the quantized levels are broadened in the well layer (i.e. when the multiple quantum well becomes a real superlattice).

5. Possibilities for applications

The preliminary results on folded phonons in the Raman spectra of chalcogenide a-ML [38] have lead to the idea that the existence of frequency gap in the phonon dispersion curves is interesting for creation of phonon filters.

More detailed investigations have been done [19,39-41] on the possibilities for electrophotographic applications of Se/Se-Te and Se/CdSe ML. These ML have some properties which make them appropriate for including in photoreceptors:

(i) the dark conductivity in a direction perpendicular to the layers planes is lower than the conductivity of the constituent materials, (ii) the stability against thermal and light treatments is good and increases with decreasing sublayer thickness and (iii) the spectral response of the Se-based photoreceprors may be shift towards the red region in order to be used in some diode laser printer, without decreasing their stability. The best results which we have received [41] are based on including Se/Se-Te ML. Most modern photocopying machines employ a-$Se_{1-x}Te_x$ alloy films with a Te content up to 5 at % usually, because the Te alloying increases the spectral sensitivity in the red region but reduces the charge acceptance and increases the dark decay rate. The designed new xerographic photoreceptor is compared in Fig. 12 with a photoreceptor of $Se_{95}Te_5$. One can see that if we use $Se_{95}Te_5$ as a transport layer only and include a photo-

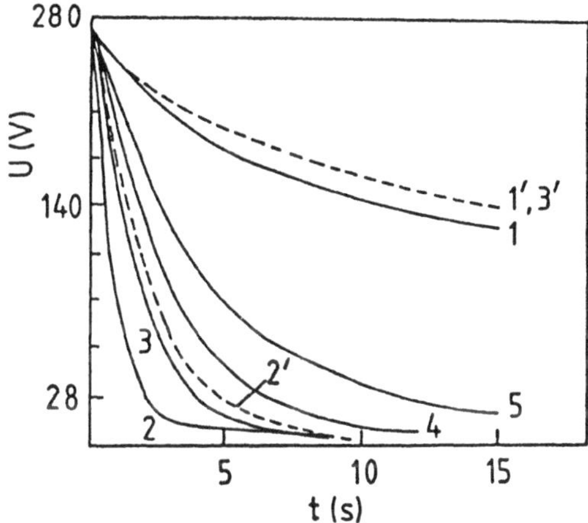

Figure 12. Xerographic characteristics of a single layer of $Se_{95}Te_5$ (4 μm) (dashed lines) and photoreceptor with $Se_{95}Te_5$ (4 μm) transport layer and a photogeneration layer of $Se_{87}Te_{13}$ (0.07 μm) single layer combined with Se/$Se_{87}Te_{13}$ ML (solid lines): 1, 1' - dark discharge, 2, 2' - photodischarge with λ = 600 nm, 3, 3' - λ = 750 nm, 4 - λ = 800 nm, 5 - λ = 925 nm [41].

generation layer of $Se_{87}Te_{13}$ combined with Se/$Se_{87}Te_{13}$ ML on top of the structure, a fast photoinduced discharge under illumination not only with λ = 750 nm but also with λ = 800 nm and even at λ = 925 nm can be seen. The sensitivity at 750 nm is higher than that of As_2Se_3 photoreceptors and at 800 nm about two times higher than that of a-Si:H photoreceptors. Besides no changes were observed in the structures for two years after their preparation.

6. Conclusion

High quality amorphous multilayers and superlattices of chalcogenide materials can be prepared by suitable deposition methods. They open the way for new fundamental investigations in the classic and quantum fields in the disordered materials physics. Many new phenomena have been observed which raise new questions for discussion. The possibilities for electrophotographic and other optoelectronic applications may also be a subject for wider investigations.

7. Acknowledgements

The work is supported by Bulgarian National Scientific Foundation under Contract F505. The author is very grateful to Dr Nesheva for useful discussions.

References

1. Esaki, L. and Tsu, R. (1970) Superlattice and negative differential conductivity, IBM J. Res. Develop. **14**, 61-65.
2. Weisbuch, C. and Vinter, B. (1991) Quantum semiconductor structures, Academic Press, San Diego.
3. Abeles, B. and Tiedje, T. (1983) Amorphous Semiconductor Superlattices, Phys. Rev. Lett. **5 1**, 2003-2006.
4. Vateva, E. (1988) Photoinduced effects in disordered materials, in M. Borisov, N. Kirov and A. Vavrek (eds.), Disordered Systems and New materials, World Scientific Publishing, Singapore, pp.226-289.
5. Vateva, E., Nesheva, D., Georgieva, I., Arsova, D., Ionov, R. and Chaushev, G. (1989-1990) Electrical and photoelectrical properties of amorphous multilayers bsed on Selenium, University Ann. Techn. Phys. **26**, 95-100.
6. Vateva, E. and Georgieva, I. (1989) Amorphous Se/Se-Te multilayers, J. Non-Cryst. Solids **114**, 124-126.
7. Vateva, E. and Georgieva, I. (1989) Amorphous superlattices of vitreous chalcogenide semiconductors in V.V. Himinetz and N.I. Dovgoshey (eds) Physical phenomena in non-crystlline semiconductors, UGU, Uzhgorod, v. 2, pp.22-24.
8. Maruyama, E. (1982) Amorphous built-in-field effect photoreceptors, Jpn. J. Appl. Phys. **21**, 213-223.
9. Ogino, T. and Mizushima, Y. (1983) Long-range interaction in multilayer amorphous film structure, Jpn. J. Appl. Phys. **22**, 1647-1651.
10. Shirai, H., Oda, Sh., Nakamura, T. and Shimizu, I. (1987) Preparation method and optoelectrical properties of a-Se/$CdSe_{1-x}$ multilayer films, Jpn. J. Appl. Phys. **26**, 991-995.
11. Vateva, E. and Nesheva, D. (1993) ISSP Jubilee Collection, in press.
12. Zhang, Z., Cheng, R. and Fritzsche, H. (1987) Electrical properties of a-Si:H/a-SiC_x:H multilayers, J. Non-Cryst. Solids **97-98**, 923-926.
13. Vateva, E. and Chaushev, G., to be published.
14. Ionov, R. and Nesheva, D. (1992) Preparation and characterization of amorphous SeTe/CdSe superlattices and their consistent thin layers, Thin Solid Films **213**, 230-234.
15. Ionov, R., Nesheva, D. and Arsova, D. (1991) Electrical and electrophotographic properties of CdSe/SeTe and CdSe/Se multilayers, J. Non-Cryst. Solids **137-138**, 1151-1154.
16. Vateva, E., Ionov, R., Nesheva, D. and Arsova, D. (1991) X-ray diffraction investigation on chalcogenide amorphous superlattices, Phys. Stat. Solidi (a) **128**, K23-K25.
17. Popescu, M., Sava, F., Lörinczi, A., Kodi, P.-J., Gutberlet, T., Uebach, W., Bradaczek, H., Vateva, E. and Nesheva, D. (to be published) Thermal stability of Se/CdSe multilayers.
18. Nesheva, D., Arsova, D. and Ionov, R. (1993) Thin and superthin photoconductive CdSe films deposited at room substrate temperature, J. Mater. Sci. **28**, 2183-2186.
19. Vateva, E. (1995) On the dispersive transport in electrophotographic structures from Se and CdSe/Se multilayers, Dokl. BAN **48**, 37-40.
20. Nesheva, D., Arsova, D. and Levi, Z. (1994) Band and subband absorption of Se/CdSe amorphous multilayers, Phil. Mag. B **70**, 205-213.
21. Vateva, E. and Nesheva, D. (1995) Small-angle X-ray diffraction studies on interface sharpness of a-Se/CdSe superlattices, J. Non-Cryst. Solids **191**, 205-208.
22. Santos, P.V., Hundhausen, M., Ley, L. and Viozian, C. (1991) Structure of interfaces in a-Si:H/a-SiN_x:H superlattices, J. Appl. Physics **69**, 778-785.
23. Hattori, K., Mori, T., Okamoto, H. and Hamakawa, J. (1988) Differential absorption spectroscopy on a-Si quantum well structures, Amorphous Silicon and Related Materials, World Sci. Publ., p. 957-978.

24. Nesheva, D., Kotsalas, I.P., Raptis, C. and Vateva, E. (to be published) On the structural stability of amorphous Se/CdSe multilayers: a Raman study.
25. Lukovsky, G. and Galeener, F.L. (1980) Intermediate range order in amorphous solids, J. Non-Cryst. Solids **35-36**, 1209-1214.
26. Kikineshy, A. (1995) Structural phototransformations and optical recording in Se/As_2S_3 multilayer films, Optical Engineering **34**, 1040-1043.
27. Nesheva, D., Vateva, E., Levi, Z. and Arsova, D. (1995) Interface and structural disorder changes in Se/CdSe multilayers, Phil. Mag. B **72**, 67-73.
28. Ionov, R. and Nesheva, D. (1992) Characterization of amorphous selenium sublayers in chalcogenide superlattices by Raman scattering, Solid State Communications **82**, 959-964.
29. Bernhard, N., Frank, B., Movaghar, B. and Bauer, G.H. (1994) Irregularities in current-voltage characteristics of hydrogenated-amorphous-silicon-based barrier structures, Phil. Mag. B **70**, 1139-1157.
30. Tsu, R. (1989) Phase coherence and damping in amorphous quantum wells, J. Non-Cryst. Solids **114**, 708-710.
31. Vateva, E. and Georgieva, I. (1993) Optical properties of Se/$Se_{85}Te_{15}$ multilayers, in J.M. Marshall, N. Kirov and A. Vavrek (eds), Electronic and Optoelectronic Materials for the 21st Sentury, World Scientific Publishing, Singapore, pp. 319-322.
32. Nesheva, D., Arsova, D. and Skordeva, E. (1993) Influence of the sublayer thickness on some properties of Se/CdSe multilayers, ibid, pp. 394-397.
33. Vateva, E. and Georgieva, I. (1993) Optically-induced stepwise absorption in Se/$Se_{85}Te_{15}$ multilayers, J. Non-Cryst. Solids **164-166**, 865-868.
34. Bernard, N., Dittrich, H. and Bauer, G.H. (1991) The optical band gap of multilayers and alloys, J. Non-Cryst. Solids **137-138**, 1103-1106.
35. Nesheva, D. (1993) Thickness dependence of the refractive index of Se/CdSe and Se-Te/CdSe multilayers, Phys. Stat. Sol. (a) **139**, K65-K67.
36. Ionov, R. and Nesheva, D. (1992) Stepwise optical absorption in amorphous Se-Te/CdSe superlattices, Superlattices and Microstructures **11**, 439-443.
37. Hamanaka, H., Konagai, S., Murayama, K., Yamaguchi, M. and Morigaki, K. (1996) Optical properties of amorphous As_2Se_3/As_2S_8 multilayers, J. Non-Cryst. Solids **198-200**, 808-812.
38. Ionov, R.I. (1992) Folded acoustic phonons in amorphous SeTe/CdSe superlattices, Europhys. Lett. **19**, 317-322.
39. Ionov, R., Nesheva, D. and Arsova, D. (1991) Electrical and electrophotographic properties of CdSe/SeTe and CdSe/Se multilayers, J. Non-Cryst. Solids **137-138**, 1151-1154.
40. Arsova, D., Nesheva, D. and Vateva, E. (1995) Xerographic Se photoreceptor including Se-Te/Se multilayer, in J.M. Marshall, N. Kirov and A. Vavrek (eds), Electronic, Optoelectronic and Magnetic Thin Films, Research Studies Press LTD, England, pp. 299-302.
41. Nesheva, D., Arsova, D. and Vateva, E., to be published.

DENSITY OF LOCALIZED STATES IN THE GAP OF NON-CRYSTALLINE SEMICONDUCTORS

GUY J. ADRIAENSSENS AND ASTRID ELIAT
Laboratorium voor Halfgeleiderfysica
Katholieke Universiteit Leuven
Celestijnenlaan 200D, B-3001 Heverlee, Belgium.

1. Introduction

It is a natural consequence of the inherent lattice disorder of a non-crystalline semiconductor, that a distribution of electronic site energies will be generated, as well as a certain number of lattice defects. Anderson [1] and Mott [2] showed that the site energy distribution results in the appearance of bandtails at the edges of the electronic energy bands, and in the localization of the electronic wavefunctions representing those tail states. For semiconductors, where the phenomena of interest are related to the energy gap between the top of the valence band and the bottom of the conduction band, that gap will consequently be filled-in by a varying density of such localized states. A further contribution to the localized-state density in the gap will result from lattice defects. In amorphous structures these are in fact the coordination defects such as the silicon dangling bond or the over- or undercoordinated chalcogenide negative-U centers. While it is easy to show in general terms that such defects will give rise to electronic energy levels in the gap, deducing the actual energy position and density of those levels from experimental data has been less straightforward.

There are several reasons for this difficulty. One very obvious reason is the low density of tail-state and defect levels in comparison with the total density of electronic states in the bands. By using appropriate spectroscopic techniques with special sensitivity for the localized states, this problem can readily be solved. But other, more troublesome ones remain. They are, in general, related to the fact that the localized-state distribution in disordered semiconductors is not static, but subject to both spontaneous and induced changes. Random potential fluctuations due to dispersive carrier

A. Andriesh and M. Bertolotti (eds.),
Physics and Applications of Non-Crystalline Semiconductors in Optoelectronics, 77–91.

motion are responsible for some of these changes [3], but also lattice relaxations in response to longer-term carrier trapping will contribute. While such relaxations are still being debated [4] for hydrogenated amorphous silicon (a-Si:H), they are the essential ingredient of the negative effective correlation energy model [5] for the chalcogenide semiconductors. A further complication will be the occurrence of spatial potential fluctuations that introduce a degree of inhomogeneity in the system. This contributes to the distribution of system parameters, but also influences the recombination behaviour of excited carriers.

Finally, there is the problem of the interdependence of tail- and gap-state densities. The idea has now been generally accepted that, for a-Si:H and its alloys, equilibration between the most strained bonds and the broken bonds will take place after deposition, or whenever external conditions change. This is the so-called weak to dangling bond conversion [6]. It is a key element in allowing the defect pool model [7] to provide a unified description for the defect-state distribution in doped and undoped material. For chalcogenide materials, a somewhat analogous link between electronic states in the bandtail and in the gap can be found in the concept of self-localization of electron pairs that has been introduced [8] to model the negative-U behaviour.

It is well-known that the total density of the localized states (DOS), as well as their energy distribution in the gap, are of primary importance for assessing the electrical properties of semiconductors. Also the specific electron affinity of the gap states – *i.e.* whether they exhibit a positive or negative effective correlation energy U – does determine many of these characteristics. However, a detailed study of the localized gap states is also important for evaluating the optoelectronic properties of non-crystalline semiconductors, since many of those states are either light-sensitive, or can be photo-induced. Amongst the best-known examples are the Staebler-Wronski effect in a-Si:H, and the photo-structural changes in chalcogenide glasses. But equally important is the contribution of gap states to phenomena such as photoinduced birefringence which cannot be ignored for optoelectronic applications.

2. Experimental observations

It will be clear from the above considerations that gap-state spectroscopies should avoid changing the distribution of localized states that is being measured. Small-signal excitation will hence be desirable. In addition, the resulting energy distributions can only be averages over the sample. The latter means that not all resolved energy levels are necessarily available in all parts of the sample; especially when analyzing electronic transport properties, this fact should not be overlooked.

2.1. TECHNIQUES

To probe the energy distribution of localized gap states, three basic types of spectroscopy have been most useful. They are based on, respectively, optical transitions, junction capacitance, and multiple-trapping transport. Of these, we will not discuss the capacitive methods here. They involve a metal-semiconductor junction where a change in applied voltage will change the depletion region and force gap states to either give up or accept (depending on the direction of change) charge carriers. Given the special nature of the depletion region, data acquisition and analysis require special care.

2.1.1. *Optical Spectroscopy*

Optical spectroscopy, in its usual transmission/reflection mode, is undoubtedly most widely used in non-crystalline semiconductors to obtain estimates for the width of the bandgap from the energy dependence of the absorption coefficient $\alpha(E)$, but it also has given rise to the one parameter that is now almost universally used to characterize the degree of lattice disorder in non-crystalline semiconductors: the Urbach slope E_0. This quantity describes the exponential slope of the optical absorption

$$\alpha(E) = \alpha_0 \exp[(E^* - E)/E_0], \quad (1)$$

that is observed below the band gap in all amorphous semiconductors. Low values of E_0 indicate minimal disorder. Values about 50 meV have been measured for both the best-quality a-Si:H and annealed a-As_2Se_3 glass. It was shown for a-Si:H by Stutzmann [9] that the value of E_0 also correlates with the dangling bond density of the material, and that it thus can be used as an overall indicator of lattice disorder.

The straightforward optical transmission spectroscopy is not very sensitive in the region of low absorption caused by the states in the gap. It is therefore supplemented by techniques such as photo-thermal deflection spectroscopy (PDS) or the constant-photocurrent method (CPM) which measure the absorption coefficient indirectly. But whatever the technique used, absorption coefficients always contain information on both initial and final state of the optical transition, and hence lead to a so-called *joined* density of states (JDS). Deconvolution of the JDS is necessary to obtain the desired DOS, a process that requires that some assumptions be made about at least one of the bands. For example, it has become accepted that the exponential parts of $\alpha(E)$ curves represent the joint density of valence and conduction band tail states, and thus in fact the distribution $g_v(E)$ of valence band tail states. This follows from the fact that the conduction band tail is invariably found to be steeper, and that the conduction band can hence be thought of as a step function.

Optical techniques which are specifically suited for studying gap state energies are photoinduced absorption (PA) [10] and the more encompassing optical modulation spectroscopy (OMS) [11], which measure changes in optical transmission caused by additional (mostly across-gap) illumination. The energies whereby photoinduced absorption or bleaching is resolved, can be related to energy levels in the gap. Changes in the defect density due to changing preparation conditions can easily be followed [12]. Unfortunately, it is not possible to deduce quantitative information on the *absolute* density of gap states in this way.

2.1.2. *Multiple-Trapping Transport*

Electrical transport in amorphous semiconductors is trap-limited. Due to the large number of localized states in the gap, free carriers will be very quickly trapped in this distribution of gap states, and remain trapped until they are thermally released after an average time of

$$t_r = \nu_0^{-1} \exp(E_t/kT), \tag{2}$$

where ν_0 is the attempt-to-escape frequency (on the order of 10^{12}s^{-1} for gap states), E_t is the energy depth of the trap as measured from the band edge into the gap, k is the Boltzmann constant, and T is the absolute temperature. Once free, the carrier will move with free-carrier mobility μ_0 until trapped again. For electrons in a-Si:H, $\mu_0 \simeq 10\text{cm}^2\text{V}^{-1}\text{s}^{-1}$ has become the accepted value; no good estimates for this quantity are available for any of the chalcogenides. The experimentally accessible quantity is the drift mobility μ_d, which differs from the free-carrier mobility by the ratio of free to trapped charge, and which can be obtained in a time-of-flight transient photoconductivity experiment [13] through the transit time t_T , the sample length L, and the applied field F as

$$\mu_d = L/t_T F. \tag{3}$$

As the average carrier will have been trapped many times before completing its path in an amorphous semiconductor, this current mode is called multiple-trapping (MT) transport. In MT mode each individual carrier will have interacted with a specific subset of the traps during its transit, and therefore stay in the sample for a long time if it encountered a deep trap, or exit quickly if it met shallow traps only, but a sufficiently large ensemble of carriers will contain information about a representative sample of localized states. To extract that information from measured currents, two types of photoconductive spectroscopy have been developed, one in the time and one in the frequency domain. In this text, we will mainly use information obtained from time-domain measurements. When not specifically stated otherwise, these spectroscopies always assume that charge capture is instantaneous on the time scale of the experiment (such that – apart from the transit time – only the release time from the traps need be considered), and that all traps have an equal capture probability, i.e. that the capture cross-section is energy-independent.

2.2. TAIL STATES

The density and distribution of tail states should provide an immediate indication of the degree of disorder in the material. Even for a perfect *Continuous Random Network*, bandtails follow unavoidably from the random distribution of the electronic site energies.

2.2.1. *Hydrogenated Amorphous Silicon*

As mentioned earlier, a deconvolution of the optical absorption Urbach slope can provide an estimate for the distribution of valence band tail states, but it will not give any information on the conduction band tail. Time-of-flight (TOF) experiments on the other hand can be used to study either side of the bandgap. The measurements are carried out in sandwich-configuration samples where electron-hole pairs are optically excited just below one of the (blocking) contacts, and where then, depending on the sign of the applied electric field, either electrons or holes are made to move across the sample. The transit time used in Eq.(3) will correspond to carriers which have been trapped in shallow states only [14], and studies of the drift mobility μ_d can therefore be used to estimate the distribution of tail states. Traditionally this is done by modeling the temperature and electric-field dependence of μ_d data sets. It should be noted that the use of Eq.(2) for determining the DOS implies that no information can be obtained from localized states with energies between the mobility edge and $E_m = kT\ln(\nu_0 t_m)$, where t_m is the earliest time for which the transient photocurrent can be reliably measured. For a-Si:H E_m is on the order of 0.1 eV.

Since the temperature dependence of the drift mobility, at 'moderate' temperatures and constant applied field, shows a more-or-less activated behaviour (*e.g.* with an activation energy of ~0.13 eV for electrons in a-Si:H), early TOF studies [15] interpreted this as evidence for a discrete gap state. However, it has since been recognized that the activation depends on the applied field, and is representative for the depth to which a typical carrier will have sampled the continuous distribution of gap states by the time t_T [14, 16].

Various sets of TOF data have been analyzed under varying assumptions by different research teams over the past years, with as a consequence that a number of specific, but often conflicting, proposals has been offered for the tail-state distribution. Unfortunately, there is a good deal of uncertainty about many of the model parameters that have to be assumed in elaborating the various models. Therefore, the general use of the exponential distributions $g(E) = g_0 \exp(-E/kT_0)$ adopted by Street [17], should be advocated as an expedient. They consist of characteristic temperatures of $T_0 \simeq 250$ K for the conduction band tail, some 400-450 K for the valence band side, and $g_0 \simeq 5 \times 10^{21}$ cm^{-3}eV^{-1} as estimated by Jackson *et al.* [18] from photoemission spectroscopy. A detailed discussion of some of the issues involved in evaluating TOF data, may be found in the Proceedings of the 1996 NATO Advanced Study Institute *Amorphous Insulators & Semiconductors* [19].

The parameter values cited above are valid for the so-called device-quality a-Si:H. Lesser-quality material, as well as Si alloys, will exhibit higher densities of localized states and wider band tails. The drift mobility data from C and S alloys, presented in Fig. 1, show evidence for these trends. Indeed, the increasingly lower values of the drift mobility, as larger amounts of C are added to the Si network, indicate that increasingly larger fractions of the photogenerated carriers are immobilized at any given moment. Increased disorder which leads to an increased number of localized states accounts for such additional trapping. Measurements of the optical absorption coefficients in a series of a-Si:C:H samples do show that the Urbach parameter becomes larger with growing C content [20], in agreement with the TOF results. Fig. 1 further shows that the drift mobility activation energy is also a function of the C content. Higher activation energy values can result from a recession of the mobility edge into the band due to the increased disorder, as well as from a shift into the gap of the localized states distribution in accordance with the widening of the tails indicated by the optical measurements. A final point of interest in Fig. 1, is provided by the a-Si:S:H data. It is seen that the addition of 1.5% of S causes the same deterioration in the material quality as is caused by 8.8% of C. This discrepancy is a logical consequence of the fact that C can replace a four-fold coordinated Si atom in the network without too much problems, while two-fold coordinated S sites will induce more distortions and strains.

2.2.2. *Chalcogenide Glasses*

Direct determination of the tail-state density of the chalcogenide glasses is more difficult than it is for the a-Si:H-type materials. As the overall defect density is much higher, the conductivity is much lower, and more amplification of the transient signals is needed; this in turn limits the short-time

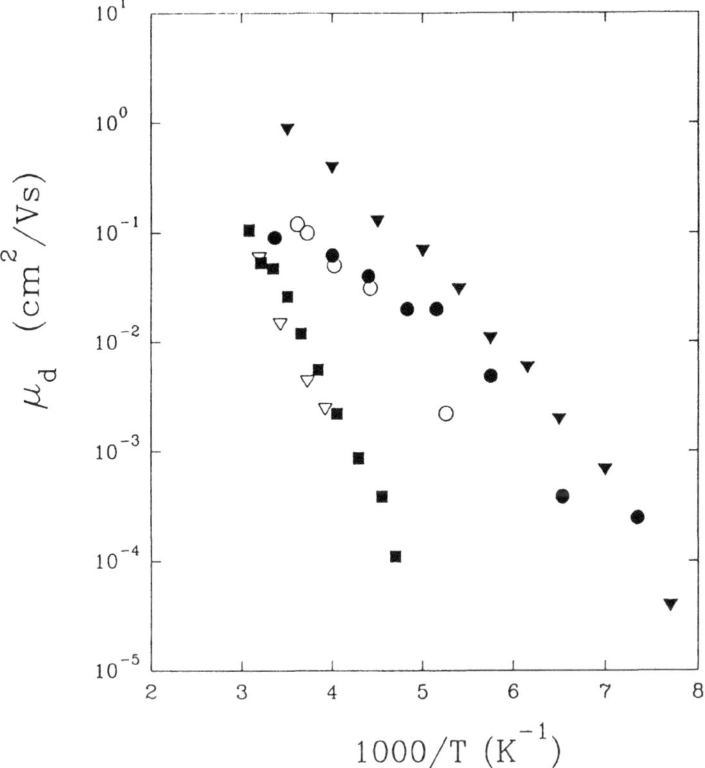

Figure 1. Temperature dependence of the drift mobility in a-Si:H (full triangles), a-Si:C:H samples with C contents of 2.2% (full circles), 3.9% (open circles), and 8.8% (full squares), and a a-Si:S:H sample with 1.5% of S (open triangles).

resolution. Since holes are the mobile species in chalcogenides, it is the valence band which will now be most accessible. However, that is also more or less the information which can be deduced from Urbach parameter values. Consequently, there is no real incentive to attempt further investigations.

As in the case of a-Si:H, some early TOF experiments, notably by Spear [13] or Marshall and Owen [21], again tended to assume that the observed activation energy indicated some discrete level. For the case of a-As_2Se_3 it has since been shown [22] that, analogous to the a-Si:H results, the TOF drift mobility activation energy depends on the applied electric field, and is indicative of a distribution of states. Evidence for a very wide *exponential* distribution of states was obtained, again for a-As_2Se_3, by Monroe and Kastner [23] by means of transient photocurrent measurements. They found a power-law decay $I_{ph}(t) \propto t^{-0.48}$ extending over nine decades of time at

$T = 295$ K. Standard interpretation of this result, where $-0.48 = -1+T/T_0$ with T_0 the characteristic temperature of an exponential DOS, leads to a $T_0 \simeq 565$ K exponential extending from 0.3 to 0.86 eV above the valence-band mobility edge. The question may be asked whether this can still be considered as a distribution of *tail* states. Given the fact that with the above T_0, the energy $E_0 = kT_0 \simeq 49$ meV corresponds to the measured a-As_2Se_3 Urbach slope, the answer apparently has to be positive.

An actual set of TOF data, obtained from 0.55 As_2S_3 : 0.45 Sb_2S_3 layers at different electric fields, was analyzed by Arkhipov, Iovu, Rudenko and Shutov [24] in their pioneering dispersive transport paper. They resolved $T_0 = 588$ K for this alloy, the wider bandtail suggesting a somewhat more disordered structure than the pure a-As_2Se_3 one. Much stronger lattice disorder should then be expected in a series of $(Ge_2Se_7)_{88}Bi_xSb_{12-x}$ alloys, since they are far removed from any stoichiometric composition. Transient photocurrents [25] did indeed decay quickly, indicating a slowly-decreasing DOS. The DOS of those alloys does in fact decrease slower than an exponential with $T_0 \simeq 1000$ K would do.

2.3. GAP STATES

While conceptually the difference between tail states and gap states is clear, even though equilibration amongst them may make the deviding line fluid, spectroscopically the distinction can often not be resolved. In practice, most spectroscopic measurements produce only some broad features which require a degree of additional 'modeling' to reveal the underlying defect states. Fairly direct access to the DOS is obtained through either transient or frequency-dependent photoconductivity. If the transient photocurrent is obtained in the TOF configuration, a simple analysis of either hole or electron post-transit current does give the DOS on valence or conduction band side respectively [26, 27]. Transient photocurrents from gap cells will give DOS information on one side of the gap only. The DOS can then be extracted after a Laplace [28] or Fourier transform [29] of the original data. In the latter case, the transform data are used as input for a frequency-domain analysis. Examination of DOS determinations, based on the use of these and other methods, learns that until this moment, there is no spectroscopic evidence whatsoever for the so-called 'real' DOS distributions with a fair number of well-defined defect bands in the gap, that were proposed in the early 1970's [30].

As the above discussion of tail-state distributions already indicated, it will not be obvious how to devide tail from gap states in the chalcogenides. The valence band tail states seem to extend all the way to mid-gap. We propose to start our discussion of gap states at this point, and discuss the somewhat clearer case of a-Si:H and its alloys afterwards.

2.3.1. *Chalcogenides*

An exponential distribution of localized states in the a-As_2Se_3 gap which matches the tail-state distribution has also been confirmed by other experimental techniques, such as modulated photoconductivity (MPC) [31]. At the same time, there has been firm evidence for well-defined recombination levels in the gap from the earliest studies of steady-state conductivity on [32]. As summarized in [22], the position of those levels in the bandgap, and other spectroscopic evidence from absorption, induced absorption or photoluminescence studies, all fit into the energy level scheme that is to be expected on the basis of the charged-defects model for the negative effective correlation energy concept. Normally, the signature of such discrete defect levels should be seen in the transient current traces as soon as they rise above the background density. In a-As_2Se_3, at 0.6 eV above the valence band which is the position of the lower recombination level, the tail-state density should be down to $\sim 10^{16}$ $cm^{-3}eV^{-1}$, and therefore about an order of magnitude below the accepted density of negative-U centers. Yet no trace of the discrete levels is found in the transient measurements (apart from a weak influence on the MPC phase factor [31]?). One possible explanation [33] may be that the lattice relaxation which moves the negatively charged center from just above the valence band to its 0.6 eV position occurs too slowly for the new position to be seen in a transient experiment. Some support for this point of view may be found in the very slow dynamics of induction or reorientation of photoinduced anisotropy in chalcogenide glasses; processes which also depend on charged-defect realignments if our interpretation [34] is correct.

2.3.2. *Amorphous Silicon*

There are no comparable difficulties in identifying the gap-state distribution in the hydrogenated amorphous silicon materials. The transient photocurrent and MPC methods have been used [35], and easily resolve the D^- band of occupied Si dangling bonds. Again the TOF technique proves useful when distributions on both the valence and conduction band sides are required. It is then necessary – as with the gap-cell transient photocurrent methods – to measure the transient current over many decades of time, as indicated in Fig. 2(a). The release-time approximation of Eq.(2) can now be used to deduce the distribution of gap states, if released charge leaves the sample rather than being retrapped, (*i.e.* for times large with respect to t_T), and provided charge trapping is proportional to the DOS [19, 26]. The second condition corresponds to the assumption of energy-independent capture cross-sections.

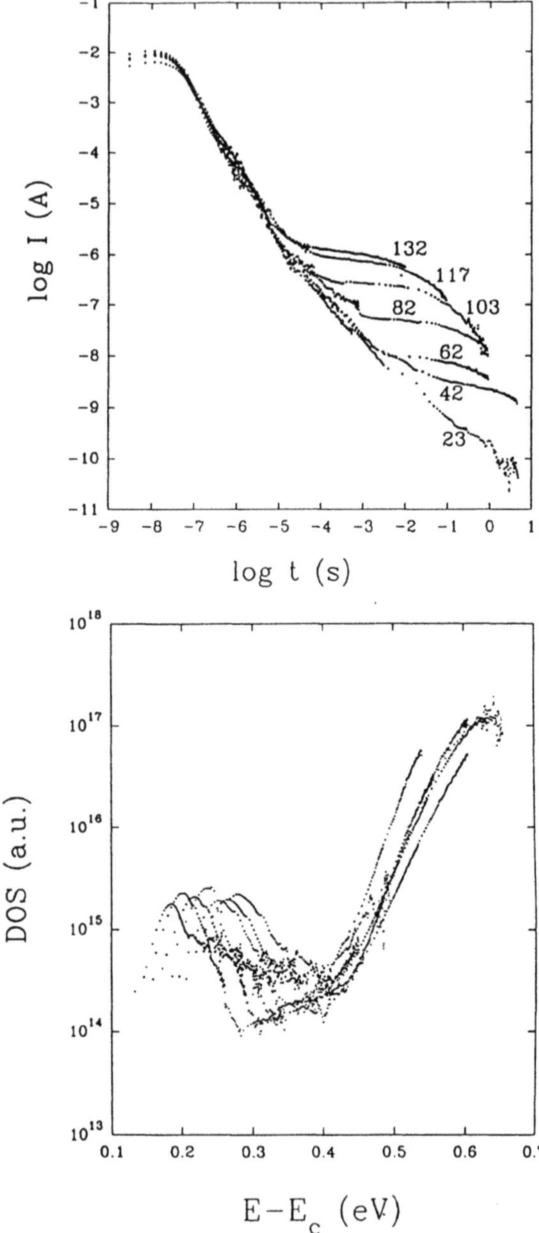

Figure 2. (a) Post-transit photocurrent from a Cr/a-Si:C:H/Cr sample with [C]$\simeq$ 2.5 at%, at temperatures indicated (in °C) near the curves; (b) DOS calculated on the basis of those curves.

The DOS is then proportional to the product of the photocurrent and the time, with the proportionality factor depending on both the material and the intensity of the laser pulse, and the energy scale set by the release time approximation:

$$g(E) \propto I_{ph}(t).t \quad , \qquad E = kT\ln(\nu_0 t). \tag{4}$$

The method can be used with a-Si:H Schottky or p-i-n diodes (solar cells for instance), but will only work for alloys as long as the defect density remains low enough to allow for a resolvable transit time. Otherwise no times $t \gg t_T$ can be defined, and only current decay due to deep trapping is observed.

Fig. 2(a) shows a series of post-transit curves, measured at various temperatures, from an a-Si:C:H Schottky diode containing 2.5 at% C. It is seen that the steep decline of the photocurrent is halted at the moment that charge emission from the D^- states becomes important. That point is reached sooner at more elevated temperatures. Transformation of the $I_{ph}(t)$ curves into DOS information, as shown in Fig. 2(b), reveals a very strong D^- band in the C-containing material. In pure (annealed) a-Si:H that band is considerably weaker. The arbitrary units (a.u.) shown on Fig. 2(b) are in fact $cm^{-3}eV^{-1}$ up to a poorly known multiplier of order one.

While no systematic variation with temperature of the resolved DOS is present in the curves of Fig. 2(b), this is no longer the case when measurements are carried out below room temperature [36], or when the gap state density of the sample becomes large [37]. One example of such behaviour is reproduced in Fig. 3 (from [38]). That the apparent temperature dependence of the DOS is not an artifact of the TOF post-transit analysis, is demonstrated by the observation of the same behaviour in MPC results of Zhong and Cohen [39] and in the Fourier transform of photocurrent transients by Bayley *et al.* [40]. Recent calculations by Arkhipov have allowed us to understand these results [3].

Up to now it was always assumed that the DOS could be considered stationary, unless some very specific excitation would induce specific environments to relax (as *e.g.* when a negative-U center is occupied). It turns out however that random fluctuations of the localized-state energies play an important role in carrier release from deep traps. It is shown in [3] that a variation in trap energy according to

$$E(t) = E_t + (kT\Delta_0/f)\sin(2\pi ft), \tag{5}$$

where E_t is the average energy of the trap, Δ_0 a characteristic constant, and f the frequency, will lead to the observation of a distribution of *effective*

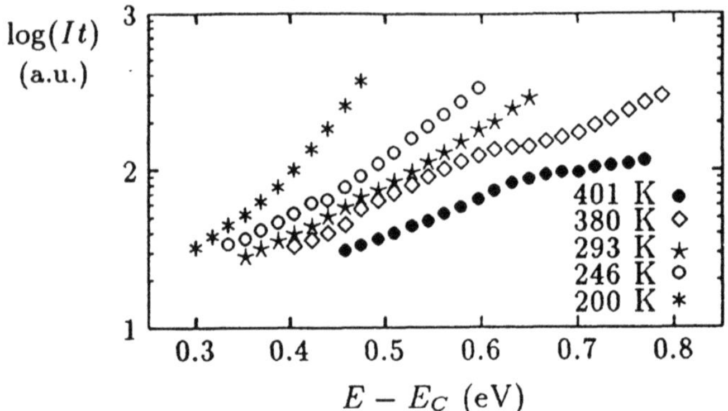

Figure 3. Temperature dependence of the $I_{ph}.t$ product of a poor-quality Cr/a-Si:S:H/Cr sample containing ~ 1.5 at% S.

activation energies

$$g_{eff}(E_a) = g[E(E_a)]\left[1 + \frac{\Delta_0}{\nu_0}\exp\left(\frac{E_a}{kT}\right)\right], \tag{6}$$

instead of the anticipated $g(E)$. The activation energy E_a is defined implicitly through

$$E_a = E - (kT\Delta_0/\nu_0)\exp(E_a/kT). \tag{7}$$

It follows from Eq.(6) that when $1 \gg (\Delta_0/\nu_0)\exp(E_a/kT)$, *i.e.* at high temperatures, a constant $g(E)$ will be measured, but that at lower temperatures a temperature dependent DOS will result. The value of the characteristic parameter Δ_0, needed to match measured and calculated $g_{eff}(E_a)$ distributions, is on the order of 1 Hz. This low frequency, as well as the choice of the $1/f$ in the amplitude factor in Eq.(5), do indicate that the fluctuations we envision and the standard semiconductor $1/f$ noise may very well be one and the same. It has in fact been shown [41] in general, that the dispersive motion of charge which results from the multiple-trapping process in disordered semiconductors, necessarily generates $1/f$ noise.

It will have become obvious from this discussion that, to study the distribution of localized states in the gap of amorphous semiconductors, one should ascertain that the results are not distorted by the dynamical aspects of the disorder. For good-quality a-Si:H, measurements at or above room temperature will provide a correct DOS profile. But when the defect density is high, as is the case for the a-Si:S:H sample used for Fig. 3, this is clearly not the case.

3. Discussion

Localized states in the gap of amorphous semiconductors have always been undesirable, and serious efforts have hence been made to reduce their number and narrow their distributions. A well-known and spectacular success in this area has been the removal from the gap of most of the silicon dangling bonds by hydrogenation. The use of *hydrogenated* amorphous silicon is now so self-evident that the need is not always felt to specify this qualification explicitly when hydrogenated material is being used. Still, there are between 10^{15} and 10^{16} dangling bonds per cm^3 and eV remaining in the material, and it is becoming increasingly clear that there exists a dynamic balance, which cannot be circumvented, between those remaining defects and strained silicon-silicon bonds. So, it will not be possible to produce amorphous silicon with a 'clear' gap. The situation will not be fundamentally different for other semiconductors since it is the dynamic equilibration between strained bonds and defect configurations that stabilizes the amorphous lattice.

In addition, any attempt to change material properties by going off-stoichiometry, or by introducing other elements, leads to an increased density in both tail and gap states. The much-cited 8-N rule explains how the chalcogenide glasses manage to accommodate all intended dopants, rather than being doped by them, but that does not prevent those foreign elements from adding to the lattice disorder and thus raising the number of localized states. In fact, in the case of Bi-modified germaniumselenides where a sign reversal of the conductivity can actually be obtained, the material has become so disordered in the process, and the localized-state density so high, that it is essentially useless.

Similar effects are observed when amorphous silicon is being modified. The addition of CH_4 to the SiH_4 that is mostly used to produce a-Si:H will increase the optical bandgap of the material (a-Si:C:H for solar-cell window layers), but it will at the same time widen the band tails and raise the deep-state density. A similar widening of the optical gap can be achieved by adding H_2S or H_2Se to the silane to produce a-Si:S(Se):H, but then the increase in localized state densities is much more significant. The addition of C evidently causes less lattice distortion in silicon than the addition of the chalcogen elements.

4. Conclusions

The *prototype* materials for each group of amorphous semiconductors, a-Si:H for those with positive U and Se or As_2Se_3 for the negative-U group, do show the lowest localized-state densities, when properly prepared. Alloying, or modification aimed at improving specific characteristics always seems to generate higher gap-state densities. This is, for instance, observed with p- or n-type doped a-Si:H, with wide-gap a-Si:C:H layers, or with Bi- or Ga-modified Ge-S glasses. Whenever a composition is farther removed from one of the simpler amorphous structures, the disorder is obviously increased, and the localized-state density raised.

Acknowledgements

Financial support for our studies of disordered materials by the Belgian *Fonds voor Wetenschappelijk Onderzoek* is gratefully acknowledged.

References

1. P.W. Anderson (1958) Phys. Rev. **109**, 1492.
2. N.F. Mott and E.A. Davis (1979) *Electronic Processes in Non-Crystalline Materials*, 2nd ed., Clarendon Press, Oxford.
3. V.I. Arkhipov, G.J. Adriaenssens and B. Yan (1996) Solid State Commun. **100**, 471.
4. W.B. Jackson and N.M. Johnson (1996) J. Non-Cryst. Solids **198-200**, 517; K. Lips, T. Unold, Y. Xu and R.S. Crandall, *ibid.*, 525; D. Kwon, A. Gardner and J.D. Cohen, *ibid.*, 530.
5. P.W. Anderson (1975) Phys. Rev. Lett. **34**, 953; R.A. Street and N.F. Mott (1975) Phys. Rev. Lett. **35**, 1293.
6. M. Stutzmann (1987) Phil. Mag. B **56**, 63.
7. Z.E. Smith and S. Wagner (1987) Phys. Rev. Lett. **59**, 688; K. Winer (1990) Phys. Rev. B **41**, 12150.
8. V.G. Karpov (1983) Sov. Phys. JETP **58**, 592.
9. M. Stutzmann (1989) Phil. Mag. B **60**, 531.
10. J. Orenstein and M. Kastner (1979) Phys. Rev. Lett. **43**, 161.
11. Z. Vardeny, T.X. Zhou and J. Tauc (1989) in *Amorphous Silicon and Related Materials*, H. Fritzsche (ed.), World Scientific, Singapore, p.513.
12. J. Dauwen and W. Grevendonk (1989) J. Non-Cryst. Solids **114**, 295; W. Grevendonk, M. Verluyten, J. Dauwen, G.J. Adriaenssens and J. Bezemer (1990) Phil. Mag. B **61**, 393.
13. A full desciption of the time-of-flight experiment for amorphous semiconductors will be found in W.E. Spear (1969) J. Non-Cryst. Solids **1**, 197.
14. G. Seynhaeve, G.J. Adriaenssens, H. Michiel and H. Overhof (1988) Phil. Mag. B **58**, 421.
15. W.E. Spear (1983) J. Non-Cryst. Solids **59-60**, 1.
16. H. Michiel, G.J. Adriaenssens and E.A. Davis (1986) Phys. Rev. B **34**, 2486.
17. R.A. Street (1991) *Hydrogenated Amorphous Silicon* , Cambridge University Press, Cambridge, U.K.
18. W.B. Jackson, S.M. Kelso, C.C. Tsai, J.W. Allen and S.-J. Oh (1985) Phys. Rev. B **31**, 5187.
19. G.J. Adriaenssens (1996) in *Amorphous Insulators & Semiconductors*, M. Thorpe and M. Mitkova (eds.), Kluwer, Dordrecht, in press.
20. Ö. Öktü, W. Lauwerens, S. Usala, G.J. Adriaenssens, O.B. Verbeke, A. Eray and H. Tolunay (1992) Mat. Sci. Eng. B **11**, 47.
21. J.M. Marshall and A.E. Owen (1971) Phil. Mag. **24**, 1281.
22. G.J. Adriaenssens (1990) Phil. Mag. B **62**, 79.
23. D. Monroe and M.A. Kastner (1986) Phys. Rev. B **33**, 8881.

24. V.I. Arkhipov, M.S. Iovu, A.I. Rudenko and S.D. Shutov (1979) phys. stat. sol. (a) **54**, 67.
25. G.J. Adriaenssens and P. Nagels (1989) J. Non-Cryst. Solids **114**, 100.
26. G. Seynhaeve, R.P. Barclay, G.J. Adriaenssens and J.M. Marshall (1989) Phys. Rev. B **39**, 10196.
27. B. Yan, D. Han and G.J. Adriaenssens (1996) J. Appl. Phys. **79**, 3597.
28. H. Naito and M. Okuda (1995) J. Appl. Phys. **77**, 3541.
29. C. Main, R. Brüggeman, D.P. Webb and S. Reynolds (1992) Solid State Commun. **83**, 401.
30. For an example, see the 'Edinburgh group model' (Figure 4c) in J.M. Marshall (1983) Rep. Prog. Phys. **46**, 1235.
31. C. Main, D.P. Webb, R. Brüggeman and S. Reynolds (1991) J. Non-Cryst. Solids **137&138**, 951.
32. C. Main and A.E. Owen (1973) in *Electronic and structural properties of amorphous semiconductors*, G.G. Lecomber and J. Mort (eds.), Academic Press, London, p. 527; G.W. Taylor and J.G. Simmons (1974) J. Phys. C: Solid State Phys. **7**, 3067.
33. V.I. Arkhipov (1995) private communication.
34. G.J. Adriaenssens and V.K. Tikhomirov (1996) in *Future Directions In Thin Film Science and Technology*, J.M.Marshall *et al.* (eds.), World Scientific, Singapore, in press; V.K. Tikhomirov, G.J. Adriaenssens and S.R. Elliott (1996) Phys. Rev. B, in press.
35. C. Main, R. Brüggeman, D.P. Webb and S. Reynolds (1993) J. Non-Cryst. Solids **164–166**, 481.
36. B. Yan and G.J. Adriaenssens (1995) J. Appl. Phys. **77**, 5661.
37. M. Nesládek, G.J. Adriaenssens and A.S. Volkov (1991) J. Non-Cryst. Solids **137-138**, 443.
38. A. Eliat, J. Jansen, S. Usala and G.J. Adriaenssens (1993) J. Non-Cryst. Solids **164-166**, 1093.
39. F. Zhong and J.D. Cohen (1992) Mat. Res. Soc. Symp. Proc. **258**, 813.
40. P.A. Bayley, J.M. Marshall, C. Main, D.P. Webb, R. Van Swaaij and J. Bezemer (1996) J. Non-Cryst. Solids **198-200**, 161.
41. K.L. Ngai and F.-S. Liu (1981) Phys. Rev. B **24**, 1049.

PHOTO STRUCTURAL PHASE TRANSITIONS IN AMORPHOUS CHALCOGENIDES: BASIC PRINCIPLES AND APPLICATIONS IN HOLOGRAPHY AND OPTICAL INFORMATION PROCESSING

C. H. Dietrich

*Deutsche Thomson Brandt GmbH**
Post Box 1307, D-78003 VS-Villingen, Germany
**During the reported work the author was with:*
Department of Computer Science 5
University of Mannheim
B6, D-68131 Mannheim, Germany

1. Introduction

Since their discovery in the mid 50th by Kolomiets and Goryunova amorphous chalcogenides have gained an increasing attention during the last decades [1]. They exist with widely varying stoichiometry in binary A_xB_{1-x} and more complex compounds. Due to their common layered structure amorphous chalcogenides are known as Se-like systems. They can be produced by cooling molten solids (bulk glasses) or thermal evaporation (thin films). The degree of disorder determined by the medium range structure is higher in thin films than in bulk glasses. Most of the amorphous chalcogenide semiconductors show photoinduced phase transitions and thermal annealing. Therefore, three states are distinguished. The freshly quenched or evaporated, the annealed and the photodarkened state. The photosensitivity and the photoinduced changes of the optical parameters like the refractive index etc. are about one order of magnitude larger for the fresh state than the annealed one. Nevertheless annealed materials and the reversal photodarkening are better investigated due to the greater determination of the annealed state and the independence of different production parameters. Scientists from all over the world used many different methods to learn more about the attractive photostructural phase transitions in amorphous chalcogenides. Treacy et al. used nuclear quadrupole resonance to detect certain molecular units in thin films and bulk glasses of As_2S_3 and As_2Se_3 [2]. Raman-spectroscopy was employed by Fumar et al. [3]. At the University of Raleigh, North Carolina, the dependence of reversal photodarkening on the intermediate range order were investigated by means of several x-ray scattering experiments [4-7]. Elliot used related methods to get more information about the significance of the first-sharp-diffraction-peak for a medium range order in different glasses [8]. X-ray diffraction and Raman spectroscopic measurements were also undertaken by Tanaka [9, 10]. Spotyuk used IR-Fourier-spectroscopy to investigate the stabilization of $As_4^+S_1^-$ co-ordination defects [11]. To get further information about the surface microstructure, electron microscopy and atomic force microscopy were applied by Starbov et al. and Nochte et al. [12, 13]. Different optical anisotropy phenomena like linear and circular dichroism and polarized photodoping were investigated by Lyubin, Kolobov and Tikhomirov [14-17].

A. Andriesh and M. Bertolotti (eds.),
Physics and Applications of Non-Crystalline Semiconductors in Optoelectronics, 93–108.

A slightly different approach to get more information about the photostructural phase transition is the used one. Starting with Keneman many investigators used holographic and non coherent recording on amorphous films to measure the reachable diffraction efficiency, the spatial resolution, changes of the refractive index and other material parameters [18-24]. Progressing to the next step Kikineshy influenced the recording parameters of amorphous chalcogenide films by producing multilayered nano structures [25]. Focusing on all this information about photostructural changes in amorphous chalcogenides mainly Tanaka, Paesler and Pfeiffer et al., Fritzsche, Kolobov and Adriaenssens elaborated models of the physics behind [26-32].
Due to our major working field – optical neural nets – our first interest in amorphous chalcogenides was using them as holographic recording material. Prerequisites to do so, were the investigation and quantitative determination of many material parameters of amorphous chalcogenide films – in our case especially a-As_2S_3 films. In the following, starting with a review of some theoretical models of the photostructural changes in amorphous chalcogenides different experiments are described, covering measurements of photoinduced phase shift and refractive index change, of diffraction efficiencies and dark self enhancement of holographic gratings in a-As_2S_3 films. In the final part of this article a specially to chalcogenide films adopted architecture of an optical vector-matrix-multiplier is introduced. Experimental results show the performance and the applicability for optical neural nets.

2. Theoretical models

Due to the high spatial resolution, the bandgap in the visible spectrum and the high photoinduced refractive index change ($\Delta n \geq 0.1$) especially non annealed a-As_2S_3 films are very interesting for holographic applications. As stated above the minor part of investigators worked with this kind of materials, because the photostructural changes are not reversible and freshly evaporated films do vary due to their varying preparation conditions. Nevertheless non annealed As_2S_3 films join the basic structure with annealed films and other binary amorphous chalcogenides. The molecular structure unit is the tetrahedral AsS_3. Three of five valence electrons of As form covalent bonds to S. Sulfur as the other chalcogens build two covalent bonds with their p-electrons and have one non bonding lone p-electron pair. Crystalline As_2S_3 has a layered structure with covalent intra-layer bonds and van der Waals inter-layer bonds. Similar to the crystalline state amorphous As_2S_3 shows a tendency to a layered structure, but without long range order. The existence of medium range order could be confirmed by extended x-ray absorption fine structure measurements [6, 7]. Additionally non-annealed films consist of molecular clusters and voids. Treacy et al. showed the existence of As_4S_4 and other molecules in amorphous films by nuclear quadrupole resonance [33]. The photoinduced dissolution of these molecular clusters accompanied by the reduction of the number of voids is connected with a transition to a phase of greater order and smaller volume. Vice versa, the photoinduced transition from the annealed to the photodarkened phase is connected with a volume increase and a reduction of the medium range order.
To explain these different phase transitions Fritzsche gives the lone p-electron pair the key-rule. Photons can excite these electrons into the conduction band. Caused by the vitreous structure they are fast localized by tail states in the bandgap. The localized charges influence the generation of valence-alternation defect pairs (VADP). In the case of the reversible phase transition the generation of VADP is followed by slight changes of the bonding angles. In the case of non reversible phase transition, single atoms can

fold down and fill voids. Eventually the charged defect pairs either neutralize each other while the changes in the lattice remain or the charged defects move inside the lattice and become stabilized, while the number of homopolar bonds is rising.
This model explains photoinduced phase transitions of annealed and non annealed chalcogenide glasses taking into account different physical processes like defect pair generation, bond angle changes and bond switching. On the other side Paesler, Pfeiffer, Zhou and Fumar make a single process, the generation of homopolar bonds, responsible for the photoinduced changes [3, 4, 7]. Kolobov and Adriaenssens take all the photoinduced effects and changes into account and explain them by means of mechanical strain caused by the different properties of amorphous chalcogenide films and its substrates [30].

3. Experiments

The experiments described in this section were motivated by the objective to build up a holographic vector-matrix-multiplier for an optical neural network. Holographic interconnection had to be realized quantitatively with the highest possible dynamic. With this background the experiments focused on the photographic material properties. In the wide field of amorphous chalcogenides the here reported work was limited to a-As_2S_3 films. Due to its bandgap of about 2.4 eV amorphous As_2S_3 shows a spectral sensitivity to green and blue light. Since 1971 Keneman had recorded holographic gratings with diffraction efficiencies up to 80% into a-As_2S_3 films their applicability as holographic recording material has been known [18]. Later, Schwartz et al. measured the remarkable spatial resolution of more than 10 000 lines/mm [19]. Together with the important property of direct photoinduced phase transition, that makes a direct recording without any chemical processing possible, these properties of a-As_2S_3 films were responsible for the limitation on that specific material. In the following experiments the a-As_2S_3 films were illuminated by green light (λ_1 = 514 nm) during recording and red light (λ_2 = 633 nm) was used to read out the photoinduced changes in a phase sensitive way.
The first experiments aimed on the photoinduced phaseshift with respect to its dependence on the writing intensity. Then, the diffraction efficiency of holographic gratings was measured. Its dependence on different exposure parameters like film thickness, intensity and grating period etc. was determined. To estimate the dynamics of a-As_2S_3 films the known effect of dark self enhancement of holographic grating in a-As_2S_3 films was investigated for technical relevant efficiencies up to 32%. Finally, white light interference and atomic force microscopy were used to determine the surface structure of exposed and fresh evaporated films.

3.1. PHASE SHIFT MEASUREMENTS

A precondition for quantitative holographic recording is the knowledge of the phase shift in the holographic material as a function of exposure and its dependence on the recording intensity. The phase shift was measured for films with a thickness from 1,3 μm to 11 μm. An interferometric measurement technique in combination with computer added data post processing was used. The complete experimental set-up and the measurement technique is described in [23]

3.1.1. Intensity Dependence

The investigation of the intensity dependence was motivated by technical and physics aspects. First, due to different experimental conditions the recording intensities cannot generally be kept constant, on the other side a dependence of the recording process on the intensity – namely an intensity threshold – gives a hint about the physics behind the photoinduced phase transition. Table 1 lists the used exposure intensities and the related exposure times.

TABLE 1. Exposure settings

I_{max} [W/sqcm]	3.0	1.0	0.3	0.1	0.03	0.003	0.0003
t_{max}	30 s	90 s	5 min	15 min	50 min	8 h 20	41 h 40

Since we mainly were interested in applying a-As_2S_3 films for holographic recording the exposures were stopped at 90 J/sqcm, before the total saturation was reached. In Figure 1 the results of the phase shift measurements are plotted for a 5.2 μm thick film.

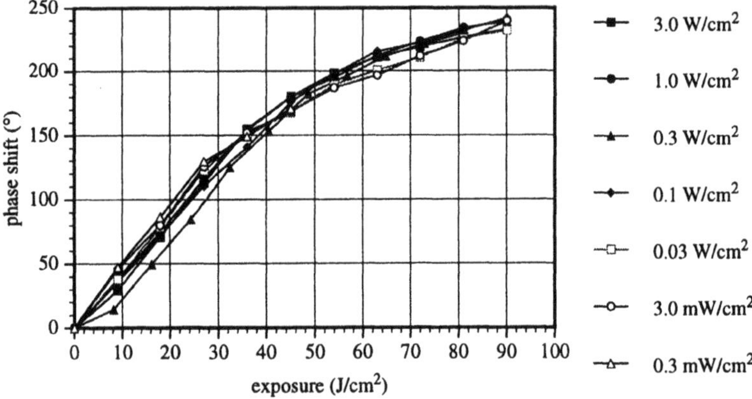

Figure 1. Intensitiy dependence of the photoinduced phase shift in sample E2 (5.2 μm)

The course of all curves is nearly the same. It shows a linear dependence of the phase shift of the exposure between 0 J/sqcm and 40 J/sqcm. A linear phase shift of 150° was reached. The stop value is about 240° at 90 J/sqcm. No significant dependence on the illumination intensity could be found. That is remarkable due to a variation of the intensity over four orders of magnitude (0.3 mW/sqcm – 3 W/sqcm).

Figure 2 shows the hole range of the phase shift measurements performed with the same experimental set-up. Amorphous films of different thicknesses and from different manufacturers were tested. Starting with a maximal phase shift of 100° for the 1.8 μm sample M965 the highest value of 410° was reached with a 11μm thick film. In between the curves of the samples E7, Q10 and Q12 are plotted. All of them were said to have the same thickness of about 5 μm, but Q10 and Q12 were samples made by a different laboratory to sample E7. Different manufacturing conditions, different ages and an inexact thickness evaluation during the evaporation process could be made responsible for the large variation of the results.All in all, figure 2 shows the common linear phase shift response for exposures up to 50 J/sqcm for rather different samples and the dependence of the results on the production conditions.

Additionally to the phase shift measurements the intensity dependence of the photoinduced processes in amorphous As_2S_3 films was investigated by simple transmission measurements. Covering the intensity interval between 0.1 W/sqcm and 1000 W/sqcm the results are plotted in figure 3.

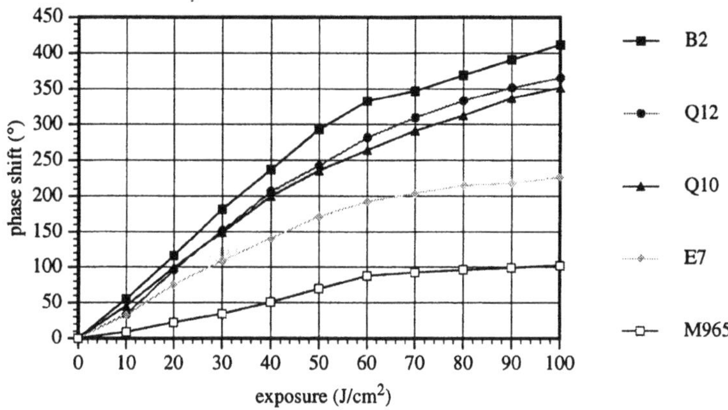

Figure 2. Sample and thickness dependence of the photoinduced phase shift

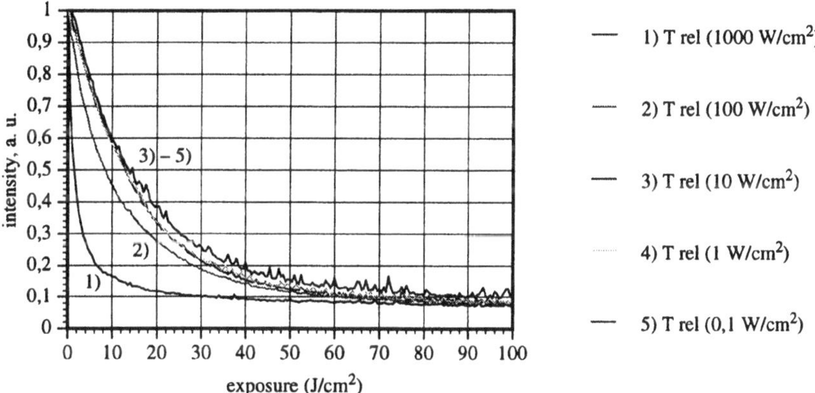

Figure 3. Transmission measurements – Ploted are the transmitted intensity of the recording beam (λ_1 = 514 nm)

The normalized curves have nearly the same course for intensities up to 10 W/sqcm. The necessary exposure of half value transmission was about 14 J/sqcm. Significant smaller values have been obtained for 100 W/sqcm and 1000 W/sqcm. The related half values were 8 J/sqcm and 2J/sqcm. This behavior strongly indicates an additional thermal induced effect, that makes the films more sensitive to light. Thus, a threshold for the additional thermal photoinduced phase transitions could be estimated between 10 W/sqcm and 100 W/sqcm. Despite the results of Salminen et al. [21], who measured the diffractive efficiency of holographic gratings in a-As_2S_3 films with recording intensities down to 0.01 mW/sqcm, the phase shift measurements, reported above, gave no indication for the existence of an intensity threshold at low intensities (<1 W/sqcm). We therefore conclude, that photoinduced phase transitions in non annealed a-As_2S_3 films must be explained by pure photon-electron processes up to the recording intensity of about 10 W/sqcm.

3.2. DIFFRACTION EFFICIENCY

Diffraction efficiency measurements were first used to get a cross check on the phase shift results. Further, the dependence of the efficiency on the intensity ratio of the two recording beams was investigated, to evaluate the optimal recording conditions. Then, the maximal diffraction efficiency was measured. Finally, the so called dark self enhancement, an effect reported earlier by Schwartz et al. [19], was investigated. The reachable signal to noise ratio of the holographic interconnects should be estimated.

3.2.1. Cross Check on the Phase Shift Measurements

The measured dependence of the photoinduced phase shift and the known theoretical dependence of the diffraction efficiency on the modulation of a planar phase grating were combined and compared with diffraction efficiency measurements of recorded gratings. The gratings had a spatial wavelength of 14.7 μm and were recorded by two beams with an intensity of 0.4 W/sqcm. Figure 4 shows the experimental set-up used. The result of the theoretically calculated and the measured efficiencies are plotted in figure 5.

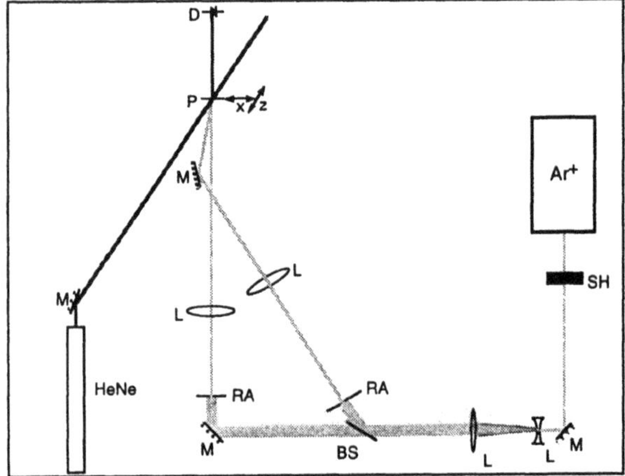

Figure 4. Experimental set-up for recording planar diffraction grating

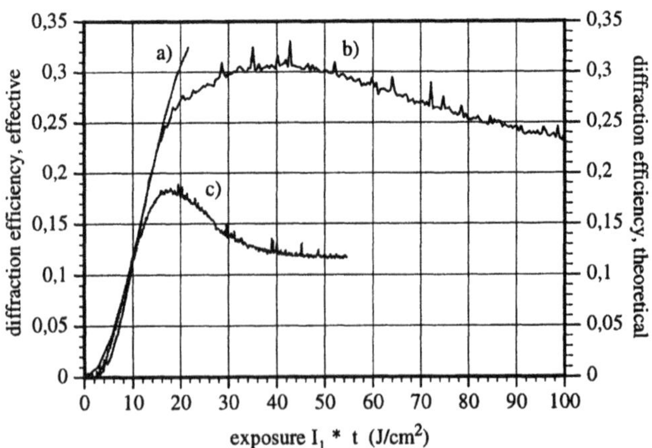

Figure 5. Comparison of theoretical and experimental diffraction efficiencies

The theoretical (smooth) and the two experimental curves are shown. The two measured curves differ in the ratio of the intensities of the two recording beams. Curve b) was obtained by using beam of equal intensity, whereas the intensity ratio of object and reference beam amounted to 14 during recording c). At this ratio the intensity of the constant bias I_0 and the amplitude of the sinusoidal intensity modulation I_1 is 2:1. Since the plots are shown versus the modulation exposure I_1*t curve c) reached only half the modulation of curve b), before saturation effects dominated the course. Nevertheless curve c) match the theoretical one up to an exposure of 10 J/sqcm. The deviations of curve b) from the theoretical course for exposures of up to 10 J/sqcm can be explained by mechanical stress inside the lattice between the bright and dark stripes of the recorded phase grating. This stress could decrease the photoinduced phase shift compared with the values received during homogenous illumination. For higher exposures of up to approximately 20 J/sqcm the experimental and theoretical curves matched; thus the phase shift measurements could be confirmed by the diffraction efficiency measurements.

3.2.2. Maximal Diffraction Efficiency and Volume Phase Gratings

The diffraction efficiency of a planar sinusoidal phase grating can be calculated by the squared first order Bessel-function:

$$\eta = J_1^2(2\pi/\lambda\, \Delta n\, d) \qquad (1)$$

Here λ stands for the wavelength, Δn for the modulation of the refractive index and d for the film thickness. The maximum of this function is 0.34, i. e. the highest diffraction efficiency reached by planar phase grating is 34%. Therefore higher efficiencies can only be explained by volume gratings. Due to Kogelniks coupled-wave-theory 100% diffraction efficiency can be achieved by volume phase gratings [34]. Klein and later Kaspers elaborated criteria classifying holograms into thin or thick ones [35, 36]. For instance, a 10 μm grating in a 5 μm thick film is a planar hologram, whereas a 0.36 μm grating in the same film is a volume hologram true to both criteria. In-between lies a 1μm grating in a 5 μm thick film,which cannot be classified by either criteria.

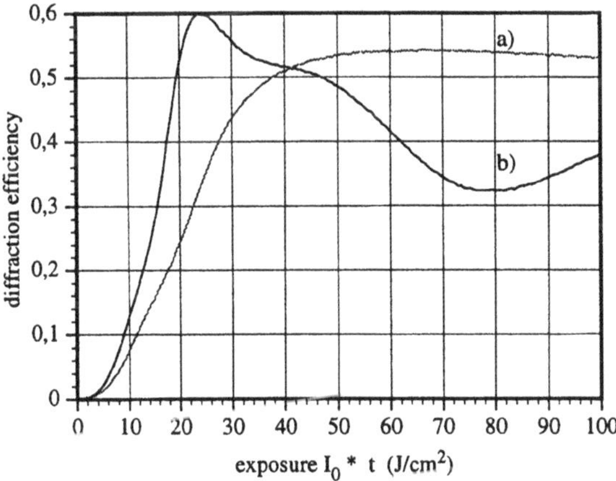

Figure 6. Diffraction efficiencies of 1 μm gratings

Figure 6 shows the result of two diffraction efficiency measurements. The real diffraction efficiency, that is not corrected regarding the absorption losses, is plotted as a function of the bias exposure. Both interfering beams had the same intensity. Curve a) belongs to a 1 μm grating exposed in a 5.2 μm film and curve b) was measured recording a 1 μm grating in a 11 μm thick amorphous film. The absolute maximal values of both curves were about the same, $\eta_{max} = 0.6$, but the course was different caused by their clear and less clear attribution to the class of volume hologram. The real volume grating (curve b)) shows a steeper slope of the diffraction efficiency and the beginning of the $\sin^2 x$-modulation decrease after the peak before saturation effects dominated.

3.2.3. Dark Self Enhancement

Dark self enhancement of the holographic gratings in amorphous As_2S_3 films was first investigated by Schwartz et al. [18]. They measured that effect by recording gratings with an efficiency of about 1%.

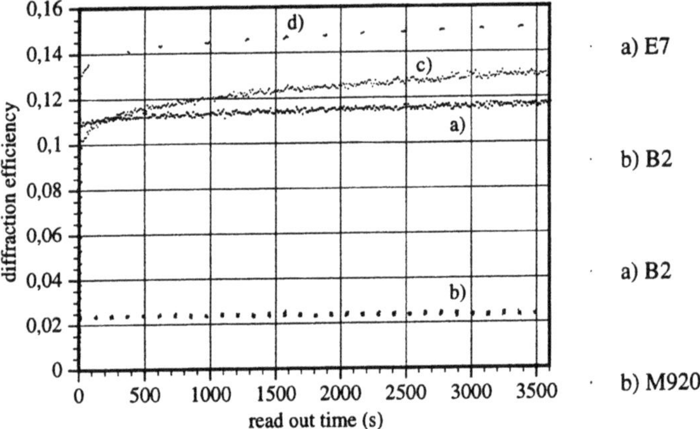

Figure 7. Diffraction efficiency measurements during 1 hour after exposure

Due to the planned application of this material for recording quantitatively optical interconnects this effect had to be investigated for the hole range of diffraction efficiencies and grating periods. Four different experiments were done. The diffraction efficiency was measured during 1 hour and during 15 hours after recording. The read out intensity ($\lambda_2 = 633$ nm) was set to 1.5 mW/sqcm. Additionally, measurements with a read out laser periodically switched on and off were made during the two same time periods. In this way the influence of the read out light was investigated. Figure 7 shows the results of four 1 hour-measurements. Continous and periodically switched data are plotted. The curves a) and b) belong to a 5 μm thick film and show an increase of the diffraction efficiency of less than 1%. The curves c) and d) were measured reading out gratings recorded in a 11 μm thick film. Here the dark self enhancement was more obvious and amounted to approximately 3%. Comparing the continous measurements with the periodically switched ones no significant difference could be found. The same result was got for the 15 hours measurements. Table 2 lists the results of all measurements. The results show three main tendencies. First, the dark self enhancement is increasing with the thickness of the films. Second and third, it is decreasing with increasing grating period and increasing value of the pre-recorded diffraction efficiency. The highest absolute value amounted to 3.5%. The connection of the effect with the grating period could be explained by the mechanical stress between the exposed and non exposed stripes of a grating, that is increasing with the density of transitions between the exposed and unexposed stripes. The other two correlations are connected with the saturation grade of the photoinduced transitions between the freshly evaporated and the photodarkened phase.

TABLE 2: Dark self enhancement of diffraction gratings in different samples

10 μm - gratings						
sample	thickn. [μm]	η_{begin} [%]	η_{end} [%]	read time [h]	Δ_{abs} [%]	Δ_{rel} [%]
B2	11	12.5	15	1	2.5	20
E8	5.2	19	20	15	1	5.3
E7	5.2	2.3	2.4	1	0.1	4.3
E7	5.2	12	13	1	1	8.3
O8	5	10	11.5	14	1.5	15
Q10	>5	44	45	1	1	2.2
Q10	>5	14	15	1	1	7.1
M920	5	1.1	1.15	15	0.05	4.5
M920	5	8.5	9	1	0.5	5.9
1 μm - gratings						
sample	thickn. [μm]	η_{begin} [%]	η_{end} [%]	read time [h]	Δ_{abs} [%]	Δ_{rel} [%]
B2	11	10	13	1	3	30
E8	5.2	32.5	34	1	1.5	4.6
E7	5.2	25	27	1	2	8
O8	5	23	24.5	1	1.5	6.5
Q10	>5	41	43	1	2	4.9
M920	5	11	11.5	1	2.5	4.5
0.36 μm - gratings						
sample	thickn. [μm]	η_{begin} [%]	η_{end} [%]	read time [h]	Δ_{abs} [%]	Δ_{rel} [%]
B2	11	6.5	8	1	1.5	23
E6	5.2	21	22	1	1	4.8
Q10	>5	15.5	19	1	3.5	22.6

3.3. WHITE LIGHT INTERFERENCE AND ATOMIC FORCE MICROSCOPIC MEASUREMENTS

White light interference and atomic force microscopy allow the investigation of surfaces with an horizontal resolution of 0.1 nm and higher. Thus, interesting surface and thickness measurements could be performed.

Figure 8 shows the result of an interference microscopy measurement of a 3 μm grating exposed in a 5.2 μm thick a-As_2S_3 film by two beam holography with an intensity of 0.01 W/sqcm. Due to the poor black and white reproduction of the colored original the scales are hard to recognize. The most interesting part of the figure is the profile plot (mid-left plot). It shows a mean value of about 25 nm for the depth of the surface grating and about 10 nm for the surface roughness. The same measurement of a 3 μm grating in a 2 μm thick film showed a depth of 5 nm and a surface roughness of about 6 nm.

Figure 9 shows a profile plot of a homogeneously illuminated region, that was exposed during the phase shift measurements up to an exposure of 100 J/sqcm. Again, the surface roughness is approximately 5 nm and the depth of the 0.4 mm wide ditch is about 15 nm. These results could be checked by atomic force microscopic measurements, shown in figure 10. A surface roughness of the freshly evaporated films of 5 nm could be confirmed.

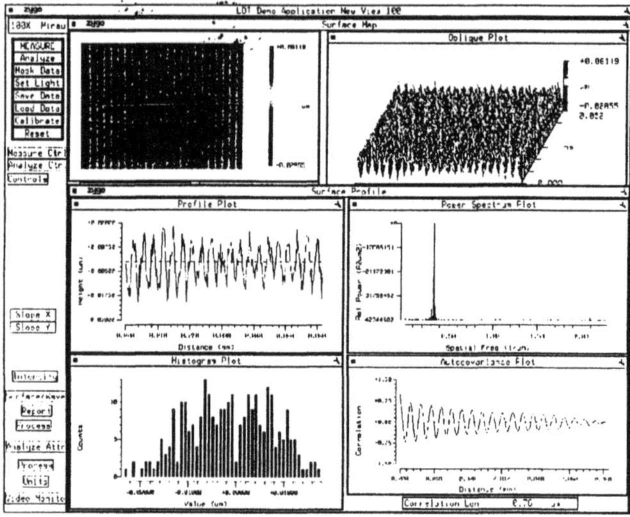

Figure 8. White light interference microscpic measurement of a 3 μm grating

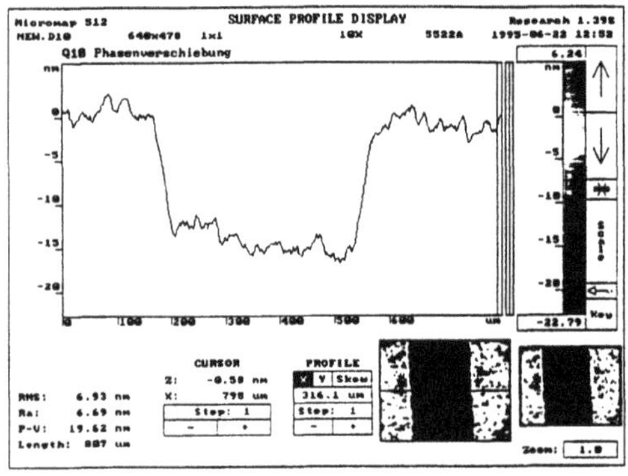

Figure 9. White light interference microscopic profile plot

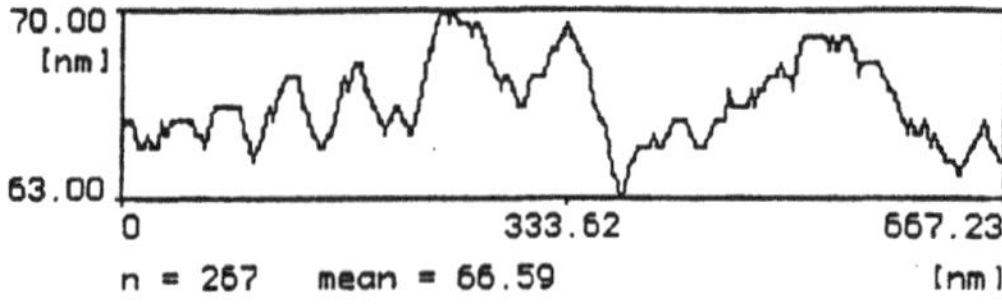

Figure 10. Atomic force microscopic profile plot

The picture of a 3 μm grating (exposed by lithographic recording) and a surface profile is shown by figure 11. Because of the high spot intensity of $6*10^4$ W/sqcm thermal effects dominated the recording process. Thus, the peak to peak value of the surface grating is higher than for the two beam holographic gratings, it amounts to 95 nm.

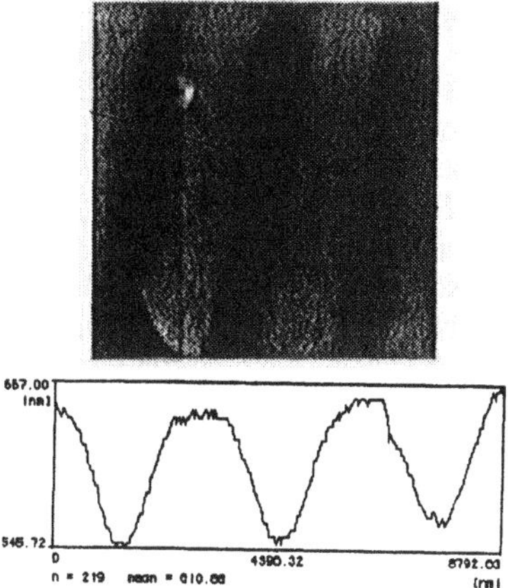

Figure 11. Atomic force microscpic picture of a 3 μm grating

Apart from the surface profiles, thickness measurements were undertaken with the interference microscope. Combining the results of these measurements with the previous phase shift measurements the real photoinduced refractive index change could be evaluated. Table 3 lists the results for three different substrates.

TABLE 3: Photoinduced refractive index change

sample	B2	E7	M965
n	2.5 ± 0.1	2.5 ± 0.1	2.5 ± 0.1
d [μm]	9.90 ± 0.05	5.00 ± 0.05	1.90 ± 0.05
Δd [μm]	0.025 ± 0.005	0.013 ± 0.005	0.005 ± 0.005
ΔOPL_{514} [μm]	0.918 ± 0.055	0.556 ± 0.045	0.278 ± 0.036
ΔOPL_{633} [μm]	0.721 ± 0.018	0.401 ± 0.018	0.185 ± 0.018
Δn_{514}	0.086 ± 0.006	0.099 ± 0.009	0.140 ± 0.020
Δn_{633}	0.067 ± 0.002	0.0686 ± 0.004	0.091 ± 0.012

The starting value of the refractive index n was taken from Keneman and Schwartz [18, 19]. The thickness and its change were measured by interference microscopy. The photoinduced change of the optical path length delta OPL were obtained by evaluating the phase shift measurements. It should be stated, that the listed results correspond to a non saturated state of the photodarkened phase.
The striking difference of the results between sample B2 and E2 on one side and sample M965 on the other side could be explained by different ages of the samples and the influence of the so called temporal annealing or temporal relaxation of the fresh evaporated films during the first months.

4. Applications in Optical Information Processing

Since Keneman 1971 showed the applicability of a-As_2S_3 films for holographic storage [18], different applications of the photoinduced phase transition in amorphous As_2S_3 films were proposed. Suhara et al. used a layer of a-As_2S_3 on a glass substrate to build a waveguide hologram [37]. Photoinduced anisotropy a-As_2S_3 films were used by Kwak

et al. to realize non linear image processing like image addition and subtraction [38]. Later he applied the same effect to expose Dammann gratings [39].
As said above the investigations of a-As_2S_3 films reported in this article were motivated by the objective to realize an optical vector-matrix-multiplier using a-As_2S_3 films as holographic recording material. Different approaches were made to build the optical vector-matrix-multiplier [40]. Starting with an architecture similar to the original one proposed by Goodman two architectures were realized using holographic interconnections [41]. One was based on spatial multiplexed Fourier holograms; the other used an array of simple holographic gratings, each of them representing just a single interconnection. Although advantages of a holographic implementation were lost – the multiplexing of many interconnections in one single hologram – all features of a-As_2S_3 films as holographic recording material could be exploited by the second holographic approach. Thus, this article focuses on the optical vector-matrix-multiplier based on an array of simple gratings.

4.1. EXPERIMENTAL SET-UP

The experimental set-up is shown in Figure 12.

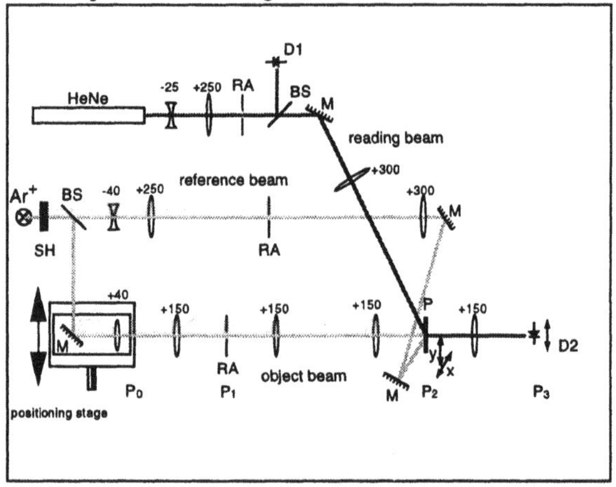

Figure 12. Experimental set-up for the recording of a diffraction grating array

It was used to record 81 gratings interconnecting 9 inputs and 9 outputs. The two beam recording set-up ($\lambda_1 = 514$ nm) and the real-time reconstruction set-up ($\lambda_2 = 633$ nm) is presented graphically. A double f-f-f-f optical Fourier system was used to image a point spot source of plain P_0 to plain P_3 A x-positioning stage was employed to adjust the spot at the nine input/output-positions. A rectangular aperture located in plain P_1 and imaged to plain P2 limited the beam in the recording plain to an area of 0.3 x 0.3 sqmm. Two other rectangular apertures imaged to the same plain P_2 limited the reference beam and the reconstruction beam to the same sharp recording area. A motor driven x,y-stage was taken to adjust the position of the recording area the a-As_2S_3 film after each recording step. The hole experimental set-up was computer controlled. The diffraction efficiency of the single gratings was measured during the recording process. Having reached the programmed efficiency the exposure was stopped automatically and the x,y-stage was driven to the position of the next grating.
Figure 13 shows a picture of the 81 exposed gratings, each grating had a size of 0.3 x 0.3 sqmm and the hole recording area amounted to 3 x 3 sqmm.

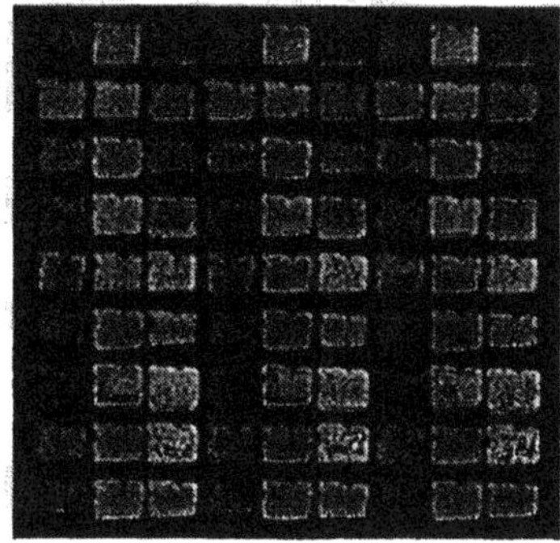

Figure 13. Picture of an array of 81 simple diffraction gratings

To facilitate the analysis only 9 different interconnection efficiencies were recorded. Therefore, the 3 x 3 matrix shown by table 4 was written nine times; in this way each input had the same interconnection pattern with the nine outputs.

TABLE 4. Optical interconnection matrix

1.2	0.75	0.15
1	0.3	0.6
0.075	0.5	0.375

4.2. RESULTS

The evaluation of the recorded grating array was done by reading all nine 3 x 3 matrices by the same homogeneous beam and measuring the total intensity of all interconnections at each output. Table 5 lists the results of two measurements, the target-values and the absolute error for output O_1 - O_9.

TABLE 5. Intensities at the nine outputs of the simple grating VMM

Output	O_1	O_2	O_3	O_4	O_5	O_6	O_7	O_8	O_9
target [%]	7.58	10.10	1.52	13.10	6.06	20.20	3.03	15.15	24.24
1. measure. [%]	7.74	10.68	2.31	13.72	7.60	18.46	3.73	14.68	21.10
2. measure. [%]	7.67	10.83	2.03	14.52	6.80	16.80	3.00	13.86	24.47
mean [%]	7.71	10.74	2.17	14.12	7.20	17.63	3.37	14.27	22.78
error, abs. [%]	0.13	0.64	0.65	1.02	1.14	-2.57	0.34	-0.88	-1.46

Taking the highest diffraction efficiency of 22.8% at output O_1 and the highest absolute error of 2.57% at output O_4 a dynamics range of 8:1 was estimated. This value did not fit the target dynamics of 16:1, because this estimation was a worst case one due to the fact, that the errors of smaller efficiencies were smaller than that of output O_4. Furthermore, the dark self enhancement effect was not yet taken into account at these grating exposures.

4.3. OUTLOOK

Without any optimization a dynamic of 3 bits could be realized by the grating array vector-matrix-multiplier. This is an excellent result due to the fact that the signal-to-noise ratio can easily be increased by adjusting the size of the output detectors to fit the whole coherently added output channels. Additionally the crosstalk could be reduced by another output geometry [40]. Further and this is a unique feature of this architecture in combination with a-As_2S_3 films as recording material, with thicker films diffraction efficiencies up 70% could be realized and dark self enhancement effects could be taken into account already during the recording process. This becomes possible due to the real time recording process in a-As_2S_3 films, that allows a real time control of the recorded efficiencies. The use of thick films for hologram recording is not possible realizing the optical interconnections by higher integrated holograms, because the necessary switch between the writing and the reading wavelength reduces the quality of complex thick holograms.
To check the potential of this architecture a realistic application was simulated. To realize the vector-matrix-multiplier of a classifying optical neural net with 512 inputs and 36 outputs – these values are derived from a realistic application in neural chromosome classification – 16384 interconnects had to be realized. With a fully automated recording set-up these interconnects could be exposed taking 24 hours. Recording an area of 275 x 275 sqμm for each grating the size of the hole grating array would be 25 x 25 sqmm. To realize 36 bipolar outputs 64 unipolar output detectors would be needed with a size of 0.7 x 0.7 sqmm each, to fit the size of the first diffraction order spots in at f = 150 mm optical Fourier system. Using a chessboard like arrangement of the detectors the output plain amounted to about 10 x 10 sqmm and the possible dynamics regarding cross talk etc. could be 4-5 bit.

5. Conclusion

Starting with a brief view over the state of the work regarding the photoinduced phase transitions in amorphous chalcogenides, this article focused on the application of a-As_2S_3 films as holographic recording material. The photoinduced phase shift of the reading beam was measured as a function of exposure. No dependence on the recording intensity could be found for the intensities <10 J/sqcm. This let to the assumption that a pure photo-electron process is responsible for the photo induced phase transition of a-As_2S_3 in this intensity range. Transmission measurements with exposure intensities >100 W/sqcm showed an increased sensitivity of a-As_2S_3 films due to additional thermal driven phase transitions. Diffraction efficiency measurements confirmed the previous phase shift measurements and the high potential of a-As_2S_3 films as holographic recording material. The dynamics of a-As_2S_3 films was investigated measuring the dark self enhancement for different gratings, different film thickness and diffraction efficiencies, which are relevant for applications. Eventually white light interference and atomic force microscopy were used to investigate the surface properties and the photo induced volume reduction of a-As_2S_3 films. It could be shown, that the relative thickness reduction is less than 0.5% for two beam recording. For lithographic recording this effect can be more serious and depends strongly on the exact recording intensity. Further the microscopic measurements were exploited for evaluating the photoinduced refractive index change. The measured value amounts to $\Delta n = 0.1$ and fits the values reported before.
As an application of a-As_2S_3 films an optical vector-matrix multiplier was realized. The

architecture employed, based on an array of simple gratings turned out to be excellently suited to exploit the advantages of a-As_2S_3 films as holographic recording material.

Acknowledgments

The author wants to thank Dr. S. Noehte for fruitful discussions. Prof. K. K. Schwartz and Prof. M. Mitkowa made this work posssible providing us with excellent samples of amorphous As_2S_3 films.

References

1. Goryunova, N.A. and Kolomiets, B.T. (1956) Izv. AN SSSR Ser. fiz. **20**(12), 1496-1500.
2. Treacy, D.J., Storm, U., Klein, P.B, Taylor, P.C. and Martin, T.P. (1980), Photostructural effects in glassy As_2Se_3 and As_2S_3, J. of Non-Crystalline Solids **35&36**, 1035-1039.
3. Fumar, M., Firth, A.P. and Owen, A.E. (1983), A model for photostructural changes in amorphous As-S systems, J. of Non-Crystalline Solids **59&60**, 921-924.
4. Pfeiffer, G., Brabec, C.J., Jefferys, S.R. and Paesler, M.A. (1989), Structural models of glassy As_2S_3,: intermediate range order and photostructural changes, Physic Review B **39**(17), 12861-12871.
5. Yang, C.Y., Paesler, M.A. and Sayers, D.E. (1998), Chemical order in glassy As_xS_{1-x} systems: An x-ray-absorption spectroscopy study, Physic Review **39**(14), 10342-10351.
6. Zhou, W., Paesler, M.A. and Sayers, D.E. (1992), Structure and photostructural changes in non stoichiometric As_xS_{1-x} : A study by x-ray absorption fine structure, Physic Review B **46**(7), 3817-3825.
7. Zhou, W., Sayers, D.E., Paesler, M.A., Bouchet-Fabre, B., Ma, Q. and Raoux, D. (1993), Structure and photoinduced structural changes in a-As_2S_3,films: A study differential anomalous x-ray scattering, Physic Review B **47**(2), 686-693.
8. Elliot, S.R. (1991), Origin of the first sharp diffraction peak in the structure factor of covalent glasses, Physical Review Letters **67**(6), 711-714.
9. Tanaka, K. (1980), Reversible photostructural changes: mechanics, properties and applications, J. of Non-Crystalline Solids **35&36**, 1023-1034.
10. Tanaka K. (1987), Chemical and medium range order in As_2S_3 glass, Physic Review B **36**(18), 9746-9752.
11. Shpotyuk, O.I. (1994), Mechanism of radiation-structural transformations in amorphous As_2S_3,, Radiation Effects and Defects in Solids **132**, 393-396.
12. Starbov, N., Starbova, K. and Dikova, J. (1992), Surface microstructure and growth morphology of vacuum deposited a-As_2S_3 thin films, J. of Non-Crystalline Solids **139**, 222-230.
13. Noehte, S., Ackermann, J., Schwartz, K. K., Dietrich, C. H. and Männer, R. (1994), Optical grating studies in a-As_2S_3 films using scanning force microscopy, Abstrats of the 7the Europhysical Conference, Lyon.
14. Lyubin, V. M. and Tikhomirov, V. K. (1989), Photodarkening and photoinduced anisometry in chalcogenide vitreous semiconductor films, J. of Non Crystalline Solids **114**, 133-135.
15. Lyubin, V. M. and Tikhomirov, V. K. (1990), Photoinduced dichroism in glassy chalcogenide semiconductor films, Sov. Phys. Solid State **32**(6), 1069-1074.
16. Kolobov, A. V., Lyubin, V. M. and Tikhomirov, V. K. (1992), Polarized photodoping of As_2S_3 films by silver, Philosophical Magazine Letters **65**(1), 67-69.
17. Lyubin, V. M. and Klebanov, M. (1996), Photoinduced generation and reorientation of linear dichroism in AsSe glassy films, Physic Review B, **53**(18), 924-926.
18. Keneman, S. A. (1971), Hologram storage in arsenic trisulfide thin films, Appl. Phys. Lett. **19**(6), 205-207.
19. Schwartz, K.K., Reinfelde, M.J. and Teteris, J.M. (1991), Holographic recording processes in amorphous As_2S_3 films, Internal report of the Letvian Academy of Science, Riga, Salaspils.

20. Ozols A., Salminen O. and Reinfelde M. (1994), Relaxational self-enhancement of holographic gratings in amorphous As_2S_3 films, J. Appl. Phys. **75**(7), 3326-3334.
21. Salminen O:, Ozols A., Riihola, P. and Mönkkönen, P. (1995), Intensity threshold for holographic recording in amorphous As_2S_3 films, J. Appl. Phys. **78**(2), 718-722.
22. Salminen, O., Nordman, N., Riihola, P. and Ozols, A. (1995), Holographic recording and photocontraction of amorphous As_2S_3 films by 488 nm and 513 nm laser light illumination, Optics Communications **116**, 310-315.
23. Dietrich, C.H., Noehte, S., Männer, R. and Schwartz, K.K. (1994), Linear phase shift response with high dynamic range for holographic recording in As_2S_3, Opt. Rev. **1**(1), 36-38.
24. Marquez, E., Ramirez-Malo, J., Villares, P., Jiménez-Garay, R., Ewen, P.J.S. and Owen, A.E. (1992), Calculation of the thickness and optical constants of amorphous arsenic sulfide films form their transmission spectra, J. Phys. D: Appl. Phys. **25**, 535-541.
25. Kikineshy A. (1995), Structural phototransformations and optical recording in Se/As_2S_3 multilayered films, Optical Engineering **34**(4), 1040-1043.
26. Pfeiffer, G., Paesler, M.A. and Agerwal, S.C. (1991), Reversible photodarkening in amorphous arsenic chalcogen, J. of Non-Crystalline Solids **130**, 111-143.
27. Paesler, M.A. and Pfeiffer G. (1991), Modeling the structure and photostructural changes in amorphous arsenic sulfide, J. of Non-Crystalline Solids **137&138**, 967-972.
28. Fritzsche, H. (1993), The origin of reversible and irreversible photostructural changes in chalcogenide glasses, Phyilosophical Magazine B.
29. Fritzsche, H. (1993), The origin of photoinduced optical anisometries in chalcogenide glasses, ICAS Proceedings of the 15th International Conference on Amorphous Semiconductors.
30. Kolobov, A.V. and Adriaenssens, G.J. (1994), On the mechanism of photostructural changes in As-based vitrious chalcogenides: Microscopic, dynamic and electronic aspects, Philosophical Magazine B **69**(1), 21-30.
31. Tanaka, K. (1983), Mechanisms of photodarkening in amorphous chalcogenides, J. of Non-Crystalline Solids **59&60**, 925-928.
32. Tanaka, K. (1980), Reversible photostructural changes: mechanisms, properties and applications, J. of Non-Crystalline Solids **35&36**, 1023-1034.
33. Traecy, D.J., Strom, U., Klein, P.B., Taylor, P. C. and Martin, T. P. (1980), Photostructural effects in glassy As_2Se_3 and As_2S_3, J. of Non-Crystalline Solids **35&36**, 1035-1039.
34. Kogelnik, H. (1969), Coupled wave theory for thick hologram gratings, The Bell system Tech. J., **48**(9), 2909-2947.
35. Klein, W.R. and Cook (1967), Unifield approach to ultrasonic light diffraction, IEEE, SU 14, 123-134.
36. Kaspers, F. G. (1973), Diffraction by thick, periodically , stratified gratings with complex dielectric constant, J. Opt. Soc. Am. **63**, 37-45.
37. Suhara, T., Nishihara, H. and Koyama, J. (1976), Waveguide holograms: a new approach to hologram integration, Optics Communications **19**(3), 353-358.
38. Kwak, C.H., Kim J.T. and Lee, S.S. (1989), Non linear optical image processing in photoanisometric amorphous As_2S_3 thin film, Appl. Optics **28**(4), 737-739.
39. Kwak, C.H., Park, S.Y., Kim, H.M., Lee, E.-H. and Kim, C.M. (1992), Dammann gratings for multispot array generation by using photoinduced anisotropic materials, Optics Communications **88**, 249-257.
40. Dietrich, C.H. (1995), Optische Vektor-Matrix-Multiplizierer für hybride Neuronale Netze basierend auf As_2S_3 als holographisches Photomaterial, Doctoral thesis, Department of Computer Science 5, University of Mannheim, Germany.
41. Goodman, J.W., Dias, A.R. and Woody, L.M. (1978), Fully parllel high speed incoherent optical method for performing discrete Fourier-transforms, Optics Lett., **2**(1), 1-3.

THERMALLY AND PHOTO INDUCED PHENOMENA IN AMORPHOUS CHALCOGENIDES.

P. NAGELS
RUCA, University of Antwerp
B-2020 Antwerpen, Belgium

Abstract

Films of amorphous Ge_xSe_{100-x} were prepared by plasma-enhanced vapour deposition (PECVD) using the hydrides GeH_4 and H_2Se as precursor gases and by standard thermal evaporation. Information concerning the structure of the films was obtained from infrared and Raman spectroscopy. The bandgap values of the virgin evaporated films were much lower in the range x=40 to 75 than those of PECVD films, the difference being due to the incorporation of hydrogen in the PECVD films. Thermal annealing and illumination induced irreversible bleaching in the evaporated films. The PECVD films showed a more complex behaviour of the optical shift upon annealing due to the loss of hydrogen. An IR and Raman investigation demonstrated that the changes in the optical transmission are accompanied by an increase of ordering in the local structure.

The photo-induced reversible shift in the optical gap (photodarkening) was studied at three temperatures (13, 77 and 300 K). Our measurements at 13 and 77 K showed that the magnitude of the optical shift increases with an increasing Se concentration in the films. An increase of localized states at the band edges could be responsible for the observed photodarkening at low temperature.

1. Introduction

Amorphous chalcogenides exhibit a wide variety of changes in their structural and physical properties under the action of light or temperature. The changes can be either irreversible or reversible, in the sense that they are permanent or that they can be removed by annealing. Most studies have dealt with thin films prepared by rapid quenching from the vapour phase. These films usually have a more or less disordered network containing a substantial density of wrong chemical bonds. Upon thermal annealing or illumination with bandgap light, bond rearrangements can take place, leading to irreversible changes in many physical properties, e.g. optical absorption

A. Andriesh and M. Bertolotti (eds.),
Physics and Applications of Non-Crystalline Semiconductors in Optoelectronics, 109–121.

edge shifts (photodarkening or photobleaching).

The relaxed state is usually the starting one for the study of photo-induced reversible changes which have a different physical origin. The latter have mainly been studied on thin films in two groups of materials : arsenic chalcogenides which represent a 3:2 coordinated system and germanium chalcogenides which belong to a 4:2 coordinated system. We chose Ge_xSe_{100-x} thin films for the reason that the irreversible process (first response of the film to the action of heating or light) results in bleaching, whereas the reversible process induced by bandgap illumination results in darkening. So, it is easier to separate the reversible darkening component from the irreversible one.

In our study we have prepared films of amorphous Ge_xSe_{100-x} by plasma-enhanced chemical vapour deposition and by thermal evaporation. This contribution consists of two parts. First we report on thermally and photo-induced changes in the optical transmission of amorphous Ge_xSe_{100-x}. Here, we used Raman and IR spectroscopy, since these are powerful techniques for observing even subtle structural changes. The second part describes a study of reversible photodarkening performed at different temperatures (13, 77 and 300 K). The aim is to show that the magnitude of the red shift of the gap largely depends on the temperature at which the material is illuminated.

2. Material preparation

Thin films of amorphous Ge_xSe_{100-x} were prepared by plasma-enhanced chemical vapour deposition in a stainless steel reactor [1]. The precursor gases were GeH_4 and H_2Se diluted in hydrogen (15 vol.% of the hydrides). A low pressure plasma (total gas pressure from 0.1 to 1 mbar) was created by an rf discharge (13.56 MHz) between parallel plate electrodes. In all experiments the separation between the powered (top) and grounded electrode (bottom) was kept constant at 3 cm. The rf power coupled into the reactor varied from 20 to 80 W. Depositions were made without additional heating of the substrates by a furnace, but due to the plasma heating the temperature was approximately 50°C. The GeH_4/H_2Se hydride ratio expressed in volumes was varied from 1/2 to 1/24. The chemical composition of the films was determined by electron microprobe analysis. Films of a-Se were also deposited by PECVD using pure H_2Se. In addition, films of amorphous Ge_xSe_{100-x} were prepared by thermal evaporation of glasses obtained by a standard melting technique, with Ge concentrations x=0, 15, 25, 30, 33 and 40.

Annealing of the as-deposited films was carried out in nitrogen atmosphere at ambient pressure for periods up to 2 h at temperatures below the glass-transition temperature. The latter, determined by means of a thermal mechanical analyser, was found to increase linearly from ~ 50°C for Se to ~ 300°C for $Ge_{30}Se_{70}$, followed by a jump to ~ 400° C for stoichiometric $GeSe_2$.

3. Thermally and photo-induced irreversible changes

3.1. RESULTS

This investigation was described by us in Ref. [2]. Three types of measurements were carried out. Optical transmission and reflection spectra were measured in the wavelength range from 1100 to 500 nm using a Beckman DU-640 UV-VIS spectrometer. Infrared transmission spectra were recorded in the wavenumber range from 4000 to 150 cm^{-1} on films deposited onto polished crystalline Si wafers using a Beckman 4240 and a Bruker IFV 113 V spectrometer. Raman measurements were made with the help of a Bruker 66 V spectrometer in a reflection mode using a 1.06 μm YAG:Nd^{+3} laser source.

Measurements of optical transmission and reflection were analyzed in the standard way to yield the optical absorption coefficient α as a function of the photon energy $h\nu$. In the high-absorption range, the optical gap was evaluated from Tauc's equation $\alpha h\nu = B(h\nu - E_g)^2$, where E_g is the energy distance between the valence and conduction band mobility edges. The parameter B given by the slope of the plots is an interesting parameter, since it can be taken as a measure of the disorder. Figure 1 shows plots of $(\alpha h\nu)^{1/2}$ versus photon energy $h\nu$ for virgin Ge_xSe_{100-x} films prepared by PECVD. The characteristic optical parameters, E_g (eV) and $B^{1/2}$ ($cm^{-1/2}$ $eV^{-1/2}$) of PECVD and evaporated films are presented in Table 1. In Figure 2 the compositional dependence of the optical gap is shown for as-deposited evaporated and PECVD films.

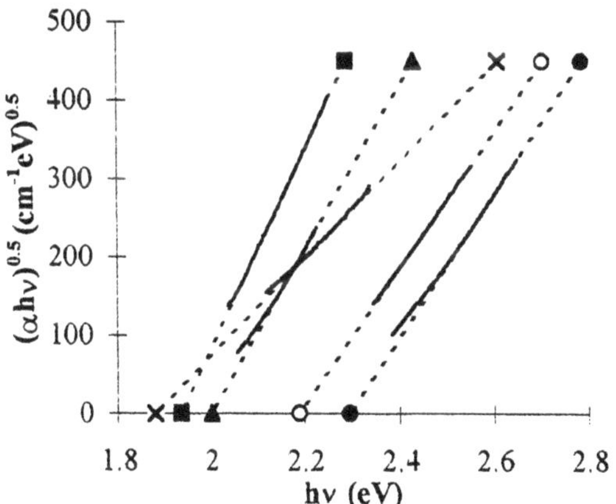

Figure 1. Plot of $(\alpha h\nu)^{1/2}$ vs. photon energy $h\nu$ according to Tauc's formula for virgin PECVD Ge_xSe_{100-x} films; composition in at.% Se : ■ 100, ▲ 85, o 75, • 66, x 60.

The evaporated films, subjected to thermal annealing for 2 h below the glass-transition temperature, displayed bleaching for all compositions, except a-Se. The blue shift of the optical absorption edge substantially increased for higher Ge concentrations. The E_g and $B^{1/2}$ values of the annealed materials are also summarized in Table 1. Annealing of PECVD films showed a more complex behaviour of the optical transmission change. Upon annealing to 175°C, the compositions $Ge_{25}Se_{75}$ and $Ge_{33}Se_{67}$ displayed gradual bleaching similar to that observed in evaporated films, changing to darkening, however, above 175°C (see Table 1). On the contrary, annealing of a $Ge_{40}Se_{60}$ film at 125° C resulted in darkening, followed by a change to bleaching for T_{ann} > 150° C (see Figure 3).

Table 1
Optical parameters E_g (eV) and $B^{1/2}$ ($cm^{-1/2}$ $eV^{-1/2}$) for Ge_xSe_{100-x} films prepared by PECVD and evaporation (EV), before and after annealing.

		PECVD		EV	
		E_g	$B^{1/2}$	E_g	$B^{1/2}$
Se	(Virgin)	1.936	1280	1.945	1297
$Ge_{15}Se_{85}$	(Virgin)	2.000	1053	2.017	945
	(100° C)	2.003	1058	2.029	947
$Ge_{25}Se_{75}$	(Virgin)	2.090	816	2.044	827
	(175° C)	2.118	827		
	(250° C)	2.094	816	2.115	839
$GeSe_2$	(Virgin)	2.266	836	2.073	913
	(175° C)	2.326	891		
	(250° C)	2.269	854	2233	1040
$Ge_{40}Se_{60}$	(Virgin)	1.881	621	1.452	552
	(125° C)	1.779	605		
	(300° C)	2.047	742	1.699	637

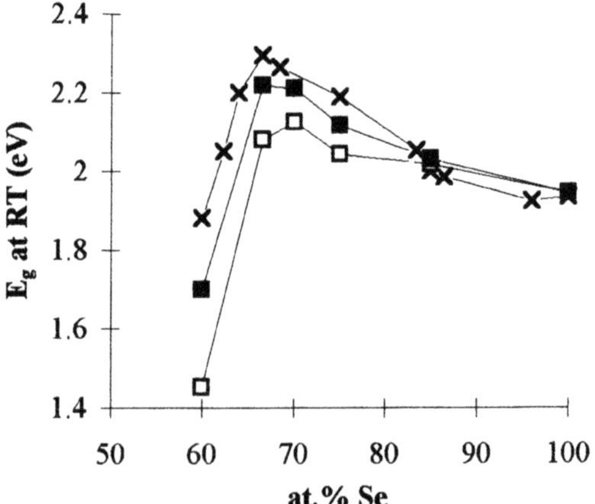

Figure 2. Compositional dependence of the optical gap E_g for amorphous Ge_xSe_{100-x} films: (X) PECVD, (□) virgin evaporated, (■) annealed evaporated. Lines are drawn as guides for the eye.

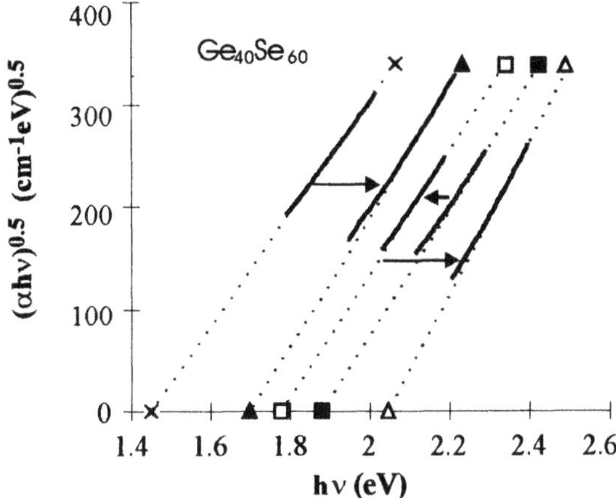

Figure 3. Plot of $(\alpha h\nu)^{1/2}$ versus hν for $Ge_{40}Se_{60}$ films; evaporated:(X) virgin, (▲) annealed at 300° C; PECVD: (■) virgin,(□) annealed at 125°C, (Δ) annealed at 300° C.

Illumination by white light induced a blue shift of the optical gap in the evaporated films of composition x = 25, 30 and 33 but no effect was observed in the Se- and Ge-rich films (x=15 and 40). The shift was maximum in the stoichiometric material (ΔE_g = 0.057 eV) but bleaching by illumination was less pronounced than the thermally-induced one. All compositions of PECVD films did not show any measurable shift of the optical absorption edge, indicating that these materials are closer to the ordered state than the evaporated ones.

3.2. DISCUSSION

From Figure 2 it can be seen that the optical bandgap of the PECVD films exhibits a maximum (E_g=2.27 eV) at the stoichiometric composition (x=33) with a rapid decrease in the Ge-rich region and a slower one in the Se-rich region. The bandgap values of the virgin evaporated films were much lower in the range x=40 to x=75 and the maximum was slightly shifted to the Se-rich side. The difference can be explained by the incorporation of hydrogen in the PECVD films during the deposition in a hydrogen-rich plasma. This was evidenced by the presence of a Raman band at 2030 cm^{-1}, arising from Se-H vibrations, in stoichiometric and Se-rich samples.

Figure 2 also shows that the E_g values of the annealed evaporated films are still lower than those of the non-annealed PECVD films. After annealing, the maximum in the compositional dependence of the optical gap is shifted to the stoichiometric composition $Ge_{33}Se_{67}$. In Figure 3 the Tauc plots of evaporated and PECVD $Ge_{40}Se_{60}$, before and after annealing for two hours, are shown. The striking difference between both films can be explained by the loss of hydrogen when PECVD films were subjected to heat treatment. Direct evidence came from Raman spectra, recorded on the three different compositions after annealing at increasing temperature. In $Ge_{40}Se_{60}$ a Raman band at 2030 cm^{-1}, due to Ge-H vibrations, completely disappeared in the temperature range 125 to 150°C. In $Ge_{33}Se_{67}$ and $Ge_{25}Se_{75}$ the hydrogen evolution occurred at higher temperature, as indicated by the gradual decrease of the intensity of a Raman band at 2240 cm^{-1} (associated with Se-H vibrations) between 150°C and 225°C. These observations indicated that darkening is a consequence of the disappearance of Se-H or Ge-H bonds with the formation of new Se-Se bonds in the Se-rich material and Ge-Ge bonds in the Ge-rich one. The Raman spectra of evaporated films before and after annealing are shown in Figure 4. In $Ge_{25}Se_{75}$ the intensity of the vibration mode at 260 cm^{-1}, related to Se-Se bonds, decreased. The relative intensity of the companion peak at 215 cm^{-1} to the peak at 200 cm^{-1} slightly decreased, pointing to an increased number of corner-sharing tetrahedra. The spectra of stoichiometric $GeSe_2$, clearly indicated that annealing reduced the intensity of the bands at 260 cm^{-1} and 175 cm^{-1}, originating from Se-Se and Ge-Ge bonds. Hence, the density of homopolar bonds decreased, resulting in bond rearrangements with the formation of Ge-Se bonds. Further evidence for an increase in the density of heteropolar Ge-Se bonds was found in the slight increase in the intensity of the absorption band located at 260 cm^{-1} in the far-infrared spectrum of $GeSe_2$, annealed at 255°C for 180 min (see Figure 5). In $Ge_{40}Se_{60}$ the Raman band at 175 cm^{-1} became sharper and better defined, reaching the shape of the same band observed in a PECVD sample. This is a sign of an increase in the short-range order.

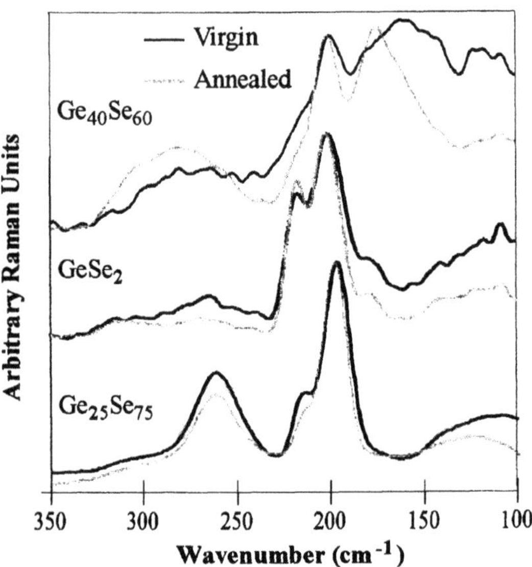

Figure 4. Raman spectra of evaporated Ge_xSe_{100-x} films before and after annealing

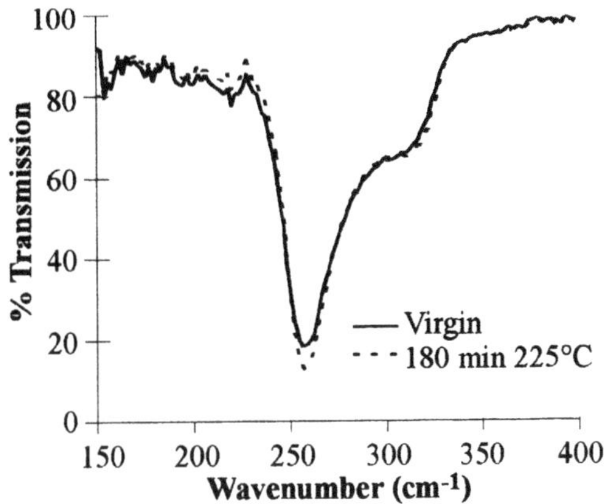

Figure 5. IR transmission spectra of evaporated $Ge_{33}Se_{67}$ before and after annealing at 225° C

3.3. CONCLUSIONS

From the Raman and IR measurements we can conclude that a certain degree of ordering in the local structure is the main cause of the irreversible blue shift of the optical gap in Ge_xSe_{100-x} films. In general, the films of different compositions prepared either by evaporation or by PECVD exhibit similar structural changes but the behaviour of the optical change is more complex in PECVD films than in evaporated ones, due to loss of hydrogen during heating.

4. **Photo-induced reversible changes**

4.1. RESULTS

In order to suppress a possible irreversible component the films were first annealed by heating them close to the glass transition temperature. The optical transmission was then measured using an optical cryostat mounted in the sample compartment of the spectrometer. All samples were illuminated at 77 and 300 K by a halogen lamp equipped with an infrared-cut filter. During illumination, the lamp (power density 80 mW cm^{-2}) was fixed in the sample compartment. In addition illumination was also carried out at 13 K on two samples : Se and one rich Se sample, $Ge_{15}Se_{85}$.

It is well known that the photo-induced reversible shift of the absorption edge increases with decreasing temperature. In the case of amorphous Se, photodarkening has only been observed, when the film is illuminated at low temperature (77 K).

4.1.1. *Photo-induced changes at room temperature.*
We shall first discuss the results of photodarkening obtained on evaporated and PECVD samples illuminated at room temperature. The light-saturated state can be practically reached after 2 h of illumination. The light-induced darkening of the virgin films is not accompanied only by a parallel shift of the band edge since the slope of the Tauc curve $B^{1/2}$ ($(\alpha h\nu)^{1/2} = B^{1/2}(h\nu - E_g)$) slighly changes. This is shown in Figure 6, where the Tauc curve before and after illumination is given for an annealed film with composition $Ge_{25}Se_{75}$ prepared by PECVD. The results of the measured saturated optical shift $|\Delta E_g|$ and of the change in $B^{1/2}$ are presented in Table 2 for Ge_xSe_{100-x} samples prepared by thermal evaporation and by PECVD.

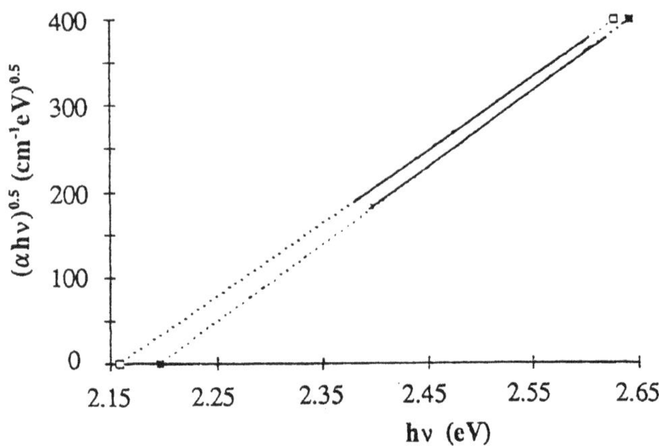

Figure 6. Tauc plots before (■)and after (□) illumination of an annealed film with composition $Ge_{25}Se_{75}$ prepared by PECVD.

The dependence of the optical shift $|\Delta E_g|$ on the chemical composition is also shown in Figure 7. The values of the optical shift are comparable in magnitude for both preparation methods. For $Ge_{30}Se_{70}$ and $GeSe_2$ the PECVD films showed a smaller ΔE_g.

4.1.2. *Photo-induced changes at 77 K and 13 K.*
Since our interest in this work was the role of the temperature in the photo-induced shift of the gap we also studied the temperature shift of the gap $E_g(T)$ in the range 13-300 K. The results obtained on a-Se are summarized in Figure 8. The solid curve represents a fit to the experimental data (open squares) using Fan's one-phonen approximation :

$$E_g(T) = E_g(0) - A[\exp(h\nu/kt) - 1]^{-1}$$

where $E_g(0)$=2.12 eV, A=0.22 eV and $h\nu$=0.021 eV.
The dashed line in Figure 8 indicates a fit of the experimental data to the standard linear equation $E_g(T)=E_g(0)-\beta T$ with $\beta=8.4\times10^{-4}$eV K^{-1}.

Table 2. Photodarkening observed at room temperature on evaporated (EV), and PECVD Ge_xSe_{100-x} films. $|\Delta E_g|$ in eV and $|\Delta B^{1/2}|$ in $cm^{-1/2}\ eV^{-1/2}$.

Comp.	EV		PECVD	
	$\|\Delta E_g\|$	$\|\Delta B^{1/2}\|$	$\|\Delta E_g\|$	$\|\Delta B^{1/2}\|$
Se	0	0	0	0
$Ge_{15}Se_{85}$	0.011	19	0.012	15
$Ge_{20}Se_{80}$			0.022	17
$Ge_{25}Se_{75}$	0.038	34	0.037	43
$Ge_{30}Se_{70}$	0.026	26	0.018	32
$GeSe_2$	0.026	33	0.015	23
$Ge_{40}Se_{60}$	0	0	0	0

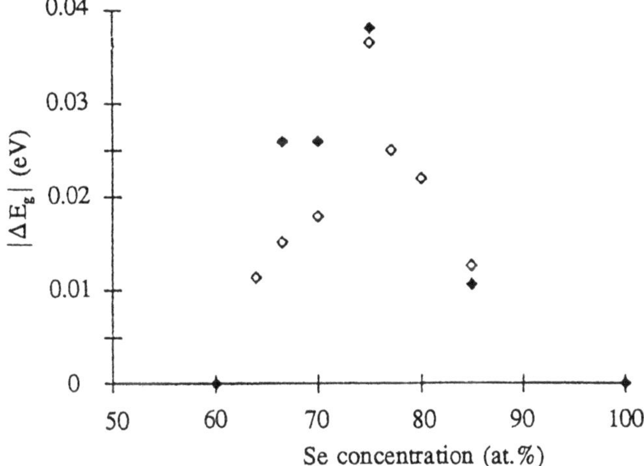

Figure 7. Dependence of $|\Delta E_g|$ on the Se concentration upon illumination at room temperature, (◆) evaporated films; (◇) PECVD films

For the case of low-temperature illumination, we examined the influence of illumination at liquid nitrogen temperature of both virgin (V) and well-annealed (A) Ge_xSe_{100-x} films (40<x<0) and at 13 K for Se and $Ge_{15}Se_{85}$. The Tauc curves obtained on a-Se at 13 K before and after illumination up to 120 min are shown in Figure 9. The saturated value of the shift was equal to 0.157 eV. This change in the Tauc gap upon illumination does not result from a simple parallel shift of the absorption edge, but is also accompanied by a strong decrease in the slope of the Tauc curve.

The experimental $|B^{1/2}|$ value was equal to 1944 and 1315 $cm^{-1/2}\ eV^{-1/2}$ before and after illumination, respectively.

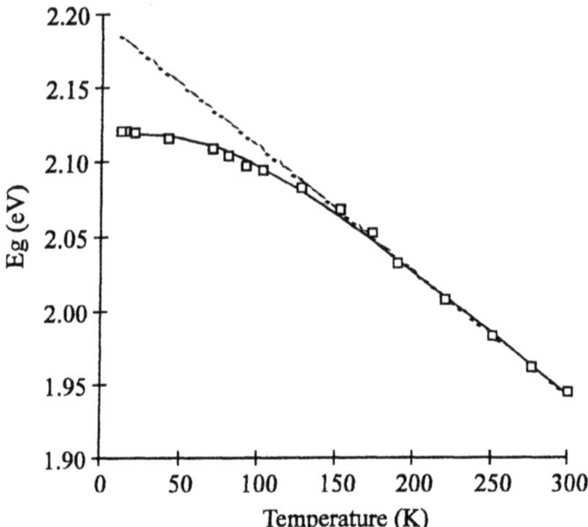

Figure 8. Temperature dependence of the optical gap E_g of a-Se

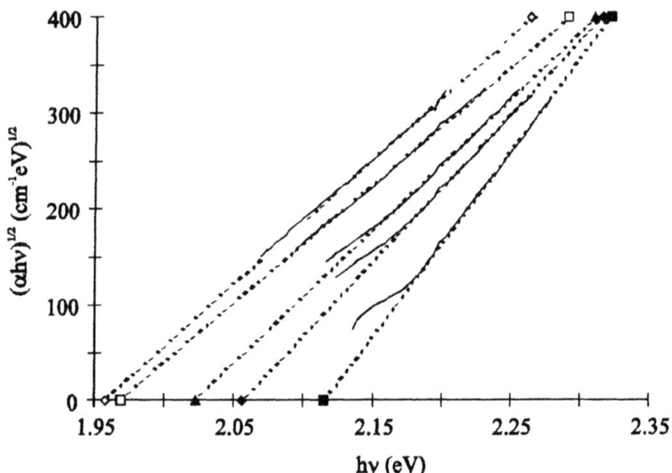

Figure 9. Tauc curves of a-Se at 13 K before (■) and after illumination: (◆) 5 min., (▲) 15 min., (□) 60 min., (◇) 120 min.

The $|\Delta E_g|$ values of the different compositions measured at 13 K and 77 K are listed in Table 3. By comparing the $|\Delta E_g|$ values in Tables 2 and 3, one can see that much higher optical shifts are found at low temperature than at room temperature.

Table 3.
$|\Delta E_g|$ values of evaporated and PECVD Ge_xSe_{100-x} films after illumination at 13 K and 77 K- (V) virgin films; (A) annealed films.

$|\Delta Eg|$ (eV)	Evap.(V) 13 K	Evap. (V) 77 K	Evap. (A) 77 K	PECVD (V) 77 K
Se	0.157	0.108		0.105
$Ge_{15}Se_{85}$	0.150	0.122	0.122	0.118
$Ge_{20}Se_{80}$				0.109
$Ge_{25}Se_{75}$		0.102	0.104	0.081
$Ge_{30}Se_{70}$				
$GeSe_2$		0.078	0.027	0.031
$Ge_{40}Se_{60}$		0.042	0.026	0.025
		0	0	0

Our measurements at 13 K and 77 K showed that $|\Delta E_g|$ increases with an increasing Se concentration in the films. At room temperature, however, the maximum of $|\Delta Eg|$ was found for $Ge_{25}Se_{75}$. We suppose that the process responsible for photodarkening is the same at room temperature, 77 and 13 K, but that during illumination at higher temperature a part of the metastable states already converts back to the ground state. From our results it follows that the temperature at which the film is illuminated is an important factor in the study of the photo-induced changes and that measurements at room temperature do not give the right information, because it is not possible to measure the total effect at this temperature.

An amorphous Se film photodarkened at 13 and 77 K had its initial abosption restored when heated up to room temperature. The process was completely reversible. An isochromal annealing experiment using a constant heating rate yielded a low activation energy for the bleaching process of the order of 0.07 eV. The ease of recovery at temperatures below 300 K can be understood if one takes into account that the twofold-coordinated Se matrix is very flexible and that the glass transition temperature is around 50°C.

The change in the slope of the Tauc curve $|\Delta B^{1/2}|$ upon illumination increased with decreasing temperature for all compositions. It was the largest for pure Se, with $|\Delta B^{1/2}|$ equal to 316 and 629 $cm^{-1/2}$ $eV^{-1/2}$ at 77 and 13 K, respectively. In the case of $Ge_{15}Se_{85}$, $|\Delta B^{1/2}|$ was found equal to 143 and 186 $cm^{-1/2}eV^{-1/2}$ at the same temperatures. Since in the Mott-Davis interpretation [3] B is inversely proportional to the width of localized states at band edges, a decrease of $B^{1/2}$ values induced by illumination indicates an increase of localized states at the band edges. The band tailing parameter in Tauc's formula is given by the equation [3] :

$$B = \frac{4\pi \, \sigma_{min}}{nc\,\Delta E}$$

where σ_{min} is the minimum metallic conductivity, n the refractive index, c the light velocity and ΔE the width of the band tails.

Let us consider the case of amorphous Se. The experimental $B^{1/2}$ value varied from 1944 to 1315 $cm^{-1/2}eV^{-1/2}$ before and after illumination at 13 K. Assuming $\sigma_{min} = 300$ $ohm^{-1}cm^{-1}$ and $\bar{n} = 2.5$, we derive $\Delta E = 0.15$ eV and 0.32 eV before and after illumination, respectively. The difference is 0.17 eV, which is very close to the value found for the photo-induced optical shift at 13 K.

Several models have been proposed for explaining the photodarkening process. The models fall into two categories : 1. breaking and reforming of covalent bonds; 2. changes in intermediate-range order involving the rotation of molecular units.

In an elemental material as a-Se the occurrence of wrong homopolar bonds is excluded. Therefore, models based on changes in short-range order involving breaking of homopolar bonds and formation of heteropolar bonds (applied for As_2Se_3) do not provide an adequate description.

Two important observations follow from our low temperature photodarkening experiments : 1. the optical shift increases with an increasing concentration of Se-Se bonds in the Ge_xSe_{100-x} samples; 2. the slope of the Tauc plots decreases upon illumination, which gives evidence for a broadening of the tails of localized states at the band edges. The influence of the change of the refractive index n upon illumination ($\Delta n = 2$ %) on B is very small. Therefore, models based on changes of refractive index cannot explain our experimental results.

5. Conclusions

Darkening induced by illumination is accompanied by an increase of localized states at band edges. Suitable defects responsible for the observed photodarkening could be self-trapped exciton (STE)-like states. The role of STE in the process of the photodarkening of amorphous chalcogenides was proposed by Street [4] and discussed in connection with possible models of photodarkening by Eliott [5]. It could also be possible that STE occur at centres where p lone-pair electronic states of Se atoms strongly interact [6]. Such a situation can also occur if dihedral angle distortions give rise to localized states to in the gap close to the band edgs [7], the process proposed by Nagels et al. [8] as a possible origin of photodarkening in a-Se.

6. References

[1] Sleeckx, E., Nagels, P., Callaerts, R., and Van Roy, M. (1993) Plasma-enhanced chemical vapour deposition of amorphous Ge_xSe_{100-x} films, *J. Non-Cryst. Solids* **164-166**, 1195-1198.

[2] Sleeckx, E., Tichý, L., Nagels, P., and Callaerts, R. (1996) Thermally and photo-induced irreversible changes in the optical properties of amorphous Ge_xSe_{100-x} films, *J. Non-Cryst. Solids* **198-200**, 723-727.

[3] Mott, N.F., and Davis, E.A. (1979) *Electronic Processes in Non-Crystalline Materials*, Clarendon Press, Oxford p. 289.

[4] Street, R.A. (1977) Non-radiative recombination in chalcogenide glasses, *Solid State Commun.* **24**, 363-365.

[5] Eliott, S.R. (1986) A unified model for reversible photostructural effects in chalcogenide glasses. *J. Non-Cryst. Solids* **81**, 71-98.

[6] Liu, J.Z., and Taylor, P.C. (1987) Absence of photodarkening in bulk, glassy As_2S_3 and As_2Se_3 alloyed with copper, *Phys. Rev. Lett.* **59**, 1938-1941.

[7] Wong, C.K., Lucovsky, G., and Bernholc, J. (1987) Intrinsic localized defect states in a-Se associated with dihedral angle distortions, *J. Non-Cryst. Solids* **97-98**, 1171-1174.

[8] Nagels, P., Sleeckx, E., Callaerts, R., and Tichý, L. (1995) Structural and optical properties of amorphous selenium prepared by plasma-enhanced CVD, *Solid State Commun.* **94**, 49-52.

PHOTOINDUCED EFFECTS IN AMORPHOUS CHALCOGENIDES

M. FRUMAR, Z. POLÁK, Z. ČERNOŠEK, M. VLČEK AND B. FRUMAROVÁ*
Department of General and Inorganic Chemistry, Faculty of Chemical Technology, University of Pardubice, 53210 Pardubice, Czech Republic
** Joint Laboratory of Solid State Chemistry of Czech Acad. Sci. and of Univ. Pardubice, Studentská 84, 53009 Pardubice, Czech Rep.*

Abstract

New results of the study of photoinduced effects in amorphous chalcogenide layers and glasses are reviewed. The main accent is given to the changes of short range order, to the mechanisms of optically induced effects, to the optically induced changes of reactivity of chalcogenide layers. Both irreversible and reversible effects are discussed. Attention is paid also to different materials in which the photoinduced effects were studied recently.

The study of Raman and IR spectra revealed the changes of short range order in As-S layers not only for irreversible but also for reversible photoinduced changes.

The exposed and unexposed parts of the surface attract some organic sorbats in different way and, as a result, the dissolution rates of layers in alkaline solutions are strongly changed. It is concluded that the As-rich parts of As-S layers are responsible for increased sorption of the adsorbats.

The review stems from the results of authors' group and covers also the other groups' main results published recently.

Besides the basic importance of the results of the interaction of amorphous solids with radiation and light, the photoinduced changes of structure and properties can be applied in high resolution litography, for preparation of optical memories, photoresists, optical diffraction elements, planar optoelectronic parts and circuits. Some of these applications are also mentioned.

1. Introduction

The illumination of many amorphous and glassy chalcogenides changes their structure and many physicochemical properties. The resulting photoinduced effects (PE) in chalcogenide layers and glasses have been studied in many laboratories for many years since the description of these effects in papers of Keneman, Berkes and de Neufville in early seventieth. The subject, or its parts, were already reviewed several times (in the last time see e.g. [1-9]). In spite of such a large research effort, many photoinduced

A. Andriesh and M. Bertolotti (eds.),
Physics and Applications of Non-Crystalline Semiconductors in Optoelectronics, 123–139.

effects are still not clear, new photoinduced phenomena such as photoinduced anisotropy, gyrotropy etc. have emerged [6-16]. The number of studied materials is also expanding (see e.g. [16-18]).

Such a large effort stems from the need of description and understanding of physical and chemical properties of amorphous chalcogenides and their interaction with light. There are also application reasons: the photoinduced effects in chalcogenide glasses and layers can be applied for production of submicron photoresistors, optoelectronic elements and devices, e.g. optical memories and diffraction elements (see e. g. [6,7,16 19-27]). Very high resolution can be achieved (>10^4lines/mm [7]); the predicted resolution limit is 10nm [24] and the resolution of 50nm was already demonstrated [26]. The PE are very often accompanied by large changes of refraction index (up to 10%; [7]; for Ag photodoping even more), they can be used also for preparation of optical light-guides, planar optical circuits, etc..

The PE are reversible or irreversible [5]. The fresh evaporated layers of a-chalcogenide are very often inhomogeneous, the illumination increases their homogeneity. Such a change is irreversible, it cannot be reversed by annealing or by illumination. The reversible PE can be produced by illumination of bulk glasses or well-annealed layers.

Unpolarised excitation light produces optically isotropic exposed layers with "isotropic" PE. When the polarised light (either band-gap or even sub-band-gap) is used for excitation, the layers can become optically anisotropic, the effect is called anisotropic PE or vectorial one.

The sensitivity of PE in amorphous layers is not generally very high when compared with classical silver halides. It can be higher for Ag-photodoping [7], and for many purposes, the high sensitivity is not necessary. The sensitivity can be increased by some developers in combination with photoenhanced dissolution [28-30]. The sensitivity can be also enlarged when the excitation is accomplished by very short intensive excimer laser pulses [7]. Such pulses can increase the sensitivity to excitation light 10^3 - 10^4 times. The properties of photodarkened layers does not depend on the mode of excitation (continuous or pulses) [7], thus the nature of both changes could be identical. The origin of such an amplification effect is not clear yet. The photoinduced changes are a dynamical process [6,27,31,32], the reverse process (selfannealing) is usually slow [31-33], possibly because of its thermal origin. One can speculate that such a selfannealing might be too slow and ineffective when short pulses excitation is applied. The excited state can be then frozen down. The local "overheating" during application of short intensive pulses can also play a role.

The intensity of photostructural changes is composition sensitive [18,27,34]. The largest photoinduced changes in Ge-As-S films were found for mean coordination number of atoms in these glasses, Z = 2.67 which represents the chemical threshold in these glasses [34]. Such a coordination is connected with the largest free volume in these glasses and is favourable for atoms movement and for optically induced changes of structure [18].

In this review we will discuss new results obtained or published recently. We will paid our attention mainly to general problems of isotropic photostructural (photoinduced) effects in amorphous chalcogenides, to the changes of their structure and to the photoinduced changes of reactivity. Shortly will be mentioned the photoinduced anisotropy and new trends in application of PE. The photoinduced dissolution and diffusion of metals (photodoping) in amorphous chalcogenides will be discussed in another paper of this Workshop [35]. Some new unpublished results of our study of PEwill be also shown.

2. Isotropic photoinduced phenomena

The isotropic (scalar) PE cover a lot of physical and chemical effects. They include photoinduced changes of electron densities, of atomic structure (short range order, SRO; medium range order, MRO), of photoinduced chemical reactions (photolysis, photooxidation, photoinduced synthesis), photoenhanced dissolution and diffusion, photoinduced changes of surface state and changes of reactivity. All these effects start with changes of electron densities due to absorption of light. The excited state can relax in different ways for different effects mentioned above. The relaxation concerns electrons and in many PE also atoms, molecules and ions. The electronic effects were thoroughly discussed recently [8] and will not be mentioned here.

As a consequence of atomic and electrons relaxation, many physical and chemical properties of chalcogenide layers (CL), e.g. atomic structure, optical transmissivity and reflectivity, index of refraction, the optical anisotropy, microhardness, chemical reactivity an others, are changed.

The origin of all PE in CL is not very clear yet, because some photoinduced structural changes are small and structural and other probes are not often sensitive enough to give clear picture of these effects. Recent contribution of our group to these studies have concerned in the photoinduced changes of structure reflected by changes of Raman and IR spectra, the photoinduced amorphization, the photoinduced changes of dissolution rates of chalcogenide layers, and the photodoping processes [4,17,18, 28-32,35-44,58]. Some applied processes were also studied [29,30, 39].

2.1. PHOTOINDUCED CHANGES OF STRUCTURE

The photoinduced changes of structure and of chemical bonds redistribution in CL accompanying the irradiation of CL by UV light, electron- or γ- rays were studied intensively by Shpotyuk and co-workers [45-51]. The differential FAR IR reflection and transmission spectra revealed changes of intensity of several absorption bands, among them of the bands connected with of As-As vibrations [51]. It was supposed that the =As-As= and also the (As_2^- ; S_3^+) and (As_4^+ and S_1^-) structural units are formed due to illumination (irradiation) of fresh evaporated amorphous chalcogenides. The upper index in defect signature means the charge state, the lower one the coordination number. Such results are in accordance with our earlier model [31,32] and with our last studies of Raman spectra[1] of CL or of chalcogenide glasses (Figs. 3-8,10,11). The changes of bands intensities are quite large even for reversible part of PE (Figs. 4,7). The photoinduced changes of coordination of some atoms were not accepted formerly by many authors (see e.g. [9,52]). Now they are generally accepted ([5,6,45-51]).

---

[1]Raman spectra were measured by FT Raman spectrometer FRA106 (Bruker, Germany) using Nd:YAG laser beam (1064 nm) as excitation light.

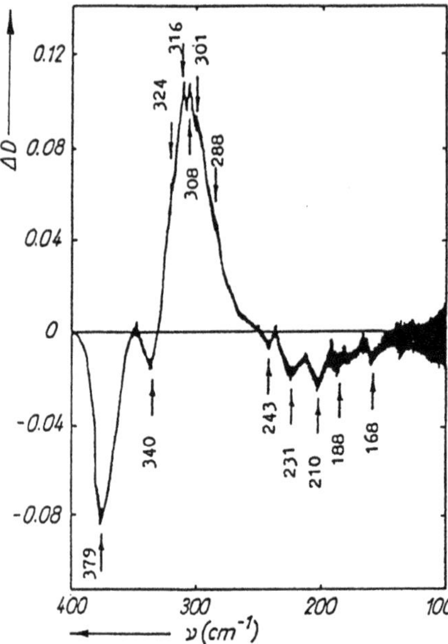

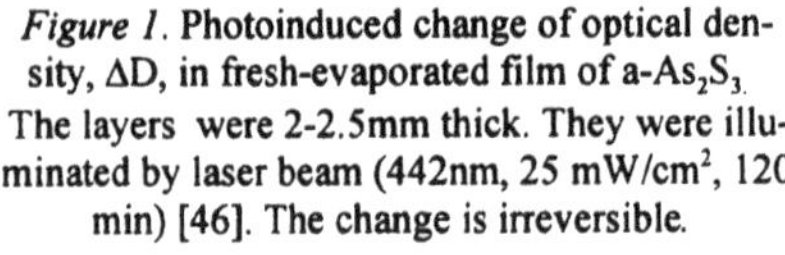

Figure 1. Photoinduced change of optical density, ΔD, in fresh-evaporated film of a-As_2S_3. The layers were 2-2.5mm thick. They were illuminated by laser beam (442nm, 25 mW/cm^2, 120 min) [46]. The change is irreversible.

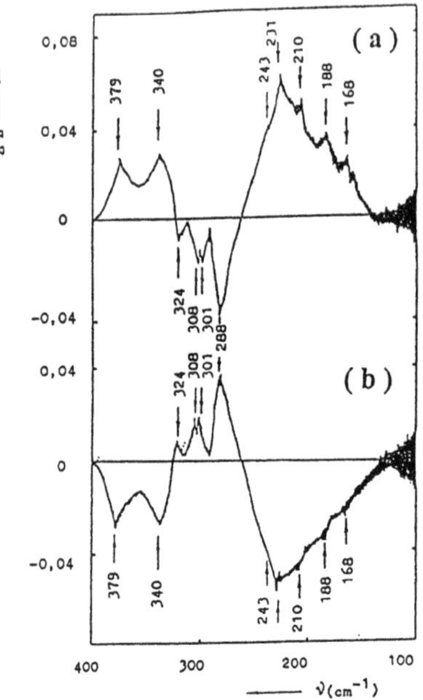

Figure 2. The change of optical density, ΔD, of a-As_2S_3 thin film induced by third cycle of photoexposure (a) and annealing (b). The layers were 2-2.5mm thick. They were illuminated by laser beam (442nm, 25 mW/cm^2, 120 min). They were annealed at 150°C for 30 min [46]. The changes are fully reversible.

The γ- and electron radiation induce also changes of optical properties, of microhardness, of acoustical properties and of LESR signal of glassy As_2S_3. The effects can be interpreted again by bond-breaking and coordination defects formation [49, 51-53]. The photoinduced changes of chemical bonds′ densities are relatively large for irreversible effects, for reversible effects are smaller. It is evaluated (in accordance with [31,32]) that approximately 6-7% of atoms are involved in photoinduced reversible changes in As-S layers [46].

The photoinduced coordination defects formation (including As-As bonds formation) starts apparently by changes of soft atomic interactions, by electron-hole formation, self-trapped exciton mechanism [1,45]. The final state (after exposure, after annealing) has different structure (at least nearby some atoms) of the amorphous layer. This fact is reflected by the changes of intensities of individual Raman bands as can be seen in Figs. 3-8.

The exposure of fresh evaporated layers changes the intensities of individual Raman bands, the system is partly shifted to the state closer to the equilibrium (Fig. 3). After annealing of the as-evaporated layers, the system approaches quasiequilibrium state and Raman spectra of annealed layers are practically identical with the spectra of bulk samples (Figs. 3-5). The photoinduced changes of Raman spectra in CL with overstoichiometry of As, e.g. in $As_{42}S_{58}$ layers are larger than in stoichiometric layers or in sulphur - rich layers [38].

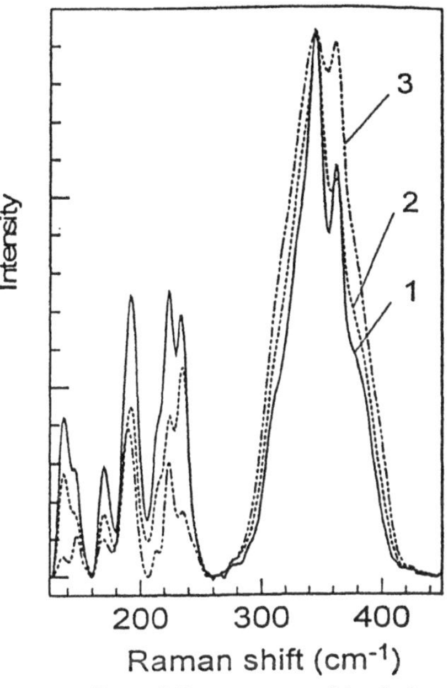

Figure 3. Raman spectra of $As_{42}S_{58}$ layers. 1 fresh evaporated layer, 2 exposed layer, 3 annealed layers (after exposure). The background of spectra was corrected, the intensities were normalized to the intensity of 345cm^{-1} band [9].

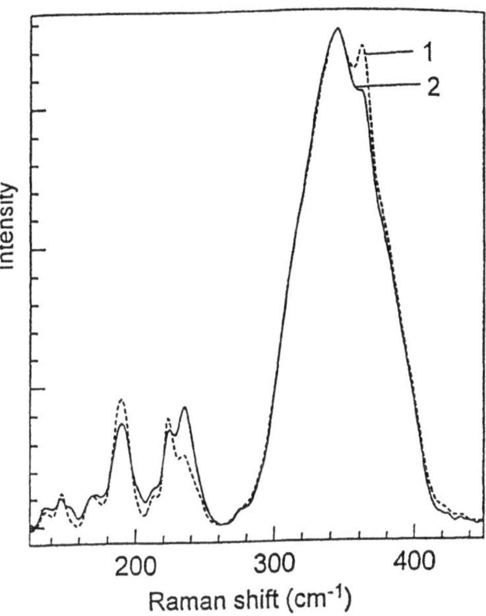

Figure 4. Raman spectra of $As_{42}S_{58}$ layers. 1 annealed layer, 2 layer exposed after annealing. The background of spectra was corrected, the intensities were normalized to the intensity of 345cm^{-1} band [9].

The photoinduced changes of Raman spectra for irreversible PE can be seen in Fig. 3, the spectra of reversible PE are in Figs. 4,7. The bands in spectra are apparently built up from several overlapping simple bands. For better understanding of structural changes, the spectra were deconvoluted into individual bands and fitted to measured curves (Figs. 5-7, [44]). The result should be taken with some care both from mathematical limitations and also due to small fluctuations of experimentally found curves. The fluctuation of experimental results was reduced very much, due to averaging of more than 700 scans for every curve. In every case, a semiquantitative evaluation of intensities of individual Raman bands can be received. The results are summarised in Fig. 8, where the relative integral areas of every band (normalised to the area of the most intensive 345 cm^{-1} band of AsS_3 pyramidal vibrations of every spectrum) is given.

It can be seen, that the intensities, half-widths and also the positions of individual bands are changed by exposure, by annealing and by following exposure as well (Figs. 3-8). The detailed discussion of these changes is going to be given elsewhere [44], here we will discuss them only qualitatively.

In the Raman spectra of fresh evaporated, exposed and annealed $As_{42}S_{58}$ layers (Fig. 3) can be seen the relatively broad band with the maximum near 345 cm^{-1} and several overlapping narrow bands in the spectral region between 125-260 cm^{-1}. These narrow bands are still present in $As_{42}S_{58}$ bulk glasses (Fig. 5), but their relative intensity (as compared with the broad band with maximum near 345 cm^{-1}) is lower. These bands are of very low intensity in $As_{40}S_{60}$ glasses or in As-S glasses with overstoichiometry of sulphur [9,45,46,54]. One can judge, therefore, that this part of spectrum is connected with the presence of overstoichiometric As: in annealed and exposed films with the presence of As_4S_4 or =As-As= structural units and, in the evaporated layers, with the presence of molecular fragments (As_4, As_4S_4 and As_4S_3). These fragments are formed during high temperature evaporation and they are cooled down during condensation of vapours on a cold substrate [54]. The exposure of fresh evaporated layers induces some photochemical reactions and polymerisation, the amount of "molecular fragments" is lowered and the relative density of As_2S_3 structural units is increased. The polymerisation manifests itself by broadening of main 345 cm^{-1} Raman band (Fig. 3).

All the vibration bands of this region (125-260 cm^{-1}) are present in both evaporated layers (Fig. 3) and in bulk samples (Fig. 5). The bands are narrower in fresh evaporated

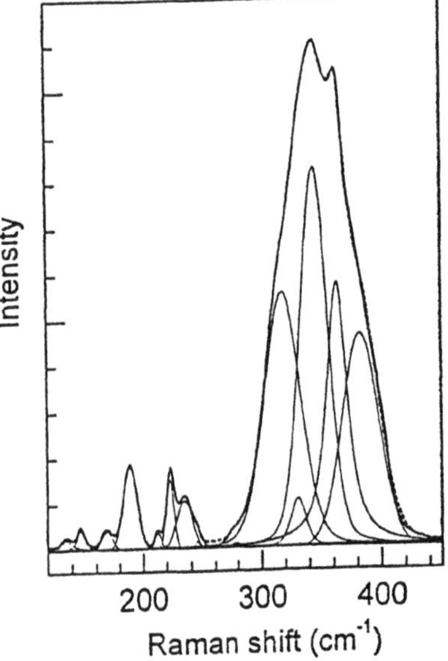

Figure 5. Raman spectrum of As_{42} S_{58} bulk glass. The complex bands were deconvoluted to single bands (thin lines). Their sum (dashed line) is practically identical with the experimental curve [44].

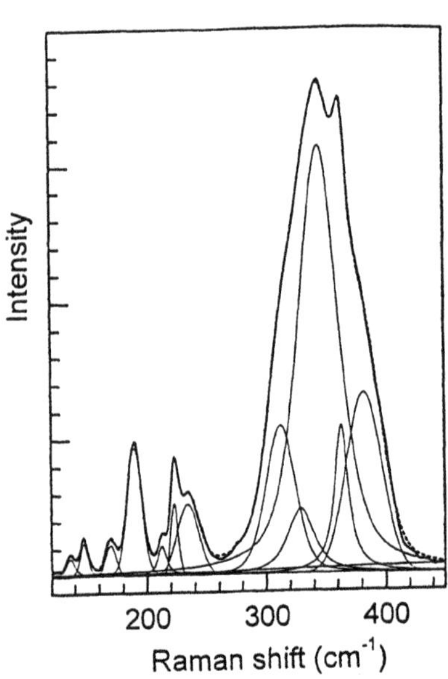

Figure 6. Raman spectrum of As_{42} S_{58} annealed layer. The complex bands were deconvoluted and to single bands (thin lines). Their sum (dashed line) is practically identical with the experimental curve [44].

layers, as it can be expected, the intensities are different.

The band near 135 cm^{-1} is present in As_2S_3 crystals [54] and also as a weak band in As_4S_4 crystals. According to [49,51,54], it is associated with As-As vibrations. The relative intensity of this band is much higher in fresh evaporated layers than in the bulk glass and in exposed or annealed layers (Figs. 3-8).

The band near 145 - 147 cm^{-1} can be seen in the spectrum of As_4S_4 crystals, where it can be assigned to As-As vibrations [51], too. It can be found also in As_2S_3 crystals and

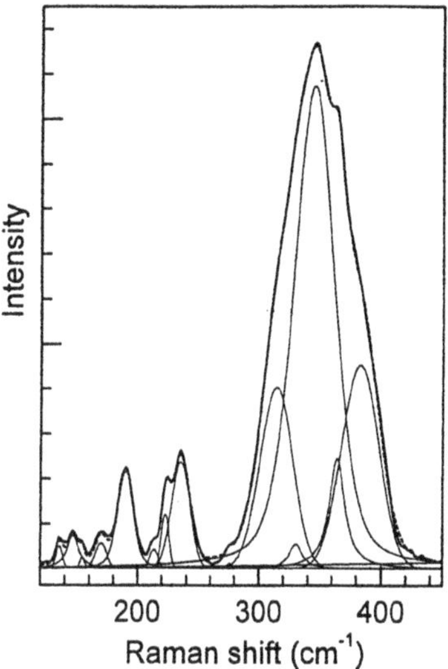

Figure 7. Raman spectrum of $As_{42}S_{58}$ annealed layer which was then exposed The complex bands were deconvolutedto single bands (thin lines). Their sum (dashed line) is practically identical with the experimental curve [44].

glasses [51], where it is assigned to As-S vibrations [54]. The intensity of this band is lowered by exposure and annealing (Figs. 3,4,8) and it is, therefore, connected with the presence of "molecular fragments".

The band near 168-170 cm^{-1} can be found in As_4S_4 and can be assigned to As-As vibrations [46,55,56]. Its relative intensity is lowered by exposure or annealing (Fig. 8).

The relatively intensive band near 187-191 cm^{-1} can correspond to a very intensive band in As_4S_4 crystals [54]. This band is probably overlapping with the band corresponding to S-S vibrations [46,49,51,56]. Its intensity is lowered again by exposure or annealing of the layers because the density of As_4S_4 and of S-S bonds is decreasing.

The next Raman band at 212-216cm^{-1} can be assigned again to As-As vibrations [46,51,56]. It is lowered a little by exposure and considerably reduced by annealing (Fig. 8).

The band near 218-222 cm^{-1} can correspond to the intensive band of realgar near 218 cm^{-1}. Also the relative intensity of this band is lowered by exposure or annealing (Fig. 8).

The band near 231-235 cm^{-1} can be assigned to As-As vibrations [31,32]. Its relative intensity is lowered by annealing (Figs. 3,8), but can be increased by exposure of fresh evaporated (Fig. 3) as well as annealed layers (Figs. 4,8).

The next group of Raman bands is centred around the most intensive band near 345 cm^{-1}. This band corresponds to symmetric vibrations of AsS_3 pyramids and is relatively broad in As_2S_3 and As-S bulk glasses. In fresh evaporated layers it is relatively sharp and narrow which is probably a result of the presence of As_4S_4 molecules in such layers, because the 334 cm^{-1} Raman band is intensive also in As_4S_4 spectra.

The shoulders of the main 345 cm^{-1} band (Figs. 3-7) are formed by four bands (312, 330, 361 and 378 cm^{-1}). The first one (~312 cm^{-1}) corresponds to the interactions between the AsS_3 pyramids and is relatively strong in $As_{42}S_{58}$ bulk glasses. Its relative intensity increases with illumination which is probably connected with polymerisation of molecular fragments. The band near 325-330 cm^{-1} is of relatively low intensity and it is not very much enhanced by annealing or illumination. Its intensity is relatively large in sulphur rich glasses [44].

The 361cm^{-1} band corresponds probably to the vibration of As_4S_4 structural units, it is the most intensive band of As_4S_4 crystals. Its relative intensity increases in irreversible process (Figs. 3,8), its intensity is lowered for reversible PE. In the second case, it probably corresponds to the incorporation of molecular units into the As-S skeleton.

The last band near 380 cm^{-1} is very broad one and can be assigned again to interaction among the AsS_3 pyramids. Its intensity is relatively high in As-S bulk glasses.

In Figs. 3-8 can be clearly seen that not only the illumination of fresh evaporated layers but also the illumination of annealed layers changes the intensities of Raman bands and, consequently, the structure of the CL in agreement with model of [31,32] and of

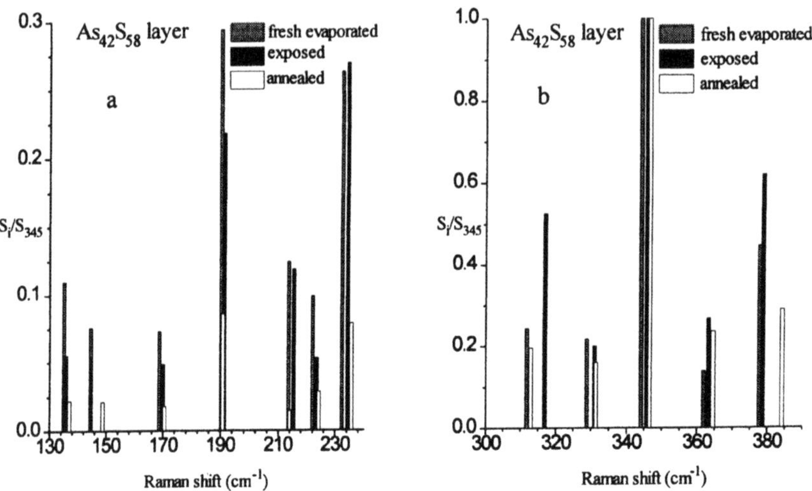

Figure 8. Integral intensities of individual single Raman bands of $As_{42}S_{58}$ layers. a - (130-235) cm^{-1} spectral region; b - (300-390) cm^{-1} region. Columns are centered at the centers of the deconvoluted bands.

[45-51]. Contrary to papers [31,32], the newly obtained spectra (e.g. Fig. 3-7) are of higher quality due to cumulation and averaging of several hundreds of spectra in FT Raman facility which reduces very effectively the noise level. From the spectra we can judge that during the exposure take place several photochemical reaction (polymerisation, photoinduced synthesis and photolysis and their combination). The photoinduced synthesis and photolysis, or at least part of them, can be described by the reaction [31,32]

$$2|\text{As-S}| \underset{h\nu_2,\, I_2,\, A}{\overset{h\nu_1,\, I_1}{\rightleftarrows}} |\text{As-As}| + |\text{S-S}|, \tag{1}$$

where $h\nu_1 \geq h\nu_2$, $I_1 > I_2$. The $h\nu_i$ is the energy of exciting light, I_i is its intensity and A is for annealing. The |X-Y| is a symbol for chemical bond between X and Y atoms.

The equilibrium of this reaction is shifted with illumination to the right hand side for reversible PE, due to annealing to the left hand side. The illumination of fresh evaporated layers causes also polymerisation of molecular species, such as As_4S_4 , As_4S_3, S_n and others.

Similarly to the models of [31,32,45-51], Shimakawa et al. [33] proposed that, the $As\text{-}As_4^+$ pairs are created by illumination of CL. Such centres are metastable and can act as additional trapping and recombination centres. They can be removed by annealing of

CL at temperatures near the glass transition temperature, T_g, [33], analogously as in the above mentioned model [51] of photoinduced effects in As_2S_3 glasses.

The PE in thin amorphous layers of As_2Se_3 (studied recently in [50] by FT- IR spectrometry) were explained again by the model of chemical bonds switching.

Contrary to above mentioned results (Figs. 1-8 and [31,32,9]), Ma and Zhou [53,57] did not found any structural changes in the first coordination sphere of As in As_2S_3 photodarkened and well-annealed layers. Using differential anomalous X-ray scattering, they noticed a large increase of the disorder and an increase in As-S-As apex angle (~2.4°) after illumination of As_2S_3 layers. They proposed [53] that such an angle increase is responsible for the reversible photodarkening observed in a-As_2S_3 films. The change of bond angles itself cannot probably explain the measured photoinduced changes of IR and Raman spectra and the photoinduced changes of reactivity as well [36]. The reversible photoinduced changes of structure (included bond switching) cover only few percent (~ 6-7) of atoms (or bonds). The X-ray scattering may be not sensitive enough to "see" such a small changes of the structure.

It is clear that the photoinduced changes of disorder can not be confined to the first coordination sphere of atoms, (SRO). They influence inevitably also further areas (intermedium range order - IRO and MRO). Unfortunately, the methods for determination of structure in further coordination spheres of individual atoms are not sensitive enough and these methods give very often the values averaged for all atoms. The relatively small changes of structure are even more lowered and a quantitative information is then in the vicinity of detection limit. The sensitivity of FT IR and FT Raman spectroscopy is probably higher (Figs. 1-8, [9,31, 32, 45-49]).

Mikla [65] has found a large differences in Raman spectra of layers of As-S and As-Se, prepared by different rates of evaporation which is not surprising, because higher rates of evaporation demand higher temperatures of evaporation, which results in larger thermal dissociation of vapours. The layer deposited on a cold substrate then contents larger amount of "molecular fragments" of the original chalcogenide. Mikla did not found any change of Raman spectra due to photodarkening of glassy samples and of well annealed layers. He concluded that the reversible part of photodarkening effect is caused by changes of medium range order. Such a result might be caused by lower sensitivity of his Raman facility.

Photo- and thermally- induced changes of the structure of in $Ge_{30}Sb_{10}S_{60}$ films were analysed in [37] using Raman, VIS and IR spectra. In accordance with the models proposed in papers [18,31,32], it is concluded that the photo- and thermally- induced effects in $Ge_{30}Sb_{10}S_{60}$ layers are caused by a chemical bonds redistribution of the type 2(M - S) = (M - M) + (S - S), where M is for metalloid (Ge,Sb), S is for sulphur or other chalcogen, (X-Y) is chemical bond between X and Y. Photodarkening of well annealed films is fully reversible and it is considered as a result of increased homopolar bonds densities.

2.2 PHOTOINDUCED CHANGES OF REACTIVITY

The illumination of CL can change not only their structure but also their reactivity, e.g. the dissolution rates in alkaline solutions (Fig. 9). The dissolution rates of CL depend also on the state and composition of the layers and on the composition of reactive solutions, e.g. on concentration of OH^- ions [28,29,58,60]. The dissolution rates of unexposed and exposed layers are generally different ([4,28,29,58-61], Fig. 9) and these rates can be even increased in the presence of some organic compounds in the etching solutions (Fig. 9,[9,29,60]).

The Raman spectra of fresh evaporated $As_{42}S_{58}$ layers contain several narrow bands in the spectral region between 125-260 cm^{-1} (Fig. 3) as we have mentioned earlier. This part of spectrum is connected with the presence of overstoichiometric As, either with As_4S_4 molecules or =As-As= structural units and, in evaporated layers, also with the presence of molecular fragments (As_4, As_4S_4 and As_4S_3). These fragments are formed during high temperature evaporation and they are frozen down during condensation of vapours on a cold substrate [54]. The exposure of fresh evaporated layers lowers the amount of "molecular fragments" and the relative content of As_2S_3 structural units is increased (Figs. 3-4, [45,46]). It is known that the crystalline counterparts of such "molecular fragments", e.g. As_4S_4 , or S_n , are dissolved in alkaline solutions more slowly than the stoichiometric glass or crystalline As_2S_3. It can be supposed that the As_4S_4 , As_4S_3 or S_n in fresh evaporated As-S layers are dissolved slowly, too, and they are responsible for a low dissolution rate of fresh evaporated layers. Such a suggestion is in agreement with the fact that the oxidising reactives (e.g. H_2O_2 , ClO_3^-, I_2, etc. and good solvents of S, which lower the concentration of these fragments on the surface), can increase substantially the rates of dissolution of fresh evaporated layers in alkaline solutions. The illumination decreases the content of these "molecular fragments" (Fig. 3) and the dissolution rate of illuminated layers is increased (Fig.9).

A large decrease of solubility of As-S layers in alkaline solutions can be found when some tetraalkylammonium salts, e.g. cetyltrimetylammonium bromide (CTAB) are present in alkaline etching solution or they are sorbed on the fresh evaporated layers before the etching. Similar influence was seen with some other cation- active surfactants especially those with quaternary (R_4N^+) ions and longer alkyl or aryl chains [28,60].

For elucidation of the interaction of cetyltrimethylammonium ion (CTA^+) or generally of R_4N^+ ions with As-S layers, the $As_{42}S_{58}$ layers (Figs. 9-11) and also powders were studied [36]. The $As_{42}S_{58}$ layers were chosen because their interaction with R_4N^+ ions is intensive. A solution containing CTAB was used as a typical representative of cation-active surfactants.

It was proved earlier that the alkyl- or aryl- ammonium salts can be selectively chemisorbed on the chalcogenide surface [58]. The chemisorbed ions (salts) slow down the dissolution rates of unexposed layers in alkaline solutions and can even protect the surface against the action of such solutions ([29,60], Fig. 9). It is considered that the positively charged alkyl- or aryl-ammonium ions can be chemisorbed by their positively charged parts (N^+) on the chalcogenide layer. The long alkyl chains or aryl groups of R_4N^+ ions then repel the hydroxyl ions during dissolution. As a result, the dissolution of the layers is slowed down or even blocked depending on pH, on concentration of organic ammonium compounds and on the layer and solution compositions. The interaction of R_4N^+ ions with negatively charged defects on As-S layers surface is probably of Coulomb type, but some Lewis′ acid-base interaction can also play some role.

The change of solubility rate of chalcogenide layers due to interaction with R_4N^+ ions is very low for $As_{38}S_{62}$ layers, it increases with increasing As content in As-S layers [60]. We shall note that with increasing of As-overstoichiometry in As-S layers, the density of As-As bonds increases, too. The formal oxidation state of As in structural units of =As-As= is lower (As^{II}) than in As_2S_3 (As^{III}). Such =As-As= groups can be then more effective electron donors (Lewis' bases) for positively charged ions (Lewis' acids, such as R_4N^+) than the As^{III}. If it was true, the sorption of R_4N^+ ions should be higher on As-rich parts of As-S layers (with =As-As= bonds) and the sorbed ions should change the intensities and vibration frequencies of absorption bands corresponding to As-As vibrations or to vibrations of As-As containing structural units.

The Raman bands intensities and frequencies should be changed also when the interaction of =As-As= units with R_4N^+ ions is only of Coulomb type. In this case, there is no full transfer of the charge and the vibrational changes should be probably smaller. It was found (Figs. 10,11) that the sorption of CTAB on $As_{42}S_{58}$ layers changes only slightly the intensities and positions of the bands which can be assigned to the vibration of As-As containing structural units (bands between 135-265 cm^{-1}, the band near 361 cm^{-1}, see above, [31,45,46,48,49,54]). Similar results were obtained for $As_{42}S_{58}$ powders, prepared from the bulk glass of the same composition [36].

From the above given data, it follows that the sorption of CTAB influences the intensities and frequencies of Raman bands connected with the presence of overstoichiometric As or As-As bonds in different structural units. We suggest that the sorption of CTA^+ proceeds mainly on such structural units. The changes of Raman frequencies and intensities are not very large and the Coulomb interaction between the CL or glass and CTA^+ is more probable.

Similar sorption of CTAB was found on the surface of silicates, where the CTA^+ ions are sorbed on negatively charged defects [62-64].

2.3. PHOTOINDUCED DISSOLUTION AND DIFFUSION OF Ag

Photoinduced dissolution and diffusion of Ag into chalcogenide layers was studied recently in many papers (see e.g. [38,39,42,67-72]. Even in this case, the origin of the effect is not fully understood. New measurement of photodoping rates [68], mobility of Ag^+ ions [69] and photoemf [70] shed some new light on these problems. New models were proposed [67,70]. The effect of photodoping has promising applications in high resolution photoresists, optoelectronic elements and devices and optical memories [67]. The problem will be discussed in another paper of this Workshop [35].

3. The photoinduced anisotropy

For a long time it was believed that the chalcogenide glasses (and glasses generally) are isotropic materials. Few years ago, it was shown that the polarised or even unpolarised light can induce a metastable anisotropy. The photoinduced anisotropy was observed in many chalcogenides (e.g. in As-S, As-Se, Sb-S, Ge-Ga-S, Ge-Pb-S and Ge-Sb-S glasses and layers [6-15,73]).

Illumination of CL and glasses by polarised light of band-gap or even of sub-band-gap energy can induce optical anisotropy (Fig. 12). The polarisation plane of CL can be moved by the change of polarisation direction of excitation light.

Both the isotropic and vectorial PE were found in thin films of As-S-Se-Te, Ge-S-Se-Sb-Pb, As-Se, As-S, Sb-S systems (see e.g. [7,10,15,73] and papers cited in). The optically induced dichroism achieved very large values especially in Sb-S and Ge-Pb-S systems [7], though the scalar photoinduced effects in these films are practically absent. The photoinduced anisotropy is usually reversible, it can be destroyed by unpolarised or

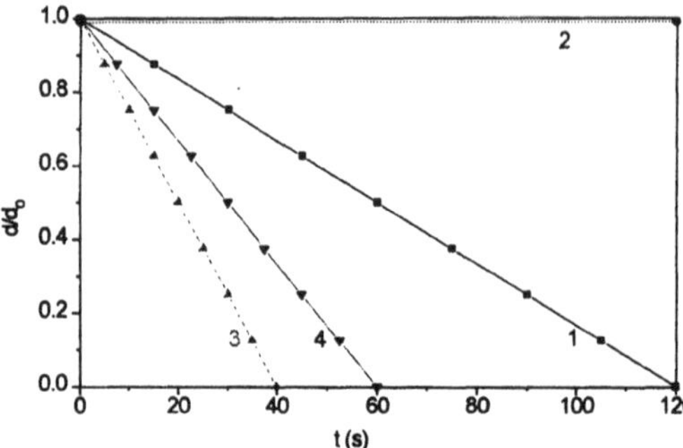

Figure 9. The change of relative thickness (d/d_0) of $As_{42}S_{58}$ layers (d_0 = 500 nm) during dissolution in alkaline solution (100g/l Na_2CO_3 + 150g/l $Na_3PO_4.12H_2O$, pH = 12.5) 1 fresh evaporated layer, 2 fresh evaporated layer which was immersed for 5 min. in CTAB solution (1g/l) before etching, 3 illuminated layer, 4 illuminated layer which was immersed for 5 min. in CTAB solution (1g/l) before etching.

circularly polarised light. Some anisotropy effects can be seen also in bulk chalcogenide samples [12]. In glassy As_2S_3 doped by In, even a stable optical anisotropy was found in [73]. This anisotropy increases with annealing at temperatures above T_g in contrast with the photoinduced anisotropy which decreases quickly at temperatures by 50-100°C below T_g [6].

The optical anisotropy can be induced also by sub-band-gap light [13] ($h\nu < E_g$), where $h\nu$ is photon energy of exciting light, E_g is energy gap of the material. The temperature dependence revealed that sub-band-gap illumination produced largest effect in As_2S_3 at temperature ~380K while the effect is very small at 430K and also at 220K. It is believed [13] that the effect can be explained proposing that in As_2S_3 glasses are present some structural fragments very similar to those present in crystalline As_2S_3. These fragments should be very small and can be generated by illumination. The formation of such fragments would need interruption of some chemical bonds and, at least the band-gap illumination. These fragments are more probably present in the glass before illumination and they can be rotated by illumination with polarised light [6].

The linearly polarised light can induce the optical anisotropy also in bulk glasses [12]. It is supposed that the microscopic mechanism comprise two parts: the optically irreversible scalar component due to creation randomly formed dipole moments, and the reversible vectorial component caused by reorientation of intrinsic dipole moments (structural units), according to the electric vector of inducing light. The detailed mechanism of optically induced anisotropy is not well known.

The band-gap light induced anisotropy was studied in [6]. Both scalar and vectorial PE are explained by photoinduced changes of chemical bonds redistribution. The surprising fact is that the vectorial effects can be found in photodarkened as well as in well annealed glasses and layers and even in materials which cannot be photodarkened. It is believed that small microvolumes of the amorphous chalcogenide are anisotropic. The absorption of polarised light creates electron-hole pairs and after their recombination, the local electronic and atomic structure is changed. This is not the only possible mechanism, because the vectorial effects can be induced also by sub-band-gap light by which the electron-hole pairs would not be created. After absorption of polarised light, the structural "microvolumes" have different orientation as compared with the former state. Large number of reoriented "microvolumes" promotes the change of the anisotropy on macroscopic scale. Changing the polarisation plane of exciting light, the orientations of individual "microvolumes" are changed and the whole macroscopic anisotropy is also changed (see Fig. 12, [74]). It is interesting to note that the vectorial effects can be induced even by unpolarized light [6]. Maximum intensity of vectorial effects can be found at temperatures near 300K [6], the intensity of scalar effects increase in many chalcogenides monotonically with decreasing temperature [14]. The vectorial effects are annealed at lower temperatures than the scalar ones (50-100° below T_g , while the scalar at $T \sim T_g$).

The attempt to describe a molecular model of photoinduced anisotropy centre is given in Fig. 13. The bonds switching and change of atomic positions causes the change of anisotropy. The model cannot describe many features of photoinduced anisotropy.

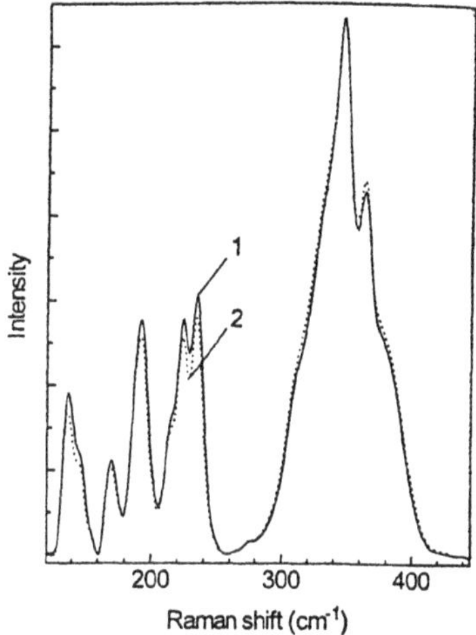

Figure 10. Raman spectra of layers 1 fresh evaporated layer, 2 fresh evaporated layer with sorbed CTAB. The background of spectra was corrected, the intensities were normalized to the intensity of 345 cm^{-1} band. The long-wavelength part (125-260 cm^{-1}) of the spectrum is given in Fig.11.

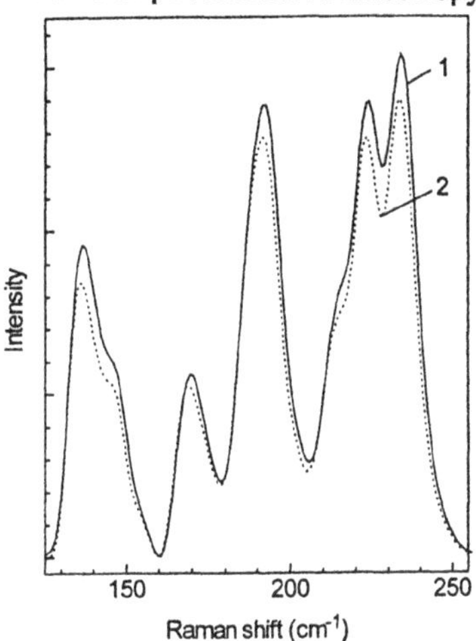

Figure 11. Raman spectra of $As_{42}S_{58}$ layers 1 fresh evaporated layer, 2 fresh evaporated layer with sorbed CTAB. The background of spectra was corrected, the intensities were normalized to the intensity of 345 cm^{-1} band.

4. Applications

There are many possible applications of photoinduced effects in chalcogenide glasses and layers. The high achievable resolution can be used for inorganic photoresist production. The exposed part of CL has different reactivity [34] and depending on solvent, positive or negative photoresist can be prepared [29]. The larger changes of the reactivity can be received with photodoped layers [75]. The effect of interdiffusion or reaction of two layers (e.g. Bi and Se [33], Ag-As_2S_3 [75], Ag-$GeSe_2$, Se-As_2S_3 etc.) can be also used for optical memories production [22]. The phase transition (crystallisation) of the glasses from the system Ge-Sb-Te and others [23] can be used in recording layers of optical discs.

Very promising seems to be the possibility to make holograms [19,27,75] or the use of holography for fabrication of optical elements, such as grids, Fresnel lenses etc. [75-77]. The advantage of these layers made from chalcogenides is their high transmission in the infrared region. Very low level of stray light can be received (10^{-6}, [76]). The diffraction efficiency can be as high as 85% [76].

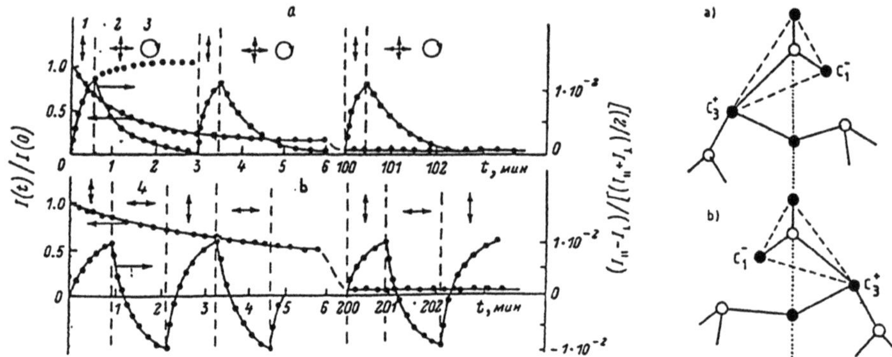

Figure 12. Kinetics of photodarkening and of photoinduced dichroism in $As_{50}Se_{50}$ layers. Intensity of excitation: 350 mW/cm^2 (a); 50 mW/cm^2 (b) [74]. The change of polarization plane of exciting light changes the dichroism to the opposite sign (b). The process can be repeated many times.

Figure 13. Schematic illustration of the photoinduced anisotropy and gyrotropy effect center in pnictogen-chalcogen systems (e.g., As-Se) before (a) and after (b) optical excitation. Open and solid circles are pnictogens and chalcogens, respectively, solid lines are bonds, dashed lines delineate the PC_3 pyramids [11].

The possibility to fabricate microlenses from CL was also described [20]. The chalcogenide glasses can be shaped under light illumination [21]. The fluidity of the glass is increased by the illumination and, under external tension, the shape of material can be changed. The mechanisms of this increased fluidity is athermal [21]. We can speculate, that the changes of orientation of chalcogenide glass microvolumes during illumination of the chalcogenides by polarised light and the increased fluidity during illumination have common physical base.

5. Conclusion

The recent results of the study of photoinduced effects in chalcogenide glasses and layers have contributed to better understanding their interaction with light. Many new and interesting effects have been studied (photostructural changes, photodoping, photoinduced anisotropy, photoinduced changes of reactivity, etc.). Many of them have very promising application in optics and optoelectronics.

Acknowledgement

The work was partly supported by grants Nos. 203/93/2090 and 203/96/0876 of Czech Grant Agency and No. ERBCIPA CT940107 of European Community (Brussels) which are gratefully acknowledged.

References

1. S.R.Elliott, *J.Non-Cryst. Solids* 81 (1986) 71.
2. Ke.Tanaka, *Rev. Solid State Sci.* 4 (1990) 641.

3. G.Pfeifer, M.A.Peasler, S.C.Agarwal, *J.Non-Cryst. Solids* 130 (1991) 111.
4. M.Frumar, M.Vlček, J.Klikorka, *Reactivity of Solids* 5 (1988) 341.
5. H.Fritzsche, *Phil. Mag.* B68 (1993) 561.
6. H.Fritzsche, *Phys. Rev.*B 52 (1995) 15854.
7. M.Klebanov, S.Shutina, I.Bar, V.Lyubin, S.Rosenwaks, V.Voltera, *Proc. SPIE* 2426 (1995) 198.
8. K.Shimakava, A.Kolobov, S.R.Elliott: *Advances in Physics* 44 (1995) 475-588.
9. M.Frumar, M.Vlček, Z.Černošek, Z.Polák and T.Wágner: Photoinduced changes of the structure and physical properties of amorphous chalcogenides. *J.Non-Cryst. Solids*, submitted.
10. V.M.Lyubin, V.K.Tikhomirov, *J.Non-Cryst. Solids* 171 (1994) 87.
11. V.K.Tikhomirov, S.R.Elliott, *Phys.Rev.* B5 (1995) 5538.
12. V.K.Tikhomirov, S.R.Elliott, *J.Phys. Condens. Matter.* 7 (1995) 1737.
13. Ke.Tanaka, M.Notani, H.Hisakuni, *Solid State Commun.* 95 (1995) 461.
14. Ke.Tanaka: *J.Non-Cryst. Solids* 59-60 (1983) 925.
15. A.V.Kolobov, V.M.Lyubin, U.K.Tikhomirov, *Phil.Mag.Lett.* 65 (1992) 67.
16. M.Mitkova, T.Petkova, R.Markovski, V.Mateev, *J.Non-Cryst. Solids* 164-166 (1993) 1203.
17. M.Vlček, C.Raptis, T.Wágner, A.Vidourek, M.Frumar, P.Kotsalas, D.Papadimitriou, *J.Non-Cryst. Solids* 192-193 (1995) 669.
18. M.Vlček, M.Frumar, A.Vidourek, *J.Non-Cryst. Solids* 90 (1987) 513.
19. O.Salminen, N.Nordman, P.Riihola, A.Ozols, *Optics Materials 116 (1995) 310.*
20. H.Hisakuni, Ke.Tanaka, *Optics Lett.* 20 (1995) 958.
21. H.Hisakuni, Ke.Tanaka, *Science* 270 (1995) 974.
22. A.Kikineshy, *Opt. Engineering* 34 (1995) 1040.
23. T.Ide, M.Suzuki, M.Okuda, *Jpn. Appl. Phys.* 34 (1995),L529.
24. Y.Mizushima, A.Yoshikawa, in: Y.Hamakawa (Ed.), *Amorphous Semiconductors, Technologies and Devices*, p.285. North-Holland, Amsterdam 1981.
25. P.J.S.Ewen, A.Zekak. C.W.Slinger, G.Dale, D.A.Pain, A.E.Owen, *J.Non-Cryst. Solids* 164-166 (1993) 1247.
26. K.Saito in: Ke.Tanaka, M.Itoh, *Optoelectronics-Devices and Technologies* , 9 (1994) 299.
27. V.I.Mikla, I.P.Mikhalko, *J.Non-Cryst. Solids* 180 (1995) 236.
28. M.Vlček, M.Frumar, M.Kubový, V.Nevšímalová, *J. Non-Cryst. Solids* 137-138 (1991) 1035.
29. M.Vlček, J.Prokop, M.Frumar, *Int. J. Electronics*, 77 (1994) 969.
30. M.Vlček, M.Frumar, T.Wágner, K.Petkov, *Int. J. Elecronics*, in press (1996).
31. M.Frumar, A.P.Firth, A.E.Owen, *Phil.Mag.B*50 (1984) 463.
32. M.Frumar, A.P.Firth, A.E.Owen, *J.Non-Cryst. Solids* 59-60 (1983) 921.
33. K.Shimakawa, S.Inami, T.Kato, S.R.Elliott, *Phys.Rev.B*46 (1992-II) 10062.
34. E.Vateva, E.Skordeva, D.Arsova, Phil.Mag.B67 (1993) 225.
35. A.E.Owen et al., in: *NATO Workshop "Physics and Application of Non-Crystalline Semiconductors in Optoelectronics*, Kishinev 1996.

36. M.Frumar, Z.Polák, M.Vlček, Z.Černošek, Photoinduced changes of the reactivity of amorphous chalcogenide layers. *J.Non-Crystalline Solids*, to be published.
37. M.Frumar, A.P.Firth, A.E.Owen, *J.Non-Cryst. Solids* 192-193 (1995) 447.
38. T.Wágner, M.Vlček, K.Nejezchleb, M.Frumar, V.Zima, V.Peřina and P.J.S.Ewen, *J.Non-Cryst. Solids*,198-200(1996) 744-748.
39. T.Wágner, M.Vlček, M.Frumar, V.Peřina, E.Rauhala, J.Scarilahti, P.J.S.Ewen, in: *Developments In Optical Components Coatings, SPIE,* Vol.2776, 1996, in press.
40. M.Frumar, T.Wágner, M.Vlček, Europ. *J.Solid State Inorg. Chem.* 28 (1991) 1193.
41. T.Wágner, M.Frumar, V.Šušková, *J.Non-Cryst. Solids* 128 (1991) 197, and papers cited in.
42. T.Wágner, M.Vlček, M.Frumar, V.Peřina, E.Rauhala, J.Saarilahti, P.J..S. Ewen: Kinetics and Rutheford backscattering study of the photo-induced solid state chemical reaction between silver and amorphous $As_{33}S_{67}$ layers. J. Non-Cryst. Solids, submitted.
43. Mir.Vlček, K.Nejezchleb, T.Wágner, M.Frumar, Mil.Vlček, A.Vidourek, P.J.S.Ewen: Structure and photoinduced changes in As-S-Te bulk glasses and amorphous layers. *Proc. 10th Internat. Conf. on Thin Films*, Salamanca, 1996, *Thin Solid Films,* to be published.
44. M.Frumar, et al., to be published.
45. O.I. Shpotyuk, A.O.Matkovskii, *J.Non-Cryst. Solids* 176 (1994) 45.
46. O.I.Shpotyuk, *Phys.Stat.Sol.(b)* 183 (1994) 365.
47. O.I.Shpotyuk, *Radiation Defects in Solids* 132 (1994) 393.
48. O.I.Shpotyuk, A.P.Kovalskii, M.M.Vakiv, O.Ya. Mrooz, *Phys.Stat.Sol.(a)* 144 (1994) 277.
49. O.I.Shpotuyk, *Phys.Stat.Sol.(a)* 145 (1994) 69.
50. O.I.Shpotyuk, *Zh.Prikl.Spektroskopi*i 61 (1994) 143.
51. O.I.Shpotyuk, *Zh.Prikl.Spektroskopii* 59 (1993) 551.
52. M.H.Peasler, J.M.Lee, in: M.Frumar, V.Černý, L.Tichý (Eds.), *Advanced Solid State Chemistry. Materials Science Monographs,* Vol. 60. Elsevier, Amsterdam 1989, pp.380-8.
53. Q.Ma, W.Zhou, D.E.Sayers, M.A.Peasler, *Phys.Rev.B*52 (1995-II) 10025.
54. S.A.Solin, G.V.Papatheodorou, *Phys. Rev. B*15 (1977) 2087.
55. V.N.Kornelyuk, I.P.Savicky, O.I.Shpotyuk, I.I.Jaskovec, *Fiz. Tverd. Tela* 31 (1989) 311.
56. O.I.Shpotyuk, V.N.Kornelyuk, *Fizikas un technisko Zinatnu*, Seria 1992 (2) 1989.
57. W.Zhou, D.E.Sayers, M.A.Peasler, B.Bouchet-Fabre, Q.Ma,D.Reoux, *Phys.Rev.B*47 (1993-II) 686.
58. M.Frumar, M.Cvrkal, M.Vlček, T.Wágner, *J.Non-Cryst.Solids* 164-166 (1993) 1243-1246.
59. K.Petkov, M.Vlček, M.Frumar, *J.Mat.Sci.* 27 (1992) 3281.
60. M.Frumar, M.Vlček, T.Wágner, *J. Non-Cryst.Solids* 164-166 (1993) 1243.
61. M.D.Michailov, S.B.Mamedov, S.T.Tsventernyi, *J. Non-Cryst. Solids* 176 (1994) 258.

62. R.Podgornik, V.L.Parsegian, *J.Phys. Chem.* 99 (1995) 9491.
63. J.L.Parker, V.V.Yaminsky, P.M.Claesson, *J.Phys. Chem.* 97 (1993) 7706.
64. S.A.Bond, W.F.Jaynes, in: A.R.Mermut (Editor), *Layer Charge Characteristics of 2:1 Silicate Clay Minerals. CMS Workshop Lectures*, Vol.6. The Clay Minerals Society, Boulder 1993, USA.
65. V.I.Mikla, *J.Phys. Condes. Matter* 8 (1996) 429.
66. E.Márquez, J.B.Ramírez-Malo, J.Fernandez-Pena, P.Villares, R.Jiménez-Garay, P.J.S.Ewen, A.E.Owen, *J.Non-Cryst. Solids* 164-166 (1993) 1223.
67. Ke.Tanaka, M.Itoh, *Optoelectronics - Devices and Technologies* 9 (1994) 299.
68. Ke. Tanaka, *J.Non-Cryst. Solids* 170 (1994) 27.
69. Y.Miyamoto, M.Itoh, Ke.Tanaka, *Solid State Commun.* 92 (1994) 895.
70. I.Z.Indutnyj, V.A.Dan'ko, A.A.Kudryatsev, F.V.Michailovskaya, V.I.Minko, *J.Non-Cryst. Solids* 185 (1995) 176.
71. T.I. Kosa, T.Wágner, P.J.S.Ewen, A.E.Owen, *Phil.Mag.B* 71 (1995) 311.
72. T.Kawaguchi, S.Maruno, *Jpn.J.Appl.Phys.* 33 (1994) 6470.
73. T.E.Mazets, E.A.Smorgonskaya and V.K.Tikhomirov, *J.Non-Cryst. Solids* 164-166 (1993) 1215.
74. V.M.Lyubin, V.K.Tikhomirov, *Fiz. Tverd. Tela,* 32 (1990) 1838.
75. C.V.Slinger, A.Zakery, P.J.S.Ewen, A.E.Owen, *Appl. Optics* 31 (1992) 2490
76. I.Z.Indutnyj, A.V.Stronski, S.A.Kostioukevitch, P.F.Romanenko, P.E.Schepeljavi, I.I.Robur, *Optical Engineering* 34 (1995) 1030.
77. M.Vlček, A.A.Stronskii, P.E.Shepeljavi, Multicomponent chalcogenide inorganic resistor properties and application in diffractive optics, *Proc. Internat. Workskop on Advanced Technologies of Multicomponent Solid Films and Structures and their Application in Photonics*. Uzhgorod, October 1996, p.81, Uzhg. Univ. 1996, Ukraine.

CW AND PULSED INVESTIGATION OF PHOTOINDUCED DARKENING IN As_2S_3 AMORPHOUS THIN FILMS

M. BERTOLOTTI[@], F. MICHELOTTI[@], A.M. ANDRIESH[&],
V.N. CHUMASH[&], M. POPESCU[#]
[@] *Università di Roma "La Sapienza", Dipartimento di Energetica, INFM and GNEQP of CNR, Via Scarpa, 16, I-00161, Roma ITALY*
Tel + 39-6-49916542; FAX +39-6-44240183
E-Mail: bertolotti@axrma.uniroma1.it
[&] *Centre of Optoelectronics, I.A.P., Academy of Sciences*
Academiei str. 1, Chisinau, 2028, Moldova
[#] *I.F.T.M., Bucharest-Magurele, P.O. Box MG-7, Romania*

Abstract

An intensity dependent complex character of the photodarkening of virgin As_2S_3 samples exposed to CW and pulsed laser radiation was observed. The excitation by CW and pulsed (picosecond) lasers for the same irradiation doses leads to significant differences in the photodarkening curves. The kinetics of the photodarkening can be decomposed into several processes related to the structural transformations in amorphous As_2S_3 material: defect healing, photopolymerization, decomposition and oxidation.

1. Introduction

Some of the most interesting phenomena occurring in chalcogenide glasses are the reversible photoinduced changes of the optical properties. They are proper to the noncrystalline state and were not observed in crystals. There was shown that photoinduced phenomena are related to the atomic scale changes in the amorphous (or glassy) network and do not imply major chemical modifications [1].

Photostructural changes of chalcogenide glasses are induced by exposing the sample to near bandgap light and are mainly characterized by shifts of the optical absorption edge (photodarkening or photobleaching). These effects are of high technological importance due to their possible applications in optical imaging, hologram recording, optical tridimensional memorics, IR windows, photo-resists [1]. The amorphous As_2S_3 is stable and can be easily produced by melt-quenching or by vacuum evaporation techniques. The damage and the aging of thc optoelectronic devices using As_2S_3 are of prime importance and further studies are needed to clarify the involved processes [2].

A. Andriesh and M. Bertolotti (eds.),
Physics and Applications of Non-Crystalline Semiconductors in Optoelectronics, 141–153.

Photo-induced edge-shift in As_2S_3 films for different intensities of incident light and exposure time was investigated in [3], and it was shown that the magnitude of the edge-shift depends only on the total energy of light. The chalcogenide films show strong nonlinear effects when illuminated by powerful bandgap light [4,5] which were confirmed by means of the Z-Scan technique for virgin As_2S_3 and As_2Se_3 films [6,7]. On the other hand, the light intensity dependence of photodarkening due to photostructural changes in amorphous As_2S_3 films was experimentally studied in [8] and it was suggested that the mechanism of the intensity dependence cannot be attributed to nonlinear optical effects and that the time averaged intensity dependencies should be dominant. If this is the case, then the photodarkening should be independent of the peak intensity of the incident light - a feature not confirmed by experiments.

In order to have a deeper insight into the photoinduced phenomena we have studied in this paper the transmission of laser beams through As_2S_3 films as a function of the exposure time and of the input intensity.

2. Experimental Results

Thin films of amorphous As_2S_3 (E_g = 2.4 eV) were prepared by thermal evaporation on glass substrates in vacuum with thickness from 0.2 μm to 3.0 μm.

The investigation of the kinetics of the photoinduced absorption change was carried out in an experimental set-up similar to that used for Z-Scan method [6,7]. The samples were put in the focal plane of a lens (focal length f = 100 mm) in a fixed position. The transmitted light was collected with a lens so that the total transmitted power is measured. Two types of lasers were used in the experiments: a CW Argon laser emitting at λ = 514.5nm ($h\nu$ = 2.41 eV) and a frequency doubled Nd: YAG CW mode-locked laser with the wavelength λ = 532 nm ($h\nu$ = 2.33 eV, pulse duration τ = 70 ps and repetition rate ϕ = 76 MHz). The focal plane spot size were ω_o = 18 μm and ω_o = 29 μm, respectively. For small beam spot-size on the film, the direct light effect on the photodarkening becomes comparable to the light-induced thermal effect [9].

The laser beam performed a dual function, serving both as a source of photons to drive the photodarkening and as a monitoring probe to measure the rate of this darkening. The experimental conditions corresponded to the situation when the change in transmittance near band-gap energies (i.e. in the region where the absorption coefficient $\alpha > 10^3$ cm^{-1}) is mainly dependent on the change of the absorption coefficient, being unaffected by changes of refractive index. In particular we were careful in checking that the behavior of the transmittance was not due to a variation of the interference conditions inside the film induced by refractive index changes.

The kinetics of the optical densification of the virgin As_2S_3 films are given, for various average input intensities I_o, in Figs. 1 for the case of CW irradiation, and in Figs. 2 for pulsed irradiation.

The change of the film transmittance is given in terms of the relative transmittance (T/T_o). The transmittance T is defined as the ratio of the transmitted intensity I to the incident one I_o at any arbitrary time t . The transmittance T_o is defined as the ratio of the transmitted to the incident flux at time t = 0 , when the experiment is initiated. Since the incident flux was maintained constant throughout all the measurement duration, the relative transmittance T/T_o is equal to the ratio of the transmitted flux at time $t > 0$ to that at $t = 0$.

In the graphs shown in Figs. 1 and 2, the logarithm of the ratio (T/T_o) is plotted versus the logarithm of the irradiation time t.

For low intensities of the CW laser beam (from 1.7 mW/cm^2 up to 15 mW/cm^2) the curves of optical densification show several characteristic regions (Fig. 1a) [16]. The first region corresponds to a small and smooth diminishing of the relative transmittance (see, for example, figure 1a, curve 1). The next one is a region with a bleaching peak. The maximum of the peak shifts toward lower values of $\log t$ with increasing the intensity according to the equation:

$$t_p \cdot I_o = \text{const} = 0.65 \text{ J/cm}^2,$$

where t_p is the time which defines the position of the bleaching peak itself and I_o is the intensity of the incident light.

The third region is characterized by a steep decrease of the transmittance, which turns into a faint shoulder (and even a kink on some curves) for higher doses of light, that's to say, for higher values of time after switching on the laser light.

For intermediate intensities (Fig. 1b) the first region of the darkening curve vanishes and the bleaching peak lowers and finally disappears for intensities higher than 50 mW/cm^2. This behavior can probably be explained by the fact that, for high intensity, these processes take place during the first 1-2 measurement points. The shoulder shifts towards shorter time when the irradiation intensity increases.

At high light doses the transmittance approaches a saturation value which is input intensity dependent. Figure 1c shows the kinetics of the photodarkening at high irradiation intensities. The shoulder is still present but it looks less pronounced and the saturation values are reached after short irradiation time [16].

In the case of picosecond laser irradiation (using the same average intensities as for CW Argon laser and taking into account the difference between spot-sizes and light absorption coefficient at 2.41 eV and at 2.33 eV) the kinetics curves show significant differences, which must be remarked (Figs. 2a to 2d).

Only for low irradiation intensities the bleaching peak does appear in the pulsed irradiation (Fig. 2a). A shoulder is more profound and persists in a larger interval of average intensities (Figs. 2b, 2c, 2d). In this case, the saturation level is reached at higher average input intensity values (Fig. 2d) with respect to the CW irradiation case (Fig. 1c).

3. Discussion

The mechanisms of the photostructural transformations are still controversial. In the case of As_2S_3 films, various processes have been suggested but a definite conclusion is lacking. The kinetics curves can help in the elucidation of the photodarkening mechanisms.

The typical kinetics curves, which have been drawn (Figs. 1 and 2) in the coordinate system $\log (T/T_o)$ versus $\log (t)$, can be divided into several regions, which prove the existence of different processes contributing to the darkening. Some of these regions can be fitted remarkably well with straight lines of different slopes. The first of

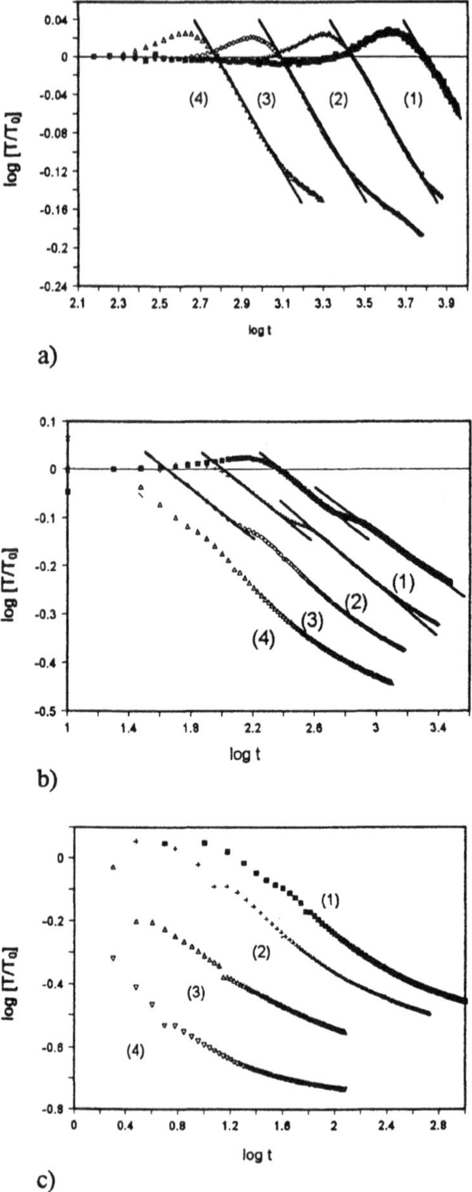

Fig.1. Kinetics of the relative transmittance through a virgin As_2S_3 film of Argon laser radiation for several input intensities (I):

a) (1) I=1.7 10^{-3} W/cm^2; (2) I=3.5 10^{-3} W/cm^2; (3) I=7 10^{-3} W/cm^2; (4) I=1.5 10^{-2} W/cm^2;

b) (1) I=2.6 10^{-2} W/cm^2; (2) I=5.2 10^{-2} W/cm^2; (3) I=0.1 W/cm^2; (4) I=0.2 W/cm^2;

c) (1) I=0.4 W/cm^2; (2) I=0.9 W/cm^2; (3) I=3.5 W/cm^2; (4) I=20 W/cm^2.

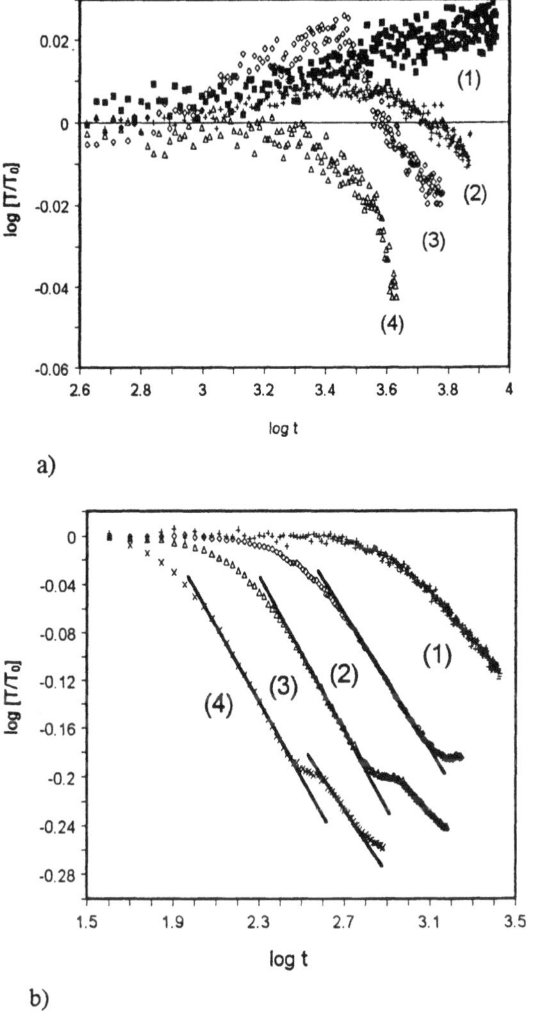

Fig.2. Kinetics of the relative transmitance through virgin As_2S_3 film of picosecond laser radiation for several input intensities (I):

a) (1) I=8 10^{-4} W/cm^2; (2) I=1.7 10^{-3} W/cm^2; (3) I=3.5 10^{-3} W/cm^2; (4) I=7 10^{-3} W/cm;

b) (1) I=2.5 10^{-2} W/cm^2; (2) I=5 10^{-2} W/cm^2; (3) I=0.1 W/cm^2; (4) I=0.2 W/cm^2;

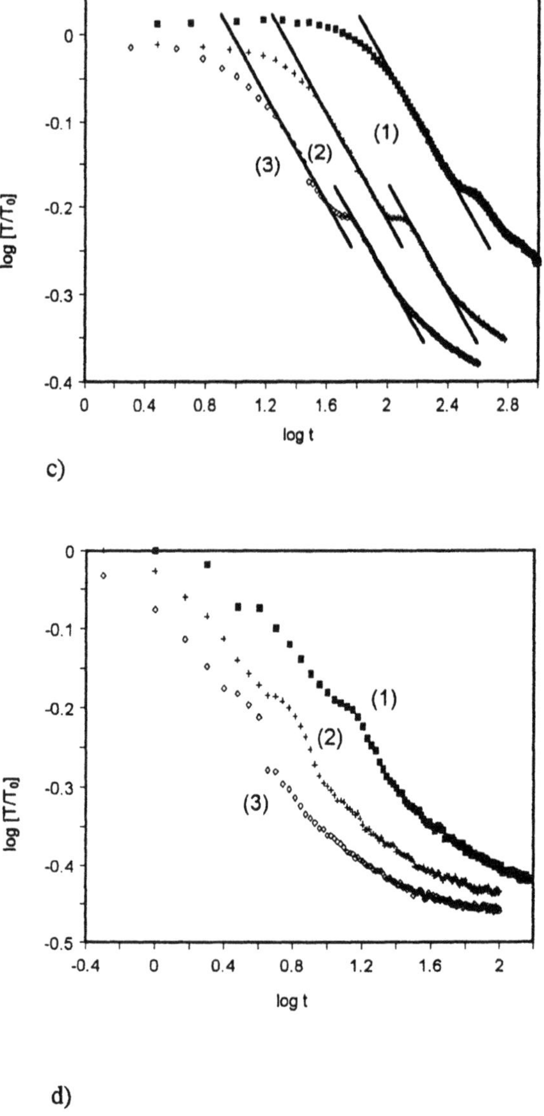

Fig.2. Kinetics of the relative transmitance through virgin As_2S_3 film of picosecond laser radiation for several input intensities (I):

c) (1) I=0.2 W/cm^2; (2) I=0.4 W/cm^2; (3) I=0.9 W/cm^2;

d) (1) I=3.5 W/cm^2; (2) I=6 W/cm^2; (3) I=10 W/cm^2.

them follows immediately after the bleaching peak, the second one follows a shoulder in the experimental curve. In some of these curves the shoulder takes the form of a relative maximum (more pronounced in linear scale). The last region, which has not a linear dependence, seems to indicate a saturation of the structural transformation process.

In the case of CW irradiation the following explanation of the kinetics curves is suggested [16]. During the first stage of irradiation, darkening appears as a consequence of the polymerization of the molecular fraction (As_4S_6 , As_4S_4) in the material [10,11]. This process gives rise to a small decrease of the relative transmittance as shown by curve 1 in Fig. 1a in the region where $\log(t) < 3.3$.

Thereafter, a bleaching peak follows, caused by processes which compete with polymerization, mainly the elimination of a part of the film disorder by defect healing. It is quite remarkable that if the films are stabilized by annealing under the glass transition temperature, T_g, or by shining with another light source, then the bleaching peak is lacking.

For longer irradiation time, a strong photodarkening is produced and a nearly linear behavior is observed from the plot $\log(T/T_o)$ versus $\log(t)$ (see Fig. 1a, b). There are two nearly linear regions separated by a shoulder. The values $\eta(1)$ and $\eta(2)$ of the slopes of the least squares fits in these two zones are reported in Table 1 for the curves drawn in Figures 1 and 2.

TABLE 1.

Figure	Curve	I (W/cm^2)	$\eta(1)$	$\eta(2)$
1a	1	1.7 10^{-3}	-0.36 ± 0.01	-
	2	3.5 10^{-3}	-0.38 ± 0.01	-
	3	7 10^{-3}	-0.37 ± 0.01	-
	4	1.5 10^{-3}	-0.35 ± 0.01	-
1b	1	2.6 10^{-2}	-0.23 ± 0.05	-0.25 ± 0.05
	2	5.2 10^{-2}	-0.25 ± 0.05	-0.28 ± 0.05
	3	0.1	-0.26 ± 0.05	-
2b	2	5 10^{-2}	-0.29 ± 0.01	-
	3	0.1	-0.30 ± 0.01	-
	4	0.2	-0.31 ± 0.01	-0.24 ± 0.05
2c	1	0.2	-0.31 ± 0.01	-
	2	0.4	-0.31 ± 0.01	-0.31 ± 0.01
	3	0.9	-0.31 ± 0.01	-0.30 ± 0.01

In the Table, $\eta(1)$ is the slope of the line which fits data immediately before the shoulder, while $\eta(2)$ is the slope of the line obtained from fitting data after the shoulder (when it is possible to obtained it). As it can be seen the slopes are mostly the same, passing from one curve to the other, confirming that the photodarkening is depending on the dose of absorbed light. The value of the slopes in Figure 1a (-0.36) is different from those in Figure 1b (-0.25) but is affected by a larger error, due to the shift towards shorter times of the shoulder, which reduces the zone in which the behavior is linear.

An induction period seems to be necessary for triggering the transition from the first nearly linear decrease of the logarithm of the transmitted intensity to the second one and this is demonstrated by the shoulder evidenced in nearly all the transmission curves. We suggest that after the end of the polymerization process a new process becomes dominant and this might be the photo-decomposition of the material, especially at the surface of the film [12,13]. Absorption of a photon of energy greater than E_g causes the dissociation of an arsenic-sulfur bond. As a result of decomposition the following reaction occurs:

$$As_2S_3 \xrightarrow{h\omega} 2As + 3S \ ,$$

and, after atom migration, sulfur rich regions appears. This is supported by our observations by optical microscopy of small yellow crystallites on the surface, ascribed to orthorhombic sulfur. The released arsenic atoms diffuse and form clusters which determine a further darkening of the material. The necessity of reaching a certain amount of decomposition before arsenic clustering, might be responsible for the induction period and for some bleaching observed in transmission curves. Small clusters of arsenic can be formed without long range diffusion of arsenic atoms [14].

A local bond reorganization will produce pairs of As-As bonds and several such reorganizations occurring in the same locality will give rise to clusters of three or four arsenic atoms. The above band-gap illumination alters the bond statistics from a distribution close to that for the chemically ordered network towards one characteristics of a random network, that is the randomness of the bond distribution is increased and , as a consequence, the darkening process is intensified. Isolated As-As bonds are energetically unfavorable but larger clusters are expected to be relatively stable. They contribute significantly to the photodarkening effect [15].

The irradiation by pulsed laser light triggers the same processes but some specific features are different. An important feature is the absence of the beaching peak for a large range of irradiation intensities (Fig. 2). This behavior indicates that the bleaching process is driven by the peak intensity of the laser beam, this speaking in favor of a nonlinear optical contribution to the process of network rebuilding.

For the case of illumination with picosecond pulses, the sub-bandgap light ($h\nu$ = 2.33 eV and E_g = 2.4 eV) and the higher light intensity (in the condition of high absorptivity of arsenic-chalcogen bond) seem to affect more the process of decomposition, accelerating it with respect to the oxidation one. This can explain the very pronounced shoulder on the photodarkening curves and the fact that the shoulder itself, evidenced in both irradiation methods, is considerably shifted towards long irradiation times (high doses) for the case of pulsed illumination. It must also be taken into account the small but not negligible difference in the quantum energy used in the experiments. Both features seem to be related to an amplification of the bond destructive effects of the radiation due to the rapid and large variations of the light intensity. It seems that the process of oxidation of arsenic is a very probable process as shown in [12,15]. Nevertheless, careful X-ray measurements performed by us on As_2S_3 films irradiated by light have not revealed any crystalline phase of As_2O_3 .

The constance of the slopes of the curves is even more evident in the pulsed measurements. In fact, the values of $\eta(1) = -0.30$ remain constant in Figures 2b and 2c.

In order to separate more clearly the processes involved in photodarkening, a plot of a tipical curve (curve 1, Fig. 2c) in the log (T/T_o) versus time scale has been drawn (Fig. 3). Three distinct processes are evidenced that are characterized by nearly linear dependence on time. Similar results can be obtained for the other curves, both in the case of pulsed and CW excitation. We ascribe these processes to photopolimerization, photodecompozition and oxidation and they become successively dominant as expected.

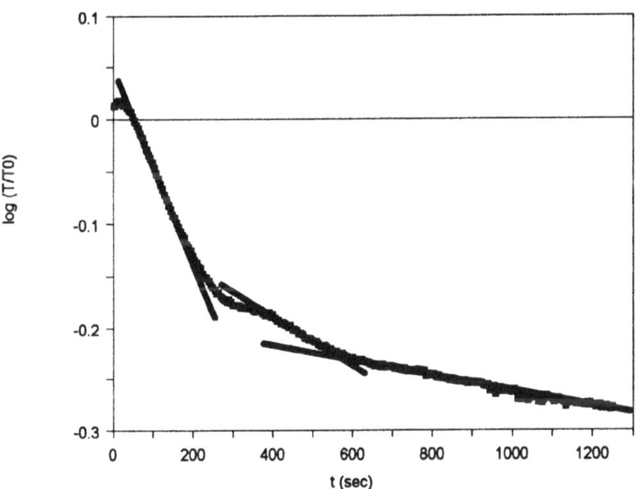

Fig.3. Kinetics of the relative transmitance through virgin As_2S_3 film (coresponding to the curve 1 in figure 2c) versus time.

The formation and the evolution of the bleaching peak revealed in the material during the first stage of irradiation can be explained in the frame of a model with two concurrent processes. The first process is the polymerization of the molecular fraction. This process leads to photodarkening effect and is a first order reaction. Let us have in the film, at time $t = 0$, "a" moles of polymerizable units. At time t after starting the irradiation, the amount of the molecules will decrease to "x". The concentration of the molecular units at time t will be:

$$c=\frac{a-x}{V}\ , \tag{1}$$

where V is the volume. According to the definition of the rate of a first order reaction:

$$\frac{dc}{dt}=-k\cdot c\ , \tag{2}$$

we have:

$$\frac{1}{V}\frac{d(a-x)}{dt}=\frac{-k(a-x)}{V}\ . \tag{3}$$

The reaction rate is proportional to the number of photons absorbed per second in the film which had the absorption coefficient α. Then $k \sim \alpha I$, where I is the flux of incident photons of energy $h\nu$.

If q is the quantum efficiency, then $k = q\,\alpha\, I$, and:

$$\begin{aligned}\frac{d(a-x)}{dt}&=-q\cdot\alpha\cdot I\cdot(a-x)\ ,\\ \frac{dx}{a-x}&=q\cdot\alpha\cdot I\cdot dt\ .\end{aligned} \tag{4}$$

By integration, considering that at $t = 0$, $x = 0$ we can write:

$$\mathrm{q}\cdot\alpha\cdot\mathrm{I}=\frac{1}{\mathrm{t}}\ln\left(\frac{\mathrm{a}}{\mathrm{a}-\mathrm{x}}\right)\ , \tag{5}$$

and finally:

$$x=a(1-e^{-q\cdot\alpha\cdot I\cdot t}) \tag{6}$$

If the energy of the photons $h\nu$, lies not too far above the absorption edge $E_g = h\nu_o$, then

$$\alpha h\nu = k\,(h\nu - h\nu_0)^2\ . \tag{7}$$

In this case it is easy to show that the following relation is valid:

$$\Delta \alpha = \frac{-2\alpha \, \Delta E_g}{h\nu - h\nu_o} = -B' \cdot \alpha \cdot x = B \cdot x \ , \tag{8}$$

where ΔE_g is the shift of the absorption edge and one admits that this shift is proportional to x, being B and B' some constants. Because the relative transmission of the film is related to the variation $\Delta\alpha$ of the absorption coefficient α during the photodarkening process by the following simple expression:

$$\frac{T}{T_O} = e^{-\Delta\alpha} \ , \tag{9}$$

one gets finally:

$$\frac{T}{T_O} = e^{-Ba(1-e^{-q \cdot \alpha \cdot I \cdot t})} \ . \tag{10}$$

The above relation can be simplified as:

$$T_1 = \frac{T}{T_O} = e^{-k_1 (1-e^{-k_2 \cdot I \cdot t})} \ . \tag{11}$$

where we have put $k_1 = Ba$, and $k_2 = q\,\alpha$.

The photobleaching process is a different process related probably to the depolymerization determined by the monophotonic processes and by defect healing. This process can be described by a similar relation:

$$T_2 = e^{\Delta\alpha} \ , \tag{12}$$

or

$$T_2 = e^{k_3 (1-e^{-k_4 \cdot I \cdot t})} \ . \tag{13}$$

where k_3 and k_4 have similar meanings as k_1 and k_2.

The total function which describes the bleaching peak is therefore: $T = T_1 T_2$, i.e.:

$$T = e^{\left[-k_1(1-e^{-k_2 I t}) + k_3(1-e^{-k_4 I t})\right]} . \tag{14}$$

For $t = 0$, $T = 1$ and for $t \to \infty$, $T \to e^{k_3 - k_1}$, with the condition: $k_3 < k_1$.

Now it is possible to find the position of the photobleaching peak by equating to zero the derivative of the function:

$$dT/dt = 0 . \tag{15}$$

After some mathematical manipulations one gets:

$$t_p = \frac{\ln \frac{k_1 k_2}{k_3 k_4}}{(k_2 - k_4) I} . \tag{16}$$

Thus, we obtained the dependence of the position of the beaching peak on the power density of the incident light I, which gives the experimental relation: $t_p \cdot I_o$ = const.

The set of constants which gives a good fit to the experimental data especially in the case of intermediate values of laser intensity is:

$$k_1 = 0.7;\; k_2 = 0.03\ \mathrm{cm^2/J};\; k_3 = 0.4;\; k_4 = 0.07\ \mathrm{cm^2/J} . \tag{17}$$

4. Conclusions

The kinetics of the photodarkening in As_2S_3 induced by laser irradiation with the energy of quanta approaching the band gap has a complex character.

A bleaching effect ascribed to the healing of the defects in the glass was revealed. A shoulder present in nearly all the transmission curves can be ascribed to the effects accompanying the chemical decomposition of the sample.

The major contribution to photodarkening in the first stages of illumination seems to be given by the photopolymerization of the molecular fraction of the material. The photodarkening in the last stages of irradiation may be related to the atomic scale modifications of the network as a result of the photodecomposition and oxidation.

In the case of irradiation with picosecond pulses the behaviour of kinetics changes points out that there is a nonlinear contribution to the process of network rebuilding.

5. Acknowledgments

The authors wish to thank Prof. C. Sibilia, Dr. E. Fazio and Dr. G. Liakhou for the frequent and useful discussions. One of us (V. C.) acknowledges the partial support of the NATO Programme for Priority Area on High Technology (Grant HTECH. EV 950772) for this research. Thanks are due to P. Cerbari for sample preparation.

6. References

1. Pfeiffer G., Paesler M. A., Agarwal S. C. (1991) Reversible photodarkening of amorphous arsenic chalcogenides, *J. Non-Crystalline Solids* **130**, 111-143.
2. Popescu M., Andries A., Ciumas V., Iovu M., Sutov S., Tiuleanu D. (1996) *Fizica Sticlelor Calcogenice*, Scientific Publisher, Bucharest & Stiinta Publisher, Chisinau, 486 pag.
3. Tanaka K., Kikuchi M., Mizuno H. (1973) Kinetics of photo-induced edge shift in optical transmission of amorphous As_2S_3 film, *Solid State Communications* **12**, 195-198.
4. Andriesh A. M., Bogdan O. I., Enaki N. A., Cojocaru I. A. and Chumash V. N. (1992) Optical hysteresis and nonlinear propagation of laser pulses in chalcogenide glasses, *Bulletin Ross. Acad. Sciences, Ser. Phys.* **56 (12)**, 96-109.
5. Chumash V., Cojocaru I., Fazio E., Michelotti F., Bertolotti M., (1996) Nonlinear propagation of strong laser pulses in chalcogenide glass films, *Progress in Optics* **36**, 1-47.
6. Michelotti F., Bertolotti M., Chumash V., Andriesh A. (1992) Chalcogenide glass thin films: Z-Scan measurements of refractive index changes, *Photonics for Computers, Neural Networks, and Memories, SPIE Proceedings* **1773**, 423-432.
7. Michelotti F., Fazio E., Senesi F., Bertolotti M., Chumash V., Andriesh A., (1993) Nonlinearity and photostructural changes in glassy As_2S_3 thin films, *Optics Communications* **101**, 74-78.
8. Tanaka K. (1988) Light intensity dependence of photodarkening in amorphous As_2S_3 films, *Thin Solid Films* **157**, 35-41.
9. Stradins P., Shvarts K., Teteris J. (1989) The relation of photo and theramal components of photoinduced changes in amorphous semiconductors, *J. Non-Crystalline Solids* **114**, 79-81.
10. Treacy I., Strom U., Kevin P. B., Taylor P. C., Martini T. P. (1980) Photostructural effects in glassy As_2Se_3 and As_2S_3 , *J. Non-Crystalline Solids* **35 & 36**, 1035-1039.
11. Onari S., Asai K., Arai T. (1985) Photo-polimerization of vacuum evaporated amorphous $(As_2S_3)_{1-x}(As_2Se_3)_x$ systems, *J. Non-Crystalline Solids* **76**, 243-251.
12. Berkes J., Ing S. W., Jr., Hillegas W. I. (1971) Photodecomposition of amorphous As_2Se_3 and As_2S_3, *J. Appl. Phys.* **42**, 4908-4916.
13. Tanaka K., Kikuchi M. (1972) Anomalous photo-induced shift of optical transmission edge observed in amorphous As_2S_3 film, *Solid State Communications* **11**, 1311-1314.
14. Frumar M., Firth A. P., Owen A. E. (1983) A model for photostructural changes in the amorphous As-S system, *J. Non-Crystalline Solids* **59 & 60**, 921-924.
15. Owen A. E., Ferth A. P., Ewen P. J. S. (1985) *Phil. Mag. B* **52**, 347-351.
16. Bertolotti M., Michelotti F., Chumash F., Cerbari P., Popescu M., Zamfira S. (1995) The kinetics of the laser induced structural changes in As_2S_3 amorphous films, *J. Non-Crystalline Solids* **192&193**, 657-660.

OPTICAL GLASSES FOR INFRARED TRANSMITTANCE

SYNTHESIS AND PROPERTIES OF CHALCOGENIDE GLASSES

B. VOIGT [a], D. LINKE [b]
[a] *VITRON Spezialwerkstoffe GmbH, Otto-Schott-Str. 13, D-07745 Jena, Germany; Phone: (493641) 616452; Fax: ( ) 616814*

[b] *Brandenburgische Technische Universität Cottbus, Fakultät 1, Lehrstuhl Anorganische Chemie, Karl-Marx-Str. 17, D-03044 Cottbus, Germany; Phone: (49355) 693622; Fax: ( ) 693090*

1. INTRODUCTION:

„Ideal" glasses for infrared imaging, wave guide and other applications should meet the following requirements:

(1) high internal transmission in the near, mid and far Infrared (IR);
(2) variability in refractive index and optical dispersion, depending on the glass composition;
(3) high chemical, mechanical and thermal durability.

For „real" glasses one has to make a compromise because some demands are mutually exclusive. For example, heavy atoms and low chemical bond energies have to be taken as a premise for mid- and far-IR transparency. This leads, however, to materials with relatively low values for hardness, elastic moduli and glass transition temperatures. Comparing IR transmitting glasses of oxide, halide or chalcogenide type, the chalcogenide glasses - well-known since the 1960s - excel above the others: They transmit to longer IR wavelengths than oxide or fluoride glasses; they are also much more resistant towards water corrosion and devitrification than chalcogen-free heavy-metal halide glasses [1 - 5].

But these and some other advantages of chalcogenide glasses can be utilized fully only, if care is exercised in carrying out the numerous steps in the preparation of the pure components and the glasses. Often, the glasses contain chemical elements for which safety measures have to be obeyed. Both aspects are probably the reason that chalcogenide glasses have a share in IR materials of only about 1 %, in comparison with 38 % for crystalline germanium and 30 % for silicon.

A. Andriesh and M. Bertolotti (eds.),
Physics and Applications of Non-Crystalline Semiconductors in Optoelectronics, 155–169.

2. COMPOSITION OF CHALCOGENIDE GLASSES AND THEIR OPTICAL PROPERTIES:

Since the early 1960s, chalcogenide glasses have been extensively researched. This had led to a great understanding of the ranges of glass formation and the interdependence of physical parameters. Because of the more than sufficient number of review papers and monographs existing now (see e.g. [2, 3, 6 - 12]) belonging to the composition of the glasses only a very brief summary introduction to the next chapters is given here.

2.1 COMPOSITION:

Chalcogenide glasses are non-oxide glasses that contain as a major component 2-fold coordinated one or more of the chalcogen elements X (X = sulfur S, selenium Se, and sometimes tellurium Te) which are bonded to one or more elements E such as germanium Ge, arsenic As, antimony Sb, more seldom gallium Ga, phosphorus P, for IR optical fibres also rare earth metals, e.g. praseodymium Pr. Its structure may be described by covalent networks formed by irregular combination of EX_n structural units. The bond energies and the connectivity of the structural units play an important role in the thermal, mechanical and optical properties of these materials.

Glasses with low-dimensional connectedness of the structural units, i.e. with a high percentage of island-like (S_8, Se_8, Se_6), chain-like (S_∞, Se_∞) or layer-like structural fragments show low levels of mechanical strength and softening temperatures and high coefficients of thermal expansion, whereas glasses rich in germanium exhibit improved thermo-mechanical properties in accordance with the three-dimensionally connected GeX_4 units existing there [13 - 15]. So it is easily understood that commercially available optical parts - apart from the glasses arsenic sulfide, v-As_2S_3, and selenide, v-As_2Se_3 - usually belong to the germanium chalcogenide based systems such as Ge-As-Se(Te) and Ge-Sb-Se (TABLE 1). For fibre applications in the near-IR germanium-sulfur glasses are also studied in order to move the short wavelength cutoff as far as possible into the visible. Such glasses with rare earth metals as dopants are interesting candidate laser materials [16 - 19].

TABLE 1. Some commercially available chalcogenide glasses (VITRON GmbH Jena)

type	composition	remarks
IG 2	$Ge_{33}As_{12}Se_{55}$	corresponds to AMTIR 1/TI 20
IG 3	$Ge_{30}As_{13}Se_{32}Te_{25}$	
IG 4	$Ge_{10}As_{40}Se_{50}$	
IG 5	$Ge_{28}Sb_{12}Se_{60}$	" · AMTIR 3/TI 1173, IRG 100
IG 6	$As_{40}Se_{60}$	
-	As-S	composition by request
-	Ge-S	composition by request
-	Ge-Ga-S	composition by request

2.2. OPTICAL PROPERTIES:

2.2.1. Infrared transmission:

Chalcogenide glasses have been used for many years as passive optical components for infrared applications up to about 16 μm. *Figure 1* compares the transmittivity of purified vitreous arsenic sulfide and of a common Ge-As-Se glass. It is very important to note that high quality optical glasses for IR transmittance can only be obtained if traces of some impurities are excluded (see 3.2).

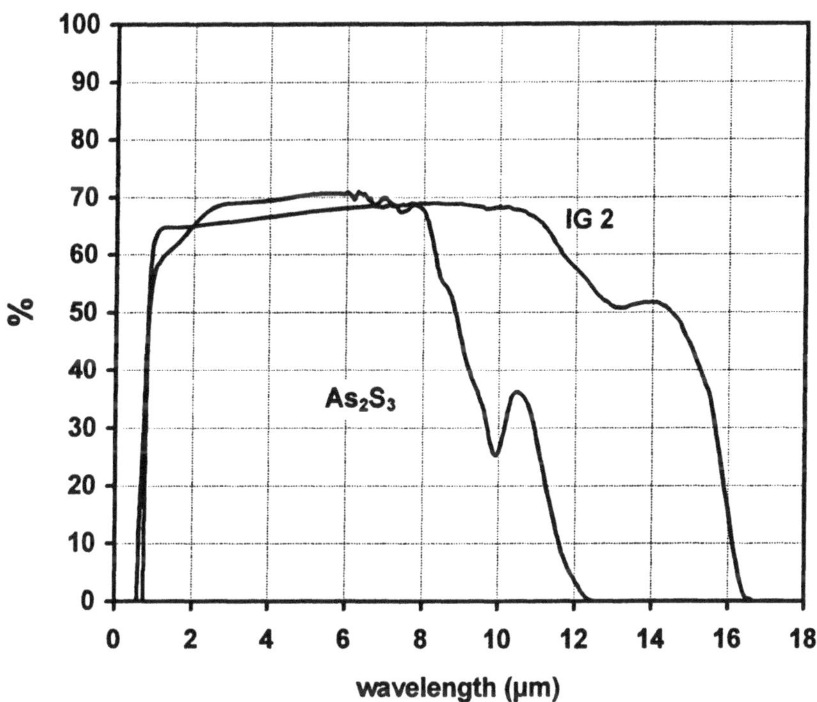

Figure 1. Transmittivity of two selected chalcogenide glasses, As_2S_3 and IG 2 ($Ge_{33}As_{12}Se_{55}$), in dependence on the wavelength (polished samples, 10 mm in thickness, surfaces non-coated)

2.2.2. Refractive indices n and Abbe numbers ν:

The refractive indices n of commercially available chalcogenide glasses vary between 2.4 and 2.8 (TABLE 2). To express the dispersion of these glasses one has to define suitable Abbe numbers ν which are adapted to the IR region under consideration. Such numbers, $\nu_{4\mu m}$ resp. $\nu_{10\mu m}$ (TABLE 3), are different enough to allow colour correct refractive optical systems of achromate or apochromate type [15].

TABLE 2. Refractive indices n of the chalcogenide glasses IG 2 - IG 6 (VITRON/Jena) at room temperature in the IR region 3 - 12 μm

wavelength (μm)	IG 2	IG 3	IG 4	IG 5	IG 6
3	2.5173	2.8111	2.6263	2.6277	2.8014
4	2.5129	2.8034	2.6210	2.6226	2.7945
5	2.5098	2.7993	2.6183	2.6187	2.7907
6	2.5072	2.7965	2.6159	2.6158	2.7880
7	2.5048	2.7941	2.6139	2.6132	2.7854
8	2.5024	2.7919	2.6121	2.6105	2.7831
9	2.4996	2.7896	2.6105	2.6075	2.7803
10	2.4967	2.7870	2.6084	2.6038	2.7775
11	2.4930	2.7841	2.6059	2.5996	2.7747
12	2.4882	2.7810	2.6029	2.5948	2.7721

TABLE 3. Abbe numbers $\nu_{4\mu m}$ resp. $\nu_{10\mu m}$ of the chalcogenide glasses IG 2 - IG 6 (VITRON/Jena);

$\nu_{4\mu m} = (n_{4\mu m} - 1) / (n_{3\mu m} - n_{5\mu m})$; $\nu_{10\mu m} = (n_{10\mu m} - 1) / (n_{8\mu m} - n_{12\mu m})$

Abbe number	IG 2	IG 3	IG 4	IG 5	IG 6
$\nu_{4\mu m}$	202	153	203	180	168
$\nu_{10\mu m}$	108	164	176	102	161

In a diagram n_{10} versus $\nu_{10\mu m}$ the glasses of the series IG 2 to IG 6 are compared with common crystalline IR materials *(Figure 2)*.

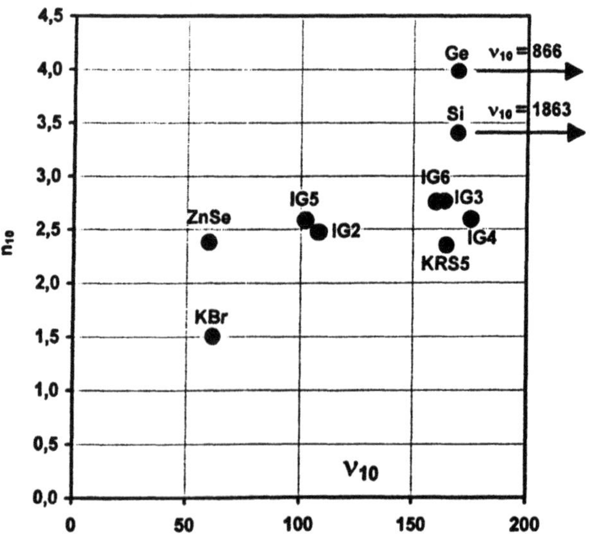

Figure 2. Abbe diagram for some glassy and crystalline IR materials (KRS 5 = 44 % TlBr, 56 % TlI)

2.2.3. Temperature coefficient of the refractive indices n, dn/dT:

If one uses IR-optical parts made of germanium IR experiments are restricted to relatively low temperatures. At elevated temperatures chalcogenide glasses remain highly transparent even at 250 °C (*Figure 3a*) whereas the transmission of germanium is drastically decreased (*Figure 3b*). In the glasses the mobilities of electrons and holes are so low that free carrier absorption is not observed at heating. The thermal changes in refractive index, dn/dT, are shown in TABLE 4.

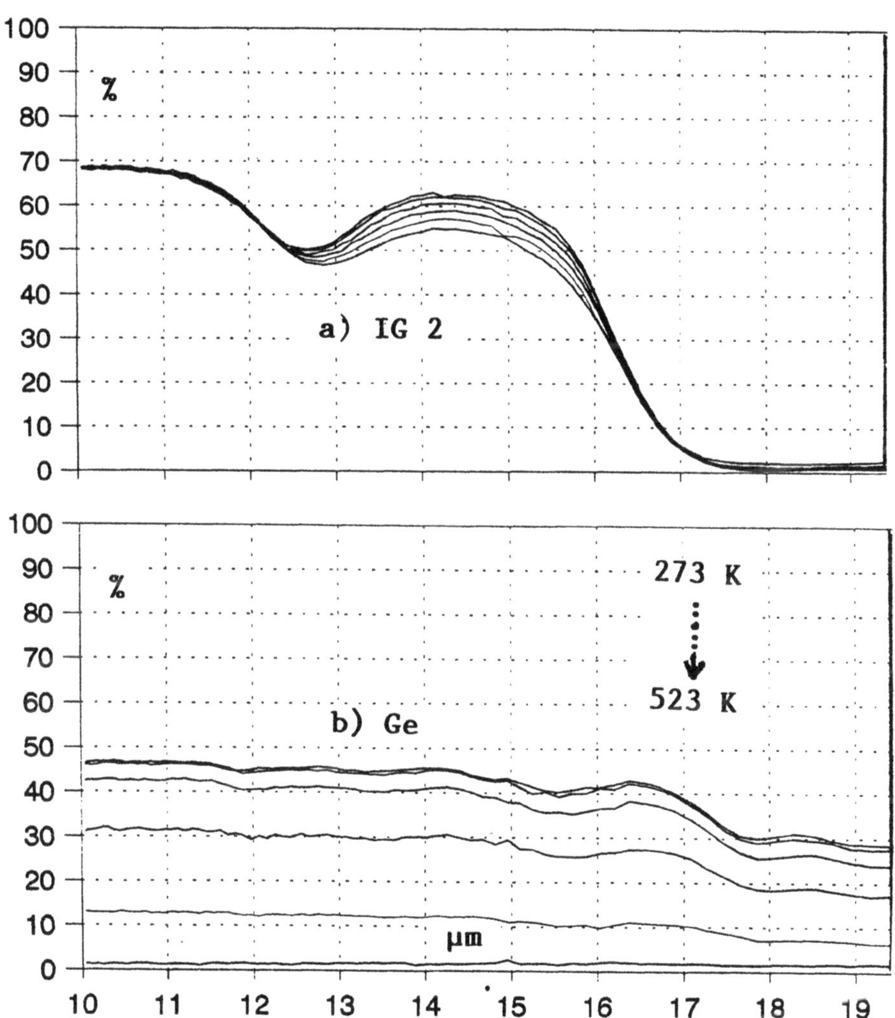

Figure 3. Influence of the temperature variation on the IR transmittivity of polished samples of
a) IG 2 (sample thickness d = 3 mm) b) crystalline germanium (d = 2 mm)
The upper curves were measured at room temperature, the lowest ones at 523 K,
the curves between at 323 K, 373 K, 423 K, 473 K

TABLE 4. Thermal change in refractive index, dn/dT, of the chalcogenide glasses IG 2 - IG 6 (VITRON/Jena) and some crystalline materials, measured for the temperature range 20 - 30 °C

10^5 dn/dT at 10.6 µm (K^{-1})	Ge	ZnSe	KRS-5	IG 2	IG 3	IG 4	IG 5	IG 6
	40	6	-23.5	6.0	14.5	3.6	9.1	4.1

2.2.4. Stress-optical coefficient C_0 (or B):

As a consequence of the „softness" of chalcogenide glasses, stress-induced birefringence attains high values. Of interest for some acoustooptic applications should be the unusually large variation of the relative stress-optical coefficient C_0 in dependence on the type of structural units existing in the chalcogenide glasses (*Figure 3*).

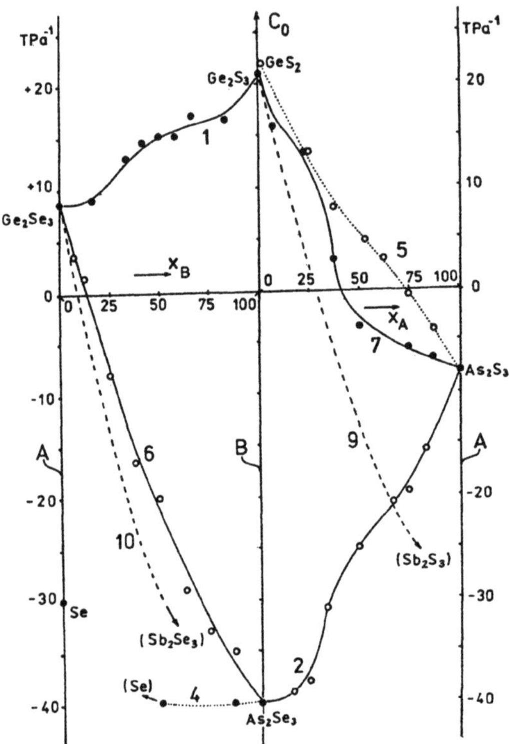

Figure 4. Variability in C_0, the relative stress-optical coefficient, as a function of composition (in mole %) for several series A - B of chalcogenide glasses [20, 21]

series	A	B	series	A	B
1	Ge_2Se_3	Ge_2S_3	2	As_2Se_3	As_2S_3
4	Se	As_2Se_3	5	GeS_2	As_2S_3
6	Ge_2Se_3	As_2Se_3	7	Ge_2S_3	As_2S_3
10	Ge_2Se_3	Sb_2Se_3	9	Ge_2S_3	Sb_2S_3

C_0 (or B) is defined as the difference of the absolute stress-optical coefficients C_1 and C_2 for the change of the refractive index parallel and perpendicular to the applied stress S_σ

$$\Delta n = \Delta n_{\parallel} - \Delta n_{\perp} = (C_1 - C_2) \cdot S_\sigma = C_0 \cdot S_\sigma. \qquad (1)$$

With values between -40 and +22 TPa^{-1} the coefficient C_0 varies on a very large scale [21, 22], much more than in the case of fluoride or oxide glasses [23]. Glasses with low connectivity of the structural units, i.e. with a chain- or layer-like linkage of these units, show high negative values B, those with three-dimensionally interconnected units yield high positive values.

Some glasses (e.g. $Ge_{22}As_{18}S_{60}$, $Ge_{36}As_4Se_{60}$, $Ge_{28.5}Sb_{11.5}S_{60}$, $Ge_{36.5}As_{3.5}Se_{60}$ [20]) do not show any optical birefringence under uniaxial pressure ($C_0 = 0$), in analogy to the well-known Pockels glasses in the system PbO-SiO_2. This could be also of value for the optimization of IR optical systems.

Measurements not only of the **relative** stress-optical coefficients C_0 but also of the **absolute** stress-optical coefficients C_1 and C_2 would allow new insights into the acousto-optical potential of chalcogenide glasses [1, 23, 24], in accordance with the equations for the relationships between the photoelastic coefficients p and q, the stress-optical coefficients C_0, C_1, C_2, the refractive index n, the Young modulus E and the Poisson ratio μ:

$$p - q = \frac{E \cdot C_0}{n^2 + (1 + \mu)} \qquad (2)$$

$$2p + q = \frac{-E \cdot (2C_2 + C_1)}{n^2 (1 - 2\mu)} \qquad (3).$$

3. METHODS OF PRODUCTION FOR CHALCOGENIDE GLASSES:

3.1. GENERAL FEATURES:

Well established procedures exist for fundamental research concerning glass formation and properties of new chalcogenide glasses:

High purity (5N-6N) elemental materials are used as starting materials for glass preparation. The batch size is usually 5-30 grams. Raw materials are weighed and loaded into a silica glass ampoule, preferably in a glove box filled with dry argon gas. The ampoule is sealed off under good vacuum (up to 10^{-8} Torr) with an oxypropane flame, and then put into a rocking or rotating furnace. There the melt is kept at temperatures between 400-1000 °C for several hours. The melting schedule has to be adapted thoroughly to the system under study in order to avoid dangerously high vapour pressures caused by non-reacted chalcogens or other volatile components *(Figure 5)*, to allow a complete reaction and a homogeneous distribution of all reaction products in the melt. It is also very important to avoid the formation of crystallites during cooling [25].

Finally, rod-like glass samples are obtained by air-quenching of the ampoule in the vertical position. For optical and other measurements the glass samples are annealed at a temperature close to the measured glass transition temperature T_g for several hours or days, in order to remove the internal strain which had been generated in the samples during quenching.

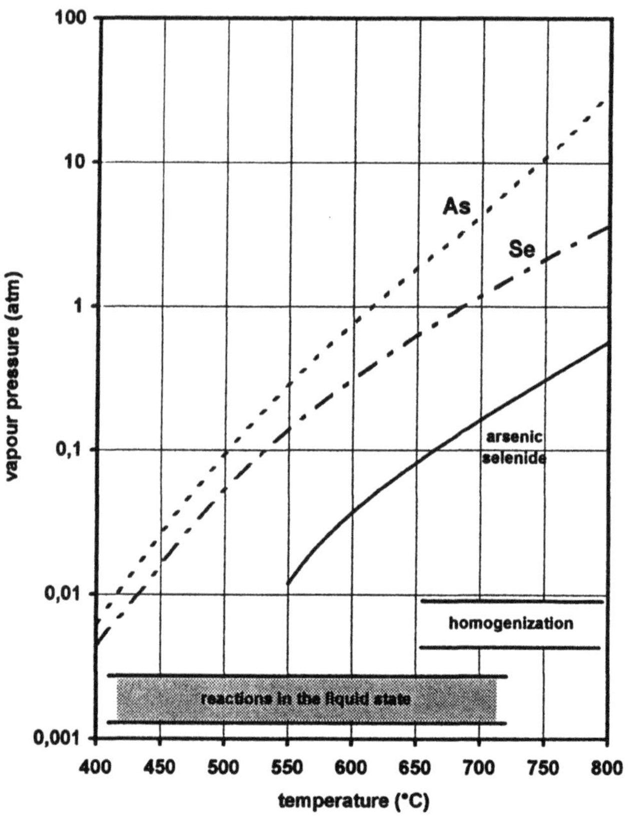

Figure 5. Vapour-pressure diagram in the system arsenic-selenium

3.2. PURITY ASPECTS:

High quality optical glasses for IR transmittance can only be obtained if during the glass preparation contaminations of elements such as oxygen, hydrogen and carbon can be minimized. These impurities carried in from the „high pure" elements, the atmosphere or the ampoules itself, cause absorption bands at specific wavelengths (TABLE 5).

TABLE 5. Wavelength λ of disturbant absorption bands A-B in the Infrared caused by trace impurities of hydrogen, carbon and oxygen

Species	O-H	S-H	Se-H	H_2O	COS	CS_2	Ge-O	As-O	Se-O
λ_{A-B} (μm)	1.4	4.0	4.1	6.3	4.9	6.3	7.9	12.5	13.6
	2.2		4.6				11.5	13.0	
	2.8						12.8	13.8	

In consequence of this the transmittivity at characteristic wavelengths is drastically reduced even by ppm amounts of such impurities *(Figure 6)*; simultaneously the absorption coefficients increase at the given wavelength [26 - 30].

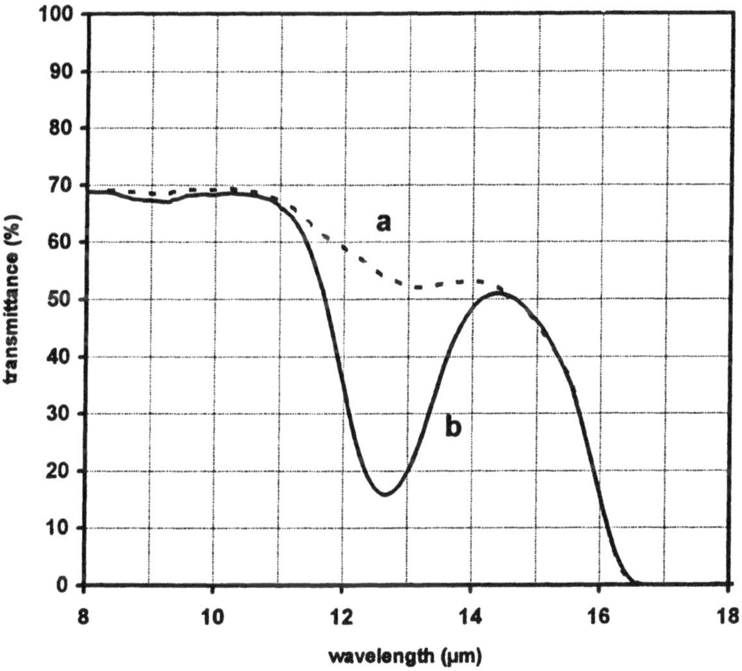

Figure 6. IR transmittance spectra of glasses $Ge_{33}As_{12}Se_{55}$ (IG 2; sample thickness 10 mm) Ge-O absorption band caused by a) 1 ppm, b) 10 ppm oxygen

Despite all the difficulties that arise from insufficient components purity and during the preparation itself, very pure selenide/telluride glasses can be produced now. Such techniques have been described in the literature in detail [11, 31]. Following the scheme given in *figure 7* batches up to 7 kg are available. The batch-to-batch variation in the refraction index n is equals 10^{-3} or lower. The optical homogeneity for variation

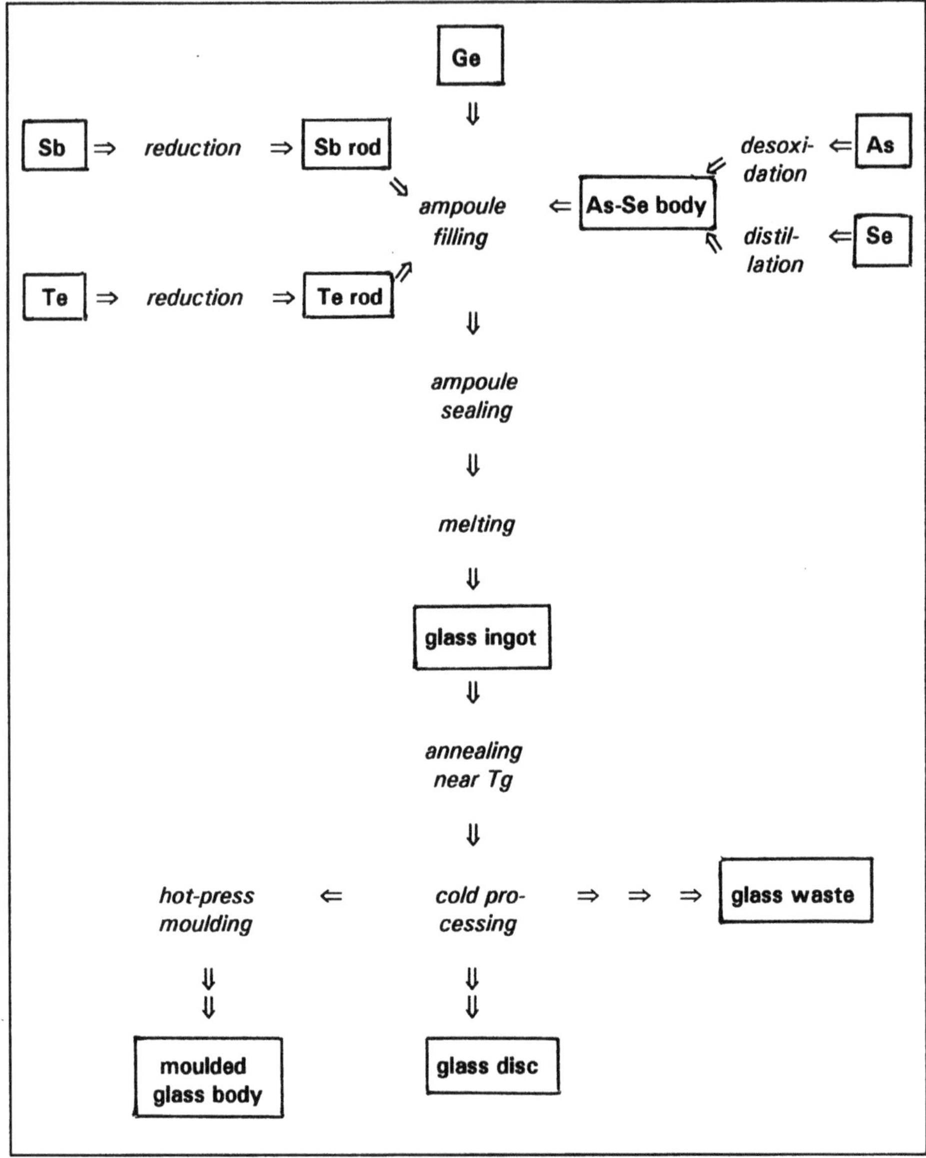

Figure 7. Technological scheme for the production of high pure chalcogenide glasses

in the refraction index n is equals 10^{-3} or lower. The optical homogeneity for a given glass plate is 10^{-4} or better. There are some subjective tests for verifying optical homogeneity (see e.g. [32, 33]. Recently, laser interferometry has been used to measure refractive index variation across a plate (*figure 8*).

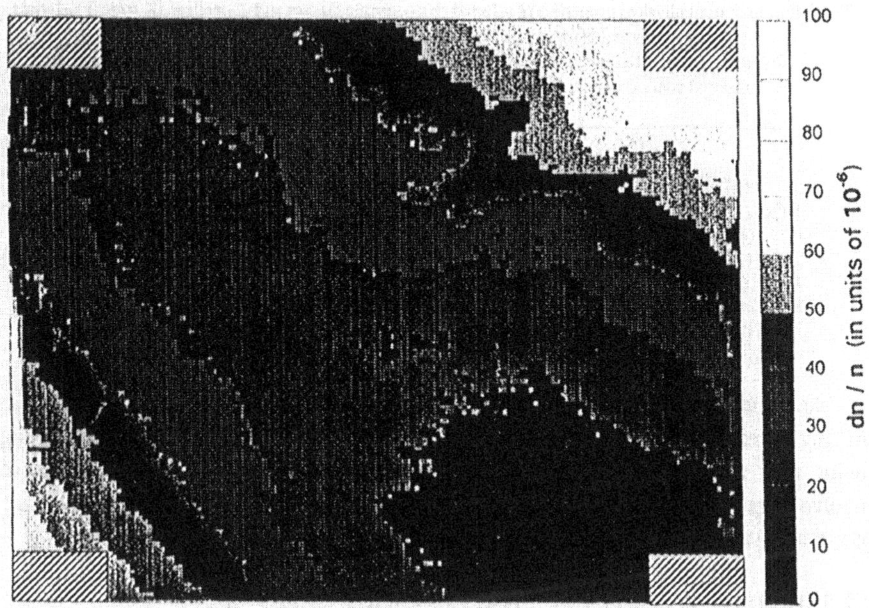

Figure 8. Refractive index homogeneity across a plate (168 x 134 x 20) mm^3 of a chalcogenide glass IG 5 (VITRON / Jena), measured interferometrically at 3.39 µm, with surface correction.
Measurement was undertaken by A. REICHMANN,
Friedrich-Schiller University Jena, Institute of Applied Optics

4. MANUFACTURING OF OPTICAL PARTS:

4.1. FINISHING OF CHALCOGENIDE GLASSES:

For the finishing of chalcogenide glasses to disks, prisms, and more complicated parts one has to take into account their relatively low values of „grinding hardness" and „thermal shock resistance", as a consequence of the relatively low mechanical parameters (Knoop microhardness MH_K, as a rough measure of abrasion hardness; Young modulus E) and thermal conductivity λ, and the high coefficients α of thermal expansion. These and some other important parameters are summarized in TABLE 6 for some glasses and crystalline IR materials.

An upper value $\Delta T = 54$ K can be estimated as tolerable temperature difference for the glass IG 2, in comparison with $\Delta T = 50$ K for optical glass BK 7, $\Delta T = 155$ K for Duran and $\Delta T = 1140$ K for vitreous silica.

TABLE 6. Important physical properties of selected chalcogenide glasses and crystalline IR optical materials {density ρ, thermal expansion α between 20 and 100 °C, specific heat capacity c_p, „thermal stability" (temperature T of melting resp. ***softening** [at viscosity $\eta = 10^{7.6}$ dPa · s]*), thermal conductivity λ at 20 °C, Knoop microhardness MH_K, Young modulus E}

	Ge	ZnSe	KRS-5	IG 2	IG 3	IG 4	IG 5	IG 6
ρ (g/cm³)	5.3	5.4	7.3	4.41	4.84	4.47	4.66	4.63
10^6 α (K^{-1})	6.1	7.1	58.0	12.1	13.4	20.4	14.0	20.7
c_p (J/g · K)	0.32	-	-	0.33	0.32	0.37	0.33	0.36
T (°C)	1210	1520	637	***445***	***360***	***310***	***350***	***235***
λ (W/m · K)	59.0	18.0	0.54	0.24	0.22	0.18	0.25	0.24
MH_K (GPa)	7.8	1.2	1.0	1.4	1.35	1.1	1.15	1.05
E (GPa)	102.7	67.2	15.9	21.5	22.0	-	22.1	18.3

Nevertheless, chalcogenide glasses can be processed by many procedures such as drilling, slicing, hot press-moulding in semifluid state, grinding, lapping, polishing, facing with mono-diamond tools, fibre drawing. Of course, some of these processes involve relatively heavy mechanical or thermal load of the glassy material. Thus, one has to be aware that dangerous chemical reactions could occur.

4.2. INHERENT TOXICITY OF CHALCOGENIDE GLASSES:

No serious danger of intoxination exists under conditions normal for the application of chalcogenide glasses [34, 35], because of the extreme low solubility of the chalcogenides in water, diluted mineral acids and organic solvents. The glass surfaces, however, are not resistant to alkalies and concentrated oxidizing acids; the resulting solutions contain toxic products such as arsenites AsO_3^{3-}, arsenates AsO_4^{3-}, selenites SeO_3^{2-} and tellurites TeO_3^{2-}.

Chalcogenide glasses are metastable against oxygen in air to temperatures up to about 150 °C. Direct combustion (dry oxidation) at high temperatures leads to oxides, e.g. to volatile arsenic trioxide, As_2O_3, or to selenium dioxide, SeO_2. Both these emissions occur also in the manufacturing of oxide glasses. Selenium and its oxidic compounds are used for example either for the decolorization of flint container glass or as a colorizing agent for the production of tinted bronze glass. Thus, the experience with such waste products can be used for chalcogenide glass production to obey the threshold limits prescribed for the emission of such substances [36].

There is no agreed opinion concerning the evaluation of waste products which remain after machining. Thus, as precautionary safety measures it is recommended to separate the working places for glass processing from the other ones, to treat all chalcogenide glass waste as toxic, to collect the different kinds of waste separately, to stock them in „special" deposits. This belongs to the red grinding slurries which contain the glass wear and likewise to all fluids used for finishing optical parts. Following contaminations were analyzed in such fluids after several weeks of utilization (in ppm): As 63 ; Se 18 ; Ge 189 ; Sb 20 ; Te < 2.5.

4.3. SURFACE IMPROVEMENT:

Although pure chalcogenide glasses are distinguished by a broad region of transmittance and low levels of the absorption coefficient, they show distinct Fresnel reflective losses, like all other IR materials with high values for the refraction index (TABLE 7). After coating with appropriate films these losses are significantly lowered. *Figure 9* shows the transmittance of a coated sample IG 2, for comparison with *Fig. 3* not only for room temperature but also for 250 ^{0}C.

TABLE 7. Reflective Fresnel losses of selected IR optical materials and related optical properties

	Ge	ZnSe	KRS-5	IG 2
Transmission range (μm)	1.8 - 23	0.5 - 22	0.5 - 45	0.8 - 16
Absorption coefficient at 10.6 μm (cm^{-1})	$2.5 \cdot 10^{-2}$	$5 \cdot 10^{-4}$	$2.2 \cdot 10^{-3}$	$7 \cdot 10^{-3}$
Refraction index at 10 μm	4.0057	2.4064	2.3707	2.4967
Fresnel loss at 10 μm (two surfaces; in %)	53	29	28.5	31.0

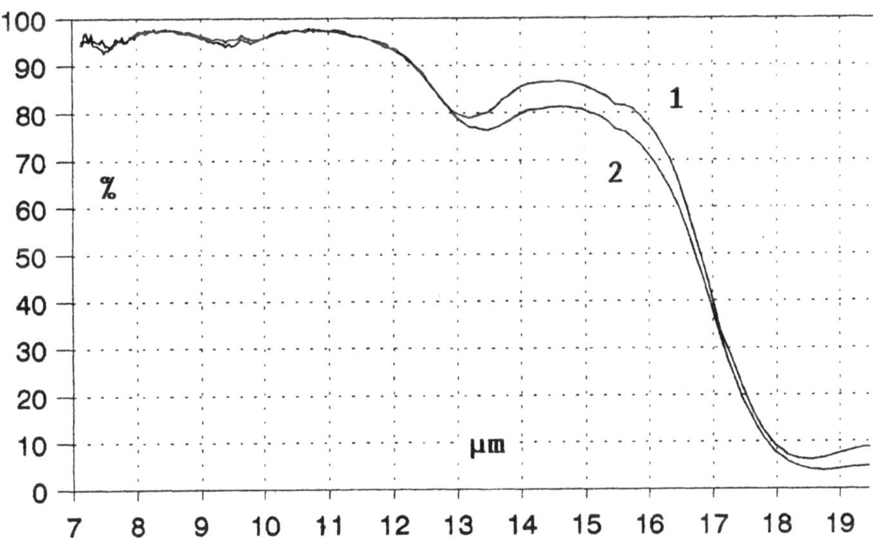

Figure 9. IR transmittivity of a polished sample of IG 2 (sample thickness d = 2 mm) surface coated with fluorides (**1** at room temperature, **2** at 523 K)

A way to improve the relatively „soft" surfaces of chalcogenide glasses consists in the deposition of diamond-like carbon layers [37]. Thereby, the surfaces can be better protected against particulate erosion.

REFERENCES:

1. Liepmann, M. J. and Neuroth, N. (1995) Infrared-Transmitting glasses, in H. Bach and N. Neuroth (eds.), *The Properties of Optical Glass*, Springer Verlag Berlin/Heidelberg/New York, pp. 299-310.

2. Feltz, A. [a] (1983) *Amorphe und glasartige anorganische Festkörper*, Akademieverlag, Berlin, 460 pp.; [b] (1986) *Amorfnuie i stekloobraznuie neorganicheskie tverduie tela* (russ.), Mir, Moscow, 556 pp.; [c] (1993) *Amorphous Inorganic Materials and Glasses*, Verlag Chemie Weinheim/New York/Basel/ Cambridge/Tokyo, 446 pp.

3. Elliott, S. R. (1991) Chalcogenide glasses, in R. W. Cahn, P. Haagen and E. J. Kramer (eds.) *Materials science and technology - A comprehensive treatment*, Vol. **9** (Vol. Ed.: J. Zarzycki) *Glasses and amorphous materials*, Verlag Chemie, Weinheim/New York/Basel/Cambridge, pp. 375-454.

4. Weber, M. J. (1991) Optical properties of glasses, in [3], pp. 619-664.

5. Lucas, J. (1991) Halide glasses, in [3], pp. 455-491.

6. Savage, J. A. and Nielsen, S. (1965) Chalcogenide glasses transmitting in the infrared between 1 and 20 μ - A state-of-the-art review, *Infrared Physics* **5**, 195-204.

7. Hilton, A. R. (1970) Optical properties of chalcogenide glasses, *J. Non-Cryst. Solids* **2**, 28-39.

8. Borisova, Z. U. (1983) *Khal'kogenidnuie poluprovodnikovuie stekla* (russ.), Publ. house Leningrad State Univ., 344 pp.

9. Vinogradova, G. Z. (1984) *Stekloobrazovanie i fazovuie ravnovesiya v khal'kogenidnuikh sistemakh*, Publ. house Mir, Moscow, 174 pp.

10. Minaev, V. S. (1991) *Stekloobraznuie poluprovodnikovuie splavui* (russ.), Publ. house Metallurgija, Moscow, 406 pp.

11. Hilton, A. R. (1993) Chalcogenide glasses for passive flir systems, *SPIE Vol.* **2018** *Passive Materials for Optical Elements II*, p. 46-57.

12. Seddon, A. B. (1995) Chalcogenide glasses: a review of their preparation, properties and applications, *J. Non-Cryst. Solids* **184**, 44-50.

13. Linke, D. (1977) Transformationstemperaturen und andere Eigenschaften von Chalkogenidgläsern und ihr Zusammenhang mit thermochemischen und strukturellen Parametern, in J. Staněk (ed.) *Proc. XIth Int. Congr. Glass, Prague*, vol. **I**, CTVS Prague, pp. 149-158.

14. Linke, D.(1978) Vzaimosvyazi sostav-svoistvo v khal'kogenidnuikh steklakh i ikh strukturno-khimicheskaya interpretaciya (russ.), in L. Štourač (ed.), *Proc. Conf. Amorphous Semiconductors '78 Pardubice*, Czechoslovak. Acad. Sci. Prague, pp. 57-64.

15. Feltz, A., Burckhardt, W., Voigt, B. and Linke, D. (1991) Optical glasses for IR transmittance, *J. Non-Cryst. Solids* **129**, 31-39.

16. Drexhage, M. G. and Moynihan, C. T. (1989) Infrarot-Lichtleiter, *Spektrum der Wissenschaft*, **1**, 104-110.

17. Ležal, D., Kasik, I., Petrovska, and M. Pospisilova (1992) Fibers for energy power delivery and chemical sensors, in A. Duran and J. M. F. Navarro (eds.) *Proc. XVIth Int. Congr. Glass, Madrid*, vol. **2**, pp. 309-313.

18. Simons, D. R., Faber, A. J. and de Waal, H. (1995) Pr-doped GeS_x-based glasses for fiber amplifiers around 1.3 μm, in H. A. Schaeffer (ed.) *Proc. 3rd Conf. Europ. Soc. Glass Science and Technol., Würzburg, May 22-24*, Verl. Dt. Glastechn. Ges. Frankfurt/Main, pp. 579-586.

19. Snitzer, E., Wie, K. (1995) Glass compositions having low energy phonon spectra and light sources fabricated therefrom, *U.S. Pat. 5,379,149*

20. Linke, D. and Wustmann, G. (1980), Opticheskii koeffitsient napryazheniya i drugie opticheskie svoistva khal'kogenidnuikh stekol, neprozrachnuikh v vidimoi oblasti spektra, in A. M. Andriesh, E. B. Ivkin, S. D. Shutov (eds.) *Physical phenomena in non-crystalline semiconductors, Amorphous semiconductors '80*, Kishinev, pp. 185-188.

21. Linke, D. (1986) Struktureinflüsse auf das spannungsoptische Verhalten von infrarotdurchlässigen Chalkogenidgläsern, *Z. anorg. allg. Chem.* **540/541**, 142-162.

22. Linke, D. and Eberhardt, G. (1979) Spannungsoptische Eigenschaften von Chalkogenidgläsern und Möglichkeiten ihrer strukturchemischen Interpretation, *Wiss. Z. Friedrich-Schiller-Univ. Jena, Math. Naturwiss. Reihe* **28**, 339-350.

23. Brückner, R. (1991) Mechanical properties of glasses, in [3], pp. 665-713.

24. Hoffmann, H.-J. (1995) Differential changes of the refractive index. Acousto-optical properties of glasses, in [1], pp. 96-129.

25. Churbanov, M. F., Shiryaev, V. S. (1994) Kristallizaciya khal'kogenidnuikh stekol, *Vuisokochistuie veshchestva* **4**, 21-33.

26. Savage, J. A., Webber, P. J., and Pitt, A. N. (1977) *Applied Optics* **16**, 2983 ff.

27. Voigt, B. and Dresler, G. (1981) Bestimmung und Abtrennung von Sauerstoffverunreinigungen in Reinst-Selen, *Anal. Chim. Acta* **127**, 87-92.

28. Voigt, B. and Wolf, M. (1982) Optical properties of vitreous GeS_2, *J. Non-Cryst. Solids* **51**, 317-322.

29. Voigt, B. and Dresler, G. (1983)) Microheterogeneities in infrared-optical selenide glasses, *J. Non-Cryst. Solids* **58**, 41-45.

30. Churbanov, M. F. (1995) High-purity chalcogenide glasses as materials for fiber optics, *J. Non-Cryst. Solids* **184**, 25-29.

31. Feltz, A., Voigt, B., Senf, L. and G. Dresler (1979) Über neue infrarotdurchlässige optische Gläser, *Wiss. Z. Friedrich-Schiller-Univ. Jena, Math. Naturwiss. Reihe* **28**, 327-338.

32. Vaško, A. (1968) *Infra-red radiation*, Iliffe Books Ltd. London, SNTL Publ. Prague, 445 pp.

33. Linke, D. and Eberhardt, G. and Dietz, G. (1975) Fotouprugie postoyannuie khal'kogenidnuikh stekol, neprozrachnuikh v vidimoi oblasti spektra (russ.), in B. T. Kolomiets (ed.) *Structure and properties of non-crystalline semiconductors, Proc. 6th Int. Conf. Amorph. Liquid Semiconductors Leningrad USSR, Nov 18-24, 1975*, pp. 251-255.

34. Murphey, W. J. (1991) New Arsenic-triselenide information, *The Bulletin of Selenium-Tellurium Development Ass.*, June, pp. 1-3.

35. Marquardt, H., and Schäfer, S. G. (1994) *Lehrbuch der Toxikologie*, Wissenschaftsverlag Mannheim.

36. Kircher, U. (1992) Emissions of selenium compounds of glass melting furnaces and state of the art of selenium waste gas treatment, in A. Duran and J. M. F. Navarro (eds.) *Proc. XVIth Int. Congr. Glass, Madrid*, vol. **6**, pp. 265-270.

37. Tosell, D. A., Costello, M. C. and Brierley, C. J. (1992) Diamond layers for the protection of infrared windows, *SPIE Vol.* **1760**, *Window and Dome Technologies and Materials III*, 268-278.

NEW CHALCOHALIDE GLASSES AS PROMISING MATERIALS FOR OPTOELECTRONICS

L.A. BAIDAKOV
St.-Petersburg Technical University
29 Politechnicheskaya str.,
195 251, St.-Petersburg
Russia

The optoelectronics and fiber optics demands stimulate search, development and preparation of new chemically stable materials, transparent in IR-area with high refractive indices. According to the general reasons, particularly, taking into account the chemical nature of elements, causing their position in the Periodical system one may hope that such glasslike systems should include heavy metals, chalcogens and halogens. However the search for new glass-forming chemical combinations and especially binary, ternary and more complex systems by trial-and-errors method in the majority of cases appear rather difficult or altogether hopeless. In the last few years the author stated an idea about the relation of glass-forming ability of melts to the quantum characteristics of free atoms and of the nature of their chemical interaction [1]. Accoding to [1] the glass-forming ability of the melt of any composition (G_t) is a function of rigidity of an electronic skeleton of the chemical bonds (RESCB), the main quantum (n) and the atomic number (Z) per averaged atom in a complex substance $A_xB_yC_z$..., for which A, B, C, ... are the components of the substance, and x, y, z, ... are the indices at the corresponding chemical element. The quantitative relation of these characteristics has the following form:

$$G_t = nK_s/ Z \qquad (1)$$

A. Andriesh and M. Bertolotti (eds.),
Physics and Applications of Non-Crystalline Semiconductors in Optoelectronics, 171–179.

Here K_s is the total RESCB, including covalent (K_c), ion (K_i),donor-acceptor (K_{d-a}), intermolecular (K_{im}) interactions, as well as the contribution of the metallization of chemical bonds (K_m) and the contribution of hydrogen bonds (K_h), if those are available in the given substance. Each of these contributions can be calculated making use of the following relationships:

$$K_c = \Sigma (U_c N_c)_i \quad (2)$$

$$K_i = \Sigma(x/y)_i (N_c N_a)_i (2Z/M)_i \quad (3)$$

$$K_{d-a} = \Sigma(2Z/M)_i (N_d \cdot N_{ac})_i (2/n_i) (n_{ac} / n_d)_i (N_p)_i \quad (4)$$

$$K_{im} = \Sigma(x/y)_i (N_c N_a)_i (M/2Z - 1)_i (S_v)_i \quad (5)$$

$$K_m = \Sigma(x/y)_i (N_c N_a)_i (M/2Z - 1)_i \quad (6)$$

$$K_h = \Sigma(Z/M)_i (N_h)_i \quad (7)$$

In the formulas (2-7) U_{ci} is the number of the valence electrons of the ith cation-former, ensuring its chemical interaction with the environment, in the majority of cases equal to the module of its oxidation degree; N_c, N_a, N_d, N_{ac}, N_h are the molar fractions of the ith cation and anionformers, donors, acceptors and atomic species of hydrogen; the x/y-is the ratio of indices of the ith cation and anionformers; Z and M is the atomic number and molar mass of the ith components; n is the main quantum number of valence electrons of the ith donor, n_{ac} and n_d are the main quantum numbers of valence electrons of the ith acceptors and donors, N_p is the number of electronic pairs of the ith donor, participating in the formation of the donor-acceptor bonds and S_v is the number of the unpared valence electrons of the ith component in the ground and exited states which are involved in linking the atomic species together. The value of the contributions of the main quantum numbers of components (n_i) and the contributions of the atomic numbers (Z_i) can be calculated by additive expressions (8) and (9):

$$n = \Sigma n_i N_i \quad (8)$$

$$Z = \Sigma Z_i N_i \quad (9)$$

Where N_i is a molar fraction of ith component. Using relationships (1-9) the calculation of the glass forming ability of concrete melts was performed, which allowed us to predict the glass forming regions in several pseudo-binary systems on the basis of heavy elements.

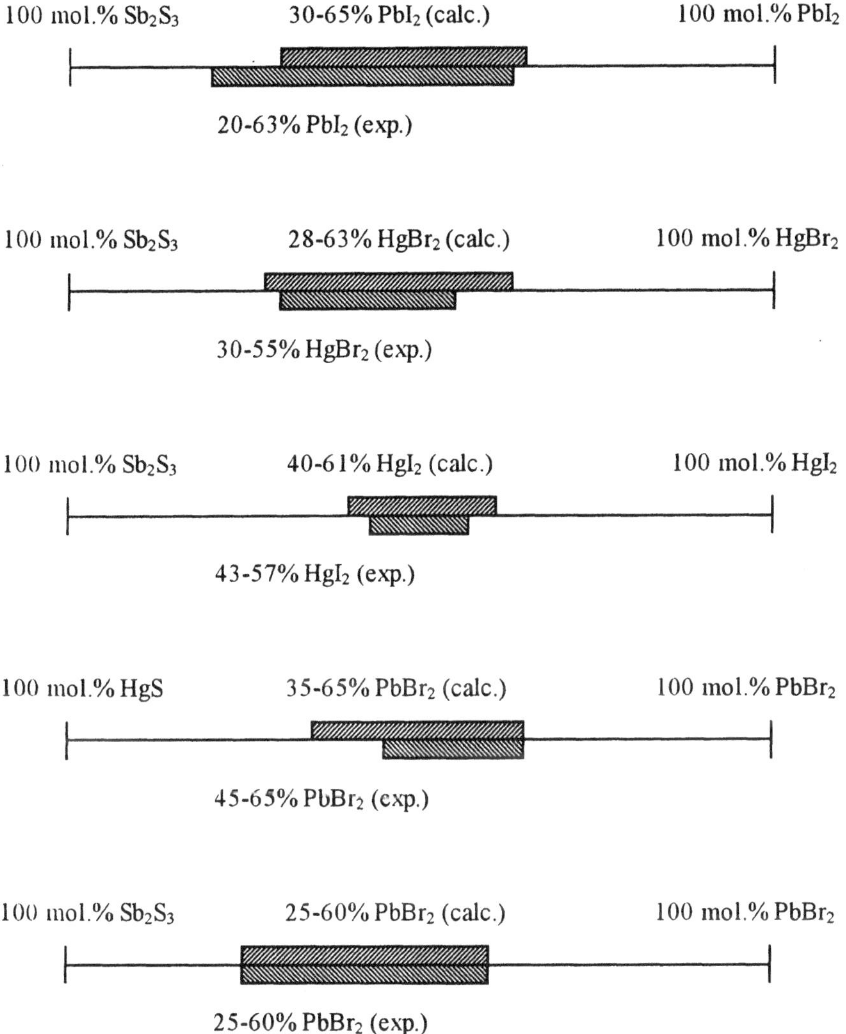

Fig.1. Calculated and experimental glass formation regions in quaternary (pseudobinary) systems: $(Sb_2S_3)_{1-x}(MeHal_2)$, where Me=Pb,Hg, Hal=Br, I, and in the system $(HgS)_{1-x}(PbBr_2)_x$.

Fig. 1 presents theoretically calculated and experimentally determined glass forming regions in the systems: $(Sb_2S_3)_{1-x}$ $(PbI_2)_x$, $(Sb_2S_3)_{1-x}$ $(HgBr_2)_x$, $(Sb_2S_3)_{1-x}$ $(HgI_2)_x$, $(HgS)_{1-x}$ $(PbBr_2)_x$ and $(Sb_2S_3)_{1-x}$ $(PbBr_2)_x$. From the figure it follows, that the calculated and experimentally determined glass forming regions are in good agreement with each other. The glasses were synthesized from previously synthesized chalcogenides and halogenides or from binary compounds of industrial manufacture of the mark "chemical pure" or "pure for analyses" in vacuum quartz ampoules in the range of temperatures, exceeding melting temperatures (T_m) of individual compounds or simple bodies by 100-200 K. The melt having the mass 2-3 g. was cooled in the air or in water with ice. The vitreous state was checked by differential thermal, X-ray diffraction analyses, optical methods and also visually. The properties of the synthesized glasses are shown in the table.

TABLE. The properties of the chalcohalide glasses

Systems	x	$d \cdot 10^3$, kg/m^3	T_g, K	T_{cr}, K	H, kg/mm^2	$-\lg\sigma_{293}$ S/cm	$\lg\sigma_0$, S/cm	E_a, eV	n, $\lambda=1\mu$
$(Sb_2S_3)_{1-x}$	0.3	4.86	-	-	-	-	-	-	2.39
$(HgBr_2)_x$	0.4	4.67	412	501	-	-	-	-	2.37
	0.5	5.01	-	-	-	-	-	-	2.36
$(Sb_2S_3)_{1-x}$	0.45	5.08	-	-	-	-	-	-	-
$(HgI_2)_x$	0.5	5.16	393	481	-	-	-	-	2.40
	0.55	5.22	-	-	-	-	-	-	-
$(Sb_2S_3)_{1-x}$	0.2	4.65	-	-	-	-	-	-	2.66
$(PbI_2)_x$	0.3	4.70	374	-	-	-	-	-	2.80
	0.4	4.85	380	-	-	-	-	-	3.10
	0.5	5.01	380	-	-	-	-	-	3.16
	0.6	5.26	386	-	-	-	-	-	3.21
$(HgS)_{1-x}$	0.5	6.69	357	-	145	9.6	1.1	0.66	-
$(PbBr_2)_x$	0.55	6.65	-	-	120	9.0	1.5	0.64	-
	0.60	6.60	346	-	115	8.6	1.7	0.58	-
	0.65	6.49	-	-	105	8.2	1.6	0.61	-

The analysis of properties of glasses shows that their density is much higher than classical chalcogenide ones and is in an interval 4.65-6.69 g/cm^3. These are rather low melting glasses. Their glass transition temperature (T_g) does not exceed 140^0 C, and glasses of system HgS-$PbBr_2$ have Tg = 70-80^0 C. It apparently will promote simplification of production technology of optical fibers for IR-region of the spectrum. Their microhardness is also rather low and changes from 105 up to 145 kG/mm^2. All glasses are semiconductors. The temperature dependence of their electroconductivity (σ) is described by the equation: $\sigma = \sigma_o \exp(-E_a/kT)$. At the increase of the contents of halides of heavy metals their conductivity slightly grows with the simultaneous increase of the frequency factor ($\log\sigma_0$) and decrease of activation energy (E_a). The rather low values of activation energy and frequency factor are usually characteristic of chalcogenide glasses with significant part of ionic conductivity. However this question requires detailed study. Fig. 2 shows temperature dependence of conductivity for glasses of system HgS-$PbBr_2$. All of them are submitted to the above law.

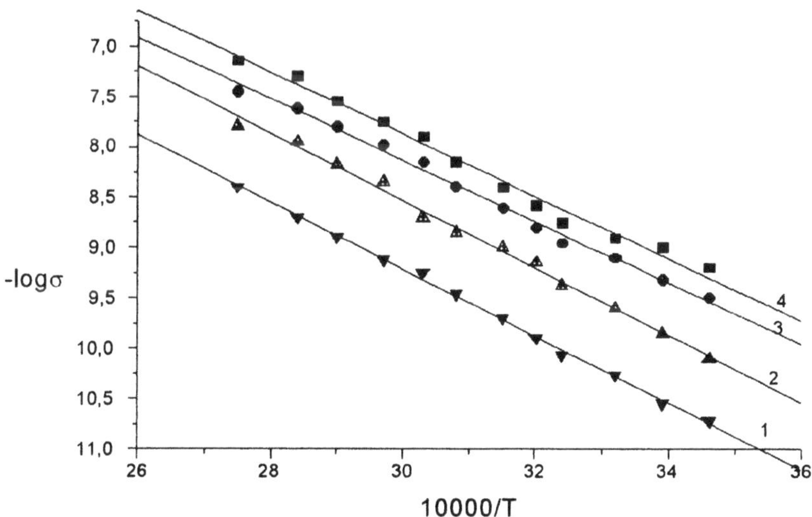

Fig. 2. Temperature dependence of electroconductivity of glasses of system HgS-$PbBr_2$.

The investigated glasses are transparent in IR-area of the spectrum. Below (see fig.3 a, b, c located under the order) IR-spectra of absorption for glasses of systems HgS-$PbBr_2$, Sb_2S_3-$HgBr_2$, Sb_2S_3-HgI_2 and Sb_2S_3-PbI_2 are resulted.

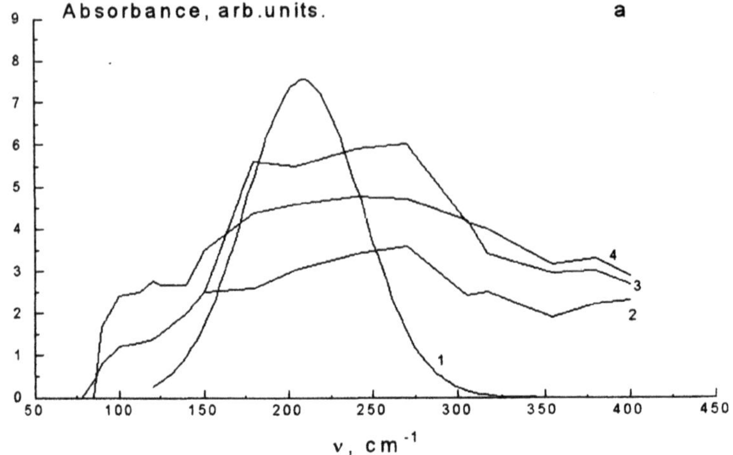
Absorbance, arb.units.
a
1
2
3
4
ν, cm⁻¹

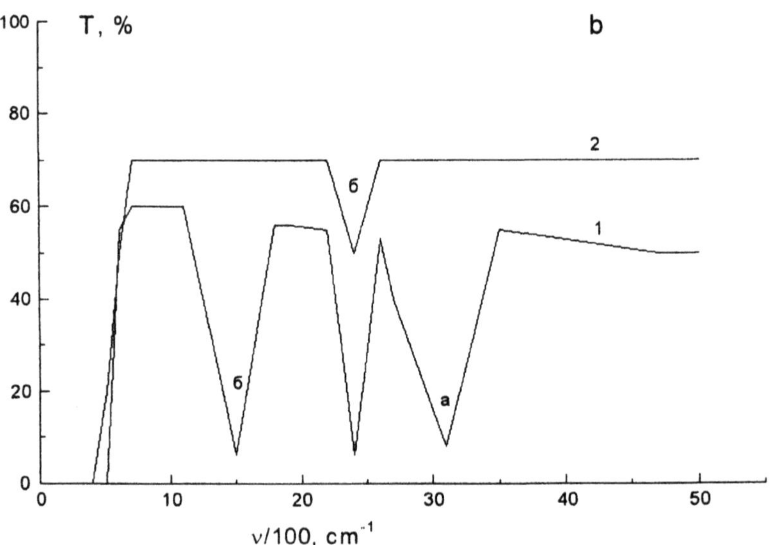
T, %
b
2
б
1
б
a
ν/100, cm⁻¹

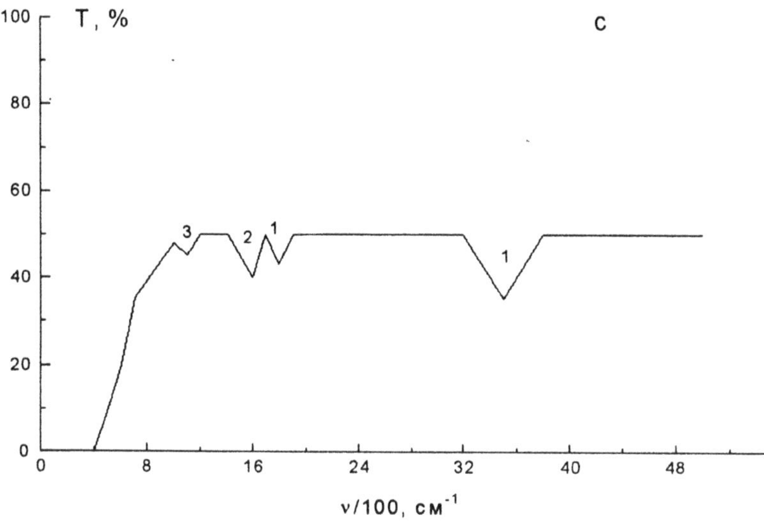

Fig. 3. Admixture and own spectra of absorption of the pseudo-binary chalcohalide glasslike systems:
a) System HgS-$PbBr_2$: 1. $PbBr_2$ (cryst.), 2. 0.47 $PbBr_2$ (glass), 3. 0.50$PbBr_2$ (glass), 4. 0.45 $PbBr_2$ (glass).
b) Glasses of the systems: 1.Sb_2S_3-$HgBr_2$ 2.Sb_2S_3 - HgI_2.A-peak O-H, b-peaks S-H.
c) Glasses of the system Sb_2S_3-PbI_2: 1- peaks O-H, 2- peak S-H,3- peak S-O.

For glasses of the system $(HgS)_{1-x}(PbBr_2)_x$ (fig.3a) in the investigated range of compositions 0.45< x< 0.50 and wave numbers 60-400 cm^{-1} rather complex and showing little information spectra of absorption were observed. For their interpretation it is necessary to turn to the spectra of appropriate crystals. In the spectrum of the crystal HgS on [2] absorption was observed at frequencies 345 and 295 cm-1. Accoding to [3] for these crystals the following bands of absorption are characteristic: 345, 280, 150 and 85 cm^{-1}. In the spectrum of the crystal $PbBr_2$ the authors [3] have fixed a band at 60 cm^{-1}. As against [3] we observed a band at frequency 200 cm^{-1}. From the analysis of spectra of glasses it follows that complete accordance IR-spectra between crystals HgS and $PbBr_2$ on the one hand, and glasses on their basis, on the other - does not exist. Nevertheless, the weak absorption in glasses in the range of frequencies 340-350 cm^{-1} is observed. Stronger, though washed out maxima of absorption in glasses are in the range of frequencies 295-85 cm^{-1}, characteristic of crystal sulfide of mercury and bromide of lead. All this allows us to assume, that motif of the structure of glasses of this system is the short range order, characteristic of crystal HgS and $PbBr_2$. It is not excluded also, that as a result of components interaction inthe melt the equilibrium is established: $HgS + PbBr_2 \leftrightarrow HgBr_2 + PbS$, that is fixed at hardening of the melt. It is possible also, that in these glasses a certain role is plaing by donor-acceptor interaction of the type > S:→ Pb < and -Br:→ Hg <, whith causes the formation of more complex structures for example ternary ones. In any case the maximum of a band absorption of glass composition 0.53HgS-0.47$PbBr_2$ coincides with a band, characteristic of $HgBr_2$ 260 cm^{-1} [3]. Such complex structure-chemical composition of glasses can also give an essentially widening and complicated an experimental spectrum. For glasses of the systems Sb_2S_3-$HgBr_2$ and Sb_2S_3-HgI_2 IR-spectra were investigated in the field of 2-25 μ. From fig. 3b it is evident, that for brominsulphide system admixture bands are characteristic, caused by vibrations of O-H (3100 cm^{-1}) and S-H (2500 and 1500 cm^{-1}) bonds [3-5]. Taking into account the character of the spectrum we can confidently speak only about the presence in glasses of significant amounts of H_2O and H_2S. As against brominesulphide-the iodinsulphide system does not practically contain bands of absorption in this area of the spectrum. So for the glass of composition 0.5Sb_2S_3- 0.5HgI_2 (see fig. 3b) one rather weak band 2500 cm^{-1} is observed only. Iodinsulphide glasses are transparent in the range of lengths of waves 2-14 μ and have 70 % transmission for patterns 2 mm thick. A similar IR-transmission spectrum in the field of 2-25 μ have also glasses of the system Sb_2S_3-PbI_2 (see fig.3c). They are also transparent in the range of lengths of waves 2-12 μ having only very weak admixture bands in the range of 2.9-9.1 μ (3450 and 1600 cm^{-1}, 1400 cm^{-1} and 1100 cm^{-1}), relating to O-H, S-H and S-O bonds accordingly [3-5]. Transmission for these glasses makes 60 %.

From observable temperatures of crystallisation one of them (500^{0}C) is close to Sb_2S_3 (550^{0}C), and a second more than likely concerns ternary structures, arising owing to donor-acceptor interactions between the components of the glass.

The refractive indices of the investigated glasslike systems change in rather wide intervals from 2.36 up to 3.21. As it was supposed, especially the high refractive indices have glasses of the system Sb_2S_3-PbI_2. Some of them exceed the maximum refractive indicses, established in [6] for chalcogenide glasses (n=3.07).

The research carried out shows that the offered method of calculation of the glass forming ability of substances any composition is rather effective and makes it possible realise computer search for new glass forming systems with properties necessary for practice. Obtained experimentally on the basis of precomputations chalcohalide glasses are stable in respect to crystallisation and also chemical steady both in the air and in water. Optimum physical-chemical and physical properties of glasslike chalcohalide systems allow us to hope, that the glasses on the basis of heavy metals, chalcogens and halogens can be successfully used for needs of optoelectronics and fiber optics in IR-region.

References

1. Baidakov, L.A. (1994) Glass-Forming Ability: A quantitative Criterion with Allowance for the Nature of Chemical Bonding, *Glass Physics and Chemistry* **20**, 232-236.

2. Jurtshenko E.N., Kustova G.N., Batsanov S.S. (1981) *Oscillation spectra of inorganic compounds,* Nauka, Novosibirsk.

3. Harfmann E., Videau J.J., Portjer J. And Tanguy B. (1992) IR-transmitting mercury and lead thiohalide glasses: effect of halide addition on the physical and structural properties, *J.de Phys. IV.Coll. C2, suppl. Au Journ. De Phys. III*, **2**, 2-239.

4. Devjatych G.G., Ulevatyi B.E., Tshurbanov M.F.,Dianov E.M, Plotnischenko V.G. (1984) Optical properties films chalcogenide glasses, deposited from unequilibrium plasma, in K. Petkov (ed.), *Proc. of the conf. Amorphous semiconductors-84,* Academy of Sciences, Gabrovo- Bulgaria, **2,** pp. 153-155.

5. Baidakov L.A., Blinov L.N. Last achievements in the field of research halcohalide of semiconductor glasses, *ibid.,* pp. 6 - 9.

6. Belych A.V., Darvoid T.V., Kanschiev Z. I., Кокоrina V.F., Miljukov V.M. (1996) Chalcogenide glass. *Author's certificate 1231814, USSR, C 08 C3/32.*

CONTRIBUTIONS TO SELF-DIFFRACTION AND OPTICAL PHASE CONJUGATION IN SEMICONDUCTORS

V. I. VLAD and A. PETRIS

Institute of Laser and Plasma Physics. Dept. of Lasers.
P.O.Box MG-36,R-76900 Bucharest, Romania

Abstract. This paper describes some of our contributions in the nonlinear optics in semiconductors: the accurate description of the light self-diffraction on the highly modulated dynamic gratings (particularly, in c-Si); the comparison of the third-order susceptibility values of semiconductors obtained by measurements in self-diffraction and phase conjugation with those produced by the theoretical models; the self-diffraction on anharmonic phase gratings generated in photorefractive materials (particularly, in sillenites) and an all-optical method for the measurement of the electro-optical coefficient using the laser light self-diffraction in the photorefractive materials; the holographic interferometry based on phase conjugation in sillenites.

1. Introduction

The dynamic gratings and the optical phase conjugation in semiconductors were strongly investigated in the past years due to their potential in the optical measurement of the microscopic material parameters, the optical processing of information and interconnects and many other modern fields [1]. The application of the dynamic gratings depends essentially on the correct description of the light diffraction in different conditions, particularly for phase gratings induced at high light intensities, when the phase modulation depth becomes large. The usual semiconductors are very interesting optical nonlinear materials with large local nonlinearities, short write/erase times and strong integration potential. The new theoretical results in the field can be easily checked on the dynamic gratings induced in semiconductor materials. This paper presents some recent contributions of the authors in the self-diffraction and optical phase conjugation in semiconductors.

The diffraction and self-diffraction on the phase gratings induced by laser light in the semiconductor materials was described by a classical model of Woerdman [2] and Eichler et al [3,4,5], in which the intensities of the diffraction orders, in the Raman-Nath regime, are proportional to the combinations of the squares of the Bessel functions with the argument provided by the nonlinear induced phase. The theoretical estimations of this simple model are not in a good agreement with the experimental results, at high (laser intensities) modulations. In order to improve the theoretical model, we have considered similar experimental observations done in the linear diffraction of light on ultrasounds [6,7] and we have solved numerically the corresponding coupled wave equations taking into account the nonlinear induced phase modulation. Moreover, in the calculations, we have introduced also the effects of the (Gaussian) laser beam profile and of the nonuniform modulation of the

A. Andriesh and M. Bertolotti (eds.),
Physics and Applications of Non-Crystalline Semiconductors in Optoelectronics, 181–200.

dynamic gratings. Our result is fitting better many previous experimental data and our experimental data obtained on c-Si with a pulsed YAG:Nd laser (up to fourth diffraction order and modulation indices of 10) [8]. The same experiments allowed us to measure more accurately the third-order nonlinear susceptibility (nonlinear refractive index, n_2), the carrier diffusion and thermal diffusion coefficients of the available c-Si samples [9].

The comparison of the third-order susceptibility values of semiconductors obtained by measurements in self-diffraction and phase conjugation with those produced by the theoretical estimations was interesting and useful. In the case of c-Si (a good test material), our results show a good agreement of all three values (4.10^{-8} esu) [9]; the small discrepancies appear due to the simple consideration of the pure phase gratings, without the effects of the (intensity dependent) absorption gratings and the actual spatial and temporal profiles of the laser beams. Despite the fact that the self-diffraction (2WM) and phase conjugation (4WM) methods are using interferometric setups, the lack of mechanical movement of optical components and the rapid and precise measurements are good arguments when they are compared to the more recently developed single beam method, the z-scan [10].

A delicate problem in the literature, the self-diffraction in photorefractive crystals on the highly modulated gratings (such as used in holography), is solved in the second section. The equation system describing the photorefractive effects in the Stepanov & Petrov model [11] can be simply and analytically solved only for small contrasts of the induced diffraction grating. Moharam and other authors [12-15] have observed that, in the case of the high contrasts and stationary conditions (of photo-ionisation, diffusion, drift and recombination), the space charge (holographic) grating developed in the crystal becomes anharmonically. Following this investigation path, we have shown that this behaviour, correlated with a given number of donor centres, imposes the limitation of the grating contrast to a well determinate maximum value (given by the donor total depletion). Furthermore, we have considered the light diffraction on different harmonic (Fourier) components of the anharmonic dynamic grating [16,17] and the incoherent summing-up rule of the effects, which was introduced by Woerdman [2] and demonstrated more rigorously by Enns & Rangnekar [18], in the Raman-Nath regime. In this regime, the result is new and interesting: the self-diffraction spectrum is asymmetrical and all diffraction orders tend to a maximum as the contrast tends to its maximum (one). In the Bragg regime, the result is plausible: the week beam is amplified by the energy transfer from the strong beam and this amplification decreases with the contrast growth to one (beam intensity equalisation). The new results were checked experimentally by the measurements of the intensity spectrum in the self-diffraction of the He-Ne laser beams on a $Bi_{12}TiO_{20}$ (BTO) crystal. Not only a good agreement with the experimental data was obtained, but a new all-optical method for the measurement of the electro-optical coefficient of the photorefractive crystal was demonstrated. For BTO, we found out $r = 5.72$ pm/V. [17], a close value to that obtained by Petrov & Stepanov [11] and Bayvel et al [19] by electro-optical methods (with electrodes on the crystals).

Real-time nondestructive optical testing using holographic interferometry (H.I.) was strongly stimulated by the advancement of the physical knowledge and the technology of electro-optic photorefractive crystals (PRC) as recording materials [20]. PRC are known to be reusable holographic materials that can be infinitely recycled and do not require additional processing (development, fixing, etc.). The erasure can be achieved optically, by uniform illumination of the crystal.

Double-exposure and time-average holographic interferometry in $Bi_{12}SiO_{20}$ (BSO) and $Bi_{12}GeO_{20}$ (BGO) crystals were successfully demonstrated by Huignard et al.[21,22] in four-wave mixing (4WM) configurations. Then, different types of two-wave mixing (2WM) (energy and polarisation self-diffraction) in PRC of the sillenite type were used in holographic interferometry by Kamshilin et al. [23,24].

This paper presents further developments in the optical testing by dynamic (real-time) holographic interferometry with PRC (mainly BTO crystals) and computer image processing. The analysis of the optical configuration of the holographic interferometer, the quantitative interpretation of the fringe patterns and the algorithms for fast and precise image processing were done. We have found that a 4WM configuration, with reduced number of optical components, in which the image holographic interferometry was implemented such as to allow the testing of large transparent objects, was optimum with respect to image quality. A very convenient matching was observed between the sensitivity of BTO crystals and the energy density yielded by the usual (low-cost) He-Ne lasers. The lifetime of holographic interferograms in PRC with continuous readout and captured by the TV camera was compatible with the acquisition time of our image digitizer controlled by the minicomputer [25]. Simple quantitative interpretation and the computer algorithms were elaborated for prismatic and lenticular objects. Some direct spatial phase reconstruction methods (in space and spatial frequency domains) were equally used [26,27].

In the optical phase conjugation, two parameters are very important: the reflectivity for the conjugated beam and the fidelity of the conjugated wavefront (mainly for the adaptive optics). We have introduced two new methods for the fidelity measurement: the synthetic aperture generated by the computer-controlled image acquisition system in the divergence measurement [28] and the direct spatial phase reconstruction in the interferometric evaluation of the conjugated wavefront (leading to Seidel and Zernike aberration coefficients)[29].

2. Accurate description of the light self-diffraction on highly modulated dynamic gratings in semiconductor materials

The interference of two laser beams in a semiconductor material yields a dynamic index (phase) grating with the spatial distribution:

$$n(x) = n_0 + \Delta n \cdot \cos Kx, \qquad (1)$$

where $\Delta n = n_2 I$, n_2 is the nonlinear refractive coefficient, I - the incident light intensity, $K = 2\pi/\Lambda$ - the spatial frequency of the grating and Λ- the grating period. For small angle between the beams, β , one can write: $\Lambda \approx \lambda/\beta$, with λ - the light wavelength.

The classical diffraction theory describes two regimes depending on the degree of the phase correlation of the diffraction orders [6,30,31]:

$$Q = 2\pi\lambda d/\left(n_0\Lambda^2\right)$$

and the depth of modulation of the phase grating: $\Delta\Phi = 2\pi\Delta nd/\lambda$
In the Raman-Nath regime, $Q \leq 0,5$ and $Q \cdot \Delta\Phi \leq 1$
and in the Bragg regime, one can write: $Q \geq 10$ and $Q/\Delta\Phi \geq 10$.
For large $\Delta\Phi$ (and Q). the Raman-Nath description of linear diffraction fails (as was found in the light diffraction on ultrasounds) and it is necessary to reconsider the theory starting from the Maxwell equations under the form of a set of coupled differential equations [6,7]:

$$2U_l'(\varphi) + U_{l+1}(\varphi) - U_{l-1}(\varphi) = i\rho l^2 U_l(\varphi). \tag{2}$$

where U_l is the amplitude of the beam diffracted in the l-th diffraction order. $\varphi = 2\pi\Delta nz/\lambda$ - light phase modulation and $\rho = Q/\Delta\Phi$.

In the self-diffraction on nonlinear materials, the Raman-Nath regime with equal intensity interfering beams, the intensity in the l-th self-diffraction order can be calculated by [1,3]:

$$I_l = T \cdot I_{inc} \cdot \left[J_l^2(\Delta\Phi) + J_{l+1}^2(\Delta\Phi)\right]. \tag{3}$$

where T is the transmission of the nonlinear material, I - the intensity of the incident beams and J_l - the Bessel function of the l-th order. In many papers, this formula is still invoked even for high incident intensities, at high phase modulation, despite the fact that its discrepancies from the experimental data become very large.

We have described accurately the self-diffraction, in the case of the highly modulated gratings (large $\Delta\Phi$) induced in nonlinear materials, by the formalism of the coupled wave equations (2), used previously for the gratings produced by linear phenomena. Moreover, we have extended this formalism to include the absorption in the materials and the nonuniformity of the interfering laser beams. This description is fitting better our experimental data obtained in self-diffraction on the phase gratings induced in crystalline silicon wafers by Nd:YAG laser beams.

The infinite set of coupled wave equations (2) can be reasonable truncated considering that all diffraction orders with the index l > N can be neglected. One can introduce [7]

$$N = 2/\sqrt{\rho} \tag{4}$$

For the usual optical nonlinearities of crystalline silicon and intensity range yielded by a Nd:YAG laser oscillator with nanosecond impulses, one can take N = 6 - 8. This truncated set of differential equations was numerically solved by the Runge-Kutta method and the results are graphically shown in Fig. 1 for the first four self-diffraction orders (the curves marked with the letter B). For comparison, on the same graphs, we have plotted also the analytical solutions given by Raman-Nath equation (3) (the curves marked with the letter A). One can remark the increasing difference between these two solutions with the increase of the phase modulation. For example, at $\Delta\Phi$ = 4, usually reached in such experiments, the differences between the two theoretical estimations are: in the first order ~15%; in the second order ~5%; in the third order ~6%; in the fourth order ~35%.

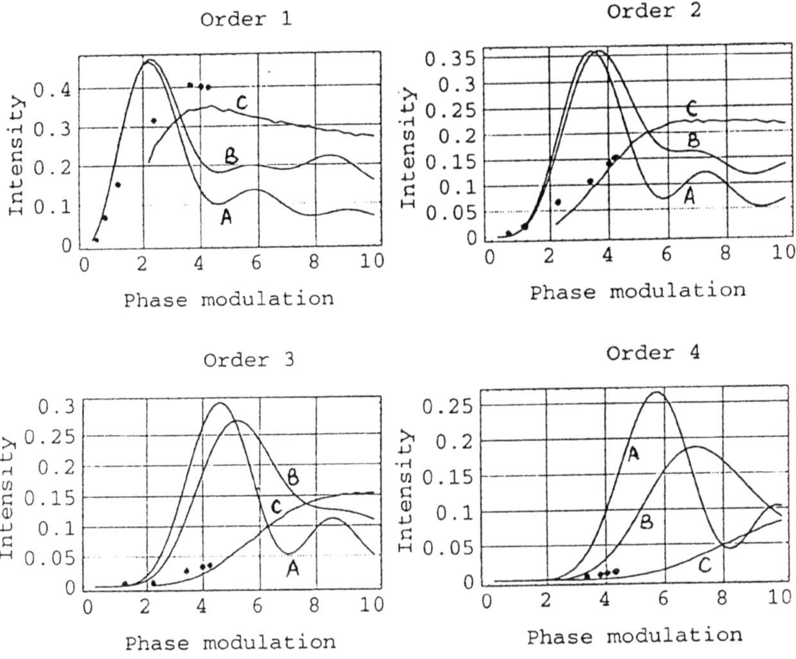

Fig.1. The theoretical (curves A - the analytical solutions given by Raman-Nath formalism; curves B - the numerical solution of the classical coupled wave equations; curves C - the numerical solution of the coupled wave equations considering the constant absorption and the gaussian profiles of the beams) and the experimental data (the dots), for a c-silicon sample with the thickness d = 0.25 mm.

In the classical theory, the assumptions of the negligible absorption α=0) and the diffraction of the plane waves are used. In many cases, including that of our experiments, the optical absorption is not small (and moreover, it is dependent on light intensity, the light losses are increasing and φ is changing nonlinearly with the propagation distance, z, which complicates seriously the computation; this dependency is no further considered). Introducing a constant absorption, the variation of the modulation index with z is:

$$\Delta n = n_2 \cdot I(z) = n_2 I_0 e^{-\alpha z} \approx n_2 I_0 (1 - \alpha z), \quad \alpha z << 1 \tag{5}$$

and we can define the integral phase modulation at certain distance inside the nonlinear material as:

$$\chi(z) = \int_0^z \frac{2\pi}{\lambda} n_1(z) dz. \tag{6}$$

With this new variable, the set of differential equations (3) becomes.

$$2U_l'(\chi) + U_{l+1}(\chi) - U_{l-1}(\chi) = il^2 \frac{Q}{\Delta\Phi} \cdot \left[1 - \frac{2\alpha d}{\Delta\Phi}\chi\right]^{-1/2} U_l(\chi) \tag{7}$$

A further step in considering the actual situation of the self-diffraction experiments consists of performing the calculations for nonuniform laser beams. For gaussian profiles of intensity for the interfering laser beams described by:

$$I(x) = I_0 \exp\left(-x^2/w^2\right), \tag{8}$$

the spatial distribution of the refractive index modulation will follow a similar nonuniform shape:

$$\Delta n(x) = n_1 \exp\left(-x^2/w^2\right), \tag{9}$$

where n_1 is the refractive index modulation on the optical axis ($x = 0$).

In this situation, the gaussian profile is commonly approximated with a suitable distribution of zones with uniform profiles. This approximation can be done for both the gaussian beam profile and the gaussian grating profile. In the case of symmetrical interference and self-diffraction at small angles, one can consider that each uniform zone of the beams is self-diffracted on the similar uniform zone of the dynamic grating. The intensity of the l-th self-diffraction order is obtained by summing up the contributions from all grating zones. In the Fig. 1, we have plotted the numerical solution of the coupled wave equations considering the absorption and the gaussian profile of the beams and grating (the curves marked C, with $\alpha = 10$ cm^{-1} and $w = 1.4$ mm). The differences between the theoretical estimations, the curves A and C, are significant and the validity of the new formalism with absorption and nonuniform beam profiles can be checked in the following experiment.

The experimental setup is shown in the Fig. 2. The beam of a Q-switched Nd:YAG laser ($\lambda = 1.06$ $\tau = 10$ ns) is divided in two equal intensity beams by the beam-splitter BS2. These beams are brought in a parallel position and then, superimposed on the Si sample (wafer) using the prism P, mirror M2 and lens L (F=1.2 m). The angle between the interfering beams was $\beta = 2^o10' = 28$ mrad, their energy as $W_A = W_B = 0.54$ mJ and the corresponding fluence, $I_A = I_B = 0.87$ MW/cm^2. The self-diffracted orders are measured with the photo-cell PC connected to the oscilloscope, OSC (Tektronix TDS 350).

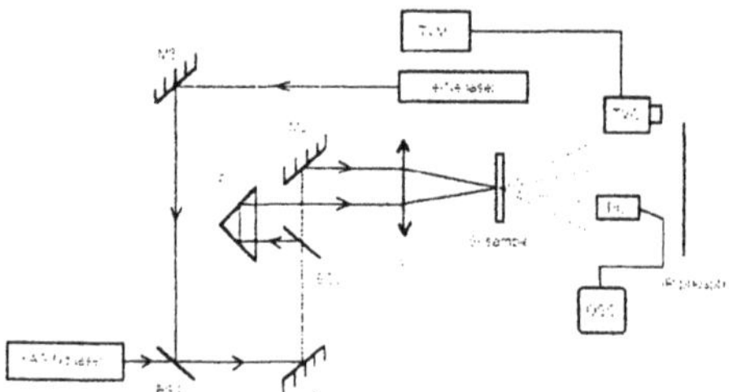

Fig. 2. The experimental setup for the measurement of the intensity spectrum in the light self-diffraction on highly modulated dynamic gratings

The nonlinear semiconductor material used in our experiments was the crystalline, nearly intrinsic silicon (n_0=3.56; α= 10 cm^{-1}; $\chi^{(3)}$ = 410^{-8} esu), in the form of plane-parallel plates with different thicknesses shown in the Table 1. The corresponding Q-parameter and the maximum diffraction order, for which the Raman-Nath regime applies, according the [7]. k_{max}, are calculated in the same table.

Table 1.

d (mm)	Q	k_{max}
0.25	0.63	3
0.42	1	2
0.87	2.12	1

The experimental results are shown in the Fig. 1 (the dots), for different diffraction orders obtained when the wafer with d = 0.25 mm was used. The agreement of these data with the theoretical predictions is much better for the curves C, the result of our numerical solution of the coupled wave equations taking into consideration the absorption and the nonuniformity of the laser beams. The agreement is better as the orders are higher.

Further corrections in order to obtain a better agreement of the experimental and theoretical data include: the dependence of the absorption coefficient on the light intensity (in the fluence profile used in our experiment, 5 - 30 cm^{-1}), the differences between the actual and gaussian profiles of the laser beams (less than 10%), the consideration of the nonuniform temporal profiles of the laser beams [4] and the influence of the thermal phase grating induced in the semiconductor material [5,9] (in our experiment, with a small effect and with a very large rise time in comparison to that of the carrier phase grating).

3. Third-order susceptibility of semiconductors obtained by measurements in self-diffraction and phase conjugation

The third order nonlinear susceptibility of the semiconductor materials can be conveniently determined by the measurements of the amplitude of the refractive index modulation in the self-diffraction (SD) spectrum (in the case of Raman-Nath regime) and the measurements of the phase conjugation (PC) reflectivity in a four wave mixing experiment. The same methods can be used also for the investigation of the electron and hole kinetics in order to calculate some transport and recombination parameters as the diffusion coefficient and recombination lifetime.

We have compared the above mentioned methods and the theoretical estimations in the determination of the third order nonlinear susceptibility of the c-Si.

The nonlinear process induced by the illumination of the c-Si sample with a YAG:Nd laser (h=1.164 eV; E_g=1.112 eV at T=295 K) consists of a free carrier plasma generation, which yields the nonlinear variation of the dielectric permitivity, ε, and implicitly, of the refractive index, n (with the laser light intensity). The change of the refractive index, Δn, is proportional to the electron-hole pair concentration, N (generated by the optical interband transition). This is, generally, a function of position and time, due to the structured illumination, diffusion and recombination.

In the case of two interfering light beams on the Si sample, the illumination varies sinusoidally across the sample and the carrier concentration, N, follows the same variation (at low light intensity).

If the illumination is done by a laser impulse, with triangular temporal shape and much shorter duration than the total carrier lifetime (due to diffusion and recombination), the effective third-order susceptibility, $\chi^{(3)}_{ef}$, (corresponding to the impulse temporal peak) can be calculated from Drude theory [30] :

$$\chi^{(3)}_{ef} = -\frac{\eta \alpha n_0 c e^2 \tau_L}{8 m^*_{eh} h \omega^3} \tag{10}$$

where: η is the quantum efficiency of the free carrier generation; α- the absorption coefficient; n_0 - the refraction index of c-Si, without this special illumination; c - the light velocity; e - the electron charge; τ_L - laser impulse duration; m^*_{eh} - the effective reduced mass of the electron-hole pair and ω - the circular frequency of the incident laser radiation. With the numerical values: $\eta = 1$, $\alpha = 10\ cm^{-1}$, $m^*_{eh} = 0.16\ m_e$, $n_0 = 3.56$, one can obtain $\chi^{(3)}_{ef} = 4.28 10^{-8}$ esu.

In the self-diffraction (two wave mixing), for the Raman-Nath regime, the light intensity in the l-th order of the diffraction spectrum can be calculated by eq. (3) or by our numerical procedure shown in the previous section. By measuring the light intensity in the orders of the diffraction spectrum, one can calculate the phase modulation index from (3) and then $\chi^{(3)}$ with the simple relation:

$$\Delta\Phi = \frac{2\pi d}{\lambda}\Delta n = \frac{2\pi d}{\lambda}\cdot\frac{\chi^{(3)}}{19 n_0^2}\cdot I. \tag{11}$$

where d is the sample thickness and I - the light intensity in the maxima of the interference pattern.

The experimental self-diffraction spectrum was obtained in a setup similar to that from Fig.1, with $I = 1.6\ MW/cm^2$. The silicon sample has d = 0.42 mm, p-type and resistivity ~100 cm. The phase modulation index was found $\Delta\Phi = 0.75$ leading to: $\Delta n = 3 10^{-4}$, $n_2 = 1.9\ 10^{-10}\ cm^2/W$ and $\chi^{(3)}_{ef} = 4.6\ 10^{-8}$ esu, in a very good agreement with the previous theoretical estimation.

In the phase conjugation by four wave mixing, three incident waves E_1, E_2, E_p interact in the sample and generate a forth wave E_c, conjugated to E_p (Fig.3a). The ratio of the intensities of the conjugate and probe waves, I_c and I_p, is defining the conjugate reflection coefficient, which can be found as [3]:

$$R^* = \frac{I_c}{I_p} = \omega^2 \frac{\mu_0^2}{\varepsilon_0^2}\frac{d^2}{n_0^4}\left[\chi^{(3)}\right]^2 I_1 I_2 \tag{12}$$

where: μ_0, ε_0 are the magnetic permeability and the dielectric permitivity of the vacuum and I_1, I_2 - the intensities of the pump waves E_1 and E_2, respectively. Using the setup from Fig.3a, we have determined the experimental dependence $R^* = f(I_2)$, (Fig.3b). According to eq. (12), the slope of this linear dependence leads to $\chi^{(3)} = 2.23 \cdot 10^{-8}$ esu (for the same samples as in the self-diffraction experiments). The difference between this value and the theoretical estimation can be attributed to some approximations in both formalisms, probably, the negligible absorption considered in eq.(12) playing a nonnegligible role in the final value of $\chi^{(3)}$ (actually, the absorption coefficient is varying between 5 cm^{-1} at the beam periphery and 30 cm^{-1} in the centre of the gaussian laser beam).

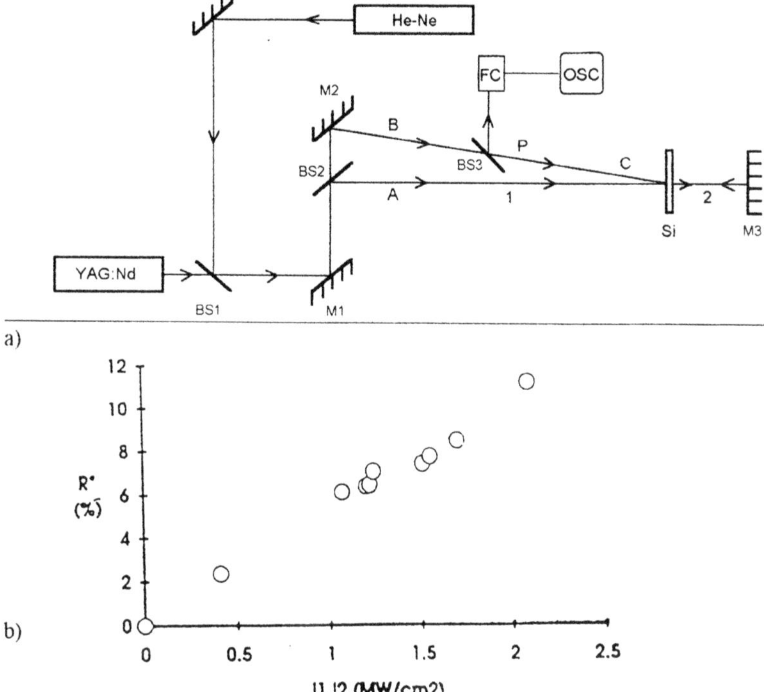

Fig.3. (a). The experimental setup for the optical phase conjugation by four wave mixing in c-Si samples. (b). The dependence of the conjugate reflection coefficient on the product of the pump intensities; the slope of this linear dependence is proportional to $\chi^{3)}$.

The same methods can be successfully used for other materials, as we have proved, for example with porous silicon [43], a-Si and GaP and A_2S_3. Moreover, they allow to determine also the lifetime of the dynamic grating in the sample (for c-Si and =30 μm, we have obtained the lifetime, τ~ 13 ns and the bipolar diffusion coefficient, D = 18 cm^2 /s, in a good agreement with the value 15 cm^2 /s, obtained by other authors [4]), as well as a map of homogeneity of the optical and electronic parameters of the sample.

4. The self-diffraction on the anharmonic phase gratings induced in photorefractive materials

Let us consider that a *harmonic* fringe pattern with the spatial light intensity distribution :

$$I(x) = I_1 + I_2 + 2\sqrt{I_1 I_2}\cos(2\pi x / \Lambda) = I_0\left[1 + m\cos(2\pi x / \Lambda)\right] \tag{13}$$

where I_1 , I_2 are the light intensities of the interfering beams; I_0=I_1+I_2 is the mean light intensity of the fringe pattern; Λ- the fringe spacing; $m = 2\sqrt{p} / (1+p)$ - the fringe contrast and $p=I_1/I_2$ - the beam intensity ratio, illuminates a photorefractive material.

The stationary space charge field generated in the photorefractive crystal (the phase grating yielded by the photogeneration, drift, diffusion and retrapping of the free charge carriers) is shown to be *anharmonic* [11-15]

$$E_{sc} = \frac{mE_D \sin(Kx)}{1+m\cos(Kx)} \tag{14}$$

where $E_D = KD/\mu = Kk_BT/e$ is the diffusion field and $K=2\pi/\Lambda$- the fringe frequency.

The case of low fringe modulation, $m \ll 1$, mostly invoked, leads to a *harmonic* space charge field distribution, with the same spatial frequency, but dephased with respect to the fringe distribution.

The space charge distribution, described by eq. (14), yields a spatial modulation of the refractive index of the photorefractive crystal by the Pockels (linear electro-optic) effect:

$$n(x) = n_0^3 r E_{sc}, \tag{15}$$

where n_0 is the refractive index of the crystal and r is the electro-optical coefficient of the photorefractive crystal. (In the above relation, we have considered the crystal orientation shown in the Fig. 4 and we ignored, for simplicity, the anisotropy of the crystal).

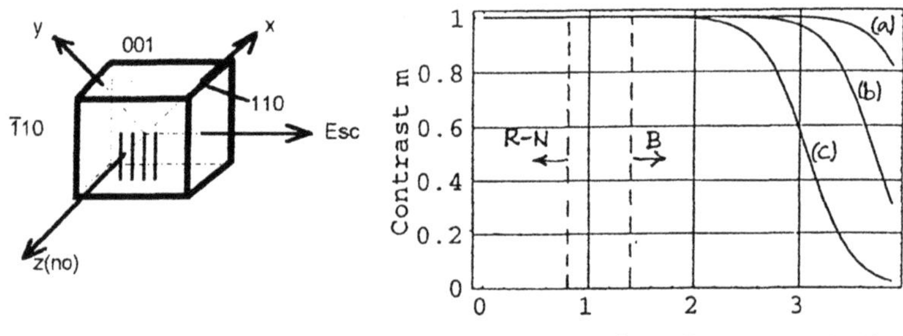

Fig. 4. The orientation of the photorefractive crystal considered in eq. (15).

Fig.5. The maximum values of the contrast in at different spatial frequencies for $N_A = 10^{17}$ (a); 10^{16} (b) and $5\cdot10^{14}$ cm^{-3} (c) (BTO crystal).

The eq. (14), written in the absence of the external field ($E_0 = 0$), can lead to local values of the space charge field which are larger than those obtained when all the donor centres from the crystal are photoionized. To consider the effect of this limitation, we have used the one-dimensional Poisson equation and the condition $\rho_{max} e N_A$, (N_A is the maximum space charge concentration) to obtain the result:

$$\varepsilon_0 \varepsilon_r \frac{k_B T}{e} \cdot \frac{4\pi^2}{\Lambda^2} \cdot \frac{m}{1-m} \le eN_A \tag{16}$$

The inequality (16) is represented in Fig. 5, from which one can determine the maximum admissible value of the fringe contrast for an arbitrary value of the fringe spacing and for different values of donor centres. One can observe, in Fig. 5, that the fringe contrast can reach the maximum value. $m_{max} = 1$, *when the fringe frequency is small*. When the fringe frequency is high, the maximum admissible value of the fringe contrast becomes smaller than 1.

Moreover, if the values of the fringe contrast and fringe spacing allow a space-charge field which can be sustained by the traps concentration of the crystal, the anharmonicity of the space charge field and, consequently that of the phase grating is determined by the eq. (14) only.

This anharmonicity can be evaluated from the Fourier series of the right-hand term of the eq.(14), i.e.:

$$E_{sc} = -2E_D \sum_{h=1}^{\infty}\left[\left(\frac{1}{m^2}-1\right)^{1/2} - \frac{1}{m}\right]^h \sin(hKx) = -2E_D \sum_{h=1}^{\infty}\left(-\sqrt{p}\right)^h \sin(hKx). \quad (17)$$

which shows that the relative amplitudes of the various Fourier components (harmonics) of the space-charge field depend on the contrast (m) only and that the anharmonicity of the space-charge field becomes significant for large values of the contrast. The corresponding refractive index modulation is:

$$\Delta n = n_0^3 r E_{sc} = \sum_{h=1}^{\infty} (\Delta n)_h \sin(hkx), \quad (18)$$

where $(\Delta n)_h = 2n_0^3 r E_D(-\sqrt{p})^h$, and the phase gratings induced in the photorefractive crystal are, for high contrasts, anharmonic too.

In the case of the photorefractive crystals, the diffraction regime is controlled by the value of Q only (for a given crystal, the numerical value of Q can be modified changing the value of Λ), due to the very small value of the electro-optical coefficient of the photorefractive crystals. In fig. 5, we pointed on the graph the values of Λ^{-1} which limit the two diffraction regimes, for a BTO crystal (with d = 8.6 mm) illuminated with a fringe pattern resulting from the interference of two He-Ne laser beams with λ = 633 nm.

In the Bragg diffraction regime, we have shown that the amplification factor reaches 1 when the ratio of the beam intensities reaches 1 and the fringe contrast is maximum [17]. This means that the energy exchange between the two beams decreases up to vanishing at high fringe contrast and the changes are difficult to be experimentally observed.

In the Raman-Nath regime, higher diffraction orders occur. In the case of the *harmonic diffraction grating*, the intensity in the l-th order in the diffraction pattern, on the side of the beam A, is:

$$I_{l,out}^{(A)} = I_{in}^{(A)} J_l^2(\Delta\varphi) + I_{in}^{(B)} J_{l+1}^2(\Delta\varphi) \quad (19)$$

where $J_l(\Delta\varphi)$ is the Bessel function of the first order. Due to the fact that $\Delta\varphi \ll 1$,

$$J_{l+1}^2(\Delta\varphi) \ll J_l^2(\Delta\varphi) \quad (20)$$

and the second term of the eq. (19) can be neglected. In the case of *the anharmonic diffraction grating*, each self-diffraction order is the superposition of some diffraction orders yielded by the diffraction of the both incident beams on the harmonics of the dynamic grating induced in the photorefractive crystal.

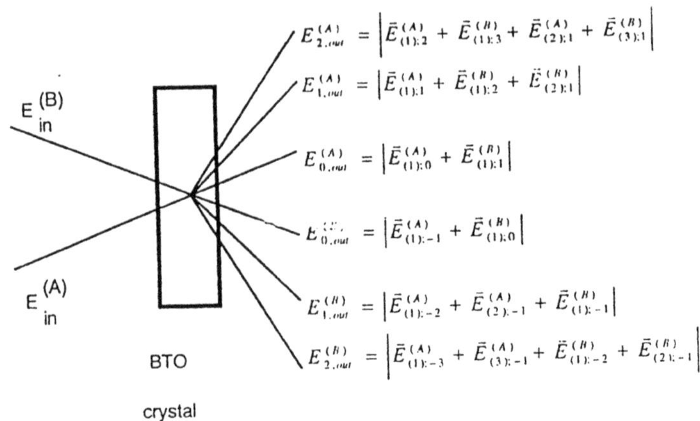

Fig. 6. The Raman-Nath diffraction regime for anharmonic gratings in photorefractive crystals

The spatial period, Λ_h, of the grating harmonics is $\Lambda_h = \Lambda/h$ leading to $Q_h = h^2 Q_1$. The Raman-Nath formalism can be used for the description of the diffraction orders with $Q_h \leq 0.5$, $h=1,2,...$

The superposition of the diffraction orders yielded by the grating harmonics is shown in the Fig. 6, considering the intensities of the incident beams $I_{in}^{(A)}$ and $I_{in}^{(B)}$ with $p = I_{in}^{(B)} / I_{in}^{(A)} < 1$ and $E_{i,j}^{(A)}$, the amplitude of light diffracted from the beam A, on the harmonics i of the phase grating, in the diffraction order j. We have calculated the resulting intensity of a self-diffraction order with the index $l > 1$ and situated on the side of the more intense beam, A:

$$I_{l,out}^{(A)} = \frac{\pi^2 d^2 n_o^6 r^2 E_D^2 p^l (1+p)^2}{\lambda^2} I_{in}^{(A)} \tag{21}$$

using again the approximation of the Bessel function of the first order by its half-argument (holding for small arguments). A similar relation can be written for the self-diffraction orders situated on the side of the beam B:

$$I_{l,out}^{(B)} = \frac{\pi^2 d^2 n_o^6 r^2 E_D^2 2p^l}{\lambda^2} I_{in}^{(B)}. \tag{22}$$

The self-diffraction spectrum becomes asymmetrical.

This accurate analysis of the self-diffraction in the photorefractive materials can provide a new all-optical method for the determination of their electro-optical coefficient. The electro-optical coefficient can be obtained from the eq.(21), for the particular case $l = 1$ and $p \ll 1$, when the diffraction efficiency is:

$$\eta_1 = I_{1,out}^{(A)} / I_{in}^{(A)} = \frac{4\pi^2 d^2 n_o^6 r^2 E_D^2 p}{\lambda^2} \tag{23}$$

In this case, the diffraction efficiency in the first self-diffraction order is a linear function of the beam ratio, p. The slope of this line is proportional to the electro-optical coefficient, r.

The setup used in our experiments is shown in Fig. 8. The linear polarised He-Ne laser beam is split in two beams which are superimposed in the BTO crystal (with the orientation and thickness given above) inducing a dynamic phase grating. Both beams are then self-diffracting on this grating. The angle between the interfering beams is β= 7' 40", leading to a spatial period of the grating Λ= 0.285 mm. The contrast of the light interference pattern can be modified by the attenuation of one of

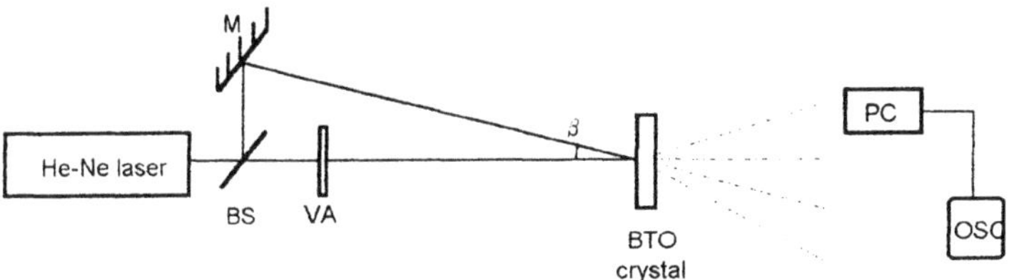

Fig. 7. The experimental setup for the determination of the electro-optic coefficient of the photorefractive materials.

the interfering beams with the calibrated variable attenuator, denoted by VA in the Fig. 8. The intensities in the self-diffraction spectrum were measured by the calibrated photo-cell, PC, coupled to the oscilloscope, OSC.

The experimental results are presented in the Fig. 8. By the best fit method, we have obtained r = 5.72 pm/V. This value is in very good agreement with the data given by Stepanov and Petrov [11] and Bayvel et al [19], namely r = 5.6 - 5.8 pm/V, using electro-optical methods.

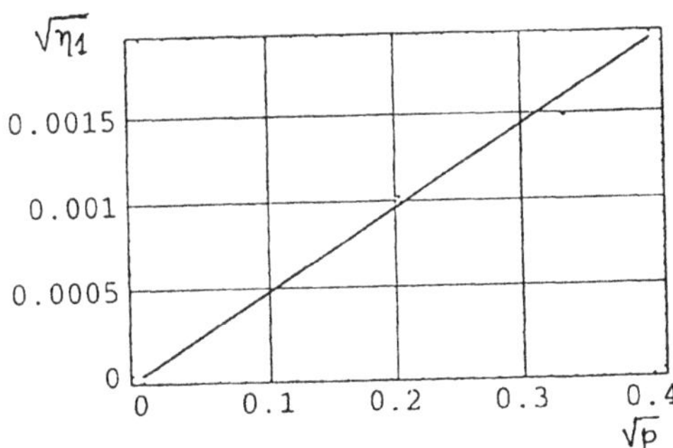

Fig. 8. The experimental dependence of the diffraction efficiency on the beam ratio. The solid line is the result of the best fit and the slope of this line is proportional to the electro-optical coefficient, r.

5. Real-time holographic interferometry using optical phase conjugation in photorefractive materials and direct spatial phase reconstruction

The structure of the holographic interferometer used in the present work is shown in Fig. 9. Both 2WM and 4WM modes of operation can be experimented in this system with simple changes. The setup without BS2 and the mirror M4, but including the polarisers P1 and P2 and the TV camera in the image plane B is easy recognised as 2WM configuration. The setup with the beam-splitter BS2 and the image plane A was used for 4WM. The lens L1 (with focal length 500 mm and an aperture of 100 mm) reduces the object beam to the crystal size (6 mm x 8 mm x 8 mm) and, together with L2 and L3, images the object plane into the sensor of the TV camera. The features of the image holographic interferometer - the fixed localisation and the quantitative evaluation of the fringe pattern - have determined essentially the design and the performance parameters of this holographic interferometer. The image collected by the TV camera is digitised and stored in a frame grabber which is interfaced to a PC computer.

In our experiments, the favourable combination between He-Ne laser power (40 mW, λ=633 nm) and the high sensitivity of BTO crystals was mainly exploited. The optical surfaces of the crystal were parallel to the (110) crystallographic planes. Without external field, the recording of holograms on sillenite crystals is mainly yielded by the diffusion of free charge carriers. The diffraction efficiency is increasing with the increase of the mean spatial frequency of the hologram. In the holographic interferometer from Fig. 9, the angle between the object and reference beam is 60 deg. leading to a mean spatial frequency of 1400 mm^{-1}. In this case, the diffraction efficiencies without and with external field are close enough justifying the hologram recording without additional sources for the external field.

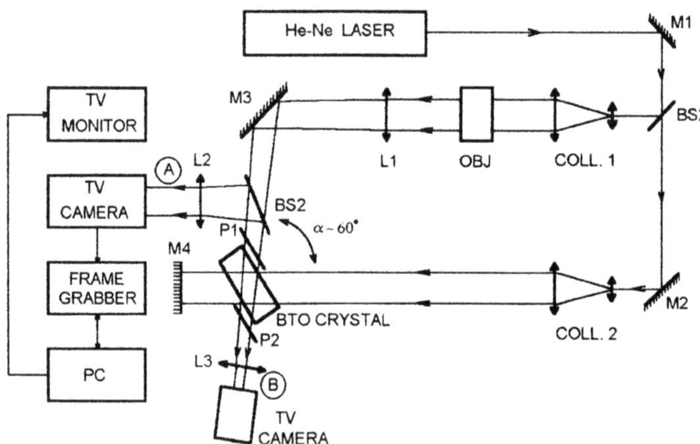

Fig. 9. The structure of the holographic interferometer using PRC and computer image processing. TV camera is in the position: A - for 4WM or B - for 2WM mode of operation.

In the 2WM mode of operation, we have taken into consideration the polarisation states of the diffracted and transmitted beams [22, 24]. To separate the diffracted light, an analyser (P2) is placed after the crystal, which transmits the diffracted beam and suppresses the object undiffracted beam emerging from the crystal. Due to the optical activity of the crystal (for BTO, 6 deg/mm), the angle between the directions of the polarisation of the incident beams and the [001] axis should be fixed properly. For an object-to-reference beam (at the crystal) of ~ 10^{-3}, the diffraction efficiency was 5%.

In the 4WM mode of operation, the counterpropagating pump (reference) beam is introduced in the crystal by the mirror M4 and the conjugate image (signal) yielded by the crystal is extracted by the beam-splitter BS2 and directed to the TV camera (image plane A). The intensity of the conjugate image was comparable with that of the diffracted image in 2WM. Disregarding some additional optical adjustments, we have found a better signal-to-noise ratio (SNR) with this configuration (due to the compensation of the stationary phase perturbations, i.e. component inhomogeneities).

The recording and erasure time in PRC can be controlled by means of the laser intensity and are, in the same conditions, approximately equal. Both the hologram recording and erasure processes (i.e. the temporal dependencies of the diffraction efficiency) can be described by exponential time variations, depending essentially on the incident light intensity. In our case, the intensity was determined mainly by the reference (pump) beam and the recording time of the hologram in BTO crystals was $\tau_H \sim 10$s (for light energy density of $4 \cdot 10^{-3}$W/cm^2).

The two exposure H.I. on PRC implies that the duration of the change of state between the exposures is much smaller than the recording time (τ_H). The hologram of the first object (state) is recorded and simultaneously read-out up to the momen t_0. At this moment, the object (state) is changed rapidly (step-like) and a hologram of this second object (state) starts to be recorded together with erasure of the previous hologram (due to the phase jump). Hence, after the moment t_0, in the crystal exist two holograms which are read-out simultaneously yielding an interference pattern between the reconstructed images. The interferogram can be visualised on the TV-monitor for a definite duration determined by the erasure of the first hologram. It is very important to remark that PRC can store in dark a recorded hologram for a definite duration, allowing such operations as delayed second exposure (important when the object evolution is slow) delayed digital acquisition of images and so on. For BTO crystals, the dark storage time of holograms is 30 to 70 min.

The class of transparent objects was particularly interesting for testing in the holographic interferometer from Fig. 9 (even we have also demonstrated its operation with reflecting objects. In this case, the laser light was enough to use all degrees of freedom given by the parameters of this setup (signal-to-noise ratio, the sensitivity of TV camera etc.). In some special cases, the transparent object testing can be conducted using fairly simple analytical expressions for processing the fringe patterns.

For homogeneous prism-like objects with the phase gradient in one direction only, the interferogram shows equally spaced parallel fringes (perpendicular to the x - direction) with a fringe period [27] (Fig. 10a)

$$i = \frac{2\pi}{a} = \frac{\lambda}{(n-1)L'_x}, \quad L' = \frac{\Delta L}{\Delta x} \tag{24}$$

In this case, an image processing algorithm can find automatically the gradient direction with an error of less than 1and the prism angle with errors of 1% in a range from several arc seconds to tenths of degrees (Fig. 10a). For polyhedral transparent objects, a more elaborated algorithm is recognising the object 3D shape by tracking automatically the plane orientation and extension.

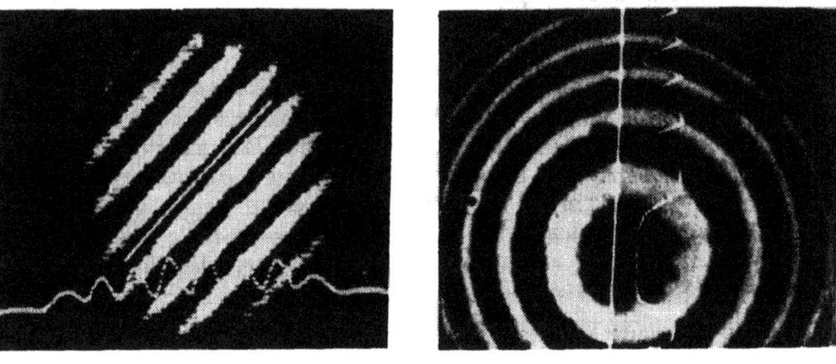

(a) (b)

Fig. 10. (a) Holographic interferogram yielded by a double-exposure of PRC with an without a prismatic object (in 4WM configuration). (b) Holographic interferogram yielded by a double-exposure of PRC with and without a lenticular object (in 4WM configuration).

For lens-like objects, denoting by f the focal length and $\rho(z)$ - the radius in cylindrical coordinates, the interferogram consists of a set of concentric circular fringes (Fig. 10b)

$$\varphi_0 + \frac{2\pi}{\lambda}\frac{\rho_k^2}{2f} = 2\pi k, \quad (k \text{ - an integer number}) \tag{25}$$

The slope of the linear dependence ρ_k^2 versus k,

$$s = \frac{\rho_j^2 - \rho_i^2}{k_j - k_i} = 2\lambda f \tag{26}$$

leads to the precise determination of the focal length. In the measurements, we have to take into consideration also the magnification factor $M = \rho_{TV}/\rho_{obj}$, the inverse ratio between the size of an object and the size of its image on the TV monitor. Thus, for a planar convex lens, in air, the radius of curvature is determined in our interferometer as [27]

$$R = s \cdot M^2 / 4\lambda = 1.64s \qquad [mm]. \tag{27}$$

In this case, the image processing algorithm finds automatically several fringe radii, ρ_j, yields an average value of the slope and finally displays the radius of curvature (focal length) of object. For lenticular arrays, the algorithm determines the position and the focal length of every element. The relative errors of the geometrical parameters digitally measured in the holographic interferometer were under 5% in the range of focal lengths of 0.5 to several hundreds meters. In order to measure precisely lenses with shorter focal length, like the geometry of the cylindrical transparent rods, we have used the object immersion into a liquid with proper refractive index.

The direct spatial reconstruction of optical phase (DSROP) was also tested for real-time holographic interferogram evaluation [26, 32-38]. These methods uses a single pattern analysis, thus, are faster than phase shifting methods [32,38,39]. In these methods, the fringe patterns (F. P. ≡ phase-modulated image) are considered equivalent to holographic recorded patterns, i.e. the result of the interference of an arbitrary optical wavefront (object) with a tilted plane reference wavefront (with a tilt given by $(\sin\Theta) / \lambda = u_0$, like in holography). The phase reconstruction is the result of the multiplication of the fringe pattern with the reference and then, by a smoothing procedure, in the space domain or by a low-pass filtering, in the Fourier domain. The difference between holographic and DSROP methods consists of the reconstruction of the 3D image in the former and of the reconstruction of the 3D phase map in the second case.

Following the holographic analogy, we can accomplish the phase demodulation by the multiplication of the fringe pattern with a reference, which can be obtained from: **a)** the external optical setup (two plane wavefront interference or a reference grating image); **b)** the recorded F. P. with a zone without phase modulation; **c)** a computer generated pattern, with the knowledge of the carrier frequency, u_0.
The product consists of two rapidly - varying terms and a slowly-varying third one. It is possible to obtain the slowly-varying term by smoothing (integration) this pattern over a space interval with the dimension $\Lambda_0 = 1/u_0$ (assuming: $u_0 = 1/3$; A, B and Φ_m = constant over the fringe period). Finally, the spatial phase distribution is obtained as the arctangent of the ratio of the imaginary and real parts of the slowly-varying term (obtained by smoothing).

We have used, in this method, digital smoothing procedures equivalent to the integration in image processing, i.e. average and median with masks of different dimensions, which are simple, fast and relatively accurate. An example is shown in Fig. 11a,b. in which the F. P. was smoothed with a 3 x 3 averaging mask (window) The computing time, with our algorithm on a PC, was ~ 5 sec. and ensured a mean error of 5%.

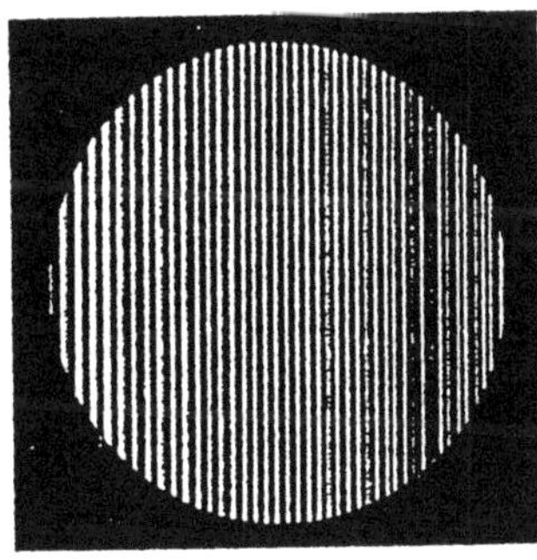

(a)

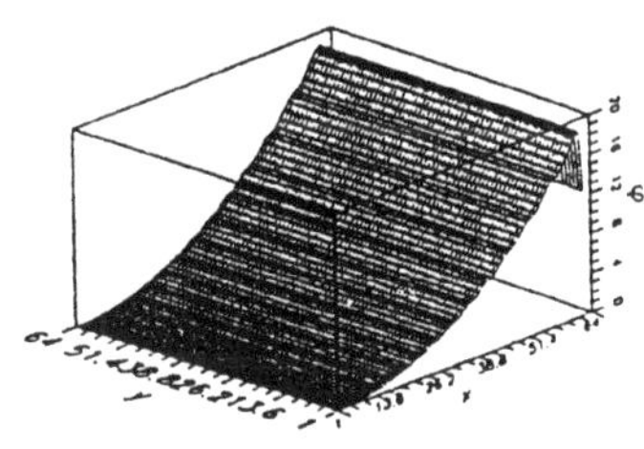

(b)

Fig.11. (a) The holographic interferogram of a cylindrical lens the DSROP in Fourier domain by the band-pass method with a Hanning window ; (b) the 3-D calculated (unwrapped) phase distribution on one side of the symmetry axis using the smoothing method with a 3 x 3 averaging window.

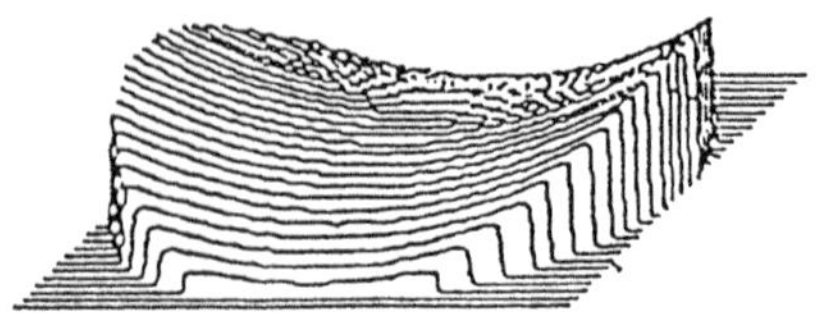

Fig.12. The 3-D phase distribution for the same holographic interferogram as in Fig.10, calculated by the band-pass method (in Fourier domain).

For comparison, we have processed the same F. P. with the band-pass method in the Fourier domain (Fig. 12). The processing time, with our algorithm on the same computer, was 10 times larger and the mean errors were also larger (due to the truncations in the two Fourier transforms and spectral band selection). Moreover, the subjective selection of the spectral band in this method yields non-reproducible results for the phase distribution (shown in Seidel and Zernike coefficients).

Our experiments with the smoothing method, in the above introduced conditions, have shown relative errors of the phase in the order of (5 - 15)%, the higher figures valid for the higher a_m and u_m (due to the approximate replacement of the integral by the simple and fast average window algorithm) [40 - 42]. Additional errors can be introduced by the unwrapping algorithms, which yield in some cases phase dislocations.

6. CONCLUSIONS

In this paper, a more accurate description of the light self-diffraction was theoretically and experimentally proved. The extension of the classical theory to the case of absorbing materials and nonuniform beams was given. We have shown that the agreement of the new formalism with the experimental data, in the case of the dynamic gratings induced in crystalline silicon, at λ=1.06 μm, is better.

The self-diffraction on phase gratings induced in photorefractive materials shows complications at higher contrasts of the light fringe pattern, more precisely, due to the anharmonicity of the space charge field developed in the material. In the Raman-Nath regime, the growth self-diffraction spectrum is asymmetric and the orders tend to a maximum with the contrast growth. In this regime, one can easily measure the diffraction efficiency in the first order and calculate precisely the electro-optical coefficient. This pure optical method is precise, general and needs no electrodes on the crystal surface. Our experimental results for the electro-optic coefficient of BTO crystals obtained from the self-diffraction measurements are in good agreement with those obtained by other methods.

A holographic interferometer using a BTO crystal as reusable recording material, an usual He-Ne laser, a computer image processing and a proper structure to work in 2WM as well as in 4WM modes of operation was described. The recording and erasure time of holograms into the crystal can be controlled by means of the laser intensity and the ambient conditions. Two-exposure H. I. were performed in this testing system. Rapid automatic processing of the fringe patterns (compatible with the recording times) was done following simple analytical results and some new methods for the direct spatial reconstruction of optical phase.

REFERENCES

1. Peyghamberian N. and Koch S.W. (1990), Semiconductor Nonlinear Materials, in *Nonlinear Photonics*, Ed. Gibbs H.M. et al, *Springer Verlag*, Berlin, 7-60.
2. Woerdman J. P. (1971), Some optical and electrical properties of a laser-generated free-carrier plasma in Si, *Philips Res. Rep. Supp.*, No. 7.
3. Eichler H. J., Gunther P., Pohl D. W. (1986), Laser Induced Dynamic Gratings, *Springer Verlag*, Berlin
4. Eichler H. J., Massman F. (1982) - Diffraction efficiency and decay times of free-carrier gratings in silicon, *J. Appl. Phys.*, **53**, No. 4, pp. 3237-3242.
5. Eichler H. J. et al.(1987), Laser-induced free-carrier and temperature gratings in silicon, *Phys. Rev. B*, **36**, No. 6, 3247-3253. .
6. Born M., Wolf E (1966), Principles of Optics, *Pergamon Press*,Oxford.
7. Berry M. V. (1986). The Diffraction of Light by Ultrasound, *Academic Press*, N.Y.
8. Petris A., Vlad V.I., Voicu L. and Negres R., (1994), Accurate description of the light self-diffraction on high-modulated dynamic gratings in semiconductor materials, *Proc. SPIE*, **2108**, 251-255.
9. Vlad V. I., Petris A., Tibuleac S. (1991)- Self-diffraction studies on Si-samples, *Rev. Roum. Phys.*, Vol. **36**, (5-6), 345-350.
10. Sheik-Bahae M., Said A.A., Wei T.H., Hagan D.J., Van Stryland E.W., (1990), Sensitive measurement of optical nonlinearities using a single beam, IEEE *J. Quantum Electronics*, **26**, 760-769.
11. Stepanov S. I. , Petrov M. P., (1988), Photorefractive materials and Their Applications, Vol. **1**, Gunter P. and Huignard J.-P. ,(1985), Eds, Springer-Verlag, Series "*Topics in Applied Physics*", Vol. **61**.
12. Moharam M. G., Gaylord T. K., Magnusson R. and Young L.,(1979), Holographic Gratings Formation in Photorefractive Crystals with Arbitrary Electron Transport Length, *J. Appl. Phys.*, **Vol. 50**, 5642-5651.
13. Petrov M. P. , Miridonov S. V., Stepanov S. I. and Kulikov V. V. , (1979), Light Diffraction and Nonlinear Image Processing in Electrooptic $Bi_{12}SiO_{20}$ Crystals, *Optics Communications*, Vol. **31**, 301-305.
14. Vachss F., Hesselnik L.,(1988), Nonlinear Photorefractive Response at High Modulation Depths, *J. Opt. Soc. Am., A*, Vol. **5**, no. 5, 690-701.
15. Au L. B., Solymar L., (1990), Higher Harmonic Gratings in Photorefractive materials at Large Modulation with Moving Fringes, *J. Opt. Soc. Am. A*, Vol. 7 (8), 1554-1561.
16. Vlad. V. I., Petris A. and Apostol I., (1994), Self-diffraction on high-modulated phase gratings induced in photorefractive crystals, *Rom. Rep. Phys.*, **46**(7-8), 589-596.
17. Petris A., Vlad V.I. and Apostol I., (1994), Determination of the electro-optic coefficient of BTO photorefractive crystal using the laser light self-diffraction, *Proc. SPIE*, **2108**, 280-284.
18. Enns R.H. and Rangnekar S.S., (1974), Diffraction by a laser induced thermal phase grating, Parts I and II, *Canadian J. Physics*, **52**, 99-109 and 562-567.
19. Bayvel P., McCall M. and Wright R.V., (1988), Continuous method for measuring the electro-optic coefficient in BSO and BGO, *Opt. Lett.*,**13**, 27-29.
20. Yeh P., (1993), Introduction of photorefractive nonlinear optics, *J. Wiley*, N.Y.
21. Huignard J. P. and Herriau J. P., (1977), Real-time double exposure interferometry with BSO crystals in transverse electrooptic configuration, *Appl. Optics* **16** (7), 1807-1809.
22. Marrakchi A., J. P. Huignard and J. P. Herriau, (1980), Application of phase conjugation in BSO crystals to mode pattern visualization in diffuse vibrating structures, *Opt. Commun.* **34** (1), 15-18.
23. Kamshilin A. A. and Mokrushina, E. V.,(1984), Photorefractive crystals for real-time holographic interferometry, *Proc. SPIE* **473**, 83-86.
24. Kamshilin A. A., Mokrushina E. V. and Petrov M. P., (1989), Adaptive holographic interferometers operating through self-diffraction of recording beams in photorefractive crystals, *Opt. Eng.***28** (6), 580- 585.
25. Vlad V. I., Popa D.,Petrov M.P., and Kamshilin A.A., (1990), Optical testing by dynamic holographic interferometry with photorefractive crystals and computer image processing, *Proc. SPIE* **1332**,pt.1, 236-245.
26. Vlad V. I. and Malacara,D., (1994), Direct spatial reconstruction of optical phase from phase-modulated images, in *Progress in Optics*, Wolf E. , Editor, Vol.**XXXIII**, 261-317.
27. Vlad V.I., Malacara D. and A. Petris, (1996), Real-time holographic interferometry using optical phase conjugation in photorefractive materials and direct spatial phase reconstruction, *Opt. Eng.*, **35**, 1383-1388.
28. Petris A., Tibuleac S. and Voicu L., (1993), Measuring the phase conjugation fidelity by computer image analysis, *Roum. J. Phys.*, **38**, 633-635 and 669-673.
29. Vlad V.I. and Petris A., (1994), Fidelity evaluation of the phase conjugation in photorefractive crystal ussing the spatial heterodyne demodulation of optical interferograms in the Digest of ICO Topical Meeting "Frontiers in Information Optics", Kyoto, Japan.

30. Jain R.K., Klein M.B., (1979), Degenerate four - wave mixing near the band gap of semiconductors, *Appl. Phys. Lett.*, **35**, 454-456.

31. Moharam M. G., Gaylord T. K., Magnusson R. (1980) - Criteria for Bragg regime diffraction by phase gratings, *Opt. Comm.*, Vol. **32**, No. 1, pp. 14 -18 and pp. 19-23.

32. Malacara D., (1992), Ed., *Optical Shop Testing*, 2nd. Edition, John Wiley and Sons, New York.

33. Robinson D. W. and Reid G. T.,(1993), Eds., *Interferogram analysis*, Institute of Physics Publ., Bristol.

34. Ichioka Y. and Inuiya M. , (1972), Direct phase detecting system, *Appl. Opt.* **11**, 1507-1514.

35. Doerband B. , Wiedman W., Wegmann U.,Kuechel W. and Freischland K. R., (1990)., Software concept for the new Zeiss interferometer, *Proc. SPIE* **1332**, 664-672.

36.Takeda M. and Tung Z.,(1985), Subfringe holographic interferometry by computer-based spatial-carrier fringe-pattern analysis, *J. Opt.* (Paris), **16**, 127-131.

37. Takeda M., (1989), Spatial carrier heterodyne techniques for precision interferometry and profilometry: An overview, *Proc. SPIE* **1121**, 73-88.

38. Creath K. , (1988), Phase measurement interferometry techniques, in *Progress in Optics*, Vol. **XXVI**, Wolf E., Ed., Elsevier Sci. Publ., Amsterdam, 349-393.

39. Georges M.P. and Lemaire Ph.C., (1995), Phase-shifting real-time holographic interferometry that uses BSO crystals, *Appl. Opt.*, **34**, 7497-7506.

40. Malacara D. , Vlad V. I. and Servin M., (1994), Spatial carrier analysis of interferograms with aspheric wavefronts, *Proc. SPIE* **2340**, 190-201.

41. Servin M., Malacara D., Malacara Z. and Vlad V. I., (1994), Sub-Nyquist null aspheric testing using a computer stored compensator, *Appl. Opt.* **33**(19), 4103-4108.

42. Vlad V. I. and Malacara D., (1995), Spatial phase demodulation in fringe images, *Proc. SPIE* **2461**, 234-244.

43. Vlad V.I., Petris A., Chumash V. and Cojocaru I., (1996), Laser induced periodic structures in porous silicon, *Appl Surface Science* (accepted for publication)

DOPING OF CHALCOGENIDE GLASSES OF SEMICONDUCTORS

S. Vikhrov
Professor at Ryazan Radio Engineering Academy
59/1, Gagarin St. Ryazan 390000, Russia

1. Introduction

It is known, that the conductivity of crystalline semiconductors displays increased sensitivity to presence of impurities, which form new levels in the band gap (donor and acceptor). Things are different at incorporation of the same elements in the non-crystalline semiconductor. Electrical activity of impure elements introduced in melt during synthesis of a disordered material is lost. For the first time the absence of impure conductivity in chalcogenide glassy semiconductors (CGS) was discovered by B.T. Kolomiets and coworkers [1], which they estimated by the temperature dependence of conductivity (the absence of characteristic bend for impure conductivity). The explanation of disordered material insensitivity to impurity is given by N. Mott [2]. The insensitivity is resulted from saturation of all valence requirements of an impurity atom due to the formation of connections with the adjacent atoms. Z. Borisova [3] explains favorable opportunities of saturation of all impurities atoms valence connections in CGS by several reasons. First, CGS practically always contains in abundance one of the components included in its structure. More often this component is chalcogen. At incorporation of impurity atoms during synthesis favorable conditions for their interaction with superfluous components of glass are available. Secondly, chalcogen materials of stehyometric compound are prepared at rather high speeds of cooling. In result the large number of defects like broken connections arise. These defects with increased activity, interact with impurities atoms neutralising their influence on properties of the material. Thirdly, more often CGS contains elements of their adjacent groups in D.Mendeleev periodic system. In this case the energy of connections between components of glass is insignificant and at addition of impurities in melt the interaction of chalcogen with introduced component turns out to be energetically favourable.

For a long time insensitivity of disordered semiconductors to various sorts of impurities was considered as one of the advantages of this class of materials in comparison with crystals. However, subsequent experiments have shown, that incorporation of impurity (as impurity of conductivity is not observed) causes change of the original material properties. The analysis of work concerning the influence of element additives of I and III groups on CGS properties was carried out by

A. Andriesh and M. Bertolotti (eds.),
Physics and Applications of Non-Crystalline Semiconductors in Optoelectronics, 201–214.

B.T.Kolomiets [4]. On an example of numerous experimental works the influence of Cu, Ag, Ga, In, Tl on electrical and photo-electric CGS properties was shown [5]. It should be noted that by incorporation of the additives during synthesis the type of CGS conductivity failed to be changed. For all chalcogenide glassy materials the dominant charge carriers were holes. To change the type of conductivity is possible, when entered additives cause the shift of Fermi level (E_F). In CGS the Fermi level is rigidly fixed near the middle of mobility band gap (forbidden band) due to large density of own defects. The concentration of defects near E_F can reach $10^{19}..10^{20}$ $см^{-3}$. In disordered semiconductors only the part of entered additives appear to be electrically active [6]. In many cases incorporation of electrically active impurities (as Ni, Fe, Co, Bi, Sn and others) causes crystallisation of the material, and small additives do not cause large shift of E_F. Hence, there was the necessity to search new components, which can be incorporated in plenty within glass formation area. Thus these additives should result in inversion of conductivity type(from hole to electronic).

2. Experimental

2.1 PREPARATION AND STRUCTURE

Our attention was drawn by bismuth, since. G.Shotmiller and other [7] discovered electronic conductivity in amorphous thin films $Se_{100-x}Bi_x$ at x=3 at. %. It was possible to receive amorphous films of selenium (Se) with contents of bismuth up to 30 at.%, which was specified by the results of electronic diffraction [7].

We made an attempt to prepare CGS of system $Se_{100-x}Bi_x$. As was shown by the results of x-ray and electronic analyses the presence of more than 2 at. % Bi in selenium causes partial crystallisation of the material. As a crystal phase x-ray analysis marks Bi_2Se_3. Specific conductivity of partially crystal samples is increased by seven orders in comparison with pure selenium, and thermopower for samples with the contents of Bi more than 3 at. % has negative sign (S_{293k}=-225 μV/K for $Se_{97}Bi_3$; S_{293K}=-210 μV/K for Se_{95} Bi_5 and S_{293K} =-124 μV/K for $Se_{90}Bi_{10}$).

The greatest quantity of bismuth in glass structure is included in systems Ge-Se-Bi and Ge-S-Bi, which has caused our detailed research of these systems. The greatest tendency to glass formation is displayed in structures $(GeSe_{3.5})_{100-x}Bi_x$ and $(GeS_{3.5})_{100-x}Bi_x$. It is possible to incorporate bismuth up to 20 at.%, into glass $(GeS_{3.5})_{100-x}Bi_x$ and in $(GeSe_{3.5})_{100-x}Bi_x$ - up to 15 at.% Bi. Subsequently these compositions were chosen basically for research of inversion of conductivity type and transport properties.

Taking into account increased tendency of compounds containing bismuth to segregation, researches on study of CGS material homogeneity were carried out. Integrated chemical composition of CGS $(GeSe_{3.5})_{100-x}Bi_x$ was supervised with the help of microprobe analysis. All prepared CGS samples had integrated chemical composition, distinguished from the rated one not more than by 0.2 at.%. The local chemical composition in various points of polished surface of bulk sample was estimated with the help of installation 'CAMECA', electronic microscope and electron scanning microscope (250Á beam diameter). Within the resolution of techniques the chemical composition of glass has appeared to be similar. Study of glass $(GeSe_{3.5})_{100-x}Bi_x$ with the help of electronic microscope at increase by 110000 granular structure was found. The results of researches on specific weight and microhardness of glass are shown in table 1. From table 1 it is clear, that specific weight and microhardness of a material$(GeSe_{3.5})_{100-x}Bi_x$ grow as the concentration of bismuth increases. Specific weight of a material grows according to the linear law up to X=16, the microhardness up to X=8 increases linearly, and at X> 8 the growth of microhardness slows down.

Table 1. Results: specific weights , microhardness and ДТА measurements for glassy samples $(GeSe_{3.5})_{100-x}Bi_x$.

#	Composition	specific weight g/cm^3	Microhardness kg/mm^2	Tg, ^{0}C	T$_{c1}$, ^{0}C	T$_{c2}$, ^{0}C	T$_m$, ^{0}C
1	$GeSe_{3.5}$	4,34	103	180	-	-	-
2	$(GeSe_{3.5})_{98}Bi_2$	4,48	128	185	-	-	-
3	$(GeSe_{3.5})_{96}Bi_4$	4,68	143	190	350	400	500
4	$(GeSe_{3.5})_{94}B_6$	4,86	168	230	330	380	520
5	$(GeSe_{3.5})_{92}Bi_8$	5,02	188	255	300	350	550
6	$(GeSe_{3.5})_{90}Bi_{10}$	5,20	203	250	300	250	555
7	$(GeSe_{3.5})_{88}Bi_{12}$	5,37	222	240	290	350	560
8	$(GeSe_{3.5})_{86}Bi_{14}$	5,56	228	235	290	345	570
9	$(GeSe_{3.5})_{84}Bi_{16}$	5,73	233	230	285	345	580

Results of ДТА at speed of temperature increase by 4,5 K / min (tab. 1) have shown, that for glass $GeSe_{3.5}$ only one endothermic peak in an interval of temperatures + 165... + 1900C, connected with glass transition, is available. Two exothermic peaks have appeared for composition with the contents of Bi more than 2 at.%. Two peaks of crystallisation confirm large tendency to stratification of glass $(GeSe_{3.5})_{100-x}Bi_x$ at heating by several binary phases. In our opinion the first exothermic peak is caused by formation of a crystal phase Bi_2Se_3, and the second by $GeSe_2$, which proves to be true by the results of x-ray analysis.

2.2. ELECTRICAL PROPERTIES

For the first time electronic conductivity in chalcogenide glassy semiconductors (the thermopower data) was found for $Ge_{20}Se_{70}Bi_{10}$ by N.Tohge and others [8]. We and others authors carried out extensive researches on revealing inversion conditions of conductivity type. As initial object system Ge-Se-Bi with the maximum contents of bismuth $(GeSe_{3.5})_{100-x}Bi_x$ has been chosen. Thus the ratio between germanium and selenium for all structures remained constant.

Temperature dependencies of conductivity (σ) were measured by the method of Van-der-Pauw. The results of researches $\lg\sigma=f(1/T)$ for composition

$(GeSe_{3.5})_{100-x}Bi_x$ are shown in fig. 1 and in table 2. From fig. 1 it is clear, that increase of the contents of Bi results in growth of conductivity and reduction of activation energy (E_σ). In investigated interval of temperatures for all structures the dependencies $lg(\sigma)=f(1/T)$ are represented as direct line.

Table 2. Electrophysical parameters CGS of system Ge (Si) -Se (Te) -Bi (Sb)

#	Composition	$\sigma_{293K}, 10^{-10}$ $(\Omega\bullet cm)^{-1}$	E_σ, eV	$\sigma_0, 10^3$ $(\Omega\bullet cm)^{-1}$	S_{293K} mV/K	E_s, eV	E_{opt}, eV
1	$GeSe_{3.5}$	10^{-6} xx	1,18	17,8	-	-	1,99
2	$(GeSe_{3.5})_{96}Bi_4$	3,9x10^{-6}	1,12	5,0	+0,8	-	1,46
3	$(GeSe_{3.5})_{94}Bi_6$	2,3x10^{-3}	1,05	4,9	+0,32^x	-	1,40
4	$(GeSe_{3.5})_{92}Bi_8$	1,47x10^{-1}	0,92	4,4	-0,16^x	-	1,25
5	$(GeSe_{3.5})_{90}Bi_{10}$	25,8	0,65	0,15	-1,78	0,76	1,22
6	$(GeSe_{3.5})_{88}Bi_{12}$	62	0,65	3,1	-1,97	0,69	1,21
7	$(GeSe_{3.5})_{86}Bi_{14}$	1,8x10^2	0,66	1,7	-2,67	0,71	1,20
8	$(GeSe_{3.25}Te_{0.25})_{90}Bi_{10}$	83	0,64	0,33	-2,42	0,87	-
9	$(GeSe_3Te_{0.5})_{90}Bi_{10}$	9,4x10^{-2}	0,80	0,13	+0,50	-	-
10	$Si_{15}Te_{75}Bi_{10}$	99,6	0,69	0,4	-0,46^x	-	-
11	$(GeSe_{3.5})_{88}Sb_{12}$	8,9x10^{-5}	1,00	1,3	+1,1^x	-	1,87

x - is determined at temperature 450 K, xx- is determined by approximation.

Due to high resistance of samples it was possible to measure temperature dependencies of thermopower (S) in a narrow interval of high temperatures only for four different composition of structures (fig. 2). For other samples the thermopower was measured at temperature 450 K. From fig. 2 it is clear, that all glassy materials have negative thermopower sign and the dependence $S=f(1/T)$ is rather well described by direct line. The transition from p-type conductivity to electronic one occurs for $(GeSe_{3.5})_{100-x}Bi_x$ at $x \geq 8$ (see tab. 2) [9,10].

For CGS distinction between thermopower sign and Hall coefficient is the most striking feature. As a rule the sign of thermopower is positive and Hall coefficient is negative. In this connection study of Hall effect on glassy semiconductors with electronic type of conductivity (thermopower data) represents certain interest. The temperature dependencies of Hall coefficient (R_H) and Hall mobility (μ_H) are represented in fig. 3a,b. The sign of Hall coefficient is negative, i.e. the sign R_H coincides with results of thermopower measurement. Hall mobility for $(GeSe_{3.5})_{88}Bi_{12}$ is equal to $\approx 10^{-3}$ $cm^2/V\bullet s$ at 370 K [10].

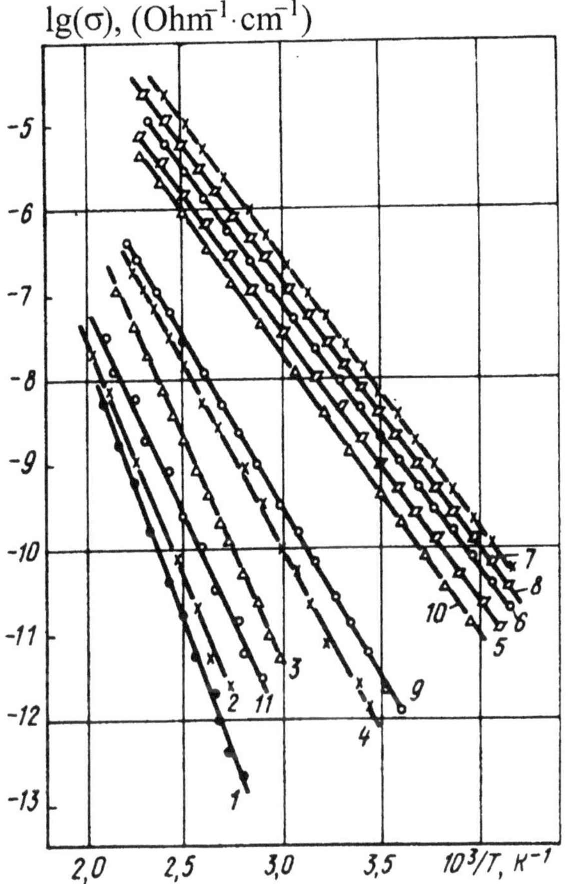

Fig. 1 Temperature dependence of electrical conductivity of CGS (numbers correspond to the Table 2)

Reduction of conductivity activation energy $(GeSe_{3.5})_{100-x}Bi_x$ at increase of bismuth concentration can occur both due to the change of optical band gap width (E_{opt}), and as a result shift of the Fermi level (E_F) shift closer to the edge of mobility gap (fig. 4). To answer this question measurements of optical band gap were carried out. E_{opt} was determined from dependencies of absorption coefficient versus photon energy at the level $\approx 10^3$.

Study of structure and physical properties of $(GeSe_{3.5})_{100-x}Bi_x$ do not allow to reveal the conditions and nature of inversion of conductivity type. The further experiments were directed on finding - out the inversion conditions of conductivity type by serial replacements of one glass components with other components, as far as possible in the same amount. To carry out the last condition strictly it is not always possible because of distinction between areas of glass formation at replacement of system components were used. The type of conductivity which was determined by measuring the thermopower had been chosen as a criterion for material properties. Results of materials properties analysis have shown, that the replacement of Ge by Si and Se on S results in reduction of conductivity and thermopower measured at high temperatures has a negative sign. Replacement of Se by Te has caused complete crystallisation of the material and crystal $Ge_{20}Te_{70}Bi_{10}$ has conduction of p-type. Incorporation of antimony instead of bismuth causes reduction of specific conductivity and change of negative thermopower sign for positive. Thus only in materials containing such components as Bi and Se or Bi and S, electronic conductivity is observed. Researches of thermopower for $(GeSe_{3.5})_{100-x}Bi_x$ c x > 15 have shown, that S_{293K} decreases, and the sign remains negative, i.e. the phase state of the material does not influence on the type of conductivity.

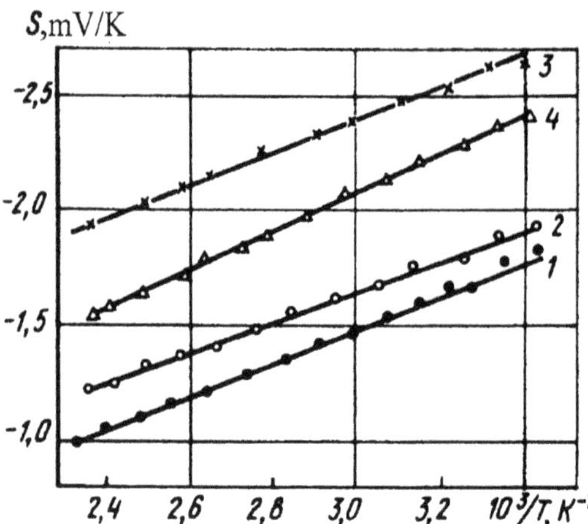

Fig.2 Thermopower versus reciprocal temperature of some composition: 1- $(GeSe_{3.5})_{90}Bi_{10}$, 2 - $(GeSe_{3.5})_{88}Bi_{12}$, 3 - $(GeSe_{3.5})_{86}Bi_{14}$, 4 - $(GeSe_3Te_{0.5})_{90}Bi_{10}$.

The replacement of selenium with sulfur does not cause the change of conductivity type. In this case study of transport properties and inversion conditions of conductivity type in Ge-S-Bi system was of great interest. As the object for researches compositions $(GeS_{3.5})_{100-x}$ Bi_x and $(GeS_3)_{100-x}$ Bi_x have been chosen. The dependencies $lg\sigma=f(1/T)$ in investigated temperature interval are rather well described by simple direct line. The high sensitivity σ and E_σ are observed at incorporation of small additives of bismuth [11]. Due to high resistance of the compositions $(GeS_3)_{100-x}$ Bi_x the thermopower was measured only for few samples near the transition from p-type to n-type of conductivity at high temperatures. Nevertheless, the received results have shown, that the inversion of conductivity type for $(GeS_3)_{100-x}$ Bi_x occurs at x> 8.

The further experiments were directed to revealing the nature of inversion of conductivity type in compositions $(GeSe_{3,5})_{100-x}$ Bi_x and $(GeS_{3,5})_{100-x}$ Bi_x. On mobility and transport properties of disordered semiconductors large influence was exercised by defects, which can form acceptor or donor levels. The new levels in mobility band gap of disordered semiconductors may act as traps for free charge carriers reducing their mobility. These additional levels in a mobility band gap occur due to introduction of the additives (for example Bi) and as a result of conditions of disordered materials preparation(speed of melt cooling).

In order to check the above stated assumptions on conditions of inversion of conductivity type the researches of temperature dependencies of conductivity and thermopower were carried out for amorphous thin films $(GeSe_{3,5})_{100-x}$ Bi_x. The results of conductivity temperature dependence measurement for film samples are given in fig. 5. From fig. 5 it is clear, that the temperature dependencies of conductivity of amorphous films do not comply with exponential law, and have more complicated behaviour. In modified amorphous thin films the conductivity is rather well described by expression $\sigma= \sigma_0 \exp [(T_0/T)^{1/4}]$. In this connection the dependencies were constructed in co-ordinates $lg(\sigma)=f(1/T)^{1/4}$. However even in these co-ordinates the results do not go in one direct line. In film samples the same behaviour σ_{293K}

depending on the contents of Bi is kept also, for many bulk samples, i.e. with growth of bismuth contents σ_{293K} increases. In this case σ_{293K} for film samples is by two orders more, than σ_{293K} for bulk samples of the same composition (table 3).

Table 3. Conductivity and thermopower for bulk and thin films samples of system Ge-Se-Bi.

Composition	$\sigma_{293K}, 10^{-8}$ $(\Omega \bullet cm)^{-1}$ film	$\sigma_{293K}, 10^{-10}$ $(\Omega \bullet cm)^{-1}$ bulk	S_{450K}, μV/K film	S_{450K}, μV/K bulk
$(GeSe_{3.5})_{94}Bi_6$	0,9	$2,3x10^{-3}$	+120	+320
$(GeSe_{3.5})_{90}Bi_{10}$	7,98	25,8	-56	-340
$(GeSe_{3.5})_{88}Bi_{12}$	50	62	-140	-460
$(GeSe_3Te_{0,5})_{90}Bi_{10}$	11	86	-	-350

Due to high resistance of the samples it was possible to measure the thermopower only for some samples, which are represented in tab. 3. It should be noted, that transition from a positive sign of thermopower to negative one occurs at the contents of bismuth 6-10 at.%, i.e. the inversion of conductivity type of films and bulk samples is observed practically at the same bismuth concentrations.

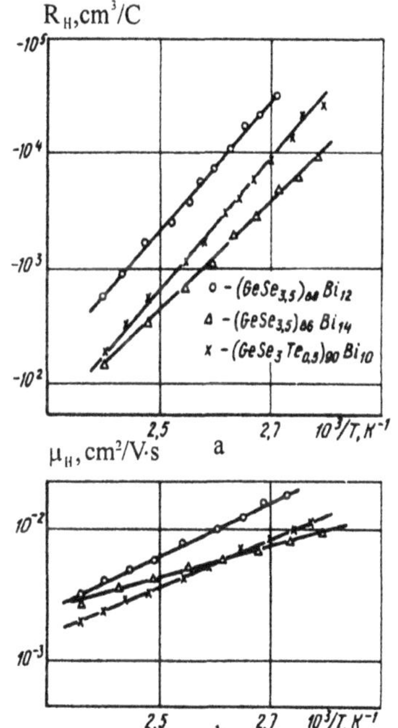

Fig. 3 Temperature dependence of the Hall coefficient (a) and the Hall mobility (b) of the CGS with n-type conductivity

Thus, the large concentration of defects in amorphous thin films, which is determined by synthesis conditions, does not exercise determining influence on conditions of inversion of conductivity type, but essentially changes transport properties. Change of transport properties of film samples are specified by large value of σ_{293K} and complicated behaviour of dependence $\lg(\sigma)=f(1/T)$.

Another approach to explanation of inversion of conductivity type is based on the fact, that in the matrix of glass Ge-Se or Ge-S microcrystalline phase Bi_2Se_3 or Bi_2S_3 with n-type conductivity are accumulated. It is connected with the fact that the incorporation of bismuth in $GeSe_{3,5}$ and $GeS_{3,5}$ significantly raises tendency of the material to crystallisation.

For revealing the prevailing role of

crystal phase in inversion of conductivity type in disordered materials $(GeSe_{3,5})_{100-x}Bi_x$ the researches on conductivity and thermopower were carried out for melt phase.

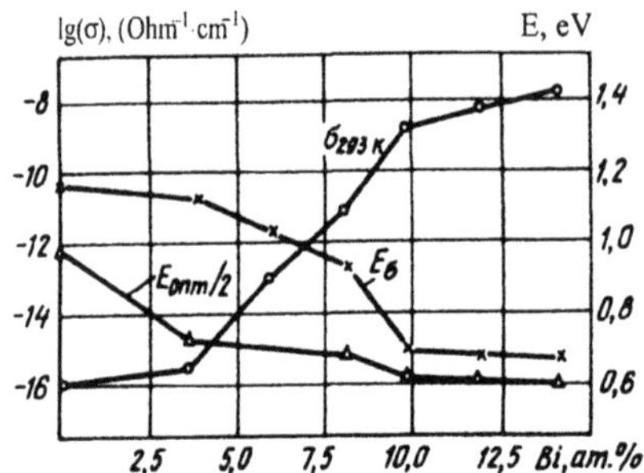

Fig.4 Concentration dependencies of electrical conductivity (σ_{293K}), activation energy of electrical conductivity (E_σ) and half width of an optical gap (E_{opt}) of CGS system $(GeSe_{3.5})_{100-x}Bi_x$.

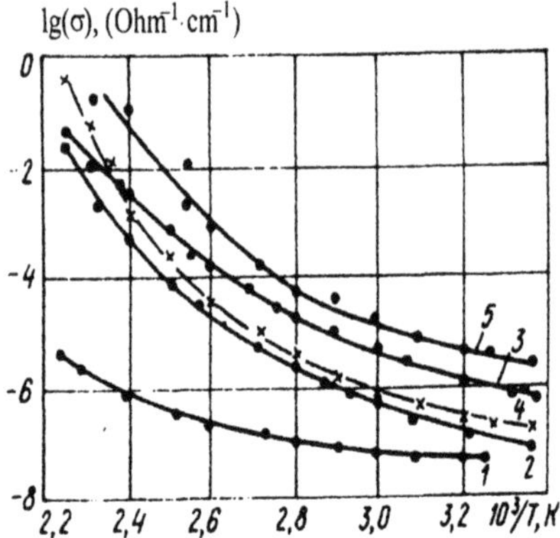

Fig.5 Temperature dependence of electrical conductivity of amorhous thin films of compositions: 1 - $(GeSe_{3,5})_{94}Bi_6$, 2 - $(GeSe)_{90}Bi_{10}$, 3 - $(GeSe)_{88}Bi_{12}$, 4 - $(GeSe_3Te_{0,5})_{90}Bi_{10}$, 5 - $Ge_{24}Se_{55}Te_{15}Bi_{10}$.

As the object for experiments compositions near the transition of inversion of conductivity type have been chosen. The peculiarity of the technique consists in the fact that after the load of glass powder in special ceramic measuring cells and heating to the temperature above melting, materials after cooling turned into a crystal phase. Results of conductivity and thermopower measurement in a wide interval of temperatures are represented in fig 6 a,b. Steep increase of conductivity appears during melting of the material. The further increasing of melt temperature up to 1000 K is also accompanied by increase of σ with the subsequent saturation. It should be noted, that in the whole measured interval of temperatures the thermopower for composition $(GeSe_{3,5})_{94}Bi_6$ has a positive sign, and for $(GeSe_{3,5})_{92}Bi_8$ and $(GeS_3)_{92}Bi_8$ - a negative one (fig. 6b). The most important thing about the received results is , that the change of thermopower sign from positive to negative in melted materials $(GeSe_{3,5})_{100-x}Bi_x$ and $(GeS_3)_{100-x}Bi_x$ occurs at x= 8, like in glassy state. So it is experimentally defined, that the transition in disordered solid state (glassy , amorphous) and melted state occurs at the same concentration of bismuth.

Another approach for revealing a determining role of a crystalline phase in the inversion of conductivity type includes the following. At the presence of microcrystalline phase with n-type conductivity homogeneously distributed in a glass matrix, material can be presented as heterojunctions. For systems $(GeSe_{3,5})_{100-x}Bi_x$ and $(GeS_{3,5})_{100-x}Bi_x$ glassy phase on the basis of Ge-Se or Ge-S is a width-band semiconductor with a p-type conductivity, and crystalline phase Bi_2Se_3 (Bi_2S_3) is a narrow-band one with electronic conductivity. In this case the electron injection from a crystaline phase into a matrix glass can stipulate electronic conductivity in a material. To check the above stated assumption concerning the transport mechanism measurements such as photoconductivity and value of photogeneration "red" border were carried out. A type of photoconductivity was determined by specific shift of carriers [12].The drift mobility (μ) was measured on film samples of 1,35 μm thick (d), received by thermal evaporation in vacuum. Aluminium or drop of water were used as electrode material. For the given structure the time of flight of electrons (t_n) at applied voltage (U) equal 3V is 5μs and consequently, the drift mobility, being determined by the formula $\mu_n = d^2/t_n \bullet U$, is equal to $1{,}2 \bullet 10^{-3}$ $cm^2/V \bullet s$ at temperature 300 K.

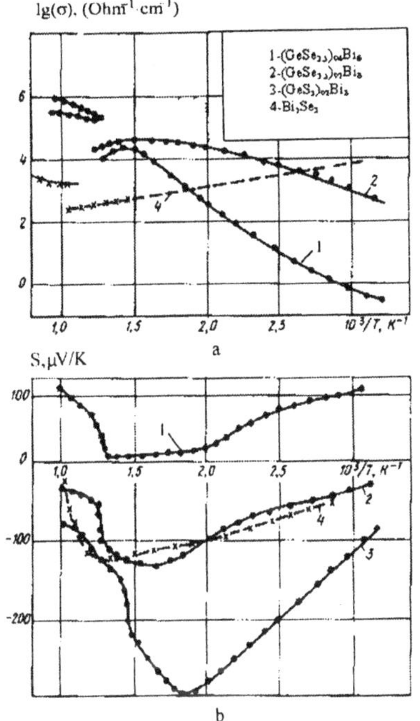

Fig. 6 Temperature dependencies of electrical conductivity (a) and thermopower (b) of solid (crystalline) and melted states

The specific shift of carriers ($\mu\tau$ is a product of mobility on time of life of change carriers) could be measured only for some compositions. In the samples $(GeSe_{3,5})_{88}Bi_{12}$, $(GeSe_{3,5})_{90}Bi_{10}$ and $(GeS_3)_{90}Bi_{10}$ specific shift for electrons considerably greater than for holes ($\mu_n\tau_n >> \mu_p\tau_p$), which correlates with measurement data of thermopower sign. For $(GeSe_{3,5})_{94}Bi_6$ specific shift for holes is more, than for electrons. The value of photogeneration red border was determined by appearance of specific shift signal at change of radiation wave length. The values of photogeneration red border change from 0,95 eV up to 1,3 eV. These values are closer to E_{opt} for the appropriate composition, which specifies photogeneration of carriers in the matrix of glassy phase. Since electronic conductivity is caused by injection of carriers from a crystalline phase Bi_2Se_3 or Bi_2S_3, the red border of photogeneration should not exceed the half of E_{opt} of the material. Crystalline Bi_2Se_3 is a narrow-band semiconductor with width of forbidden band 0,28 eV, and E_{opt} for Bi_2S_3 makes 1,3 eV. In this case the height of the barrier on the border of glassy and crystalline phases should be different for Ge-Se-Bi and Ge-S-Bi systems. These experiments have not established the similar law. Thus the alternative opportunity of an explanation of a n-type photoconductivity due to the electron injection from a assumed crystalline phase of Bi_2Se_3 (Bi_2S_3) is excluded.

3. Discussion

Thus the analysis of experimental results on studying the properties of disordered and crystalline materials with bismuth, specifies, that the inversion of conductivity type in these substances is not connected with the shift of E_F, as it occurs in a-SiH and partially in modified amorphous thin films. At incorporation of Bi in CGS the Fermi level remains practically in the middle of optical forbidden band and bends are not observed on temperature dependencies $lg(\sigma)=f(1/T)$. It could be assumed, that bismuth causes sharp increase of density of states near the middle of mobility gap and the hopping mechanism near E_F is the dominant transport character, similarly to modified amorphous thin films. Due to the large value of E_σ and E_S, linear behaviour of dependence $lg(\sigma)=f(1/T)$ and large value of thermopower (S=-1...2eV/K) completely eliminate the possibility of variable range hopping near E_F. Hence, the unique approach for an explanation of n-type conductivity in CGS with bismuth remains the mechanism of redistribution of hole and electron mobility ratio. It was established experimentally, that the addition of Bi in Se causes reduction of hole mobility to greater extent than of electron one [13] and at certain concentrations of bismuth mobilities are levelled, and then $\mu_n>\mu_p$. The larger value of electron mobility, than of hole mobility is a condition of inversion of thermopower sign.

It is necessary to find out, what causes the change in mobility of carriers at addition of Bi in CGS. The incorporation of any component in a disordered material causes increase of density of localised states in mobility gap. Bismuth atoms cannot form donor defects (negatively charged) because of large value of Se electronegativity. The electroneutrality of defects is provided as a result of occurrence of positively charged atoms of Bi_4^+, i.e. it is possible to present reaction as: $Se + Bi = Se_1^- + Bi_4^+$

Another alternative opportunity for keeping greater mobility of holes in CGS in disordered semiconductors is connected to that fact, that the deformation influences to greater extent on bonding and antibonding orbitals of electrons than on lone-pair orbitals. The energy levels of valence band are formed by lone-pairs and bonding orbitals, but conduction band is formed by antibondig orbitals. In this case the conduction band is disordered more than the valence band near mobility edge, because bonding electrons are located deeper relatively to lone - pair energy levels. Incorporation of additives in CGS decreases both holes and electron mobility but with different speed. As a result large concentrations of additives are required. To greater degree the additives, forming the bond with chalcogen will destroy energy levels near the edge of mobility. The energy bond of additive atom with chalcogen may be the same as energy of lone-pair electrons so in this way it is possible to explain the fact that the additives incorporated in CGS decrease μ_p more than μ_n. The upper part of the valence band near the edge of mobility is formed from levels of lone-pairs for CGS of systems $(GeS_3)_{100-x}Bi_x$ and $(GeSe_{3,5})_{100-x}Bi_x$, selenium and sulfur respectively. The additives of Bi in $GeSe_{3,5}$ interact with Se, forming chemical bond Bi-Se. The energy of bond Bi-Se is much less, than Ge-Se and Se-Se [14], according to the stated assumption concerning the increase of the disorder in $GeSe_{3,5}$ at introduction of Bi. Such approach explains obtained results of the experiment. So, the inversion of conductivity type in disordered materials $(GeS_3)_{100-x}Bi_x$ and $(GeSe_{3,5})_{100-x}Bi_x$ is observed at x> 8 in glassy, amorphous (thin film) and melt states. In each separate case the concentration of own defects should be different and one could think, that for their compensation, different quantity of bismuth is necessary. Hence, the concentration of own defects is not determinative during inversion of conductivity type, while the valence band in glassy, amorphous and melt states of CGS cannot change essentially and, accordingly, the dominant hole mobility in these states is kept. Only the additives of Bi cause changes in valence band, resulting in reduction of hole mobility. The transition from dominant hole conduction to electronic one is satisfactorily described by the percolation. For the first time the theory of percolation for the description of concentration dependencies of CGS conductivity of $Ge_{20}Se_{80-x}$, Bi_x, $Ge_{20}S_{80-x}Bi_x$, $(GeS_2)_x(Bi_2S_3)_{1-x}$ systems is used by P.Nagels [15].

Other additives, which can result in inversion of conductivity type should satisfy the following conditions. First,, the bond energy of this additive with chalcogen atom should be much less, than bond energy between other atoms, included in glass structure. Secondly, the incorporation of these additives in melt should not cause appearance of acceptor levels, as it takes place at addition of Mn and Cu. Thirdly, these additives should not result in crystallisation of glassy materials.

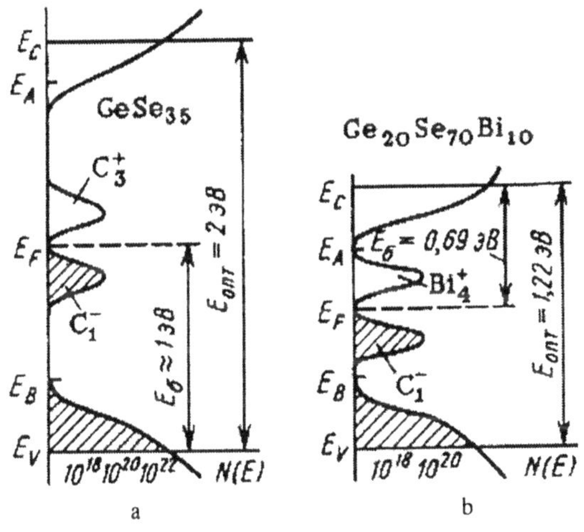

Fig. 7 Schematic density of states diagrams for two glasses.

To carry out all listed above conditions proves to be rather difficult. One should expect dominant electronic conductivity in CGS with the additives Ni and Fe. However, introduction of nickel and iron in melted chalcogenide material significantly raises tendency of the substance to crystallisation at cooling. As a result under normal conditions of cooling (10^2 K/s) one manages to incorporate only 2 ...3 at. % Fe [16], 1 at. % Ni in glassy material [17]. In this connection the inversion of conductivity type is already observed in glass-crystalline state. It is much easier to incorporate transitive metals in amorphous thin films in quantity by the method of sputtering (modification).

Experimental results obtained let us estimate the changes in energy spectrum of glass $(GeSe_{3,5})_{100-x}Bi_x$ and $(GeS_3)_{100-x}Bi_x$. For example in fig. 7 the qualitative diagrams of density of states for CGS $(GeSe_{3,5})_{100-x}Bi_x$ are shown at x=0 and x=10. These diagrams are constructed on the basis of experimental data. The diagrams of density of states for CGS $(GeSe_{3,5})_{100-x}Bi_x$ given in fig.7 allow to explain transport properties of the disordered materials. The detailed analysis of experimental results and the possible transport mechanisms are presented in [18].

We shall further discuss some results concerning the influence of different additives on properties of CGS. First we shall consider the additives, which display valence equal to unit. In the first place to such additives it is necessary to attribute the elements of the first group Li, K, Na. The transport properties of disordered materials with the additives Li, Na and K were studied in [19, 20]. Analysing these works it is possible to conclude the following: the elements of the first group in glass compensate positively charged defects and act as donor levels. Electronic conductivity at small concentration of the additives (less than 1 at. %) Li and Na are found in [20] and [21] respectively. It is clear, that the introduction of Li, Na and K causes appearance of donor level shifting the E_F closer to the E_C and increases density of states in tails and in mobility gap.

Such additives as Ga, In, Tl, Zn incorporated in glassy As_2Se_3 during synthesis [4, 5, 21] exercise the same influence on transport properties and the changes of the zone diagram of energy spectrum. Given additives do not change a p-type conductivity of As_2Se_3, but significantly change width of optical forbidden gap and reduce energy of activation . It is possible, to explain such results if we assume that the additives as Ga, In, Tl, Zn cause increase of density of states both in tails near the edge of the conduction and in valence band, and near the middle of mobility gap. Thus the reduction of E_{opt} occurs due to the delocalization of states near the edge of mobility valence band, and the shift of E_F to E_V is caused by two factors simultaneously by expansion of tails area and density of states fixing Fermi level.

The influence of transitive metals (Fe, Ni, Mn) and intransitive (Cu) on transport properties of CGS was already partially discussed earlier. We shall remind, that all these additives cause insignificant reduction of E_{opt}; increase of conductivity; sharp reduction of E_σ; the temperature dependencies of conductivity at low temperatures are rather well described by dependence $\lg(\sigma)=f\ (T)^{-1/4}$; the value of thermopower is about 50 μV/K; a sign of thermopower is positive for the additives Mn and Cu and negative for the additives Ni and Fe [16, 22-24]. Like in case of modified thin films, additive Ni, Fe, Mn and Cu, incorporated in melt in quantity up to 10 at.% cause increase of density of states in the field of tails (reducing E_{opt}), in mobility gap near E_F and appearance of new doping levels (reduction of E_σ). Donor and acceptor levels were determined according to the sign of thermopower, though the nature of acceptor levels remains unclear.

4. Concluding remarks

Thus the analysis of experiments and other scientific publications, has shown, that the incorporation of the additives in disordered materials causes simultaneous change of several elements on the diagram of density of states. As a rule, there are changes in width of an optical gap, density of states in tails and in the mobility gap. The inversion of conductivity type is stipulated in disordered materials due to

- the change of the ratio of electron to hole mobilities due expansion of the extended states or localised states in tails (for example at introduction of bismuth);
- dominant hopping transport mechanism near E_F (at modification of thin films by nickel);
- occurrence of doping levels which compensate charged states fixing E_F (Li^+ and Na^+).

References

1. Kolomiets, B.T. and Nazarova, T.F. (1959). To question about the role of impurity in conduction glassy semiconductors, Fizika i tekhnika poluprovodnikov **1**, 2-6.
2. Mott, N.F. and Davis, E.A. (1979) Electronic Processes in Noncrystalline Materials, Clarendon Press, Oxford.
3. Borisova, Z.U. (1976) Glassformation in chalcogenide systems and periodic system of elements, in B.T. Kolomiets (ed.), Proc. of the 6th Int. Conf. On. Amorph. and Liquid Semicond. Leningrad pp.2-6.
4. Kolomiets, B.T. (1976) Impurities and properties of chalcogenide glassy semiconductors, in B.T. Kolomiets (ed.), Proc. of the 6th Int. Conf. On. Amorph. and Liquid Srmicond. Leningrad pp. 23-34.
5. Kolomiets, B.T., Rukhlyadev, Yu. V., and Shilo, V.F. (1971) The effect of gallium, indium and thallium on the conductivity and photoelectric properties of glassy arsenic selenide, J. Non-Crystalline Solids **5**, 402-414.
6. Kastner, M. (1985) Chalcogenide glasses: Solutions and problem, J. Non-Crystalline Solids **77-78**, 1173-1182.
7. Schottmiller, J.C., Bowman, D.L. and Wood C. (1968) New vitreons semiconductors, J. Applied. Physics **39**, 1163-1169.
8. Tohge, N., Yamomoto, Y., Minami, T. and Tanaka, M. (1979) Preparation of n-tipe semiconducting $Ge_{20}Bi_{10}Se_{70}$ glass , J. Applied. Physics Letters **10**, 640-641.
9. Vikhrov, S.P. and Ampilogov, V.N. (1987) Electronic conductiviti of $(GeSe_{3,5})_{100-x}Bi_x$ and $(GeS_{3,5})_{100-x}Bi_x$ in solid and liquid states , J. Non-Crystalline Solids **90**, 441-444.
10. Vikhrov, S.P., Nagels, P. and Bhat, P.K. (1981) N-type conduction in cholcogenide glasses of the Ge-Se-Bi system, in J. T. Devrece, L.F. Lemmens, V.E. Van Doren and J. Van Royen (eds.), Recent Development in Condensed Matter Physics v.2 Metals, Disordered Systems, Surfaces and Interface, N.Y. and London: Plenum Press, 333-340.
11. Vikhrov, S.P., Ampilogov, V.N., Kengerlinski, L.Yu. and Himinets, V.V. (1984) N - type in glasses Ge-S-Bi system, Neorganicheskie materialy **9**, 1459-1461.

12. Vikhrov, S.P., Juska, G. and Ampilogov, V.N. (1984) To nature of inversion the type conductivity in CGS of Ge-Se-Bi and Ge-S-Bi systems, Fizika i tekhnika poluprovodnikov **2**, 348-350.

13. Takahashi, T. (1981) Two types of drift mobilities in amorphous Se-Bi and Se-As-Te-Bi systems, J. Non-Crystalline Solids **2-3**, 239-247.

14. Tohge, N., Minami, T. and Tanaka, M. (1980) Preparation and conductivity mechanism of n-type semiconducting chalcogenide glasses chemically modified by bismuth, J. Non-Crystalline Solids **38-39**, 283-289.

15. Nagels, P., Tichy L., Triska A. and Ticha, H., (1985) Physical properties of $(GeS_2)_x$ $(Bi_2S_3)_{1-x}$ glasses, J. Non-Crystalline Solids **77-78**, 1265-1268.

16. Bychkov, E.A., Vlasov, Yu.G. and Borisova, Z.U. (1978) Influence of additives of iron on physical and chemical properties of chalcogenide glasses, J. Fizika i khimiya stekla **3**, 335-339.

17. Averyanov, V.L. (1984) Modification of chalcogenide glassy semiconductors, Proc. Int. Conf. "Amorphous Semiconductors - 84", Gabrovo 155-159.

18. Aivazov, A.A., Budagyan, B.G., Vikhrov S.P. and Popov A.I. (1995) Non-Crystalline Semiconductors, Vysshaya Shkola Press, Moscow.

19.Endo, H. (1983) Electronic and thermodynamic properties of liquid chalcogenides, J. Non-Crystalline Solids **59-60**, 1047-1054.

20. Yao, M., Hosokawa S. and Endo, H. (1983) The effects of charged additives on the conductivity and the thermopower in liquid selenium, J. Non-Crystalline Solids **59-60**, 1083-1086.

21. Pfister, G. and Morgan, M. (1980) I The influence of thermally induced defects on transport in a - As_2Se_3, II The effect of metallic impurities on the transport properties of a-As_2Se_3, J. Philosophical Magazine **B.2**, 191-207, 209-234.

22. Shimizu, T., Watamable, I. and Shiomi, S. (1981) Effects of transition additives on Ge-S glasses, J. Solids State Communication **38**, 483-488.

23. Borisova, Z.U. (1982) Influence of mettalic impurities on physical and chemical properties of chalcogenide glasses, Proc. Int. Conf. "Amorphous Semiconductors - 82", Bucharest, 8-13.

24. Averyanov, V.L. and Tsendin, K.D. (1985) Doping glassy and amorphous semiconductors, "Fiziko - Tekhnicheski Institut A.F. Ioffe" Press. Leningrad.

MEDIUM RANGE ORDER IN CHALCOGENIDE GLASSES

M. POPESCU

Institute of Physics and Technology of Materials.
Bucharest-Magurele, P.O. Box MG7, Romania

1. Introduction

In contrast with the crystalline state characterized by *long range order* (LRO), i.e. by the existence of correlations between the positions of every two atoms situated as far as possible one from another, the non-crystalline state is characterized by the absence of the LRO. The remnant is not a total disorder but a certain limited order called *short range order* (SRO) defined by the inter-atomic correlations in the first coordination spheres of an arbitrary atom, i.e. up to the maximum distance where the bonding forces are acting.

The SRO in crystals and in non-crystalline materials of identical composition shows both resemblances and differences, the last ones representing the energy cost of the loss of LRO. The SRO is reflected in the radial distribution function (RDF) calculated from X-ray, neutron or electron scattering.

In many non-crystalline materials and, especially, in chalcogenide glasses the order extends up to larger inter-atomic distances. On this basis a new type of order was defined: the *medium range order* (MRO) or the *intermediate range order* (IRO).

The SRO is strongly related to the physico-chemical properties of the materials. MRO plays an essential role in the particular properties of chalcogenide glasses as e.g. the photo-induced modifications of various physical parameters and the recently found stable anisotropy.

Lucovsky [1] defined SRO in terms of inter-atomic correlations and local symmetries and proposed as limit for this order the distance to the third order neighbours (5-6 Å). The correlations which extend to a larger number of atomic spheres define the MRO.

The structures which exhibit MRO do not give well defined spectroscopic signatures. The vibration spectra and, particularly, the Raman spectra can be in some circumstances sensible to some details in MRO [2]. The most powerful methods for the detection of MRO are the diffraction methods (X-rays, neutrons, electrons). The structural modelling approaches are of great help in revealing subtle details of atomic arrangements.

The main information on MRO is obtained from the diffraction pattern which exhibits either a pre-peak on the low angle side of the main diffraction peak or a more or less developed *first sharp diffraction peak* (FSDP) in the low angles region and, even, a

A. Andriesh and M. Bertolotti (eds.),
Physics and Applications of Non-Crystalline Semiconductors in Optoelectronics, 215–232.

complex structure in the vicinity of FSDP. The details in RDF at high distances are related to MRO but very accurate curves are rarely available.

This paper aims to shed more light on MRO in non-crystalline chalcogenides, to discuss critically various models of MRO and to suggest a general model.

2. MRO in non-crystalline chalcogens.

The SRO in non-crystalline chalcogens (S, Se, Te) is now satisfactorily known.

Sulphur exhibits predominantly 8-atom rings while tellurium has the tendency to form atom chains. The mixture chains-rings seems to be proper to various forms of amorphous selenium. Nevertheless, there are experimental data which cannot be explained by assuming only SRO. Therefore MRO must be taken into account.

One of the most accurate structural investigations of non-crystalline sulphur was made by Tompson and Gingrich [3]. The X-ray diffraction pattern of amorphous sulphur exhibits a pronounced shoulder on the low angle side of the first (main) diffraction peak (Fig. 1a). When the temperature is raised the shoulder increases and gradually becomes a true peak. In the liquid phase this peak continue to increase with the temperature so that at 300°C it exceeds the height of the main peak. These details are, undoubtedly, related to MRO. The RDFs exhibit at large distances poor structural features related to MRO (Fig. 1b).

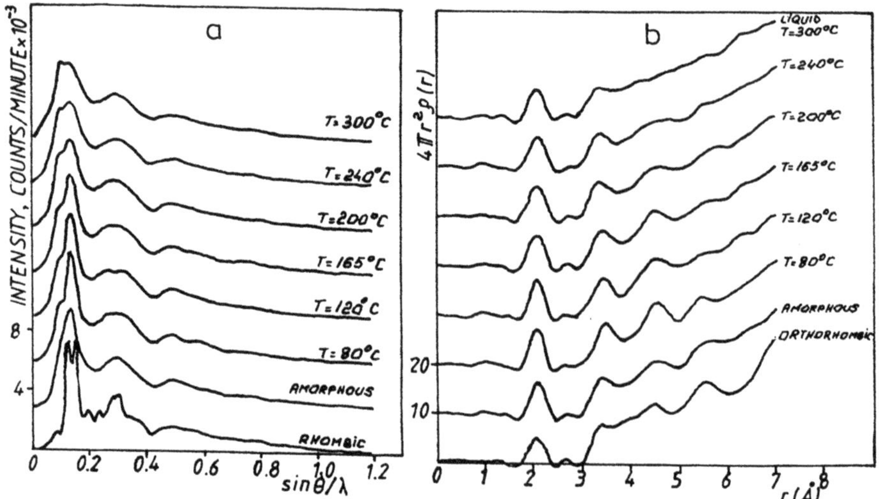

Figure 1 a. X-ray diffraction patterns for amorphous sulfur at 4°C, liquid sulfur at 80°C(supercooled) and at higher temperatures [3].

b. The RDFs obtained from the X-ray diffraction patterns shown in (a).

Accurate structure factor and pair correlation function for amorphous selenium were obtained by Bellissent and Tourand [4] from neutron diffraction experiments (Fig. 2). The first peak in the diffraction pattern has the attributes of the shoulder-peak observed in sulphur: high sensibility against temperature, preparation method, etc... The reduced RDFs (Fig. 2b) exhibit details related to MRO up to ~7Å.

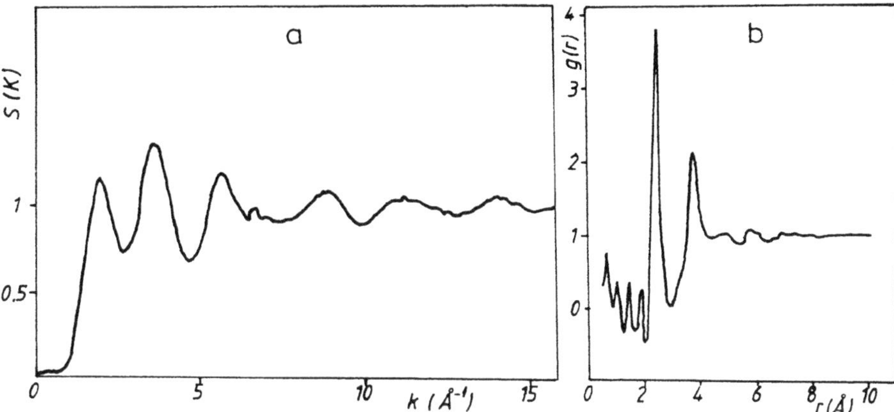

Figure 2 The structure factor (a) and the differential RDF (b) for amorphous selenium (after [4]).

As concerned amorphous tellurium, the first maximum in the X-ray diffraction pattern is very high and lacks any shoulder which fact speaks in favour of poor MRO, if any. The order limited to the first neighbours can be easily understood, because the atomic bonds of Te are far from covalent and show a strong tendency towards a metallic character. The amorphous state of Te is difficult to obtain because, probably, the order exceeding the first coordination spheres is lacking.

3. MRO in arsenic and germanium chalcogenides.

As_2S_3 and As_2Se_3 are the most investigated chalcogenides in the non-crystalline state. Due to a strong MRO signature revealed by FSDP these chalcogenides could provide the "clou" for the medium-range structure.

De Neufville *et al.* [5] have shown that fresh evaporated films of $As_2(S,Se)_3$ exhibit high, narrow FSDPs which undergo a dramatic change when the films are annealed (Fig. 3). In both type of films the FSDP decreases, broadens and shifts towards larger diffraction angles by annealing.

Big changes in MRO are revealed not only by annealing but also by transition in the liquid state, by applying high pressures and by appropriate irradiation with light of wavelength situated in the vicinity of the optical gap (Fig. 4 a-d) [6-9]. FSDP strongly increases in high temperature melts. The raising of the pressure up to 80 kbar leads to the complete vanishing of FSDP in As_2S_3. The illumination decreases the height of the FSDP, broadens this peak and shifts the peak towards higher diffraction angles.

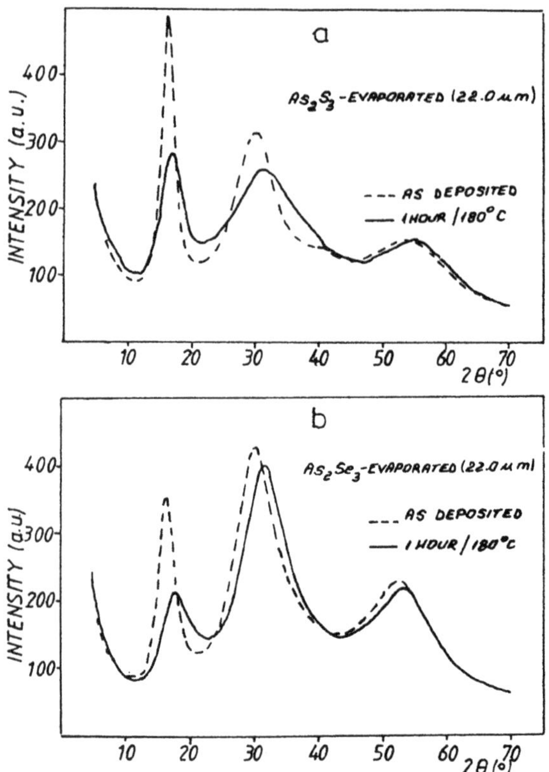

Figure 3 X-ray diffraction patterns (CuK_α) before and after annealing below Tg
a. As_2S_3 b. As_2Se_3

In the system Ge-S the best investigated amorphous composition is GeS_2. Detailed modelling of MRO in GeS_2 has been performed by Cervinka [10]. The X-ray diffraction pattern of a-GeS_2 and the corresponding RDF are given in Fig. 5 [11]. As can be easily seen a high FSDP is situated at the wavevector $k \sim 1.0 Å^{-1}$.

We have investigated three glassy compositions in the system Ge-Se: GeSe, Ge_2Se_3 and $GeSe_2$. GeSe and $GeSe_2$ can be found also as crystals while Ge_2Se_3 is a special composition situated at the boundary of the glass formation domain [12]. The X-ray diffraction patterns and the corresponding RDFs are shown in Figure 6. One observes a large variation of FSDP as a function of composition while the details in RDFs seem to remain not only less significant but also less reliable.

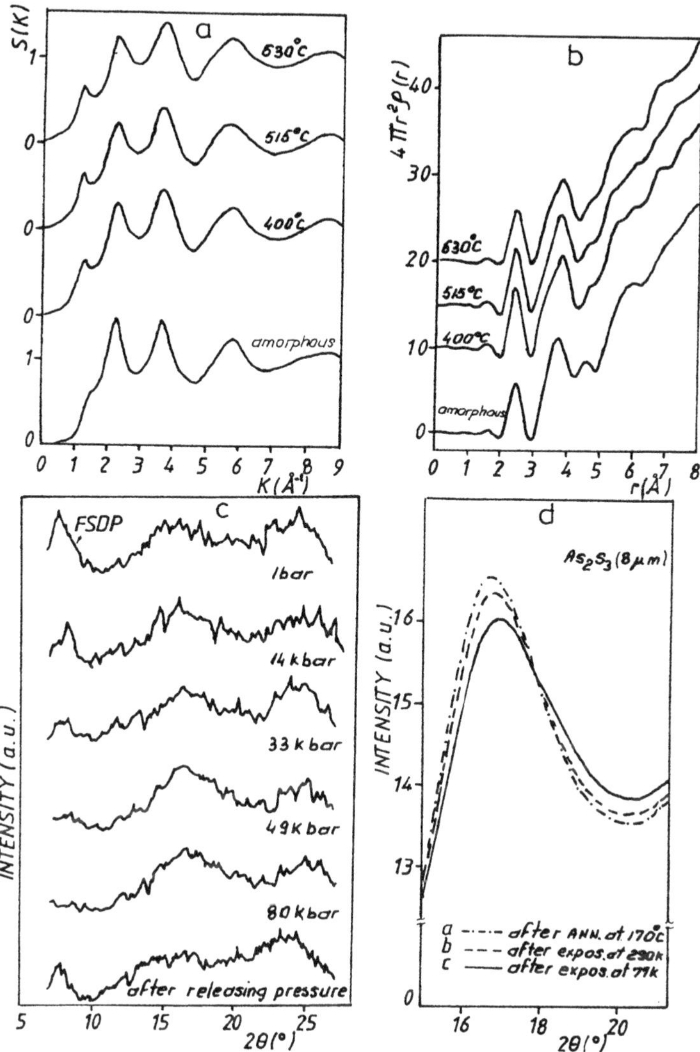

Figure 4 a. The X-ray pattern of a-As_2Se_3 at room temperature and in the liquid state at high temperatures [6]. b. The RDFs calculated from (a). c. X-ray diffraction patterns of glassy As_2S_3 at various pressures (MoK_αradiation) [7]. d. Modifications of the FSDP in As_2S_3 film by illuminations at 77K and 290K. (500W Hg arc through IR- cut filter) [8]. The a-b and a-c cycles are reversible.

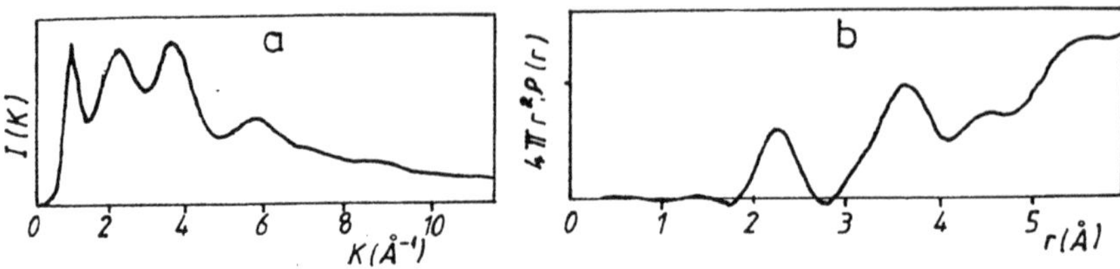

Figure 5 X-ray scattered intensity (a) and radial electron density distribution (b) in a-GeS_2 (after [11]).

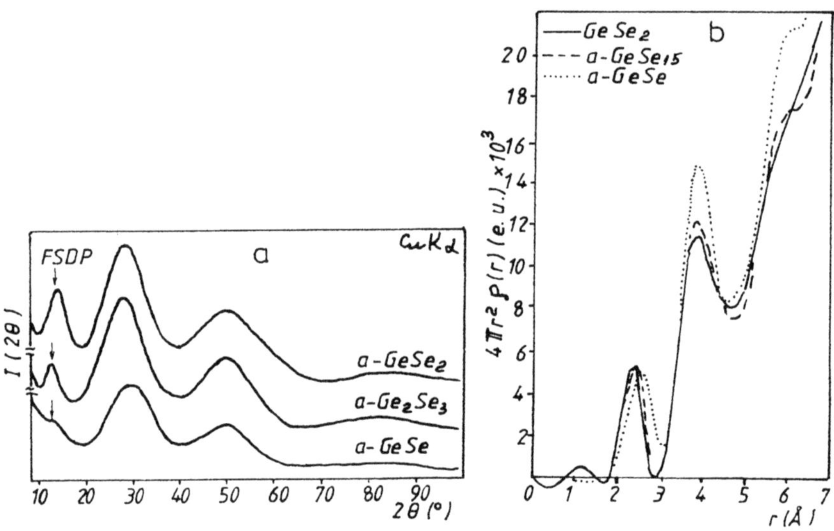

Figure 6 X-ray diffraction patterns (a) of glasses in the system Ge-Se and the corresponding RDFs (b).

In the germanium chalcogenides the FSDP exhibits significant modifications when the glasses are annealed [13], heated above the melting point [6] or subjected to pressures (Fig. 7) [14]. FSDP decreases by annealing, becomes higher in the liquid state, vanishes at high pressures and recovers by annealing at normal pressure.

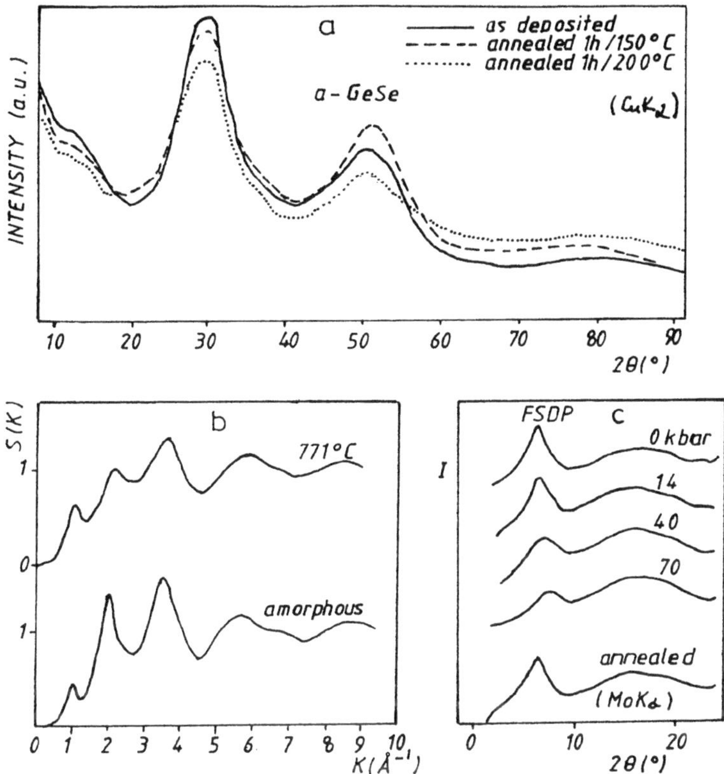

Figure 7 The modifications of FSDP in typical germanium chalcogenides
a. by annealing a GeSe glass [13]
b. by transition in the liquid state ($GeSe_2$) [6]
c. by applying high pressures (GeS_2) [14]

4. Layers against clusters: a MRO paradigm

One of the possible interpretations of FSDP is the presence of distinct layer-like features in the glass. In glasses derived from layered crystals as e.g. As, As_2S_3, As_2Se_3, the sharp and high diffraction peak at or near the reciprocal layer spacing has been interpreted as evidence for the formation of disordered layers. The FSDP is influenced by preparation and thermal treatment. X-ray measurements carried out by Tanaka [8] and Busse [15] have shown that the intensity of FSDP in well-annealed samples of As_2S_3 increases reversibly with the temperature (Fig. 4d) in contrast with the behaviour of the other peaks situated at higher diffraction angles. This anomalous behaviour has been interpreted in terms of increased ordering of the layers in the glass, made possible by the reduction in viscosity. Calculations of a quasi-crystalline model, do in fact, show the formation of a peak at the right position but of small height, impossible to correct by realistic modifications of the structural parameters.

We have performed the modelling of amorphous As_2Se_3 in the frame of a layer-like model starting from a continuous random model (CRN) with 146 atoms which was relaxed by computer. The model does not show any FSDP in the reduced interference pattern (Fig. 8) [12]. Then the CRN model has been transformed into a layered model (with three disordered layers) by changing the interatomic bonding network. The new model gives rise, surprisingly, to a FSDP of correct intensity (Fig. 8b)

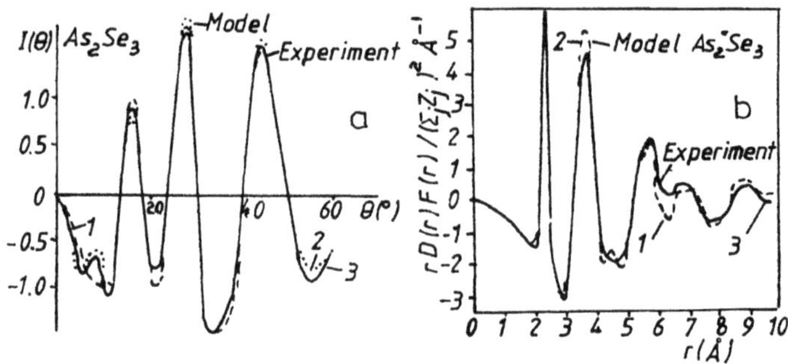

Figure 8 The reduced intensity curves (a) and the reduced RDF (b) calculated from two models of a-As_2Se_3 (1-CRN model; 2-layered model) and compared to the experimental curves (3) [12].

It is remarkable that the reduced RDF is much less sensitive to the changes in MRO. However, it is most unlikely that the quasi-crystalline picture involving interlayer correlations is generally valid. Some glasses (e.g. SiO_2) clearly exhibit FSDP, yet there is no evidence for the structure in either crystalline or amorphous state being layer-like or locally two-dimensional. Furthermore, the FSDP of e.g. $GeSe_2$ persists into the liquid state [16], seeming to rule out the microcrystalline explanation for its origin. Finally, X-ray scattering is independent of film thickness [17] and this fact could be an indication that preferential layer-like correlations probably do not exist.

On the other hand Bridenbaugh et al. [18] have presented Raman scattering data which show a companion line in $GeSe_2$ crystals similar to that found in glass. They proposed a model for $Ge(S,Se)_2$ glasses which suggests that the MRO of the network structure is basically layered than three-dimensional. With the hypothesis of layers in the glass, they succeeded to explain the origin of the companion A1 Raman line (tetrahedral breathing mode). The large clusters which represent the dominant structural elements of $GeSe_2$ (after [18]) are supposed to be made of stacked layer units of the kind shown in the Fig. 9. These are the so-called outrigger rafts and are largely discussed in the literature. The edge sharing tetrahedra stabilize the double-chain structure. The companion Raman line is associated with coupled motion of the dimerized chalcogens as indicated by the double arrows in Fig. 9.

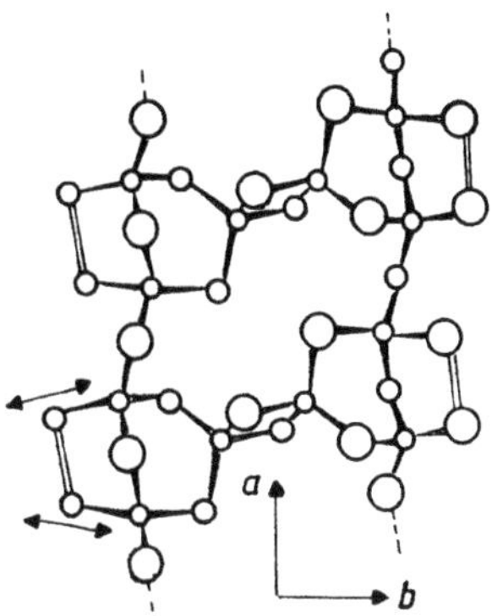

Figure 9 The outrigger raft clusters in $GeSe_2$ proposed in [18].

An alternative explanation of the origin of FSDP (and MRO) based on the packing of structural units has been largely discussed. Firstly, De Neufville *et al.* [5] proposed a structural model based on As_4S_6 (As_4Se_6) molecules for freshly deposited amorphous As_2S_3 (As_2Se_3). The samples exhibit narrow FSDPs which indicate a molecular correlation range of ~40 Å in analogy with the dense random packing of a hard sphere glass. This sharpness decreases substantially to the bulk glass value through cross-linking during annealing or by illumination.

Apling *et al.* [19] have simulated a molecular glass having as basic unit the As_4S_6 molecule (Fig. 10). The simple model is described by a random sphere packing calculated

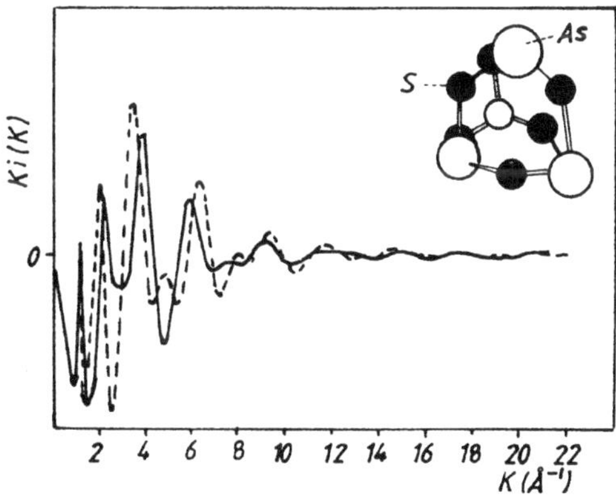

Figure 10 X-ray intensity pattern (reduced) calculated for random packing of As_4S_6 molecules: --------- model with packing fraction 0.532.
——— experimental data (inset : As_4S_6 molecule) (after [19]).

by assuming an effective molecular diameter of 6.4 Å and a packing fraction of 0.532. The model gives an excellent fit to FSDP but there are obvious deficiencies in the mid range of k. The sharpness of the FSDP in the model is easily adjusted by varying the packing fraction and altering the effective molecular diameter to maintain the overall density.

It is remarkable that many experimental and theoretical studies have shown that the FSDPof $Ge(S,Se)_2$ is determined by cation-centred correlations (Ge-Ge). The differential anomalous X-ray scattering data on a-$GeSe_2$ [17], Ge isotopic substitution neutron diffraction studies of liquid $GeSe_2$ [20] and molecular dynamic and other computer simulations of the structure of glassy and liquid $GeSe_2$ [21] all lead to the conclusion, implying that Ge-centred clusters are the structural units associated with the formation of FSDP in a-$GeSe_2$. Penfold and Salmon [20] found unambigous evidence from the partial correlation functions that cation-cation (Ge-Ge) correlations are the dominant contribution to FSDP.

5. A challenge for the MRO picture: The void correlation

It is unanimously accepted that the "signature" of the MRO in covalently bonded glasses is the FSDP or the pre-peak in the structure factor.

A new picture for MRO originated from the modelling studies of amorphous arsenic.

As early as 1979 Greaves et al. [22] pointed out that the continuous random network (CRN) model of amorphous arsenic is somewhat cavern-like. The structural relaxation and concomitant ordering of the caverns can (via a Babinet argument) produce a sharper FSDP. In the same time the comparison between the diffraction patterns of the initial a-As and the annealed a-As allows to conclude that the local order (evidenced at high diffraction angles) is not affected by longer range organization of the basic structural units. These units in a-As give rise to the value of $k_1.d_s$=2.4, similar to that of chalcogenides, but this parameter is the smallest one if compared with the same parameter of other glasses (d_s is the spacing determined from FSDP position). This fact emphasizes the openness of the a-As structure, where k_1 is associated with a spacing considerably larger than a bond length.

Elliot [23] has taken into account the FSDP to short-range ordering of interstices around cation-centred clusters. The Elliott's picture of the structure of say an AX_2 glass is thus an aggregate of soft (i.e. overlaping) cation-centred quasi-spherical clusters separated by the cation-cation distance, each cluster being surrounded by an equal number of spherical voids at a distance d_s. In this model the marked decrease in intensity of the FSDP with increasing pressure is understandable. The application of pressure causes a densification of the glass structure, i.e. a diminishing of the interstitial volume. Thus, it is expected that the FSDP intensity should decrease.

The anomalous temperature dependence of the FSDP in chalcogenide glasses can also be understood. Vashishta *et al.* [21] have shown by molecular dynamics and other computer simulations of a-$GeSe_2$ structure, that in the diffraction pattern only FSDP exhibits an anomalous increase in the peak intensity with decreasing density at constant temperature. The anomalous temperature dependence of the intensity of the FSDP is thus due to the decrease in density of the glass with increasing temperature. This behaviour is predicted by the Elliott's model since FSDP intensity is expected to scale with the amount of void volume clustered around cation-centred units in the structure.

We have compared [24] various models of a-As for the FSDP profile and for the void structure. The void radius distribution has been determined as the distribution of the radii of the largest spheres which can be introduced into the free spaces of the network.

Fig. 11 shows the results. While the CRN model of Greaves [22] gives only a very

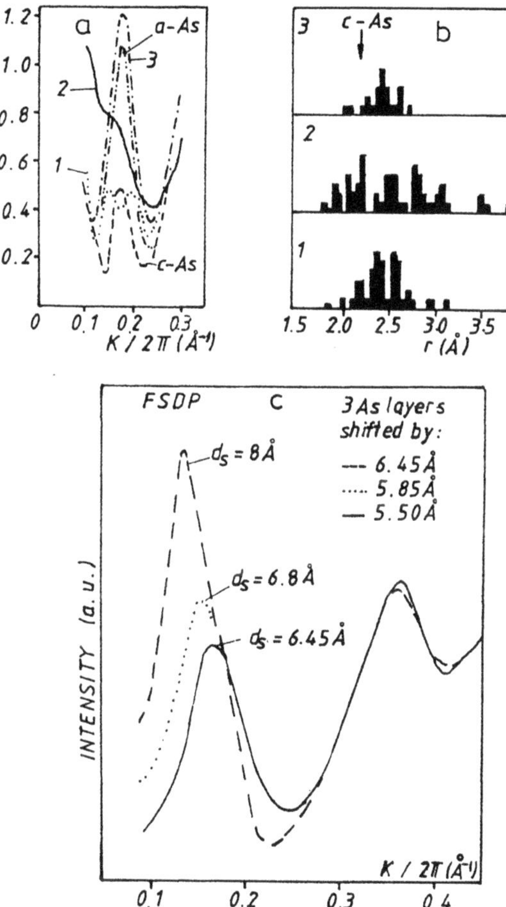

Figure 11 a. FSDP in a-As, c-As and in three models of a-As: 1. Beeman's model, 2. Greaves' model. b. The void distribution in the models. c. The evolution of the FSDP intensities when the interlayer spacing is increased in the Popescu's model.

faint FSDP and exhibits a very large void distribution, the model with disordered layer packing developed by Popescu [25] gives a very strong FSDP but exhibits a narrow void distribution. It is interesting to remark that the Beeman's model (private communication of the coordinates) is mainly CRN but, at a deeper analysis, have been found regions of very high and very low densities. This bimodal distribution of densities (and void-size) gives rise to a split FSDP. This could be regarded as a proof that void size distribution contributes to the control of the FSDP. The crystalline As is, nevertheless, intriguing by its small first diffraction peak, while the void radius is unique. In order to explain this feature we must observe that the crystal void radius is significantly smaller than the mean void radius in the models. When the layer spacing in the Popescu's model [25] is increased, then a strong increase of FSDP is observed (Fig. 11 b). A systematic increase of the FSDP intensity with the increase of the inter-cluster spacing was found also by Červinka [11] for a simulated cluster structure (the clusters were obtained by truncating the high temperature form of crystalline $GeSe_2$ (36 atoms)). Therefore the increase of the FSDP intensity of the As crystal FSDP can be regulated by the increase of the layer spacing.

6. Model with layered domains and thin interfaces: A compromise?

The new discovery of a small FSDP in carefully prepared amorphous silicon films [26] led us to suggest a model for MRO with amorphous domains (amorphites) separated by thin interfaces.

We have built a tetrahedral model with 405 atoms of the CRN type where the main feature is the separation in two equal parts through a relative uniform thin interface. A small number of distorted bonds links the two amorphous domains [27]. Figure 12 b shows the interference curves for the a-Si model with a thin interface between two amorphites (203+202 atoms) in the low angle region. One observes the formation of a small FSDP. Its intensity depends on the thickness of the interface, as proved by various simulations with positive and negative shift of the amorphous domains one from another. As regarding the chalcogenide glasses, what before was interpreted as a true layer-like structure can be now considered as a network of thin, interfaced layered domains. In the layer-like model the number of correlated layers deduced from the FSDP width is 4 for a-As_2S_3 and 5 for a-As_2Se_3 [15]. The layers in our model are not individual superstructural entities but parts of a disordered domain.

Červinka [28] has discussed in great detail the effect of the structural parallelization of the atomic configurations in two neighbouring clusters. The parallelization determines the intensification of FSDP in $Ge(S,Se)_2$. Our study gives reason to affirm that the thin interface could induce similar parallelization between boundary layers because this arrangement reduces the total free energy of the structure.

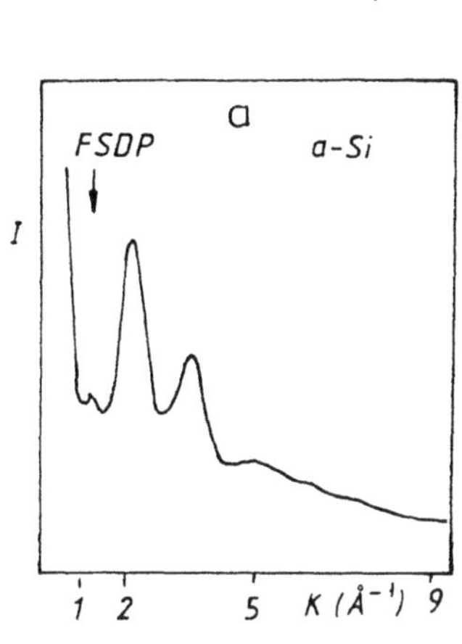

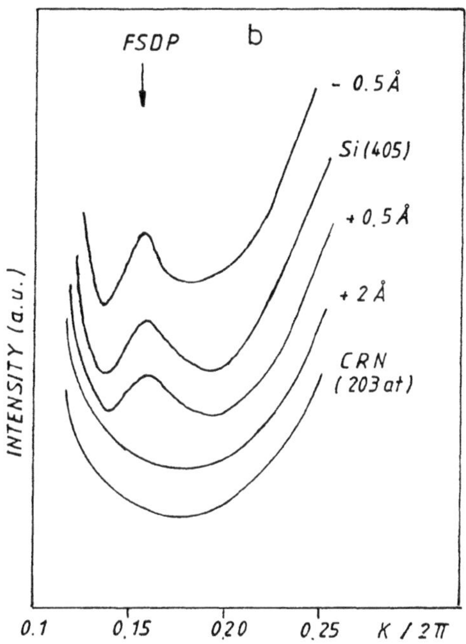

Figure 12 a. The interference pattern of a-Si obtained by Viščor [26]. A small FSDP is evidenced. b. The FSDP reproduced by the model with thin interfaces for various interface widths.

An interesting observation made by Červinka is that the inter-cluster distance (in our model the width of the interface) is a parameter which controls to a greater extent the FSDP intensity and less the position of the FSDP. Nevertheless, we have found by modelling the a-As in the frame of Popescu's model, that it is not necessary to have well defined structural entities in order to produce a large FSDP and that the position of the FSDP changes significantly and linearly with the shift of the layer spacing (Fig. 13).

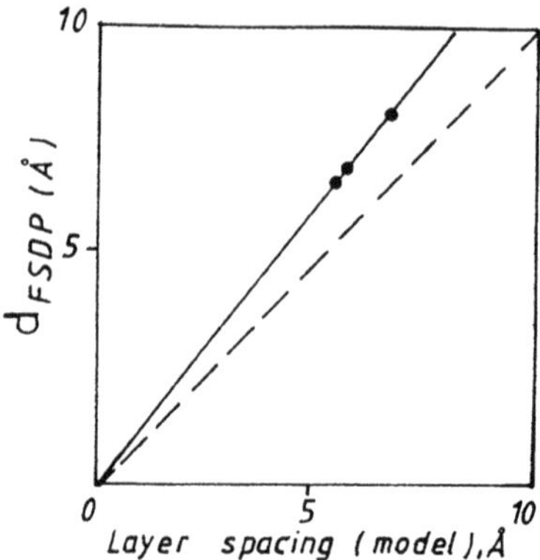

Figure 13 The linear relation between the distance extracted from the FSDP position and the layer spacing in the model with three stacked disordered layers for amorphous arsenic. (----- expected)

The simple explanation of the effects of various external parameters (temperature, pressure, illumination) on MRO, reflected in the FSDP characteristics, gives strong support to our model with layered amorphites and thin interfaces percolating through the glass volume. The increase of the temperature broadens the interface and this leads to a higher FSDP as experimentally observed. The application of high pressures narrows the interface and thus the FSDP must decrease and even vanish, as observed. The illumination with appropriate light determines the formation of bridges between amorphous domains through the interface in the points of highly distorted bonds. Thus the interface is compressed and loses its uniformity. As a consequence the FSDP peak must decrease and shift towards higher diffraction angles, while its width must increase. These features have been found experimentally and reported in the literature.

The new model, after our opinion, exhibits the main features of the layer-like and cluster model and includes, implicitly, the void structure and void ordering from the Elliott's model.

In the amorphous and liquid chalcogens the atoms can easily form stacks of rings or bundles of chains separated by interfaces.

In the non-crystalline chalcogenides the domains are probably thin, oblong, with higher lateral extension, randomly positioned. Their formation depends on the growth of the nuclei in the solid state and on the local temperature fluctuations in melts.

Quasi-molecular defects under the form of three-centre bonds, proposed by Dembovskii [29] as well as the VAPs and dangling bonds seems to be a necessary ingredient in the model due to the possible accumulation of large distortions in some points of these complex networks.

7. The complexity of MRO and its anisotropy

The first observation of the anomalous shape of a FSDP was made in 1980 [24]. In the CRN model of a-As built by Beeman we found a split FSDP.

Recently Arsova *et al.* [30] observed an assymetric shape of FSDP in Ge-As-Se bulk vitreous samples. Moreover, there was evidenced a small peak at low angles, which was called pre-FSDP (Fig. 14). This pre-peak create a new challenge for those aiming to decipher the structural order in chalcogenides. Similar results were obtained in Ge-Sb-S glasses [31] (Fig. 14 b). A complex low angle structure was revealed in the diffraction pattern of As_2S_3 bulk samples doped by indium. Besides the pre-FSDP two smaller peaks appear in the same region. The same complex structure in the FSDP region was revealed in As_2S_3 samples doped by iodine [33] (Fig. 15).

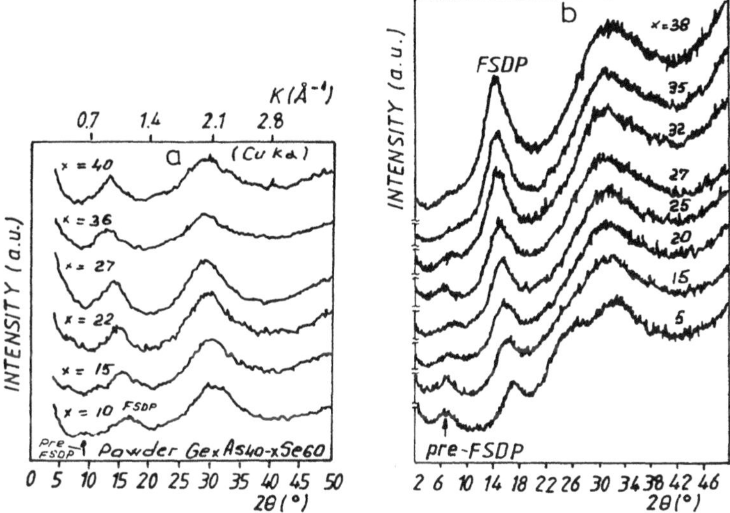

Figure 14 a. The FSDP and its pre-FSDP companion in the structor factor of $Ge_xAs_{40-x}Se_{60}$ [30] b. The region of the FSDP in the structure factor of $Ge_xSb_{40-x}S_{60}$ bulk glasses [31].

The extra features in the FSDP region of the X-ray diffraction patterns suggest that the MRO correlations in chalcogenide glasses can be very complex and can extend up to rather larger distances than those circumscribed in the classical definition of MRO.

The authors of [32,33] think that the large-scale MRO in doped chalcogenide glasses is formed by a discrete hierarchy of non-crystalline superstructural units or nano-clusters with specific dimensions. The nano-clusters are considered to be regions of Me-enriched As-S-Me solutions built in a Me-depleted matrix close to glassy As_2S_3 in composition.

In the non-doped chalcogenides the pre-FSDP is situated in a position which corresponds to a structural spacing of about twice as large as that characteristic to

FSDP. Vateva *et al.* [31] suppose that an hierarhical relationship could be assumed between the two peaks.

In the frame of our model we believe that the doors are open for more complexity in MRO because three parameters are responsible for the particular MRO features: the interface width, the amorphite size and the correlated density fluctuations during the formation of the structure with the amorphous domains.

In close relation with the discovery of the complex MRO in glassy chalcogenides a very important property of MRO seems to be revealed in some cases: the stable anisotropy. Indeed, by changing the azimuthal orientation of the bulk glassy samples in the X-ray beam, the pre-FSDP and even FSDP exhibit significant modifications of intensity. Mazets *et al.* [32] and Baidakova *et al.* [33] have observed modifications in As_2S_3:In and in As_2S_3:I (Fig. 15 a,b). Very recently Arsova *et al.* [30] have demonstrated that the stable anisotropy occurs also in the Ge-As-Se glasses (Fig. 15 c).

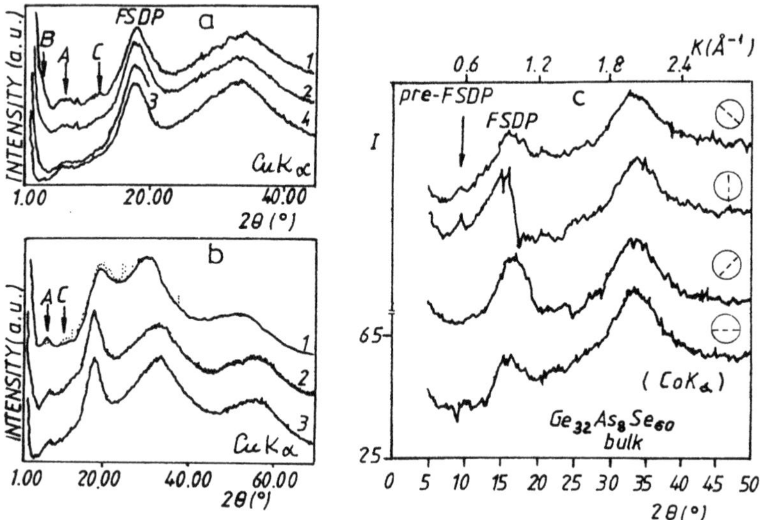

Figure 15 a. The X-ray diffraction pattern for $(As_2S_3)_{0.95}In_{0.05}$ for two orthogonal azimuthal sample orientations (1,3) and an intermediate orientation (2). The curve (4) is for pure As_2S_3.
b. The X-ray diffraction pattern for $(As_2S_3)_{0.56}I_{0.44}$. The solid and dotted lines in (1) correspond to two orthogonal azimuthal orientations of the sample. Curve (2): As_2S_3.
c. X-ray diffraction patterns for the $Ge_{32}As_8Se_{60}$ bulk glass. Four different azimuthal angles of orientation of the X-ray beam relative to the projection of the optical axis on the surface of the sample are shown.

While the complex behaviour of the FSDP can be explained satisfactorily in the frame of models with layered amorphites and interfaces, by adding new ingredients as e.g. very compressed (or stretched) regions alternated with very relaxed ones, the stable anisotropy of the glass is a challenging problem.

It is quite remarkable that 22 years ago Vancu *et al.* [34] have observed an anisotropic behaviour of the light scattering in As_2Se_3 bulk glass and demonstrated that oblong cylindrical scatterers oriented radially to the axis of the sample do exist. The measurements were performed on a cylindrical As_2Se_3 sample with the axis perpendicular to the ampoule axis). Scanning electron microscope studies revealed oblong centers with a length of a few micrometers. A possible origin of these centres could be the preparation method: the batches with the material suffer, during quenching, not only a radial stress but also a vertical contraction along their central axis.

We suppose that special elongated configurations in the glass could take a certain degree of reciprocal orientation during melt cooling thus triggering the installation of some kind of stable anisotropy.

8. Conclusions

The structural ordering at medium range distances is a general feature of the non-crystalline chalcogenide materials. MRO is reflected in the low angle range of the interference patterns as a first sharp diffraction peak (FSDP) or a more complicated structure (shoulder on the low angle side of the main peak, pre-FSDP and post-FSDP).

Several basical models for MRO have been developed more or less successfully: layer-like model, cluster model, void ordering model.

The model with amorphous, oblong, layered domains separated by thin interfaces and voids explains the FSDP signature and, moreover, offer a simple way to understand the modifications of the MRO in the chalcogenide glasses subjected to high pressures as well as those induced by light irradiation, by annealing and by melting.

The complex structure of the FSDP, recently revealed, and the discovery of the stable anisotropy are new challenges for MRO in chalcogenide glasses.

References

1. Lucovsky, G. (1987) Specification of medium-range order in amorphous materials, *J. Non-Cryst. Solids* **97&98**, 155-158.
2. Lucovsky, G. and Galeener, F.L. (1980) Local structure and vibrational spectra of a-As_2O_3, *J. Non-Cryst. Solids* **37**, 53-70.
3. Tompson, C.W. and Gingrich, N.S. (1959) Atomic distribution in liquid, plastic, and crystalline sulphur, *The Journal of Chemical Physics* **31** (6), 1598-1604.
4. Bellissent, R. and Tourand, G. (1977) Neutron diffraction study and quasicrystalline model for amorphous selenium, *Proc.of the 7th Int. Conf. on Amorphous and Liquid Semiconductors*, June 1977, ed. Spear W. E., 98-104.
5. De Neufville, J. P., Moss, S.C. and Ovshinsky S.R. (1974) Photostructural transformations in amorphous As_2Se_3 and As_2S_3 films, *J. Non-Cryst. Solids* **13**, 191-223.
6. Uemura, O., Sagara, Y., Muno, D. and Satow, T. (1978) The structure of liquid As_2Se_3 and $GeSe_2$ by neutron diffraction, *J. Non-Cryst. Solids* **30**, 155-162.
7. Tsutsu, H., Tamura, K. and Endo, H. (1984) Photodarkening in glassy As_2S_3 under pressure, *Solid State Comm.* **52**(10), 877-879.

8. Tanaka, K.(1975) Reversible photoinduced change in intermolecular distance in amorphous As_2S_3 network, *Appl. Phys. Lett.* **26**, 243-245.
9. Yang, C. Y., Paesler, M. A., and Sayers, D. E. (1987) Measurements of local structural configurations associated with reversible photostructural changes in arsenic trisulfide films, *Phys. Rev.* **B 36**, 9160-9167.
10. Červinka, L. (1987) Comments on medium-range ordering in non-crystalline solids, *J. Non-Cryst. Solids* **97&98**, 191-223.
11. Červinka, L, Smotlaha, O. and Tichý, L. (1987) A study of the structure of $(GeS_2)_{1-x}(Sb_2S_3)_x$ glasses, *J. Non-Cryst. Solids* **97&98**, 183-186.
12. Popescu, M., Andries, A., Ciumash, V., Iovu, M., Shutov, S., and Tsiulyanu, D. (1996) *Physics of Chalcogenide Glasses* (roum.), Stiintza and Stiintifica Publ. Houses, Chishinau and Bucharest.
13. Popescu, M., Stötzel, H., and Vescan, L. (1980) Structural modelling of amorphous GeSe, *Proc. Intern. Conf. "Amorph. Semic. '80"*, Kishinev, USSR, ed. Andriesh A. M., **1**, 44-47.
14. Tanaka, K. (1987) Pressure-structural and optical studies of glassy chalcogenides *J.Non-Cryst. Solids* **90**, 363-370.
15. Busse, L. E. (1984) Temperature dependence of the structures of As_2Se_3 and As_xS_{1-x} glasses near the glass transition, *Phys. Rev.* B **29**(6), 3639-3651.
16. Susman, S., Price, D.L., Volin, K.J., Dejus, R.J., and Montague, D.G. (1988) Intermediate-range order in binary chalcogenide glasses: the first sharp diffraction peak, *J. Non-Cryst. Solids* **106**, 26-29.
17. Fuoss, P.H. and Fisher-Colbrie, A. (1988) Structure of a-$GeSe_2$ from X-ray scattering measurements, *Phys. Rev.* **B 38**, 1875-1878.
18. Bridenbaugh, P.M., Espinosa, G.P., Griffiths, J.E., Phillips, J.C. and Remeika, J.P. (1979) Microscope origin of the companion A_1 Raman line in glassy $Ge(S,Se)_2$, *Phys. Rev.* **B 20**(10), 4140-4144.
19. Apling, A.J., Leadbetter, A.J. and Wright, A.C. (1977) A comparison of the structures of vapour-deposited and bulk arsenic sulphide glasses, *J. Non-Cryst. Solids* **23**, 369-384.
20. Penfold, I.T. and Solomon P.S. (1990) A neutron diffraction study on the structure of molten $GeSe_2$: the Ge coordination environment, *Journal of Physics: Condensed Matter* **2**, SA 233-SA 237.
21. Vashishta, P., Kalia, R.K., Antonio, G.A. and Ebbsjö, I. (1989) Atomic correlations and intermediate-range order in molten and amorphous $GeSe_2$, *Phys. Rev. Lett.* **62** (14), 1651-1654.
22. Greaves, G.N., Elliott, S.R. and Davis, E.A. (1979) Amorphous arsenic, *Adv. Phys.* **28**, 49-141.
23. Elliott, S.R. (1991) Origin of the First Sharp Diffraction Peak in the Structure Factor of Covalent Glasses, *Phys. Rev. Lett.* **67** (6), 711-714.
24. Popescu, M. (1980) Medium-range order in amorphous arsenic, *Proc. Intern. Conf. "Amorph. Semic. '80"*, Chisinau, p.21.
25. Popescu, M. (1978) Structural properties of low coordinated non-crystalline materials, *Proc. Intern. Conf. "Amorph. Semic. '78"*, Pardubice, p.203.
26. Viščor, P. (1988) Structure and existence of the first sharp diffraction peak in amorphous germanium prepared in UHV and measured in-situ, *J. Non-Cryst. Solids* **101**, 156-169.
27. Popescu, M. (1995) The medium-range structure of amorphous hydrogenated silicon, *J. Non-Cryst. Solids* **192&193**, 140-144.
28. Červinka, L. (1988) Medium-range order in amorphous materials, *J. Non-Cryst. Solids* **106**, 291-300.
29. Dembovsky S.A. (1989) Model of chalcogenide glassy semiconductors based on three center bonds in a rigid covalent network, *J. Non- Cryst. Solids* **114**, 115-117.
30. Arsova, D., Vateva, E., Petkov, V. and Skordeva, E. (1996) New features in the medium-range order in $Ge_xAs_{40-x}Se_{60}$ glasses, *Solid State Comm.* **98** (6), 595-598.
31. Vateva, E. and Savova, E. (1995) New medium-range order features in Ge-Sb-S glasses, *J. Non-Cryst. Solids* **192&193**, 145-148.
32. Mazets, T.F., Smorgonskaya, E.A. and Tikhomirov, V.K. (1993) Stable anisotropy in glassy As_2S_3:In, *J. Non-Cryst. Solids* **164-166**, 1215-1218.
33. Baidakova, M.V., Faleev, N.N., Mazets, T.F. and Smorgonskaya, E.A. (1995) Nano-scale medium-range order in semiconducting glassy chalcogenides, *J. Non-Cryst. Solids* **192&193**, 149-152.
34. Vancu, A., Grigorovici, R., Süptitz, P. and Brink, R. (1974), *Proc. Intern. Conf. "Amorph. Semic. '74"*, Reinhardsbrunn, **2**, 276-279.

SIMPLE THEORY OF SOFT POTENTIALS AND NEGATIVE - U CENTERS IN CHALCOGENIDES

K.D.Tsendin
A.F.Ioffe Physico-Technical Institute
194021 St.- Petersburg, Russia

1. Introduction

It is very well known, that the main part of intrinsic defects in chalcogenides are centers with negative effective correlation energy U (negative - U centers). They govern such important phenomena as radiative and non-radiative recombination, photoinduced absorption, Fermi level pinning etc.

There are two groups of negative-U centers models. Those belonging to the first group are based on a quantum chemical approach [1,2]. These models explain the negative value of U in terms of the specific electronic structure of chalcogenide atoms, which have lone - pair electrons (LP - electrons). It is the excess bonding created by LP - electrons that leads to a situation when charged defect states are favored over the neutral ones. But in these models the strong electron-phonon interaction is postulated and considered independently of LP - electrons.

In the second group models both strong local decrease in hardness coefficient (soft potential defect) and electron-phonon interaction have been supposed independently [3]. In a previos paper [4] I have shown that the last two suggestion are not independent, and the electron-phonon interaction, which takes into account the hybridization of one-electron wave functions of intrinsic defect, describes the appearance of soft, two - well potentials and negative-U simultaneously. The only thing necessary for this is a proper set of energy levels.

The present paper demonstrates the fact that chalcogenide atoms provide such a set of levels just due to the LP - electrons. It will be shown that soft and two-well potentials may be described by taking into account the electron-phonon interaction between atomic potentials, which are situated close to each other due to existence of LP-states. The form and hardness coefficient of the atomic potential will be dependent on the electronic configuration and charge state of a defect.

Thus, from the results of the present paper, it may be concluded that the above-mentioned two groups of models are not contradictory. Taking into account the chemical nature of chalcogenides and rather strong

A. Andriesh and M. Bertolotti (eds.),
Physics and Applications of Non-Crystalline Semiconductors in Optoelectronics, 233–241.

electron-phonon interaction in them one can obtain negative-U centers with soft or two-well potentials.

2. Model and calculations

The structure of the simplest chalcogenide - glassy Se is shown in the (xy) plane in Fig.1.

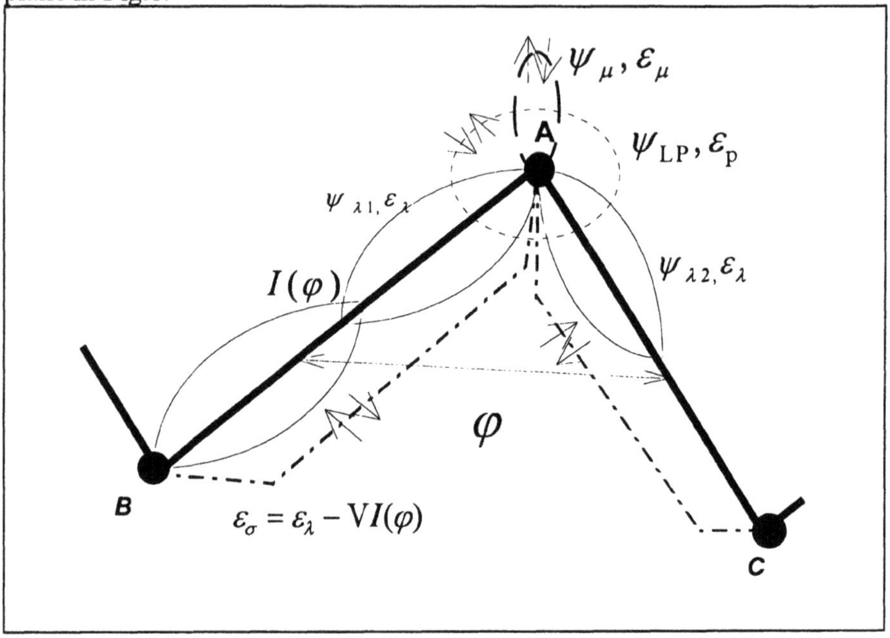

Figure 1.
Schematic picture of Se-atom chain. (ψ_μ , ε_μ), ($\psi_{\lambda 1,2}$, ε_λ) and (ψ_{LP}, ε_p) - wave functions and energy levels. I(φ) - overlap integral between two ψ_λ - wave functions centered at neighbouring atoms A and B. $\varepsilon_\sigma = \varepsilon_\lambda$ - VI - bonding energy.ε

Let us consider angle φ between two bonds of atom A with the nearest atoms B and C, as a main configuration coordinate. It is known that the equilibrium value of φ in a Se atoms chain deviates from 90^0 and equals $101\text{-}107^0$ [5]. This means that one has to construct two equivalent sp-hybridized orbitals $\psi_{\lambda 1,2}$, which are centered at atom A and directed toward atoms B and C

$$\psi_{\lambda 1,2} = (s + \lambda p_{1,2})/(1 + \lambda^2)^{0,5} \quad (1)$$

The energy of these equivalent orbitals is equal to

$$\mathcal{E}_\lambda = (\mathcal{E}_s + \lambda^2 \mathcal{E}_p)/(1 + \lambda^2) \quad (2)$$

A third non-equivalent hybridized orbital in the (xy) plane is mainly of s-type

$$\psi_\mu = (s + \mu p)/(1 + \mu^2)^{0,5} \quad (3)$$

with energy

$$\mathcal{E}_\mu = (\mathcal{E}_s + \mu^2 \mathcal{E}_p)/(1 + \mu^2) \quad (4)$$

Here ($\mathcal{E}_s$, s) and ($\mathcal{E}_p$, p) are atomic energies and wave functions of 4s and 4p electrons of Se. The overlapping of two ψ_λ orbitals, which are centered at atoms A and B causes a splitting of the energy $\mathcal{E}_\lambda$ into bonding $\mathcal{E}_\sigma$ and antibonding $\mathcal{E}_{\sigma^*}$ levels

$$\mathcal{E}_{\sigma,\sigma^*} = \mathcal{E}_\lambda \mp VI \quad (5)$$

Here the assymetry of splitting into σ and σ^* states is neglected and the matrix element which governs the splitting is supposed to be proportional to the overlap integral I. Then six electrons of each Se atom (for example atom A) with $4s^2 4p^4$ electronic configuration occupy two bonding states, that are directed toward atoms B and C, one LP-state which is a p_z-state with energy $\mathcal{E}_p$ and perpendicular to the plane of picture and a ψ_μ-state which is mainly s-type.

We shall use non-adiabatic approximation and mix one-electron wave functions by changing of configuration coordinate φ. In this case one has to require that one-electron wave functions be mutually orthogonal for any value of φ. Hence we have

$$<\psi_{\lambda 1}, \psi_{\lambda 2}> = 0 \rightarrow 1 + \lambda^2 \mathrm{Cos}\varphi = 0 \quad (6)$$

$$<\psi_\mu, \psi_\lambda> = 0 \rightarrow 1 + \lambda\mu \mathrm{Cos}(180 - \varphi/2) = 0 \quad (7)$$

These equations give us dependences of λ and μ on φ, hence one can see from (2), (3) that $\mathcal{E}_\lambda$, $\mathcal{E}_\mu$ also depend on φ. Another dependence of energy on φ is associated with the overlap integral I(φ). According to [6], the overlap integral between two equivalent ψ_λ orbitals centered at different atoms A and B and directed toward to each other along AB line depends on λ due to changing of sp-hybridization. This situation is shown in Fig.2. from [6].

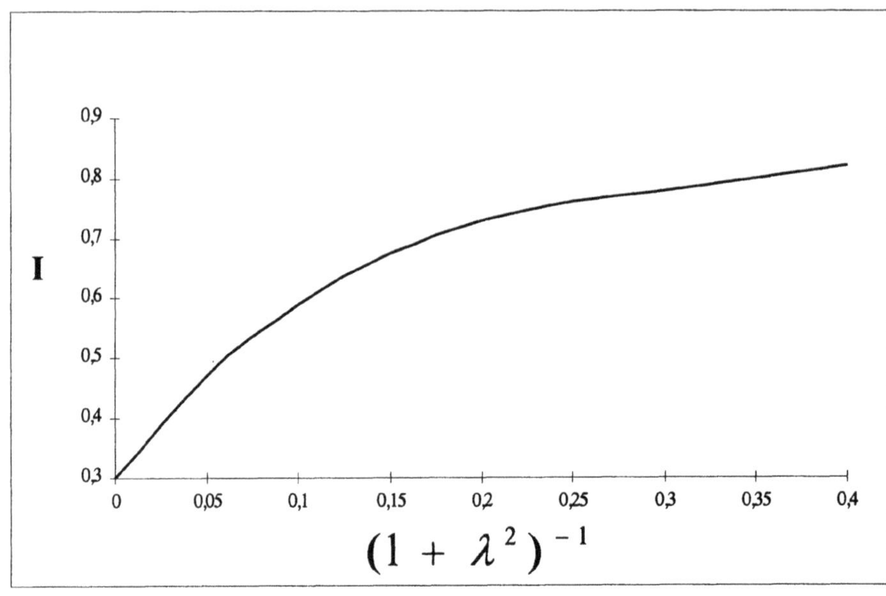

Figure 2.
Dependence of overlap integral I on λ [6].

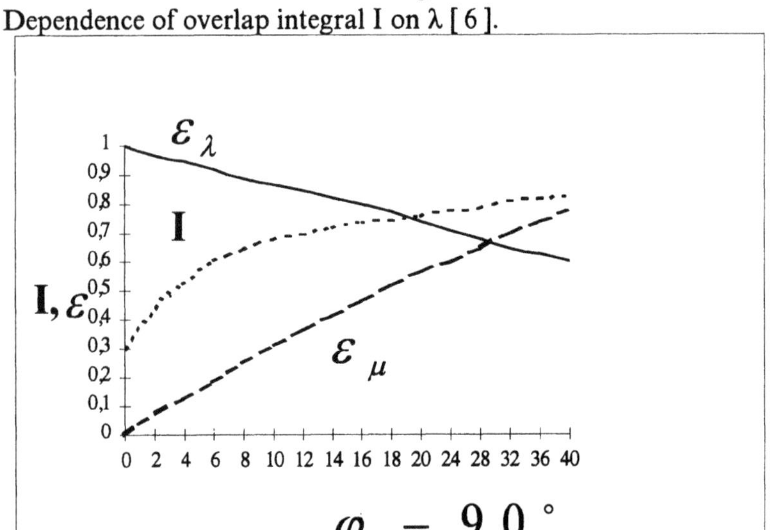

Figure 3.
Dependences of overlap integral I and energies ε_λ , ε_μ on φ. $\varepsilon_s = 0$, $\varepsilon_p = 1$.

Using the λ(φ) dependence (6) one can calculate the I(φ) dependence which is shown in Fig.3. together with $\varepsilon_\lambda(\varphi)$ and $\varepsilon_\mu(\varphi)$. Energies are given using ε_s and ε_p energies as 0 and 1 correspondingly.

Now we have everything that is necessary to calculate in our rather rough approximation the energies of different electronic and charge configurations. Several of these are depicted in Fig.4.

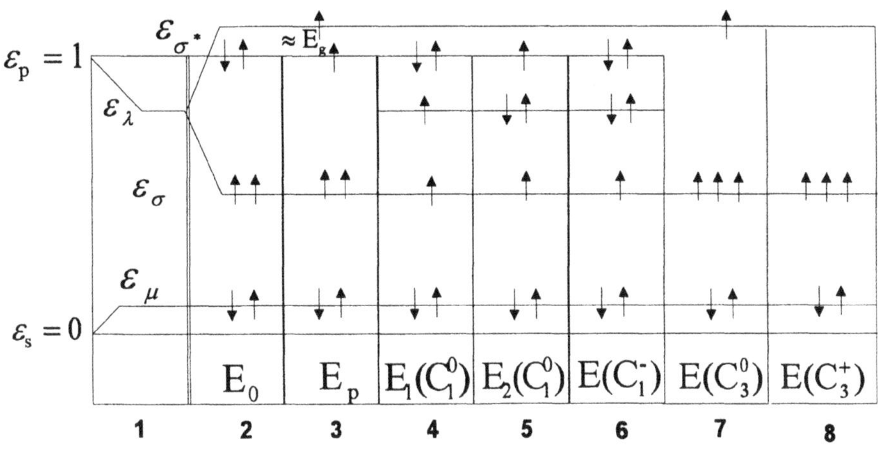

Figure 4.
Energy diagrams for hybridized states (1), basic structure unit (2), polaronic state (3), neutral dangling bond (4, 5), negative charged dangling bond (6), neutral and positive charged threefold coordinated defects (7, 8). In the latest cases the difference between splitting of the native and excess bonds are not shown.

First of all the atomic potential E_o for the basic structure unit and the polaronic potential E_p (Fig.5 a,b) have been calculated by the using the following expressions

$$E_o = 2\,\mathcal{E}_\mu + 2(\mathcal{E}_\lambda - VI) + 2\,\mathcal{E}_p \qquad (8)$$

$$E_p = 2\,\mathcal{E}_\mu + 2(\mathcal{E}_\lambda - VI) + \mathcal{E}_p + (\mathcal{E}_\lambda + VI) \qquad (9)$$

The energy difference between the minima of these potentials must be of the same order as the forbidden gap E_g of Se (see upper part of column 3 in Fig. 4). If one supposes that this difference is E_g or $2E_g$ then $V \approx 1.0$ or 0.6 must be chosen respectively. The latter value means that the splitting of levels $\mathcal{E}_{\sigma^*}$ and $\mathcal{E}_p$ into bands is taken into account. One can see that our approximation describes rather adequately the basic structure unit of Se, with equlibrium value of φ being equal to $\sim$ 100-103°. So we may conclude that $\mathcal{E}_\lambda$, $\mathcal{E}_\mu$, and I are really the three main contributions to the dependence of energy on φ. The potential E_p is more softer then E_o, but this softening is obvious, because it occurs owing to excitation of one electron.

Now we describe a much more important and pronounced softening which is due to an intraction between atomic potentials. Fig.5c shows two calculated atomic potentials for the dangling bond defect C_1^0. The two different potentials E_1 (C_1^0) and E_2 (C_1^0) of the same defect C_1^0 exist only when we take into account the hybridization (i.e. the difference between the energies $\mathcal{E}_\lambda$ and $\mathcal{E}_p$, see columns 4 and 5 in Fig.4) and coulomb repulsion U_λ of electrons occupying the ψ_λ-state of dangling bond. The energy difference between these two potentials is equal to

$$E_1(C_1^0) - E_2(C_1^0) = 1 - \mathcal{E}_\lambda - U_\lambda \qquad (10)$$

The electron-electron coulomb repulsion energy is typically 1 - 2 eV, that means in dimensionless variables $U_\lambda \approx (1 - 2)/(\mathcal{E}_p - \mathcal{E}_s) \approx 0.1 - 0.2$. Here a value of $\mathcal{E}_p - \mathcal{E}_s \approx 10$eV for Se is taken from [7].

Hence the two potentials may cross each other at $\mathcal{E}_\lambda \approx 0.9 - 0.8$, i.e., at $\varphi \approx 95 - 100^0$, according to the $\mathcal{E}_\lambda(\varphi)$ dependence (see Fig.3). It is this situation that is depicted in Fig.5c for crossing point $\varphi \approx 100^0$. But the crossing of two potentials indicates that one has to take into account the interaction between them. If we consider interaction due to the closest ($\mathcal{E}_p$, ψ_{LP}) and ($\mathcal{E}_\lambda$, ψ_λ) states then new important parameter $\upsilon = \langle \psi_\lambda | H_{int} | \psi_{LP} \rangle$ appears. Here H_{int} is Hamitonian of electron-phonon interaction. When υ is small then only slight splitting of two potentials occurs, as it is shown in Fig.5d, for small value $\upsilon = \upsilon_1$. In this case the lowest potentials has little barrier ΔE of approximately ≈ 0.2 eV. The two positions of the dangling bond which correspond to this situation are shown in Fig.6.

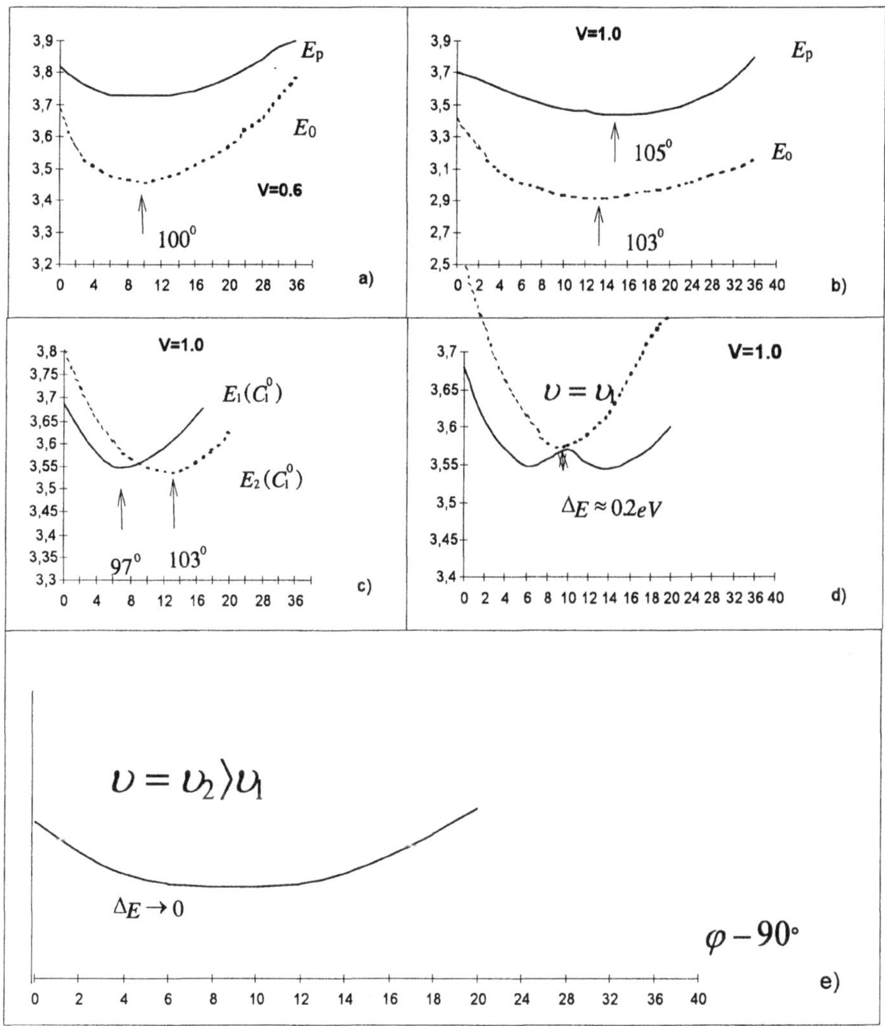

Figure 5.
Atomic potentials calculated for: the basic structure unit E_o and polaronic state E_p (a,b); for different electronic configuration of the dangling bond (C_1^0) (c). The schematic picture of the splitting of the potentials E_1 and E_2 for two values of matrix element $\upsilon = \upsilon_{1,2}$ (d, e). For the latest case ($\upsilon_2 > \upsilon_1$) only the lowest potentials is shown. All horizontal and vertical axes are angles ($\varphi - 90°$, degrees) and energies ($E/(\mathcal{E}_p - \mathcal{E}_s)$).

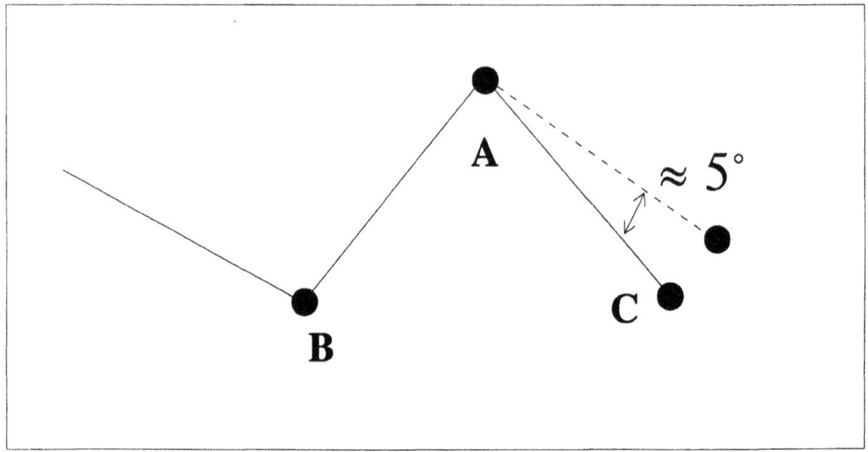

Figure 6.
Two positions of the neutral dangling bond (atom C) for the case two - well potentials ($\upsilon = \upsilon_1$).

For $\upsilon = \upsilon_2$, where υ_2 is greater by several fold then υ_1, $\Delta E \rightarrow 0$, and the lowest potential becomes flat, very soft potential in the interval ~ 5 - 10^0, as it can be seen from Fig 5e. A similar situation is realized for C_1^+ but not for the C_3^0 and C_3^+ defects. In the last two cases the set of closely spaced levels $\mathcal{E}_p$ and $\mathcal{E}_\lambda$ which provides the crossing of two potentials is absent. Using the approach of [4] and potential minima for different neutral and charged defects one can evaluate negative U. As a result the distribution of U may be obtained. Thus values of negative U are correlated both with degree of softening and with specific electronic configuration (coordination number and charge) of defects.

3. Conclusions

To summarise it should be emphasized that the quantum chemical approach [1,2] and the soft potential model [3] are not contradictory. If one takes into account the specific set of electronic states (especially LP-states) of the chalcogenide atom and the electron-phonon interaction, then negative U and soft or two-well potentials are obtainable simultaneously.

4. References

[1] Street, R.A. and Mott, N.F. (1975) States in the gap in glassy semiconductors, Phys. Rev.Lett. **35**, 1293-1296.

[2] Kastner, M., Adler, D. and Fritzsche, H. (1976) Valence - alternation model for localized gap states in lone - pair semiconductors, Phys.Rev.Lett. **37**, 1504-1507.

[3] Klinger, M.I. (1988) Glassy disordered systems: Topology, atomic dynamics and localized electron states, Phys.Repts. **165**, 275-397.

[4] Tsendin, K.D. (1992) Role of hybridization in the polaron mechanism for the formation of U^- centers, soft potentials, and two-well potentials, JETP Lett. 55, 661-665.

[5] Sobolev, V.V. and Shirokov, A.M. (1988) Electronnaya structura halcogenov, Nauka, Moskva.

[6] Coulson, C.A. (1961) Valence, Oxford university press, Oxford.

[7] Harrison, W.A. (1980) Electronic structure and properties of solids, W.H. Freeman and Company, San Francisco.

OPTICAL ABSORBABILITY IN TERNARY Ge-As-S GLASSES AND SOME POSSIBILITY OF ITS APPLICATIONS

I.V.FEKESHGAZI, K.V. MAY, V.M. MITSA, A.I. VAKARUK
Institute of Semiconductors Physics, National Academy of Sciences, prosp. Nauki 45, Kyiv-28, 252650, Ukraine

1. Introduction

Chalcogenide glasses and films are promising materials for production of different optical elements for integral and optoelectronics and laser devices. Generally it is due to the properties of the glasses that are transparent in visible and IR region of spectrum and posses different values of refractive indices and relatively low optical and acousto-optical losses. However, their relatively low damage threshold I_d is the main restriction for their wide application. As it is known the damage threshold of a material is mostly defined by the efficiency of optical absorbability, that cause to linear and nonlinear losses of laser radiation, whose values essentially depends on composition and fabrication of the glasses. Recently, it was shown that topology play important role in chalcogenide glass structure, that exhibit medium range order in atomic arrangement[1]. However, common direct methods of medium range order studying are absent now, while some information may be received by studies of the dependences of physical properties of glasses on it's concentration and average coordination number. In this connection there is important to investigate the interconnection between I_d, α and β for the various x and r of Ge-As-S glasses.

2. Experimental technique

2.1. SAMPLE PREPARATION.

All experiments have been performed on glassy samples from the ternary of Ge-As-S system along the As-GeS_2 , As-Ge_2S_3, As_2S_3-GeS_2 and As_2S_3-Ge_2S_3 sections of glass forming region[2]. The samples have been fabricated from elementary high grade purity components by synthesis in evacuated ampoule with following

A. Andriesh and M. Bertolotti (eds.),
Physics and Applications of Non-Crystalline Semiconductors in Optoelectronics, 243–248.

quenching into water. Two opposite surfaces of the plane parallel platelets with thickness of 1-6 mm have been polished to high optical quality.

2.2. METHODS OF MEASUREMENTS.

Linear α and two-photon β coefficients have been determined from the intensity dependences of sample transmittance that posses a sublinear character, while the dependences of reverse transmittance I_0/I on incident intensity I_0 were linear[3,4]. This linearity indicate the domination of two-photon nature of the absorption, that may be approximated by equation

$$I_0/I = \frac{\exp \alpha d}{(1-R)^2} + I_0 \frac{\beta(\exp \alpha d - 1)}{\alpha(1-R)}, \qquad (1)$$

were **R** is the reflection coefficient, **d** is the sample thickness, α and β are the coefficients of linear losses and two-photon absorption correspondingly. The α and β values were calculated from the experimental meaning of ordinata cutt-of and the slope of these $I_0/I=f(I_0)$ dependences correspondingly.

The optical damage threshold $\mathbf{I_d}$ was determined as a minimal power density that give rise to the appearing of bright flash and as a result to sharp decreasing of transmitted beam intensity.

The Q-switched ruby laser was used as a light source. It emitted the pulses with Gaussian spatial and time distribution at the duration of 25 ns and the energy of 0.4 J. The laser emission was focused on entrance surface of samples by the lens with focal length of 110 mm.

3. Results and discussion

The dependences of two-photon absorption constant β and linear losses coefficients α on the average coordination number values **r** and component concentration **x** are presented on Fig.1. They posses a very complicated character. The critical **r** and **x** values at which the α and β parameters reached it's extrema were determined.

For the glasses of $(GeS_2)_x(As_2S_3)_{1-x}$ pseudobinary section the minimum values of β occur at **x** value equal to 0.24 and average coordination number value equal to 2.47, while the α have a minima also at the x=0.6 (r=2.56) and x=0.85 (r=2.64). The maximum of β take place at x=0.6 (r=2.56), while the α at x=0.25 (r=2.47), x=0.85 (r=2.62) and x=1 (r=2.67).

For $As_x(GeS_2)_{1-x}$ glasses minima values of α and β are reached at x=0.07 (r=2.69) and are equal to 0.04 cm^{-1}, 0.017 cm/MW respectively. These results are due to bind of the break bonding of the tetrahedral GeS_2 chains by the As atoms that provide maximal topological homogeneity of the glass matrix and results in the increase of relative interconnection of the glass structure matrix, secure it's highest ordering[5]. The following increase of α and β for larger **r** values are connected with sharp decreasing of pseudogap width $\mathbf{E_0}$ (Fig.2b).

For glasses of $(Ge_2S_3)_x(As_2S_3)_{1-x}$ pseudobinary section the minima of β value occur at x=0.3 (r=2.52); x=0.5 (r=2.60) and x=0.65 (r=2.66), while the α at the 0.38<x<0.75 (2.5<r<2.7) and the the maxima of β at x=0.38 (r=2.55), x=0.63 (r=2.64) and x=0.9 (r=2.76) and α only at x=0.9 (r=2.76).

For $As_x(Ge_2S_3)_{1-x}$ glasses the decrease of β to minimum at x=0.03 (r=2.806) with **x** and **r** rising are observed in the region of small As concentrations due to the replacement of 4-valent Ge by 3-valent As which results in the decrease in a fraction of Ge-Ge homobonds with a low breaking energy and the increase preferential interconnection of the glass at x=0.03 (r=2.806). The increase of α and β for larger values of average coordination number may be caused by sharp decreasing of pseudogap width too (Fig.2b).

The dependences of optical damage thresholds $\mathbf{I_d}$ on the average coordination number **r** for two consecutive concentration changes of Ge-As-S glasses are shown in Fig.2a. The graph 1-2-3 belongs to the sections of $(GeS_2)_x(As_2S_3)_{1-x}$ - $As_x(GeS_2)_{1-x}$, while graph 4-5-6 belongs to $(Ge_2S_3)_x(As_2S_3)_{1-x}$ - $As_x(Ge_2S_3)_{1-x}$ ones. As can be seen they possess a very complicate and different characters. In the first case there are three maxima and two minima, while in the second one there are two maxima at quite smooth concentration dependence of $\mathbf{I_d}$. In the both cases the increasing of As concentration give rise to sharp decreasing of $\mathbf{I_d}$ due to the decrease of the glass forbidden gap (Fig.2b).

As it is known, the sharp thermal heating of illuminated region of semiconductors play a leading role in it's optical damage by the laser pulses emission of nanosecond duration[6]. The process is connected with excitation of free carriers under the linear and two-photon absorptions and subsequent thermalisation and nonradiative recombination of them. The increase of temperature $\mathbf{\Delta T}$ are determined by the equation of thermodynamic:

$$\Delta T = E/mC, \qquad (2)$$

where E - is the light energy absorbed in illuminated region, **m** - is the mass of this region, **C** - is the thermal capacity.

At the minority of decreasing the light intensity $\mathbf{\Delta I \ll I_0}$, e.g. at the $\mathbf{\alpha d \ll 1}$ and $\mathbf{\beta I d \ll 1}$, the $\mathbf{\Delta T}$ values may be appreciated from equations:

$$\Delta T_1 = \alpha I_0 \tau / c\rho \qquad (3)$$

and

$$\Delta T_2 = \beta I_0^2 \tau / c\rho \, , \qquad (4)$$

where $\mathbf{I_0}$ - is the intensity of laser radiation, τ - is the duration of laser pulse, ρ - is the density of material.

As the damage process start at the achivening of critical value of temperature gradient $\mathbf{\Delta T_{cr}}$, the optical damage threshold is inverse to values of linear α and two-photon β absorptions coefficients. As it's seen from comparison of Fig.2a

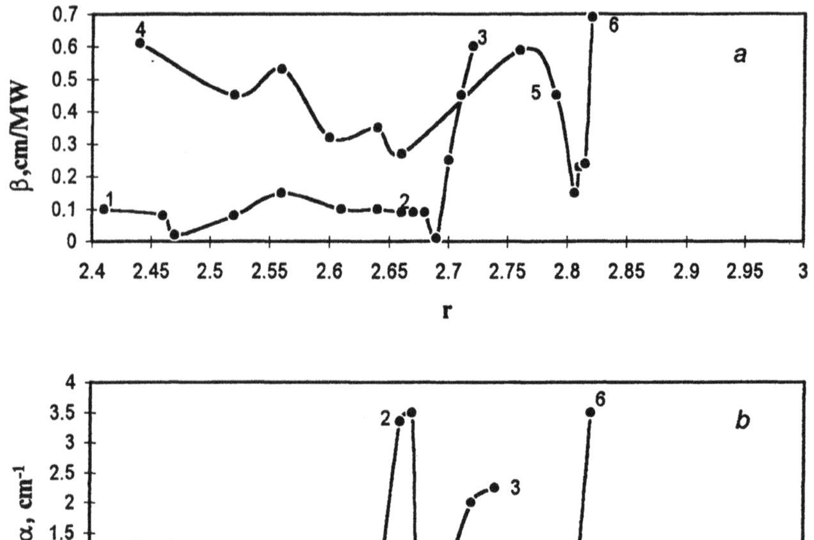

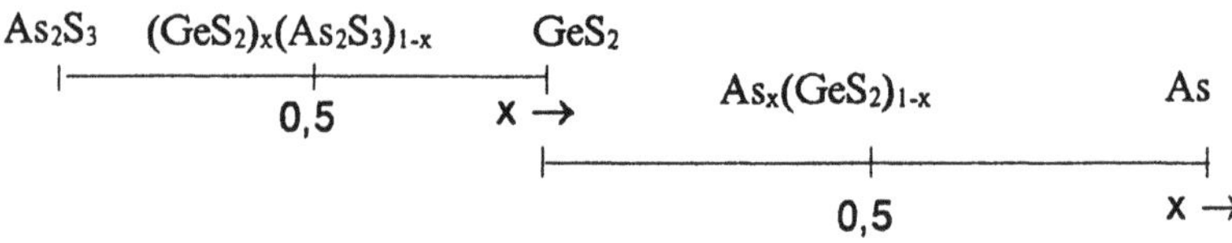

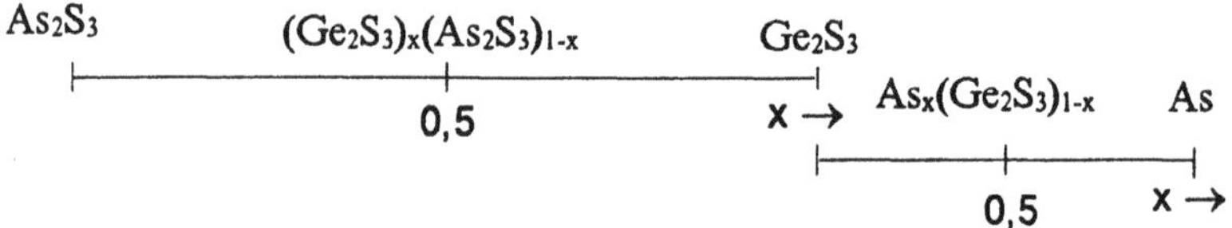

Fig.1. Dependences of the two-photon absorption coefficient β (Fig.1a) and linear losses constant α (Fig.1b) on the average coordination number **r** and concentrations **x** for ternary Ge-As-S glasses along the sections: 1-2 $(GeS_2)_x(As_2S_3)_{1-x}$, 2-3 $As_x(GeS_2)_{1-x}$, 4-5 $(Ge_2S_3)_x(As_2S_3)_{1-x}$ and 5-6 $As_x(Ge_2S_3)_{1-x}$.

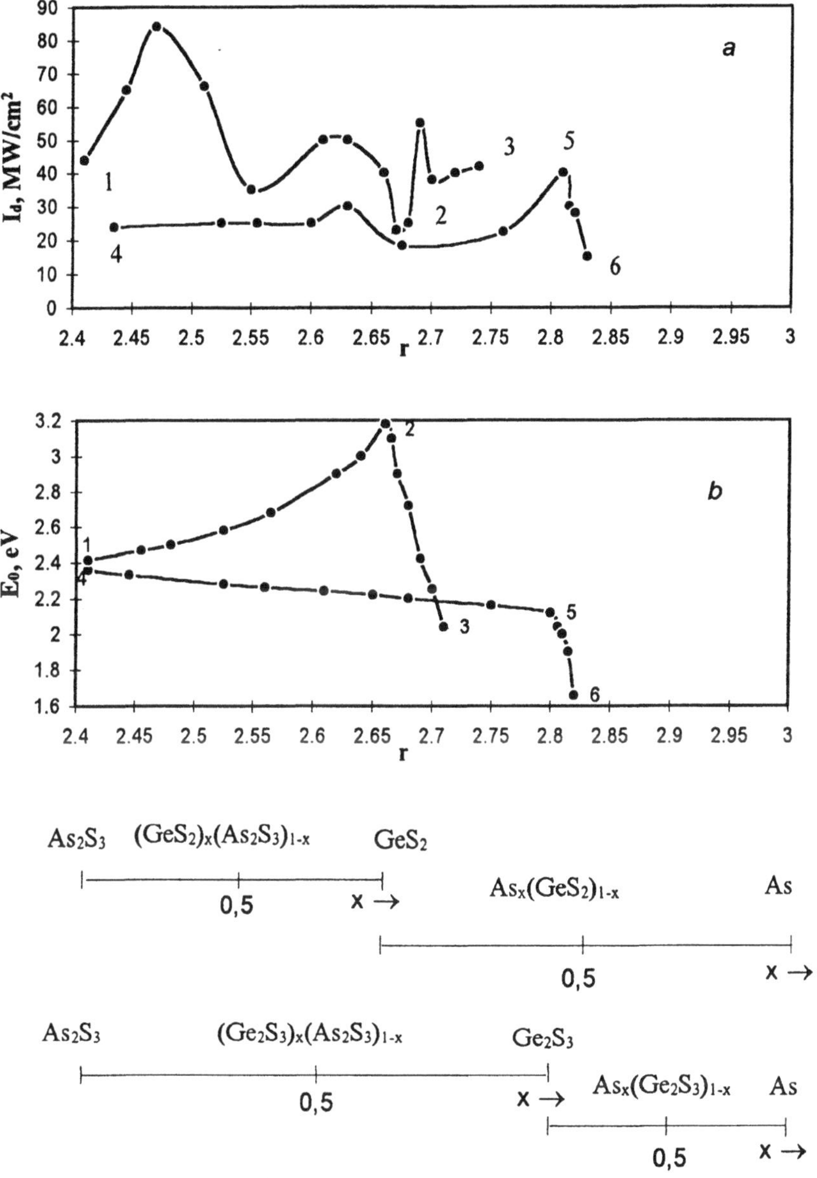

Fig.2. Dependences of the optical damage threshold $\mathbf{I_d}$ (Fig.2a) and pseudogap width $\mathbf{E_0}$ (Fig.2b) on the average coordination number **r** and concentration **x** for ternary Ge-As-S glasses along the sections: 1-2 $(GeS_2)_x(As_2S_3)_{1-x}$, 2-3 $As_x(GeS_2)_{1-x}$, 4-5 $(Ge_2S_3)_x(As_2S_3)_{1-x}$ and 5-6 $As_x(Ge_2S_3)_{1-x}$.

and Fig.1, always maxima of I_d observed at the minima α and β and vice versa. But all I_d, α and β dependences on average coordination numbers are different for 1-2-3 and 4-5-6 sections. Thus the later parameter is not very good for describing the influence of disordering on physical properties of ternary Ge-As-S glasses. Naturally, it is justifiable because one of same value of r may be received at the different concentrations of elements in these glasses.

The find glasses of compositions $(GeS_2)_{0.25}(As_2S_3)_{0.75}$, $(GeS_2)_{0.85}(As_2S_3)_{0.15}$, $As_{0.07}(GeS_2)_{0.93}$ and Ge_2S_3 with high optical damage threshold and small nonlinear absorbability are useful for fabrication of the optical elements of power optics. The glasses $(Ge_2S_3)_{0.37}(As_2S_3)_{0.63}$, $(Ge_2S_3)_{0.9}(As_2S_3)_{0.1}$ and $As_{0.1}(Ge_2S_3)_{0.9}$ with high values of two-photon absorption are used for duration control of Q-switched solid state laser pulse[7] and also may be used for fabrication of nonlinear Fabry-Perot interferometer[8] , optical power limiter and so on.

4. References

1. Phillips, J.C. (1979) Topology of covalent non-crystalline solids 1: short-range order in chalcogenide alloys, *J. Non-Cryst. Solids.* **34**, 153-181.
2. Vinogradova, G.Z. (1984) Glassing and phase equlibriume in chalcogenides systems, Nauka, M.
3. Vlasenko, Yu.V., Lisitsa, M.P., Fekeshgazi, I.V. (1985) Nonlinear optical properties of hyrotropic crystalls CdP_2 and ZnP_2 , *Quant. Electron.* **29**, 36-48.
4. Fekeshgazi, I.V., May, K.V., Mitsa, V.M., Roman, V.V. (1995) Diagnostic of glassy semiconductors by non-linear absorptive methods, *Proc. SPIE* **2648**, 257-261.
5. Arai, K., Namikawa, M. (1974) ESR in Ge-S glasses, *Solid State Commun.* **13**, 1305-1307.
6. Lisitsa,M.P., Koval, V.S., Mozol, P.E., Potykevich, I.V. (1976) Optical damage of surface of cadmium and zinc diphosphide crystalls by ruby laser emission, *Quant. Electron.* **10**, 81-84.
7. Fekeshgazi, I.V. (1986) Non-linear optical properties and applications of glassy semiconductors, *Proc. Int. Conf. "Non-crystalline semiconductors'86"*,FAA1/1-FAA1/2.
8. Bertolotti, M., Chumash, V., Fazio, E., Ferrari, A., Sibilia, C. (1993) Nonlinear Fabry-Perot cavity with chalcogenide glass thin films, *J. Appl. Phys.* **74**, 3024-3027.

PHOTOINDUCED EFFECTS IN CHALCOGENIDE GLASSES AND THEIR APPLICATION FOR OPTICAL RECORDING

A. KIKINESHI, A. MISHAK

Department of Solid-State Electronics, Uzhgorod State University
Pidhirna St. 46, 294000 Uzhgorod, UKRAINE

1. Introduction

A number of chalcogenide glasses, thin films and structures made of these inorganic polymer materials are known as photosensitive media. It means that different processes may result from the interaction of light quantum with such semiconductor material (see Table 1). These effects are described in many books [1-3], excellent review articles are given by V. Lyubin in 1984, S. R. Elliott, in 1986 [5], K. Tanaka in 1990 [5], H. Fritzsche in 1994 [7] and others [8,9]. So, further in this paper we shall make references to these ones or to the papers containing our previous results. Also we shall consider effects caused by relatively small energy quantums $h\nu \leq E_g^*$, where E_g^* is analogous to the band gap energy, but the illumination capacity may be varied in a great range from the usual for photoconductivity measurements (10^{-5}-10^{-3} W·cm^{-2}) up to the heating or even melting the material (10^{2}-10^{5} W·cm^{-2}).

First of all Type II processes of structural transformations in amorphous state are interesting for amplitude-phase recording due to the photoinduced effects of absorption ($\Delta\alpha$), refraction (Δn), reflection (ΔR) changes and polarization of the initially isotropic medium. The fundamentals of these effects, extended to the nanolayered films, and some applications for optical recording and optical elements' fabrication will be analyzed further.

Type I processes connected with charge transport and electron polarization are used for electrophotography with powder or thermoplastic development. They are rather well investigated and applied [3,9], so only some new opportunities will be mentioned. Type III processes are still under investigations nowadays, and here they will be mentioned in connection with Type II processes for certain layers and structures.

A. Andriesh and M. Bertolotti (eds.),
Physics and Applications of Non-Crystalline Semiconductors in Optoelectronics, 249–257.

TABLE 1. Photoinduced effects in chalcogenide glasses for optical recording

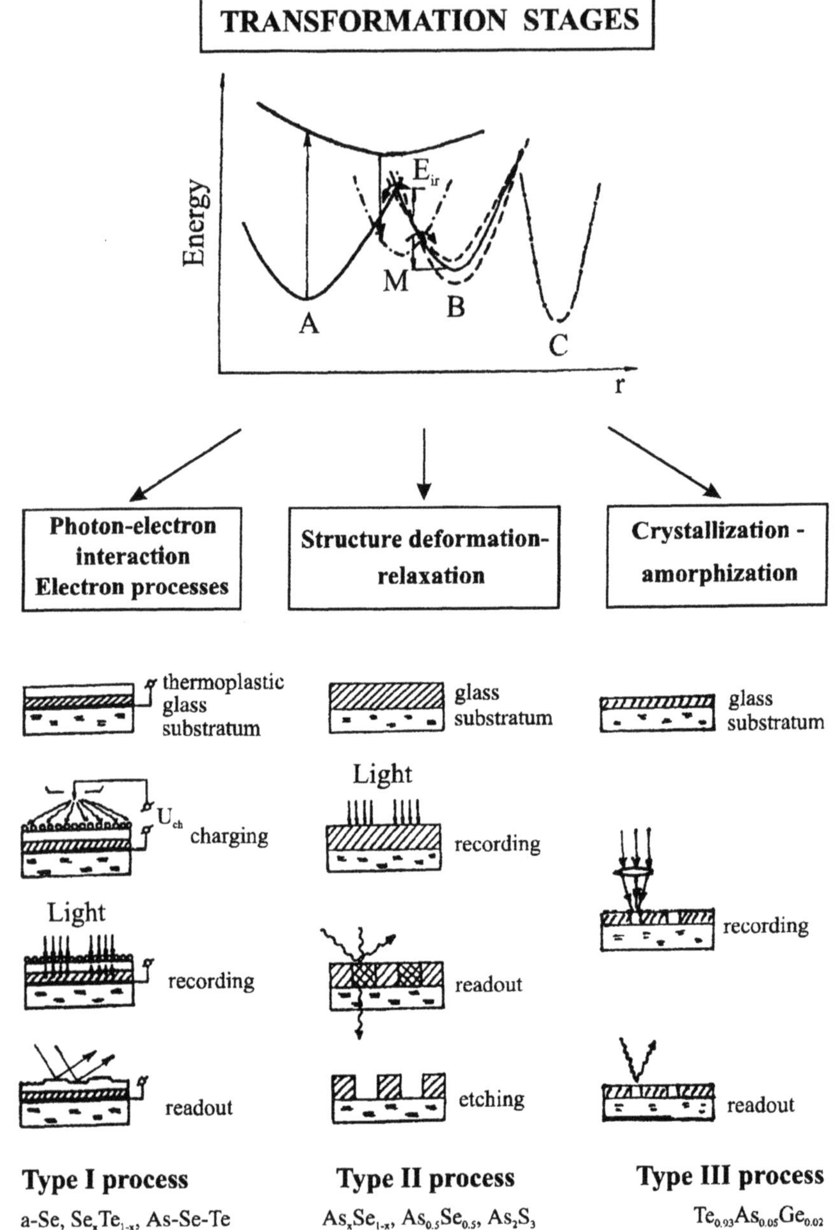

Type I process	Type II process	Type III process
a-Se, Se_xTe_{1-x}, As-Se-Te	As_xSe_{1-x}, $As_{0.5}Se_{0.5}$, As_2S_3	$Te_{0.93}As_{0.05}Ge_{0.02}$
As-Se/As-Se-Te	Ge-As-Se, As-S-I	$As_{0.10}Se_{0.67}Ge_{0.23}$
a-Se/As_xSe_{1-x}	As_2S_3/a-Se	$In_{0.35}Sb_{0.45}Se_{0.20}$

2. Fundamentals of photoinduced structural transformations

Illumination (usually laser irradiation) essentially influence optical (α, n, R) and other parameters (microhardness, solubility) of investigated layer (or sometimes bulk) materials. These changes differ in magnitude and stability, i.e. relaxation times. Typical manifestations of stable (at room temperature) changes in *AsSe* layer are shown on Figure 1. These may be divided to two types: reversible and irreversible. The last is caused by the stabilization of non-equilibrium, fresh-deposited films and in many respects is similar to the reversible one. Reversibility means thermal erasing of the induced optical changes. It is most effective at $T \leq T_g$ (T_g– softening temperature) of the given glass composition. So it is necessary to take into account the difference between recording-readout temperature T_r and T_g, when speaking about the efficiency of photoinduced changes and optical recording [6,8].

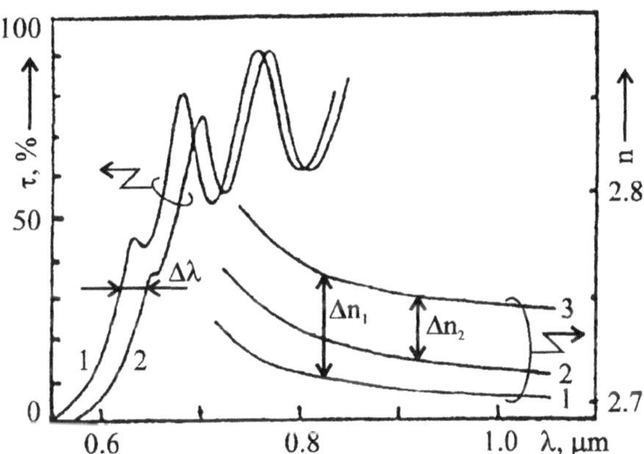

Figure 1. Transmission edge and refractive index change in *AsSe* layer under *He-Ne* laser irradiation. 1 – initial stage, 2 – first recording, 3 – stage after first thermal erasing.

The whole process of changes, i.e. optical relief formation is considered as a sequence of stages: (1) – electron-hole pairs excitation by light quantums (position *A* on a figure of Table 1); (2) – intermediate metastable localized states formation during illumination (position *M*), which may be oriented according to the light polarization; (3) – atomic rearrangement to new, more stable configurations in the vicinity of excitation, extended to the radius of deformation which is smaller than the critical radius of crystallite nucleation (in appropriate compounds) (position *B*); (4) – photocrystallization, as a final stage in the case of suitable composition and temperature conditions (position *C*).

The second stage and *M* states are characterized by rather short life-times, and in the case of low intensity short excitation the recovery of *A* positions is preferable: the electrophotographic process (Type I at the Table 1) may be repeated many times with small changes of photoreceptor's parameters. Using large light exposures Type II processes may be realized in such elements, and additional relief of necessary parameters may be formed on the plates or drums [10], enabling us to drive the parameters of electrophotographic process with different type of relief development.

The dynamical part of photoinduced changes also exists and it is smaller by one-two order of magnitude (see example on Figure 2). It manifests the existing of few components of optical recording even in annealed films or bulk glass at certain temperatures below T_g [8,11]. One of them is connected with M states, the second with B states, which are characterized by different relaxation times, according to the energy barriers E_{ir} and temperature range.

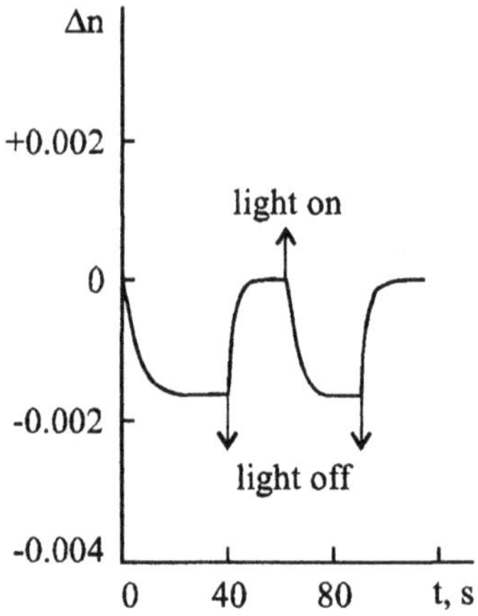

Figure 2. Dynamical component of optical recording in $As_{0.15}S_{0.85}$ layer, measured in waveguide regime (λ=0.51 μm, P=0.7 W·cm^{-2}).

As it was mentioned in few publications of different authors [6,12] and proved in our works [8,13,14], all investigated light-sensitive chalcogenide glasses possess similar characteristics of optical relief relaxation (transitions from B to A positions), which may be compared with relaxation processes in glasses and polymers [15,16]. Localized stresses are formed under photoinduced changes due to the shift of atoms (first of all bridge chalcogens in our glasses) or larger atomic groups (fragments of chains) [17]. Relaxation of stresses may be described in simple way as:

$$\delta(t) = \sum_{i=1}^{n} C_i e^{-t/\tau_i(T)}, \qquad (1)$$

where C_i – coefficient, $\tau_i(T)$ – relaxation times:

$$\tau_i(T) = B_i e^{E_{ir}/kT}, \qquad (2)$$

where B_i depends on volume and type of the relaxator.

Measuring the temperature dependences of relaxation times (diffraction efficiency of previously recorded holographic grating, see Figure 3), we have determined two main ranges of E_{ir}: 0.2-0.6 eV (below T_g) and 1.1-2.5 eV (near and above T_g). These energies may be compared with activation energies of β, β' and α-type relaxation processes in glasses (5-13 kcal/mole and 60-150 kcal/mole) [16], as far as they are connected with glass-network deformations, atomic displacements.

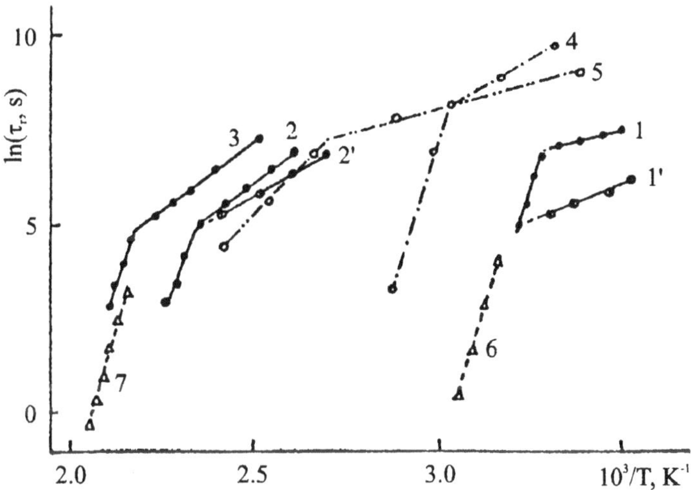

Figure 3. Optical relief relaxation in as-deposited *a-Se* (1), *AsSe* (2), As_2S_3 (3), and annealed *a-Se* (1'), *AsSe* (2') 1 μm thick uniform layers or As_2S_3/*a-Se* (4) and As_2S_3/*AsSe* (5) nanolayered films. Calculated dependences of mechanical stress relaxation in *a-Se* (6) and As_2S_3 (7).

The breaks on curves 1,2,3 are located at the vicinity of T_g for appropriate glasses: the change of the slope above these temperatures indicate the beginning of quick α-type relaxation processes, accompanied by viscous flow.

The calculated dependences of mechanical stress relaxation at such conditions (curves 6, and 7 for *a-Se* and As_2S_3) support this assumption.

The type of relaxation processes do not change in all investigated 0.5-5 μm thick layers, even in nanolayered structures, where, for example, *a-Se* or *AsSe* layers are placed between As_2S_3 layers (curves 4,5 on Figure 3). Of course, energy barriers deviate depending on the layer preparation technology, thermal treatment (1',2' for annealed *a-Se* and *AsSe* layers).

Important change of transition temperature were observed in nanolayered films (curves 4,5 on Figure 3): its correlation with T_g indicates the change of T_g under the influence of size-effects.

In general case the surface energy G_s influence the phase-transition temperature T_p in thin layers. At the equilibrium of two phases:

$$H_1 - TS_1 + G_{s1}\frac{A}{V} = H_2 - TS_2 + G_{s2}\frac{A}{V}, \tag{3}$$

where A – surface area, V – volume, S, H – entropy and enthalpy. $S_2 - S_1 = Q/T_0$, $H_2 - H_1 = Q$ – transition heat. T_0 – transition temperature in volume material.

In the thin layer: $A/V = 2d$, where d – thickness, so:

$$T_0 - T_p = \frac{2(G_{s1} - G_{s2})}{Qd}. \tag{4}$$

Depending on the correlation of G_{s1} and G_{s2}, for certain pairs of layers in our case the increase or decrease of T_g may be achieved. These cases were just proved for As_2S_3/*a-Se* and As_2S_3/*AsSe* multilayer structures. So, on the top of all this we can operate the parameters of optical relief, recorded in these structures, using Type II processes.

Type I processes in nanolayered films also are influenced because the changes of current relaxations, drift mobility of the carriers [18] and optical absorption edge shifts.

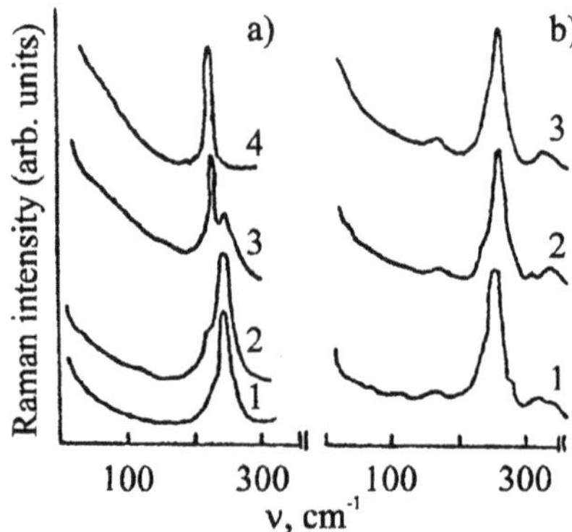

Figure 4. Raman spectra of 1 μm thick uniform *a-Se* (a) and As_2S_3/*a-Se* nanolayer films (b) for different exposures, growing from 1 to 4.

Type III processes of photo-, thermoinduced crystallization are also influenced by size effects in nanolayered films because of the above-mentioned dependences. For example, photocrystallization of the amorphous

Se layer was observed *in situ* by Raman-scattering measurement under *He-Ne* laser illumination (λ=0.63 μm, P=10 W·cm^{-2}) (see, Figure 4,a) [19]. The growing peak at 237 cm^{-1} testify the appearance of trigonal *Se* microcrystallites, which may be dissolved in amorphous phase below certain dimensions (stage of recording).

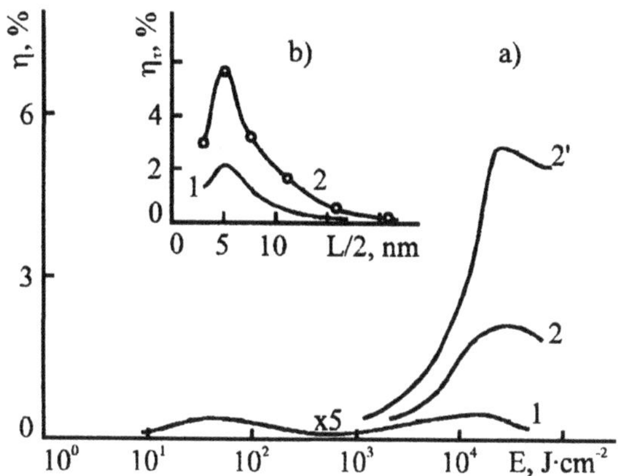

Figure 5. a) Diffraction efficiency η dependence on exposure in *a-Se* layer (1) and As_2S_3/*a-Se* multilayer (2,2') measured in transmittance (1,2) or reflectance mode (2').
b) η dependence on *a-Se* sublayer thickness in As_2S_3/*a-Se* nanolayered film at different *He-Ne* laser power density (1 – P=1 W·cm^{-2}, 2 – 30 W·cm^{-2}).

The same illumination do not change the structure of *a-Se* in As_2S_3/*a-Se* nanolayered films (Figure 4,b), in accordance with the increased T_g and restriction of crystallization processes in such films. Vice versa the photocrystallization may be enhanced in certain systems of layers and Type III optical recording process may be improved.

3. Applications of photoinduced effects for optical recording

As it was discussed above, we were interested in development of all three types of optical recording processes, recording media. The results of previous work in these directions are published in [8,14,20,21] and other papers. A number of new possibilities were achieved by introduced nanolayered films made of different pairs. of light-sensitive chalcogenide glasses, as it was also mentioned above. It is worth to say here about the new possibilities for archival holographic recording on As_2S_3/*a-Se* nanolayered films (see Figure 5). These media possess improved diffraction efficiency at the one wavelength write-readout cycle, particularly in reflective readout mode.

This work was supported by grants of the State Committee on Science and Technology of Ukraine and Ministry of Higher Education of Ukraine.

4. References

1. Mott, N.F., Davis, E.A. (1979) *Electron processes in non-crystalline materials*, Clarendon Press, Oxford.
2. Shvarts K.K. (1986) *Physics of optical recording in dielectrics and semiconductors* (in Russian), Zinatne, Riga.
3. *Nonsilver photographic processes* (in Russian). (1984) Edited by Kartuzhansky, A.L., Chimia, Leningrad.
4. Lyubin, V.M. (1984) Photographic processes on the basis of chalcogenide glass semiconductors (in Russian), in A.L. Kartuzhansky (ed.) *Nonsilver photographic processes*, Chimia, Leningrad, pp. 193-222.
5. Elliott, S.R. (1986) A unified model for reversible photostructural effects in chalcogenide glasses, *J. Non-Cryst. Solids* **81**, 71-98.
6. Tanaka K. (1990) Photoinduced changes in chalcogenide glasses, *Reviews of Solid State Science* **4**, 641-659.
7. Fritzsche, H. (1993) The origin of reversible and irreversible photostructural changes in chalcogenide glasses, *Phil. Mag. B* **68**, 561-572.
8. Kikineshi, A.A. (1989) Peculiarities of photophysical processes in the recording media based on the chalcogenide glasses (in Russian), *Kvantovaja Elektronika* (Kiev) **37**, 31-41.
9. Panasjuk, L.M. (1985) Photographic characteristics of thermoplastic materials (in Russian), *Uspekhi Nauchnoj Photographii* **23**, 187-193.
10. Vlasov, V.I., Semak, D.G., Chepur D.V. (1978) Temperature dependences of optical recording and erasing on AsSe chalcogenide glasses (in Russian), *J. Scientific and Applied Photography and Cinematography* **23**, 51-53.
11. Kikineshi, A.A., Shipljak, M.M., Spesivikh, A.A., Bentsa, V.M., Sterr, A.A. (1988) Properties of photosensitive heterostructures on the basis of AsSe glass (in Russian), *Photoelectronics* (Odessa) **2**, 91-95.
12. Jakovuk, O.A., Novoselov, S.K., Borisova, Z.U. (1976) Photostructural transformations in As-Ge-S, As-Ge-Se glasses (in Russian), *Izvestija AN USSR: Neorganicheskije Materialy* **12**, 1949-1950.
13. Kikineshi, A.A., Melnichenko, T.N. (1989) Parameters of free volume theory in the model of photoinduced changes in chalcogenide glasses, *Proc. XV International Congress on Glass*, Leningrad, **1B**, pp.66-67.
14. Kikineshi, A.A., Shipljak, M.M. (1989) Structural relaxation and photostructural changes erasing in chalcogenide vitreous semiconductors (in Russian), *Ukrainian J. of Physics* **34**, 30-34.
15. Bartenev, G.M. (1979) *Structure and relaxation properties of elastomers* (in Russian), Chimia, Moscow.
16. Sanditov, D.S., Bartenev, G.M. (1982) *Physical properties of disordered structures* (in Russian), Nauka, Novosibirsk.
17. Trunov, M.L. (1995) Structural relaxation and the phothermoplastic effect in amorphous semiconductors, *J. Non-Cryst. Solids* **192&193**, 431-434.

18. Shipljak, M., Pinzenik, V., Makauz, I., Lendel, D., Chereshnya, V. Multilayer structures with operated parameters for xerography. (1994), *Proc. International Workshop on Advanced Technologies of Multicomponent Solid Films and Structures*, Uzhgorod, pp. 87-88.
19. Kikineshi, A.A., Fedak, V.V., Stefanovich, V.A. (1986) Peculiarities of holographic recording on the layers of amorphous selenium (in Russian), *J. Scientific and Applied Photography and Cinematography* **31**, 433-437.
20. Kikineshi, A., Mishak, A., Sterr, A. (1993) Selenium based compositionally modulated recording materials for holography, *Proc. SPIE* **2108**, pp.72-75.
21. Kikineshi, A., Marjan, M., Vlasov, V., Ripka, G. (1996) Optical interconnection elements investigation in chalcogenide glass layers for integrated circuits, ISHM/NATO Advanced Research Workshop and Exhibition on Microelectronic Interconnections and Microassembly, Prague (to be published by Kluwer ed.).

COORDINATION DEFECTS FORMATION MODEL FOR REVERSIBLE PHOTOSTRUCTURAL TRANSFORMATIONS IN AMORPHOUS $As_2S(Se)_3$

O.I.SHPOTYUK
Lviv Scientific Research Institute of Materials,
Stryjska str. 202, Lviv, UA+290031, Ukraine
Physics Institute, Pedagogical University of Czestochowa,
Al. Armii Krajowej 13/15, Czestochowa, PL-42201, Poland

Photoinduced destruction-polymerization transformations in the amorphous a-$As_2S(Se)_3$ associated with coordination defects formation processes have been studied by differential IR Fourier spectroscopy method in the 400-100 cm^{-1} region. All topological variants of these processes statistically possible in the investigated samples have been taken into account for physical consideration of the real microstructural transformations. The model for the both irreversible and reversible photoinduced effects has been developed at the basis of coordination defects concept.

1. Introduction

Amorphous chalcogenide semiconductors (AChS) studied by Kolomiets and Goryunova more than forty years ago are unique solid state materials showing the effects of photoinduced changes of their physical properties [1-4]. Features of these effects have been well studied between 1970 and 1980, but now there are several views with respect to our understanding of their microstructural mechanism, especially in the case of so called reversible photoinduced effects revealed themselves in the repeated cycles of photoexposure and thermal annealing [5-9].

This situation is considerably caused by the difficulties of direct observation of local atomic structure transformations in disordered solids using "traditional" amorphographical techniques.

The photoinduced changes are very weak, they comprise no more than 10% of the atomic sites concentration and sensitivity of the corresponding modes in the vibrational spectra is low. Optical spectroscopy methods containing useful information on the integrated signal are less informative for this purpose. The necessary

A. Andriesh and M. Bertolotti (eds.),
Physics and Applications of Non-Crystalline Semiconductors in Optoelectronics, 259–274.

information on photostructural transformations (PhST) can be accurately obtained by Fourier spectroscopy technique in the long-wavelength region (400 to 100 cm^{-1}) [8,9] .

This technique was applied to the AChS for the first time in 1988 [9]. The main aim of that experiment, the identification of irreversible short-range ordering PhST, has been successfully achieved for the a-As_2S_3 thin films.

It is expected that "differential" IR Fourier spectroscopy technique will be quite informative for the mechanism study of the reversible PhST too. However, experimental data on photoinduced changes of the AChS intermediate-range ordering structure obtained using EXAFS technique (experimental results by S.R.Elliott, M.A.Paesler and others [10-12]) must be taken into account, when deriving a unified model of the reversible PhST at the level of both short- and intermediate-range ordering.

In this paper we shall put forward a new microstructural model for the reversible PhST in the AChS based on the experimental results previously obtained using "differential" IR Fourier spectroscopy technique - the model of so called coordination defects (CD) formation or chemical bonds switchings.

2. Historical aspect of the problem

The CD concept firstly appeared in the scientific literature at the beginning of 1970s in the connection with well known problem on theoretical interpretation of the AChS energetic forbidden band gap structure. The next stages in the historical development of this definition mentioned below may be distinguished.

The numerous experimental data obtained in 1960s testified that Fermi level in the AChS was fixed near the middle of the energetic forbidden band gap and didn't change its position with changing of impurity atoms concentration (in the limits excluded strongly defined crystallisation processes, of course) [13]. This property has been explained due to "8-N" rule. Another way speaking, it has been suggested that the AChS structural network can be distorted in such way that all covalent bonds saturate themselves and local atomic coordination for each chosen atom is equal to 8-N, where N is a normal atomic valence.

At the same time, a number of experiments, such as the drift mobility investigations [14], would be successfully interpreted, if donor and acceptor defect states had existed. These defects would be diamagnetic ones at the room temperature in full agreement with ESR absence in AChS, while simultaneously they would allow to explain a photoinduced paramagnetism in these materials at the low temperatures [15]. The resolution for this contradiction had been found using fundamental Anderson's concept on effective negative correlation energy of the electronic states in the forbidden band gap of the AChS [16].

Anderson put the postulate on existing of centres with negative Habbard potential or negative U-centres in the AChS. U is the energy of two one-type carriers interaction localised at one centre. The effective attraction of these carriers is expected to be produced by a strong polaron effect, e.g. configuration-deformation disturbance of the AChS structure around negative U-centre:

$$U = U_c - 2W < 0, \tag{1}$$

where U_c is the energy of Coloumb repulsion,
2W is the polaron energetic shift.

Anderson believed that all states in the conductivity forbidden band gap of the AChS corresponded to double-paired carriers with opposite directed spins, the centres energy forming the quasicontinious spectrum in this band. Levels with $E < E_F$ (E_F - energy of Fermi level) are proper to double-paired electrons, and levels with $E > E_F$ - to double-paired holes. Thus, the ESR signal is absent in the AChS; the conductivity forbidden band gap preserves for one-particles states; the hopping conductivity is not observed because of necessity to large polaron disturbance displacements between centres; the Fermi level is frozen near the middle of the forbidden band gap; the large Stokes shift of photoluminescence is explained.

The strong polaron effect was postulated by Anderson, while its origin was not cleared. It was not understood, why the polaron shift at the centres in AChS (W = 0.6-0.8 eV) leaves so large in comparison with analogous parameter in wide-band covalent materials ($W \leq 0.1$-0.2 eV) [17]. Besides, Anderson didn't explain the nature of dispersion of centres states in the AChS forbidden band gap.

Taking into account the latter fact on discrete character of the electronic states in the band gap of the AChS and simultaneously following the Anderson's concept of negative U-centres, Mott, Davis and Street (MDS) put forward the model of D-centres or unsaturated dangling bonds [18]. They supposed, that the Stokes shift in the AChS was such, that the neutral dangling bond (D) gave out energy for two distant dangling bonds in accordance with the following reaction

$$2D \rightarrow D^{+} + D^{-}, \tag{2}$$

where D^{+} and D^{-} denote positively and negatively charged dangling bonds. The electronic energy lost in the process that brings two electrons into D^{-} defect is expected to be more than counterbalanced by the energy of distortion around D^{+} centre. As a result, there are then no unpaired spins at room temperature, as well as numerous data on drift mobility, photoconductivity and photoluminescence in the AChS [3,4] can be interpreted in a strong way.

A new point of view was presented by Kastner, Adler and Fritzsche (KAF) [19]. They used a chemical-bond model to provide a more precise description of the CD in the AChS and to explain the origin of the effective negative correlation energy. Particularly, they showed that the creation of one positively charged three-fold coordinated chalcogen C_3^{+} and one negatively charged singly coordinated chalcogen C_1^{-} left the number of bonds unchanged. The energy to overcome the Coulomb repulsion of two electrons localised at the C_1^{-} CD in the KAF model comes not from a change in the number of bonds, but rather from the conversion of two pairs of bonding and antibonding electrons into lone pairs. These CD may be characterized as valence alternation pairs (VAP) of atoms. If we take into account the Coulomb interaction between opposite charged CD, we may describe their as intimate VAP due to Kastner's model [20].

Soon after these articles Street used the concept of charged CD for explanation of the PhST mechanism in the AChS, connecting the nature of their initial origin with exciton self-trapping [21,22].

However, if we take into account only qualitative atomic configuration characteristics (chemical bond type, in the first hand) for such or other structural state, it is difficult to explain, why the photoinduced transformed bonds concentration is less than 10% of the atomic nodes [7]. The factors determining the defects formation efficiency usually don't consider in the self-trapped exciton model. So the appearing CD represent themselves as quasiatomic centres having electrical charge excess and characterized by effective negative correlation energy. The sharply defined polaron effect in this model appears owing to the strong interaction between empty-occupied positively charged centres and lone-pair electrons of neighbouring atoms (donor-acceptor interaction). Only a qualitative analysis of atomic and molecular orbitals is taken into account.

Hence more truthfully to conclude that well known reversible PhST in the AChS take place owing to electron and hole pairs excitations localized on so called soft atomic configurations created by atoms and atomic groups, having double-well potential and effective negative correlation energy [17, 23, 24]. A low quasielastic constants of the soft atomic configurations correspond in the real glass structure to the fragments with slightly bonded atoms, which concentration is near 10%. Such fragments have a low quasielastic constant and consequently a large polaron energetic shift. As a result the energetic states spectrum in the forbidden band gap of the AChS is quasicontinious with some distinguished energetic bands multiplied to $E_g/4$ in good accordance with experimental results. Unfortunately, the concrete structural complexes corresponding to the soft atomic configurations has not been established. In particular, we don't exclude that above mentioned diamagnetic charged defects of MDS, KAF, K and S may be quite accepted as structural pre-images of the soft atomic configurations, provided their initial microscopic mechanism of formation corresponds to the concept described in [23].

It must be noted that effective correlation energy for electrons of various CD in the AChS is adopted to be negative, correspondingly to the ESR signal absence. However, the consistent theoretical calculations of this parameter have not been carried out. Apart from, structural relaxation processes in the vicinity of anomalously coordinated atoms, as well as rigidity level of the structural network have not been accepted to account. Another way speaking, bonds switching processes need a certain activation barrier overcoming. The precise magnitude of this barrier left unknown, till Vardeny and Tauc defined it using optical modulation spectroscopy technique [25]. It was estimated as -1.0 eV for amorphous a-As_2S_3 and -0.7 eV for a-As_2Se_3.

It is naturally that processes of the CD formation reveal themselves at the level of not only short-range ordering, but also intermediate-range one, as AChS are characterized by strong inter-molecular interaction [5]. Some structural features of these intermediate-range ordering changes stimulated by absorbed light photoexposure have been studied in details since second half of 1980s using EXAFS technique [10-12]. Unfortunately, the limited possibilities of this technique to the investigation of the short-range ordering changes don't allow to identify the CD formation processes entirely.

3. Concept of destruction-polymerization transformations

This concept, firstly discussed in details by Zakis [26], is a basic point for consideration of a large number of essential atomic-dynamic processes in AChS (in particularly ones connected with the CD), for instance fluid viscosity, crystallization, irreversible photo- and thermopolymerization and so forth. The main principles of this concept may be shortly stated as below.

1. Chemical bonds switching, in the first approximation, belongs to not atomic but rather electronic process, inasmuch as covalent bond destruction or creation between certain atoms is determined by their electron configuration. It means that the initial elementary act of bonds switching is probably underthreshold electron excitations taking place in AChS due to electron-vibrational or electrostatic unstability mechanisms [23].

2. Equilibrium states for many atoms (atomic carrying) are changed in any bonds switching acts, being a cooperative process of configuration-deformation disturbances at the level of short- and intermediate-range ordering.

3. The process of bonds switching effectively occurs only at the some convenient configurations of atoms characterized by lower energetic barrier for metastable state. An act of "place preparation", i.e. chemical bond destruction near supposed structural zone of switching, is needed in the some of cases for this process. Hence, analysing destruction-polymerization transformation in the AChS, we must pay attention to structural fragments with small atomic compactness in comparison with remained part of normally prepared matrix.

The above mentioned comments are essentially important for understanding of the microstructural mechanism of photoinduced CD formation processes in the AChS.

4. Experimental procedure

The necessary information on induced structural transformations (the CD formation processes) in the a-As_2S_3 can be accurately obtained by the Fourier spectroscopy technique in the long-wavelength 400-100 cm^{-1} region [8,9].

For this purpose we must deduct the previously intensified absorption spectra before and after exposure or annealing, and determine the occurring processes by changes of optical density D in the region of the a-As_2S_3 main vibrational bands. Positive values of $\Delta D > 0$ correspond to structural complexes appearing under the treatment and negative ones of $\Delta D < 0$, on the contrary, correspond to complexes disappearing under the treatment.

The advantage of this technique consist in that the part of the vibrational spectrum induced by external influence is investigated, but not the whole integrated spectrum. Multiple accumulation of the useful deducted signal, when fast Fourier transformation is used, allows us to reach a sensitivity of this technique at the 1% level of breaking (switching) bonds.

The whole scheme of the experimental procedure used for the structural investigation of the CD formation processes in the AChS is presented in the Fig. 1.

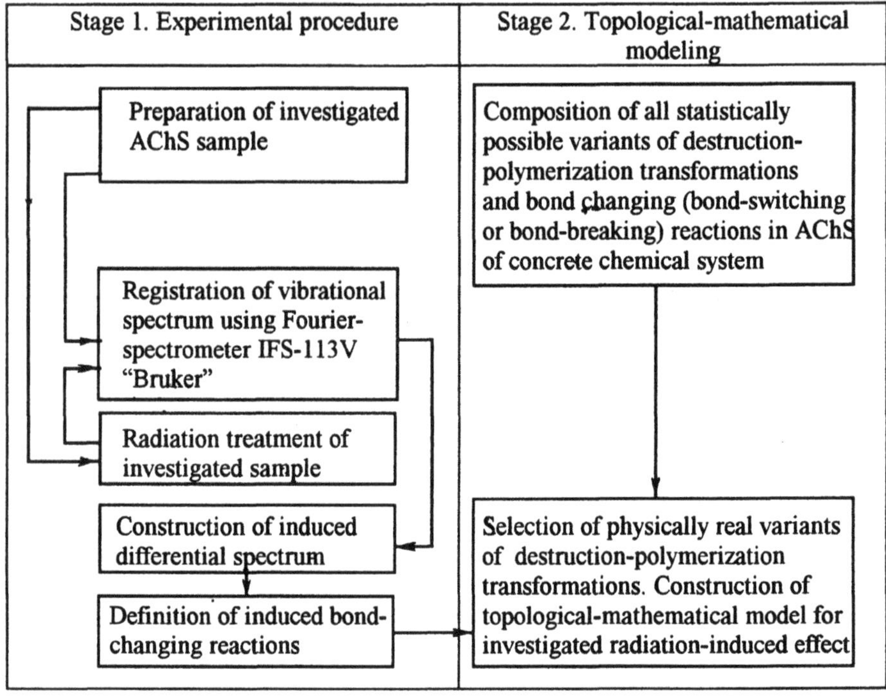

Figure 1. Scheme illustrated the conditions for experimental determination of the microstructural mechanism of the CD formation processes in AChS.

The a-As_2S_3 is concerned as a model object from the point of detection and investigation of induced structural changes. It is accounted for by a good distinction of different vibrational bands corresponding to main structural fragments, in particular the pyramidal AsS_3 units (335-285 cm^{-1}) and molecular products with "wrong" homopolar As-As (379, 340, 231, 210, 168 and 140 cm^{-1}) and S-S bonds (243 and 188 cm^{-1}) [27-30]. The factor group analysis shows that crystalline c-As_2S_3 are characterized by four different IR vibrational modes observed in the 335-285 cm^{-1} region (323.1, 307.3, 301.3 and 288.9 cm^{-1}) [29].

As to the preparation of the investigated samples and experimental details, it must be noted that correct investigation of the mechanism of the both irreversible and reversible PhST demands the strict fulfilment of the following conditions:

1) using the traditional specimens showing the PhST at the quite high level (this condition is defined by such parameters as evaporation rate, temperature, thin film thickness, molecular chemical composition and so on);

2) correct forming of the reversible channel of the PhST through subsequent cycles of the photoexposure by absorbed light at the room temperature and thermal annealing at the temperatures of the 20-30 K less than glass transition point;
3) experimental proving of the reversibility for the investigated PhST, i.e. observation of the equal but opposite to each other changes at the multiple repeated cycles of the photoexposure-thermoannealing.

It must be noted that in some cases these conditions were not kept.. Thus, it was shown in 1986 in the paper [10] that the reversible PhST in the a-As_2S_3 at the level of intermediate-range ordering have been not connected with chemical bonds switching in the first coordination sphere. Specimens in the form of 1 μm grain powder putting in the organic binder have been used for the experimental investigations. Such samples are characterized by too high light scattering and consequently a low efficiency of the reversible PhST. Besides, the photoexposure cycles have been carried out at T = 80 K, while the thermoannealing ones - only at the T = 323 K. Therefore, having changed these conditions (owing to substitution of the powder specimens on thin films), another research group in 1987 [11] observed more than 25% increase of the homopolar As-As chemical bonds concentration stimulated by photoexposure at the reversible stage.

Thin layers of the a-As_2S_3 (2-2.5 μm thickness) investigated in this work were deposited by vacuum evaporation on plates of radiation-modified polyethylene having a softening temperature over 430-440 K.

The photoexposure of the a-As_2S_3 layers was carried out by an unfocused absorbed laser beam (442 nm) at power density P = 25 mW/cm^2 during 120 min. The thermal annealing of the samples was fulfilled at the T = 423±1 K temperature during 30 min.

5. Topological variants of coordination defects formation

First of all, let's analyse all topological variants of the CD formation processes statistically possible in the a-As_2S_3, taking into account the main types of its initial structural units: heteropolar As-S bonds in the framework of pyramidal AsS_3 or bridge As-S-As complexes, homopolar As-As or S-S bonds in different fully or partially polymerised molecular fragments [4].

Since the final AChS state depends not only on the destroyed bond, but also on its nearest neighbourhood or the topology of nearest atoms distribution (it means that 2 initial units take place in the CD formation), there are 16 topological schemes of the corresponding structural transformations for 4 various initial units mentioned above. Another way speaking, the whole number of the statistically possible CD formation variants for the a-As_2S_3 is equal to the permutations of 4 taken 2 at a time.

These schemes grouped to homopolar and heteropolar chemical bonds breakings are presented at the Fig. 2 and 3.

The each scheme corresponds to one pair of the CD. The upper index in the defect signature (superscript) means the charge state of the atom, and the lower one (subscript) - the coordination number, i.e. the number of the nearest directly covalent-bonded atoms. The CD appear in the AChS matrix by pairs (negative and positive), that provide the conservation of the sample electroneutrality. The whole variety of the CD proper to the a-As_2S_3 are described by the next ones: S_1^-, S_3^+, As_2^-, As_4^+.

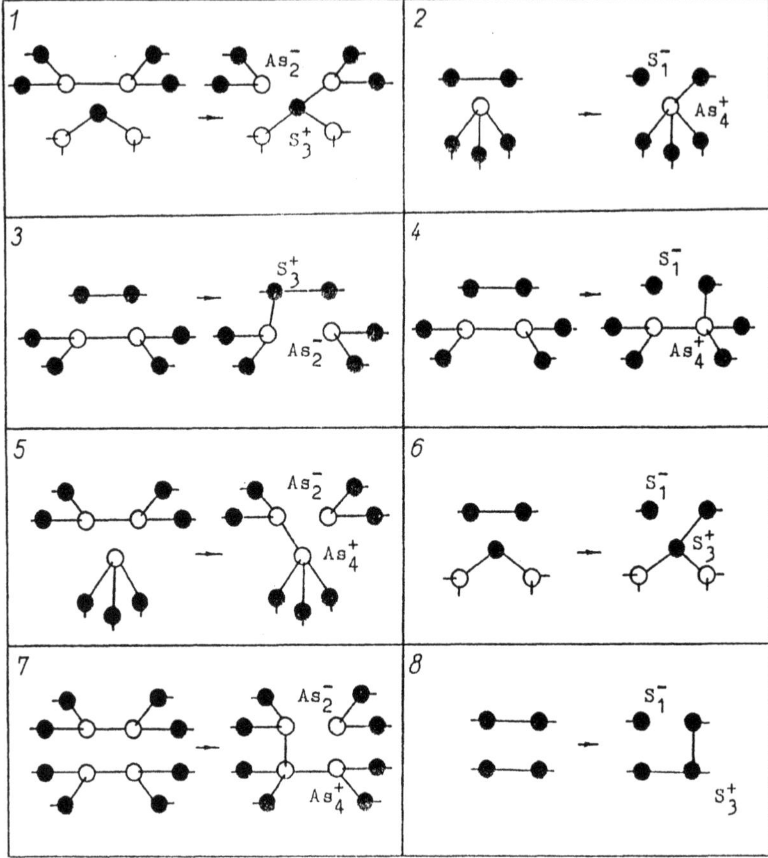

Figure 2. Defects formation in the a-As_2S_3 accompanied by homopolar bond-breaking transformations.

The another characteristic of the CD is their ordering defined by the number of "wrong" homopolar bonds in the nearest neighbourhood.

The 1-4 schemes at the Fig. 2 are connected with transformations (switchings) of homopolar covalent bonds into heteropolar ones and the 1-4 schemes at the Fig. 3, at the contrary, are connected with switchings of heteropolar bonds into homopolar ones. The 5-8 schemes at the Fig. 2 and 3 don't change the chemical bonds type (one bond is destroyed, but the same bond forms again). However, in previous case (1-4 schemes) we work with structural changes at the level of short-range ordering and wait the essential changes of the vibrational bands intensities in the deducted Fourier spectrum It is suggested, that absence of such statistical consideration for the AChS of the concrete chemical composition leads to the incorrect conclusions on the mechanism of the structural transformations, especially in the case of multiple influences such as photoexposure and thermal annealing at the reversible stage. Some topological variants of the CD formations remained out physical consideration, whereas the physically unreal processes were erroneously accepted as possible ones [5,31,32].

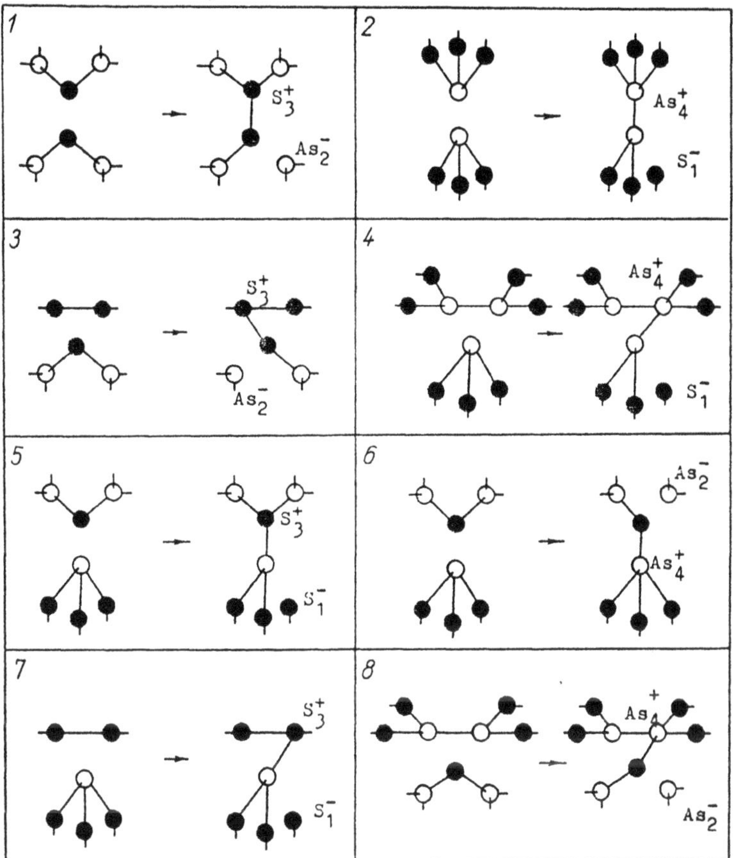

Figure 3. Defects formation in the a-As_2S_3 accompanied by heteropolar bond-breaking transformations.

6. Experimental cases

6.1. IRREVERSIBLE PhST

The spectral distribution of the additional optical density ΔD induced in the freshly evaporated a-As_2S_3 films by photoexposure (irreversible cycle) is shown in the Fig. 4.

It can be seen distinctly that absorbed light treatment leads to the decrease of the 379, 340, 243, 231, 210, 188 and 168 cm^{-1} vibrational bands and to the increase of those, corresponding to the pyramidal AsS_3 and the bridge As-S-As units (335 to 285 cm^{-1}). On the general background of the 335 to 285 cm^{-1} band one can point out weak overlapping bands at 324, 301 and 288 cm^{-1} and sharply defined peaks at 308 and 316 cm^{-1}. In our opinion, the latter (316 cm^{-1}) is caused by a superposition of modes at 324 and 308 cm^{-1} of equal intensities.

Thus, the observed irreversible PhST in the a-As_2S_3 may be conventionally represented by the next destruction-polymerization reaction:

$$(As\text{-}As) + (S\text{-}S) \rightarrow (As\text{-}S), \tag{3}$$

where the expressions in the parentheses denote the chemical bonds destroyed or formed as a result of photoexposure. Another way speaking, irreversible PhST in the a-As_2S_3 are accompanied by the transformations of the products containing homopolar S-S and As-As bonds into ones containing heteropolar As-S bonds.

A following annealing of the photodarkened a-As_2S_3 films produces the same changes as described by (3) reaction. However, the intensities of the main vibrational modes are more than twice smaller.

The photoinduced decrease of the homopolar bonds concentration in the irreversible stage (see Fig. 4) indicates that they are the initial components of the PhST. There are eight topological schemes of the CD formation for this case shown in the Fig. 2. Schemes 5 to 8 can be excluded from the further analysis since they do not change the bonds type. Schemes 1 to 4 really correspond to the obtained results and are described by the destruction-polymerization reaction (3).

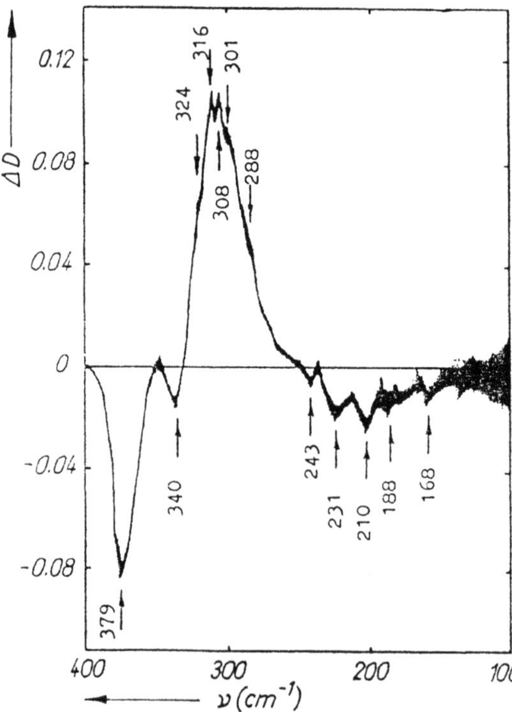

Figure 4. Photoinduced additional optical density in the freshly evaporated a-As_2S_3 thin film.

Consequently, the formation of $(As_2^-;S_3^+)$ and $(As_4^+;S_1^-)$ CD is the main process of irreversible PhST in the investigated a-As_2S_3 thin films. However, in the nearest neighbourhood of $(As_2^-;S_3^+)$ and $(As_4^+;S_1^-)$ CD, formed in accordance with schemes 1 and 2, only high-energetic heteropolar bonds exist. Such CD are thermally stable in the following treatment. But two other defects pairs (schemes 3 and 4 in the Fig. 2) having a homopolar bonds with their nearest neighbours annihilate at high temperatures in full agreement with the experimentally obtained destruction-polymerizations reaction (3). The process of annihilation may be presented as homopolar S-S (Fig. 2, scheme 3) and As-As (Fig. 2, scheme 4) bonds switching into heteropolar As-S one in the framework of AsS_3 units. This analysis explains the twofold decrease of the thermoinduced vibrational mode intensities in comparison with the photoinduced ones.

6.2. REVERSIBLE PhST

Spectral distributions of the additional optical density ΔD in the a-As_2S_3 induced by third cycle photoexposure and annealing are illustrated in the Fig. 5a,b, respectively.

The vibrational mode intensities of the structural complexes with homopolar As-As (279, 340, 231, 210, and 168 cm^{-1}) and S-S bonds (243 and 188 cm^{-1}) are increased by absorbed light beam, while the concentration of the pyramidal AsS_3 and the bridge As-S-As units containing heteropolar As-S bonds (335 to 285 cm^{-1}) is decreased.

Subsequent annealing of the investigated samples shows that this process is fully reversible (see Fig. 5b). In multiple photoexposure-thermoannealing cycles these changes of optical density ΔD may be repeated with a small irreversible component (2 to 3 %). The quantitative analysis shows that no more than 6 to 7 % of the atomic sites take place in the observed processes in good accordance with the results, obtained by the Raman spectroscopy technique [7].

We shall consider all possible variants of the reversible CD formation processes in the a-As_2S_3 described by the reaction opposite to (3). Eight CD generation schemes accompanied by the destruction of heteropolar bonds (see Fig. 3) must be put in the centre of this analysis. Schemes 5 to 8 can be excluded from the further consideration since they do not change the bonds type. Schemes 3 and 4 perhaps are characteristic of non-stoichiometric chalcogen- and arsenic-rich samples. Thus, reversible photodarkening of the a-As_2S_3 is connected with $(As_2^-;S_3^+)$ and $(As_4^+;S_1^-)$ CD formation due to the schemes 1 and 2 in the Fig. 3.

A complete topological scheme of the reversible PhST in the a-As_2S_3 thin films can be presented as a complex of destruction-polymerization transformations shown in Fig. 6. Thus, all previously described stages of the CD formation connected with photoexposure by absorbed light and thermal annealing of the a-As_2S_3 lead to the one unified scheme of the reversible structural changes.

The analogous results have been obtained for the a-As_2Se_3 thin films [33]. However, their interpretation is more complicated as not only two channels connected with heteropolar chemical bonds switching into homopolar ones take place, but also the third channel connected with mutual switchings of the heteropolar chemical bonds of various types are present in this films.

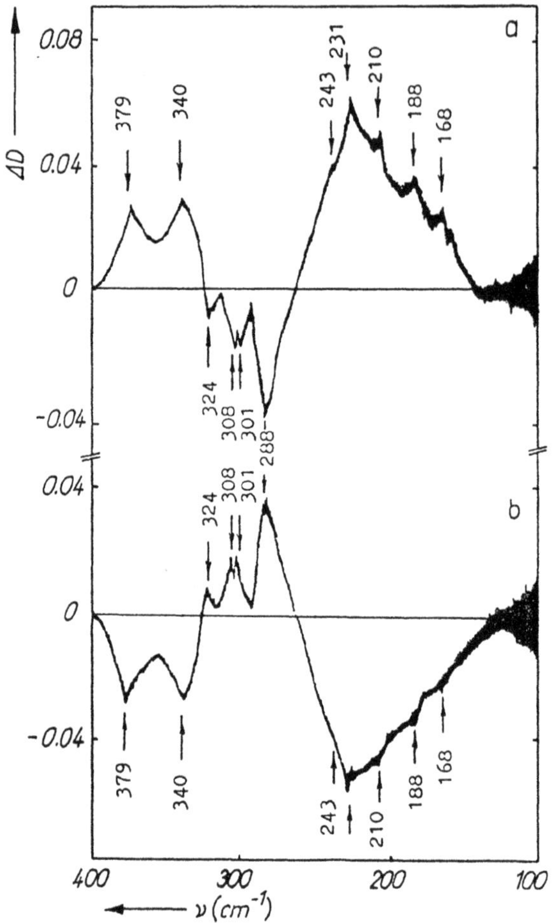

Figure 5. Additional optical density in the a-As_2S_3 thin film induced by third cycle photoexposure (a) and annealing (b).

6.3. EFFICIENCY OF THE REVERSIBLE PhST

In our model the reversible PhST are considered as interconnected processes of chemical bonds switching (short-range ordering changes) and following relaxation transformations (intermediate-range ordering changes) comprising a large space of the AChS network. As was shown in the Fig. 6, all atomic blocks taking part in the reversible PhST (pyramidal AsS_3 complexes) are formed in such nets of the structural matrix, where originally homopolar bonds existed (compare states 1 and 3). The local atomic density of changed structural blocks is much less than the average AChS density because homopolar bonds are more extended than heteropolar ones [34]. It means that the force constants of reversibly transformed complexes, first of all the

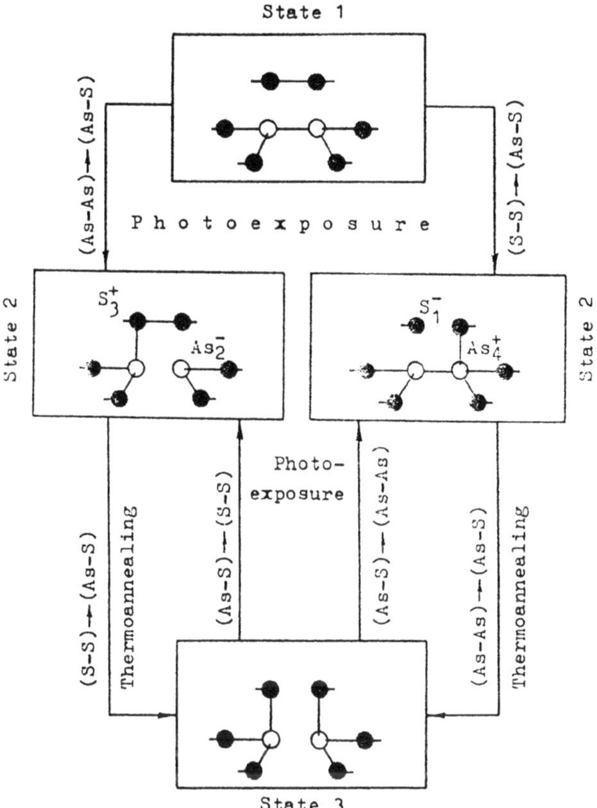

Figure 6. Topological scheme of reversible photostructural changes in the a-As_2S_3.

intermolecular AsS_3-AsS_3 interaction constants, are reduced in comparison with those lying on the other sites of the a-As_2S_3 structural matrix. Therefore, these complexes can be classified as soft atomic configurations [23] having double-well potential and negative electron correlation energy. The coincidence of the concentration of reversibly transformed bonds and soft atomic configurations is an additional confirmation of this assumption.

Network rigidity and relaxation effects in the nearest neighbourhood of appearing coordination defects are accepted in our model as decisive factors. Their quantitative characteristic is given by the local atomic density δ [7]:

$$\delta = \rho \frac{\sum \frac{A_i x_i}{\rho_i} - \frac{\sum A_i x_i}{\rho}}{\sum A_i x_i} \tag{4}$$

where A_i, x_i and ρ_i are atomic mass, atomic weight and density of the i-th structural fragment after bonds switching, ρ is average density of the sample.

The reversible PhST are more effective in the AChS with small local atomic density δ. Hence photoinduced changes of optical properties increase in the sample row "bulk-powder-thin film" in agreement with well known experimental data [5].

The existence of the photoinduced states in the AChS is conveniently connected with the CD stability. Let us introduce the parameter ε, which is equal to the difference of bonds dissociation energies before and after photoexposure. If the concentration of $(As_2^-;S_3^+)$ pairs is denoted by x_1 and the concentration of $(As_4^+;S_1^-)$ pairs - by x_2, the activation energy barrier of the reversible PhST can be written as

$$\Delta E = \frac{\varepsilon_1 x_1 + \varepsilon_2 x_2}{x_1 + x_2} \tag{5}$$

Supposing that the both reversible PhST channels are equally possible ($x_1 = x_2$), the ΔE values for the AChS films can be calculated: As_2S_3 - 20 kJ/mol, As_2Se_3 - 17,5 kJ/mol, As_2Te_3 - 7,5 kJ/mol. The photoinduced optical changes also decrease in this row [5]. The photodarkening efficiency in non-stoichiometric AChS can be estimated, taking into account the different ε_I values for the various schemes of the CD formation shown in the Fig. 6.

6.4. MODEL OF THE REVERSIBLE PhST

Thus, the microstructural mechanism of the discussed reversible PhST in the AChS can be described, taking into account main principles of the chemical bonds switching concept [30]. It contains the following photoinduced stages:

1. Excitation of electron or (and) hole pairs autolocalized at the soft atomic configurations (as initial microscopic process).
2. Weakening of the intermolecular bonds, resulting in the possibility to displacements of atomic groups.
3. Metastable state formation due to structural changes at the level of short-range (chemical bonds redistribution or the CD formation) and intermediate-range ordering (rearrangements of atomic blocks, displacements of non-bonded atoms).

These stages are interconnected and depend not only on the chemical-technological peculiarities of thin films, but also on temperature and spectral distribution of the absorbed light. Hence intra- and inter-molecular bond-breaking processes are involved in the model of the reversible PhST in the a-As_2S_3 as described in [5].

7. Conclusions

Thus, we conclude that the processes of the CD formation stimulated by absorbed light photoexposure may be studied in the AChS as destruction-polymerization transformations or chemical bonds switchings at the level of both short- and intermediate-range ordering. The stage of topological-mathematical modelling for these transformations, i.e. statistical analysis of all possible variants of the CD formation, provides a correct selection of the physically real reactions associated with concrete experimental situation.

8. References

1. DeNeufville, J.P., Moss, S.C., and Ovshinsky, S.R. (1974) Photostructural trans-formations in amorphous As_2Se_3 and As_2S_3 films, *J. Non-Cryst. Solids* **13**, 191-223.
2. Tanaka, K. (1974) Evidence for reversible photostructural change in local order of amorphous As_2S_3 film, *Solid State Commun.* **15**, 1521-1524.
3. *Amorphous Semiconductors* / Ed. Brodsky, M. (1982) Mir, Moscow.
4. Felts, A. (1986), *Amorphous and vitreous inorganic solids*, Mir, Moscow.
5. Elliott, S.R. (1986) A unified model for reversible photostructural effects in chalcogenide glasses, *J. Non-Cryst. Solids* **81**, 71-98.
6. Tanaka, K. (1983) Mechanism of photodarkening in amorphous chalcogenides, *J. Non-Cryst. Solids* **59/60**, 925-928.
7. Frumar, M., Firth, A.P., and Owen, A.E. (1983) A model for photostructural changes in the amorphous As-S system, *J. Non-Cryst. Solids* **59-60**, 921-924.
8. Shpotyuk, O.I. (1994) Reversible radiation effects in vitreous As_2S_3. 2. Mechanism of structural transformations, *Phys. Stat. Sol. A* **145**, 69-75.
9. Kornelyuk, V.N., Savytsky, I.V., Shpotyuk, O.I., and Yaskovets, I.I. (1989) Mechanism of reversible photoinduced effects in As_2S_3 thin films, *Fiz. Tverd. Tela* **31**, 311-313.
10. Lowe, A.J., Elliott, S.R., and Greaves, G.N. (1986) Extended X-ray absorption fine-structure spectroscopy study of photostructural changes in amorphous arsenic chalcogenides, *Phil. Mag. B.* **54**, 483-490.
11. Yang, C.Y., Paesler, M.A., and Sayers, D.E. (1987) Measurement of local structural configurations associated with reversible photostructural changes in arsenic trisulphide films, *Phys. Rev. B.* **36**, 9160-9167.
12. Weiqing, Z., Paesler, M.A., and Sayers, D.E. (1992) Structure and photoinduced structural changes in nonstoichiometric a-As_xS_{1-x}: A study by x-ray-absorption fine structure, *Phys. Rev. B.* **46**, 3817-3825.
13. Mott, N., and Davis, E. (1974) *Electronic processes in non-crystalline materials*, Mir, Moscow.
14. Marshall, J.M., and Owen, A.E. (1971) Drift mobility studies in vitreous arsenic triselenide, *Phil. Mag.* **24**, 1281-1305.
15. Bishop, S.G., Strom, U., and Taylor, P.C. (1976) Optically induced localized paramagnetic states in chalcogenide glasses, *Solid State Commun* **18**, 573-576.
16. Anderson, P.W. (1975) Model for the electronic structure of amorphous semiconductors, *Phys. Rev. Lett.* **34**, 953-955.
17. Baranovsky, S.D., and Karpov, V.G. (1987) Localized electronic states in vitreous semiconductors, *Phys. I Techn. Poluprov* **21**, 3-17.
18. Mott, N.F.,.Davis, E.A, and Street, R.A.(1975) States in the gap and recombination in amorphous semiconductors," *Phil. Mag.* **32**, 961-996.
19. Kastner, M.A., Adler, D, and Fritzshe, H. (1976) Valence-alternation model for localized gap states in lone-pair semiconductors, *Phys. Rev. Lett.* **37**, 1504-1507.
20. Kastner, M., (1978) Defect chemistry and states in the gap of lone-pair semiconductors, *J. Non-Cryst. Solids* **31**, 223-240.
21. Street, R.A. (1977) Non-radiative recombination in chalcogenide glasses, *Solid State Commun.* **24**, 363-365.
22. Street, R.A. (1978) Recombination in amorphous semiconductors, *Phys. Rev. B* **17**, 3984-3995.
23. Klinger, M.I. (1983) Model of electronic processes in glassy semiconductors: correlation with structural features, *Solid State Commun.* **45**, 949-953.
24. Klinger, M.I. (1988) Glassy disordered systems: topology, atomic dynamics and localised electron states, *Phys. Rep.* **165**, 275-397.
25. Vardeny, Z., and Tauc, J. (1985) Method for direct determination of the effective correlation energy of defects in semiconductors: Optical modulation spectroscopy of dangling bonds, *Phys. Rev. Lett.* **54**, 1844-1847.
26. Zakis, Yu.R. (1984) *Defects in vitreous state of substance*, Zinatne, Riga.

27. Solin, S.A., and Papatheodorou, G.N. (1977) Irreversible thermostructural transformations in amorphous As_2S_3 films: A light scattering study, *Phys. Rev. B.* **15**, 2084-2090.
28. Strom, U., and Martin, T.P. (1979) Photo-induced changes in the infrared vibrational spectrum of evaporated As_2S_3, *Solid State Commun.* **29**, 527-530.
29. Mori, T., Matsuishi, K., and Arai, T. (1984) Vibrational properties and network topology of amorphous As-S systems, *J. Non-Cryst. Solids* **65**, 269-283.
30. Scott, D.W., McCullough, J.P., and Kruse, F.H. (1964) Vibrational assignment and force constants of S_8 from a normal-coordinate treatment, *J. Molec. Spectroscopy* **13**, 313-320.
31. Fritzsche, H. (1993) The origin of reversible and irreversible photostructural changes in chalcogenide glasses, *Philosophical Magazine B* **68**, 561-572.
32. Shimakawa, K., Inami, S., and Kato, T. (1992) Origin of photoinduced metastable defects in amorphous chalcogenides, *Phys. Rev. B.* **46**, 10062-10069.
33. Shpotyuk, O.I. (1994) On the mechanism of reversible photostructural transformations in amorphous arsenic triselenide, *Zh. Prikl. Spektroskopii* **61**, 143-148.
34. Rao, K.J., and Mohan, R. (1981) Chemical bond approach to determining conductivity band gaps in amorphous chalcogenides and pnictides, *Solid State Commun.* **39**, 1065-1068.

AMORPHOGRAPHY OF CHAOS AND ORDER IN ISOTROPIC AND ANISOTROPIC GLASSES

S. A. DEMBOVSKY
Institute of General and Inorganic Chemistry of Russian Academy of Sciences, Leninsky Pr. 31, Moscow 117907, Russia. Fax: 7-095-9522382

1. ANISOTROPY, DEFECTS, AND SYMMETRY

Glass is a bulk isotropic medium consisting of anisotropic structural polyhedra. Therefore, it is not surprising that the scalar effests, such as photostructural changes accompanied by the change in density, refraction index, etc. can be induced there. However, the vectoral effects in the form of optical anisotropy were indeced too. Before the obtaining of anisotropic glasses, there were prepared anisotropic films - under the action of polarized laser illumination with the energy of $h\nu > Eg$ [1] and $h\nu < Eg$ [2]. At that time we have induced the anisotropy of viscous flow in a softening glass in magnetic [3] and electric [4] fields, and in temperature gradient [5]; the first influence being axial vector, and the second and third ones - polar vector. Later Luybin and Tichomirov, using subgap laser illumination, have obtained optical anisotropy and even gyrotropy in a narrow ($\sim 100\mu$) channel inside a bulk glass [6,7], and Tanaka have obtained the bulk optical anisotrtopy under the action of uniaxial pressure [8] - this is the polar-tensoral but not the vectoral influence. Thus, in glasses there exist scalar, vectoral and polar-tensoral effects.

In the last years these investigation have attracted a great attention from both experimental and theoretical points of view. Beginning from one of the first models anisotropy [9] the key problem remains the nature of anisotropic centers. Most of the models appeal to defects, first of all to famous VAP (valence alterbation pairs) and IVAP (intimate VAP) [10], which are considered as polar dipoles [11,12] and even chiral centes [12] responsible for anisotropy. However, there are some considerations against VAP (IVAP) as the source of anisotropy. First of all, in

A. Andriesh and M. Bertolotti (eds.),
Physics and Applications of Non-Crystalline Semiconductors in Optoelectronics, 275–290.

spite of arguments given in ref.[12], these defects, being charged and under/super-coordinated, probably have an extremely low concentration (say, about 0.001% [13]), which contradicts the macroscopic character of the effects. In addition, the scheme of reconstruction for IVAP [12] is non-realistic from the quantum chemistry point of view. Note that VAP, being based on a merely qualitative consideration, possess unknown symmetry. Therefore it is more correct to use the model of defects in the form of D^+, D^-, and D^o after [14], which is not pretend to describe the concrete defect structure/symmetry.

To introduce a more realistic centres for understanding the macroscopic effects in glasses, there were considered three-centre bonds (TCB) [15-21], a particular case of hypervalent bonds (HVB). The symmetry of TCB (HVB) was determined as the point group C_1 [17], and therefore they are polar and chiral (e.g. [22]). In contrast to VAP, the ground state of TCB is neutral and diamagnetic. For instance, in Se there are TCB-dipole (a) and TCB-quadrupole (b) in fig.1 (here — denotes covalent bond, CB, ∿∿∿ is TCB, and points are lone-pair electrons).

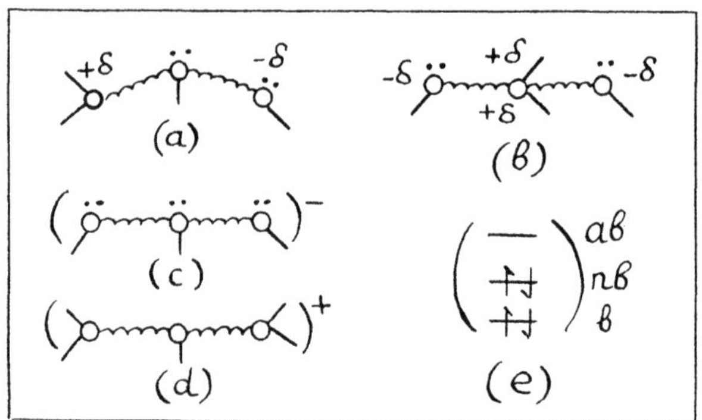

Figure 1. Examples of TCB in Se (a-d) and their molecular orbital scheme (e)

Among TCB there may be also the charged ones, e.g. (c,d) in Fig.1 having the same electronic configuration (e) from which it follows that these charged states of TCB are diamagneric too.

In contrast to VAP, the ground state of TCB (HVB) is neutral and diamagnetic, hence, concentration of TCB in glass should be much larger than concentration of VAP.

The existence of hypervalent bonds was confirmed by quantum-chemical calculations for S [23], B_2O_3 [24], GeO_2 [25], SiO_2 [26,27], BeF_2 [28], GeS_2 [29], and As_2S_3 [30], as well as by the density functional method for Se [31].

The concept of HVB is naturally connected with the theory of soft atomic configurations (SAC), which are assumed to be responsible for low-temperature anomalies, $U^{eff}<0$, quasi-continuum distribution of electonic levels in the mobility edge, etc. [32,33]. Hypervalent bond represent a concrete miscroscopic image for SAC, being able at the same time to explain the same glass features. The fact that the SAC concentration, being about 0.01-0.1 [32,33], is much higher than the VAP concentration (about 0.001) means not only [HVB(TCB)]>>[VAP(IVAP)] but also that HVB is rather not a defect but an equal in rights (with covalent bonds) structural element, and therefore one may consider glass as consisting of the HVB-subsystem and the CB-subsystem.

Since quasi-elastic constant for SAC is much lower than for rigid CB, the atoms belonging to SAC (HVB) can move at a suffisient distance without breaking of bond, and when the breaking in necessary, it takes much less energy than a direct breaking of covalent bond, Ed. For example, the switching of covalent bonds, being a necessary step at atomic transport, can proceed in a facilitated way, as it is shown below with Se as an example:

(1) formation of TCB, quasi-polymerization of chains and rings;

(2) destruction of TCB to VAP;

(3) formation of charged and diamagnetic TCB;

(4) its destruction with the translation of break (etc.)

In this scheme it is clearly seen that VAP and HVB are not alternatives, but supplement each other in the process. As a result, there is not only the break translation, but also the change of initial configurations and their mutual arrangement. Taking into account that intermediate HVB-formations are soft configurations, there arises a possibility that even sub-actions ($E<Ed$, $h\nu<Eg$) may have an orientational influence on the processes which proceed with participation of HVB, and the final configuration will reflect the nature and symmetry of external field. For example, the charged TCB in the scheme (1-4) will appear in the direction parrallel to the $\vec{E}$ vector of polarized light, and neutral and diamagnetic TCB-dipoles will reoriented in the same direction. Such a reorientation may proceed without breaking of HVB and thus the induced configuration will relax after the field off, relaxation being more or less fast depending on inherent barriers and temperature. This means coexistence of several types of relaxation, and one may consider an easily relaxed (e.g. corresponding to Arrhenius region), heavier relaxed (Kohlrausch region) and remembered orientation.

It was shown earlier [17], that TCB have a symmetry of point group C_1 which corresponds to the polar and chiral nature of the centres. Thus, if TCB were oriented under the action of linarly polarized light, this leads to the appearance of macroscopic anisotropy and chirality (gyrotropy). Owing to the HVB ability for reorintation and participation in switching of covalent bonds, one can eliminate the induced effects by non-polarized light and/or temperature.The same result one can obtain also by means of other symmetry arguments. When approaching to Tg, the specific motions, such as librations and inversions, appear in non-rigid HVB-subsystem [34]. Let us consider, for instance, the inversion of the top atom 4 beloning to the pyramidal molecule shown in fig.2. Initial figure has the symmetry C_1, while the arising composite one - C_i symmetry due to the centre of inversion. In contrast to C_1, the C_i symmetry is non-chiral and non-polar [22], then anisotropy and gyrotropy should disappear near Tg, as observed really. Thus, the symmetry arguments seems to be very useful for understanding anisotropic effects in glasses. However, there needs a more deep understanding of microscopic nature of corresponding centres than in the present day models (e.g. [11-12]).

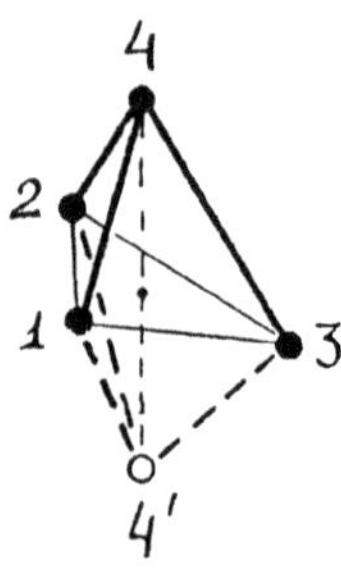

Figure 2. Example of the inversion of the C_1 pyramid

2. ORDER IN THE ATOMIC ARRANGEMENT, AND SYMMETRY

Several types of order, e.g. positional, orientational, configurational, and chemical order after [35] or cellural, topological,and continual order after [36], are considered. One should distinguish also order in static and dynamic systems. Of course, symmetry connects with order: as first approximation the higher symmetry, the higher the order in atomic arrangement. Using the language of symmetry, one can try to describe qualitatively the above types of order (and thus disorder, if Ord+Disord=1) in application to glasses:

chemical disorder - as the breaking of invariance relatively every translations. In this way one can describe the phase transition connected with violation of homogeneity of glass or violation of chemical order (see, e.g. [37]);

positional disorder - as the breaking of invariance relatively every rotations and reflections on the scales of short-, intermediate-, and long-range order; it may reveal in the decrease of number of these operations of symmetry.

This approach gives opportunity to describe order in glass in terms of symmetry, and the problem is to find, at least, a semi-quantitative measure of order in its connection with symmetry, and to desribe, using this measure, isotropic and anisotropic glasses and the transition isotropy->anisotropy itself. We shall solve this problem by means of AMORPHOGRAPHY as the science (similar to crystallography) which use symmetry as the method and amorphous (including glassy) state as the object.

The main problem of amorphography is the lack of our knowledge about symmetry in amorphous state: it is known only that the crystal symmetry, as a rule, remains on the scale of short-range order (SRO) and, sometimes, on the scale of intermediate-range order (IRO) [38]. The similarity in SRO follows from the similarity between vibrational bands in crystalline and glassy states (the bands are widen in glasses, but their position practically remains) and between the first peaks in RDF. However, a significant distinction appears on the IRO scale. For example, in glassy SiO_2 there prevails the 6-member rings which probably keep the symmetry of such rings in crystalline SiO_2 , however, there exist also 4-, 5-, and 7-member rings [39] with a very different symmetry [27]. This is the reason why a "pure" amorphography should be supplemented with a method which, being connected with symmetry, characterizes order/disorder.

Such a method in the form or informational symmetry or informational symmetry index I_{sym} was elaborated in details for molecules [40,41], but not for non-crystalline solids. I_{sym} takes into account the symmetry of molecule and of automorphous groups, from which it consists, by means of Shannon's probabalistic formulae for informtional entropy $H = -\Sigma p_i \log_2 p_i$. Recently [42] we have elaborated the non-probabalistic method for calculation of geometrical index I_{geom} which is based on the analytical geometry arguments only. The I_{geom} results correlates with the I_{sym} results, although not coinside with them strictly, therefore I_{geom} can be considered as informational index too. The geometrical informational index was suggested initially for molecules [42], however there exists a possibility to spread this method to SRO, IRO, and even to the long-range order scale in non-crystalline media, as it will be shown below.

3. GEOMETRICAL INFORMATIONAL INDEX AND ORDER

Since I_{geom} may be compared with disorder (it correlates with informational entropy connecting with the atomic arrangement), one can consider its supplemental value $(1-I_{geom})$ as negentropy, on the one hand, and as the measure of positional order, Q, on the other hand. The value of Q is calculated by means of expression [42,43]

$$Q = (3N - P)/3N = 1 - P/3N \qquad (5)$$

where N is the number of points constituting the system under consideration, and P is the number of independent parameters fixing these points. For example, to fix a separate point (N=1) one needs in P=3, and Q=0; to fix non-regular triangle (N=3) P=3x3=9, and Q=0 too; to fix regular triange P=3+2+2=7, and Q=0.22; etc.

Geometrical points mean atoms when we consider molecules. For example, let consider the pyramidal molecule of the NH_3 type with the regular triangle in the bottom. This figure has the C_{3v} symmetry. In this case N=4, P=7(3H)+1(N)=8, and Q=0.333.

When one considers not an isolated molecule but a polyhedra in condensed state, these values will change. For instance, in As_2S_3 the structural polyhedra is $AsS_{3/2}$ of the same C_{3v} symmetry. However, each S atom belongs to two neigbouring tetrahedra, and therefore N=1(As)+3(S)/2=2.5, P=7(3S)/2+1(As)=4.5, and Q=0.40, but not 0.33 as before.

4. ORDER IN THE ATOMIC ARRANGEMENT WITHIN STRUCTURAL POLYHEDRA OF THE COVALENT BOND SUBSYSTEM

From the structural point of view there are three main types of glasses: 1D, 2D, and 3D, to which the angle, trigonal (plane) or pyramidal (unplaine), and tetragonal polyhedra correspond on the scale of SRO. In accordance with the Zachariasen-Warren model (which is sufficiently improved at present - see e.g. [38]), these polyhedra are tied together by means of bridge atoms at the polyhedra vertices. Let us consider as an example the following prototype non-crystalline semiconductors: Se (1D), As_2S_3 (2D), and GeS_2 and a-Si (3D). To distinguish this case from the case of separate molecules, let us use the notation Q_1 for the order in polyhedra.

Table 1. Order (Q_1), relative order ($Q_1/1-Q_1$), and specific order ($q_1=Q_1/N$) in typical structural polyhedra

Substance	g-Se	g-As_2S_3	g-GeS_2	a-Si
Structural unit	$SeSe_{2/2}$	$AsS_{3/2}$	$GeS_{4/2}$	$Si_{4/4}$
Polyhedron		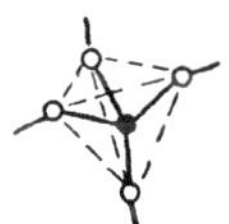		
Symmetry	$C_{2v}(4)$*	$C_{3v}(6)$	$T_d(24)$	T_d (24)
N	2	2.5	3	2
P	5.5	4.5	3.5	1.75
Q_1 [eq.(5)]	0.083	0.40	0.61	0.71
$Q_1/(1-Q_1)$	0.09	0.67	1.56	2.45
$q_1 = Q_1/N$	0.041	0.016	0.20	0.36

* The number in brackets is the number of operations of symmetry that keep invariance of the figure; this is roughly the measure of symmetry (Sym).

The values of Q_1 in Table 1 show that, like in molecules [42], Q_1=f(Sym,N). Note also a succesive increase of order with the increase of dimensionality: $Q_1(1D)<Q_1(2D)<Q_1(3D)$, a-Si being having the highest order among the 3D representatives. Additional characteristics of order - the values of relative and specific order behave in the same manner. It is interesting that the relative order, i.e. relation order/disorder, is above 1 only in the 3D case, being as high as 2.45 in a-Si, which is an especially rigid structure having about 1% of dangling bonds. In flexible chain structure of the Se type this relation is much less, that can be connected with the existence of various isomeric bonding configurations there [17,44] (see fig.1a-d), which provides excess entropy $\Delta S_o>0$ [44].

Remember the information sence of the value of disorder $(1-Q\ ) = I_{geom}$, the latter can be compated with informational entropy, and Q - with negentropy. Therefore it is not surprising that the value of $(1-Q_1)$ correlates well with the entropy of viscous flow near Tg, whose characteristic values were determined to be >200, 70-140, and 0-40 e.u. for 1D, 2D, and 3D glass formers, respectively [45]. This correlation is presented in fig.3.

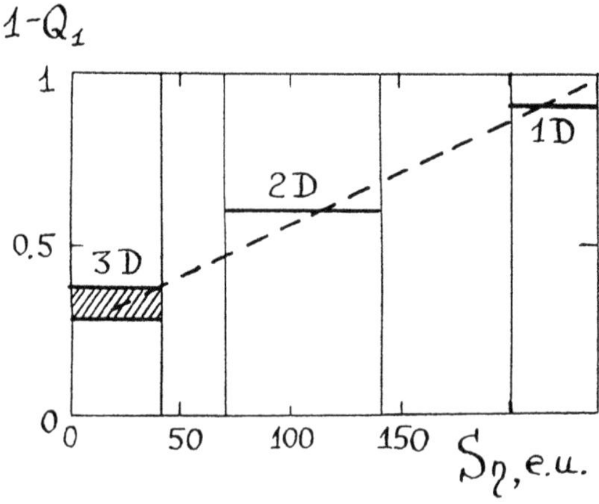

Figure 3. Correlation between geometrical disorder (from Table 1) and activation energy for viscous flow [45] in the 1D, 2D, and 3D structures.

5. ORDER ON THE SCALE OF INTERMEDIATE-RANGE ORDER AND ON THE SCALE OF LONG-RANGE ORDER IN THE SUBSYSTEM OF COVALENT BONDS WITH 1D Se GLASS AS AN EXAMPLE

The mode of joining of neighbouring polyhedra and so the resultant geometrical order on the scale of long-range order (LRO) depends not only on the dimensionality of glass (like in classical Zachariasen-Warren model)but also on the structural degrees of freedom (SDF) inside and outside the polyhedra. Let us investigate the SDF influence on the value Q_2 (new index below indicate new situation under consideration) using 1D chain-type glass like Se as an example.

It is known from structural experiment (see e.g. [46] for Se), that three structural parameters of a chain: first interatomic distance (R) or the length of covalent bond, valence angle ($\measuredangle$val), and torsional angle ($\measuredangle$tor) possess different ability to distortion, namely, in terms of force constant (β)

$$\beta(R) \gg \beta(\measuredangle val) > \beta(\measuredangle tor)$$

In Table 2 there are considered all possible variants of fixing the parameters - from the case when all three SDF realize (I) to the case of crystal (IV) that have no SDF.

Table 2. Order (Q_2) in polymeric N-atomic chains with different structural degrees of freedom

Case	I	II	III	IV
R	var	const	const	const
$\measuredangle$ val	var	var	const	const
$\measuredangle$ tor	var	var	var	const
P	3N	2N+1	N+4	8
Q_2	0	0.33-0.67/N	0.67-1.33/N	1.0-2.67/N

It is seen from Table 2 that for a sufficient large N the fixing of each new SDF leads to the increase of limQ_2 by one-third: 0 -> 0.33 -> 0.67 -> 1.0, that corrresponds to the transition from ideal gas to ideal crystal.

Situation in glass roughly corresponds to the case II (g-Se [44]) and/or the case III, and it is seen from fig.4 that the influence of additional atoms to the order depends on the system size N. One can distinguish conditionally three characteristic regions in fig.4 marked A, B, and C. At the A region, which may be compared with SRO and near IRO (2-3 coordination spheres) in terms of ref.[38], there is a strong and roughly linear dependence of order on the number of atoms, if N is represented in logarithmic coordinates. At the second B region, which corresponds to medium and far IRO [38], the development of Q_2 with the increase of logN becomes slower. Finally, the C region with Q_2 ->const corresponds to the the scale of long-range order (LRO). In conrtast to the case IV, this is not a crystalline - translational LRO but rather "chemical" LRO arising owing to regularies due to preservation of some parameters of chemical bond: R (II) or R and ∡val (III).

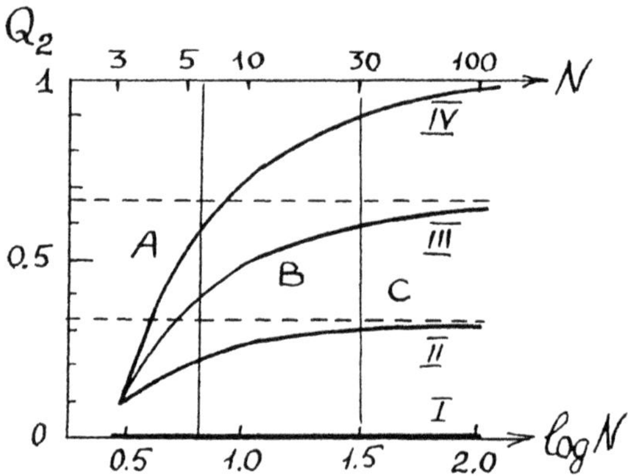

Figure 4. Order in chain structures with different SDF and varying number of atoms (see Table 2)

6. ORDER IN THE ATOMIC ARRANGEMENT IN HYPERVALENT BONDS AND DEFECTS

We have considered geometrical order in the subsystem of covalent bonds of glass. Here we begin consider the order in structural elements with violated bonding, namely, in hypervalent bond (HVB), valence alternation pairs (VAP), dangling bonds (DB), as well as in hydrogen bond (HB) in a-Si:H. In the latter case the two defect configurations (a,b) after [47] are considered.

Table 3. Order in defects (Q_3) in comparison with order in covalent polyhedra ($\Delta Q = Q_1 - Q_3$)

Substance	g-Se				a-Si:H		a-Si
Defect	HVB(TCB)	C_1^-(DB)	C_3^+		HB		Si_3° (DB)
Structure	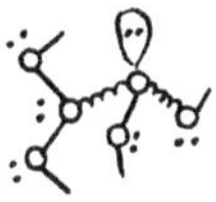	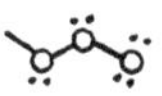	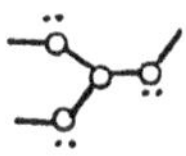		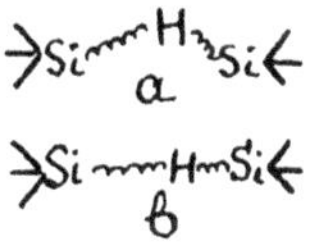 a, b		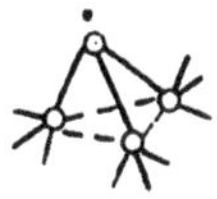
Symm.	$C_1(1)$	$C_s(2)$	$C_s(2)$	$C_1^*(1)$	$C_s(2)$	$C_{\infty v}^*$	$C_{3v}(6)$
N	1.5	2.5	2.5	2.5	1.5	1.5	1.75
P	4.5	7.5	6.5	7.5	4.5	4.25	4.0
Q	0	0	0.130	0	0	0.06	0.268
Q	0.083	0.083	-0.047	0.083	0.71	0.65	0.47

* The most probable symmetry ('a' for a-Si:H)

One should expect that the order in defects is lower than in corresponding structural polyhedra (see Table 1), and this is really the case except the situation when the unknown symmetry of the C_3^+ centre is chosen to be C_s (2). Therefore C_3^+-centre probably has the symmetry of $C_1(1)$, and hence it should be polar and chiral. The $C_s(2)$ symmetry of dangling bond C_1^- means its polar but non-chiral character.

7. ORDER IN ISOTROPIC AND ANISOTROPIC MEDIA CONSISTING OF QUASIMOLECULAR HYPERVALENT BONDS

Above we have considered the order Q_1 in structural polyhedra consisting of "normal" covalent bonds, the order Q_2 in a system of such polyhedra (with 1D chain structure as an example), and the order Q_3 in defects (HVB, VAP, etc.). Remember that hypervalent bonds, HVB, due to their high concentration ($\sim$1-10%) are rather a component of the structural network than a "defect" in it, and therefore one may consider the CB-subsystem and the HVB-subsystem of glass. The order in the CB-subsystem was, actually, considered earlier as Q_2, and, due to the rigid character of covalent bonds, external fields are hardly influence on its value. The situation ib "soft" HVB-subsystem is differ.

Calculation of order in HVB-subsystem, Q_4, is carried out by means of the basic formulae (5), in which

$$P = A + B\ (M - 1) \tag{6}$$

where P, as before, the number of indipendent parameters which fix N atom of the system (HVB-subsystem in our case), A is the number of independent parameters which fix the atoms in n-atomic group under consideration (single HVB), from which the system consists, B is the number of parameters which fix the second and followind groups (next HVB), and M=N/n. Eq.(6) was supposed earlier [43] for simple molecular systems, and here it is applied for the first time for solid/liquid state, in which the "molecules" are not isolated but embedded into the covalent network. Therefore this is a special situation of quasimolecular medium, the "quasimolecules" being HVB conjucted with covalent bonds of structural polyhedra of glass.

Let us consider, for example, two types of external influences: (i) the polar-vectoral one (PV, ↑), e.g. the electric component $\vec{E}$ of the linearly polarized light, and (ii) the polar-tensoral one (PT, ↕ or ⟟), e.g. uniaxial pressure (⟟).

From Table 4 it is seen that if the external field can act on the HVB-subsystem (e.g. electric field due to the dipole HVB - see fig.1), then the PV influence is more effective, than the PT one. Of course, the symmetry of HVB is also essential because of the Curie's principle of superposition of symmetry, which connects the symmetry of external influence, the symmetry of structure, and the symmetry of properties [48]. Thus, both the symmetry of the field and the symmetry of HVB determine the anisotropy effects.

Table 4. Order Q_4 in isotropic and anisotropic media consisting of HVB of the C_1 symmetry (N=1.5, P=4.5, Q_3 =0 - see Table 3)

Medium	Arrangement of HVB	B	P by (6)	Q_4	Fig.
isotropic	arbitrary	4.5	3N	0	5a
anisotropic (ii)	parallel to PT (Χ,↕)	3.67	0.83+2.4N	0.20-0.28/N	5b
anisotropic (i)	parallel and in direction of PV (↑)	2.83	1.67+1.89N	0.37-0.56/N	5c

8. BREAKING OF SYMMETRY AND OF DISORDER AT NON-EQUILIBRIUM PHASE TRANSITION "ISOTROPY->ANISOTROPY"

The transition from isotropic to anisotropic medium, as it is seen in Table 4, is accompanied by the increase of order and, consequently, by the decrease of disorder. Corresponding breaking of the symmetry of the medium is illustrated in fig.4, in which the case (a) is isotropic medium consisting of arbitrary oriented dipoles (→) and quadripoles (≻≺,↔), the case (b) is anisotropic medium arisiug under the action of, say, uniaxial pressure ($\vec{P}$), and the case (c) is anisotropic medium arising under the action of, say, the linearly polarized irradiation ($\vec{E}$).

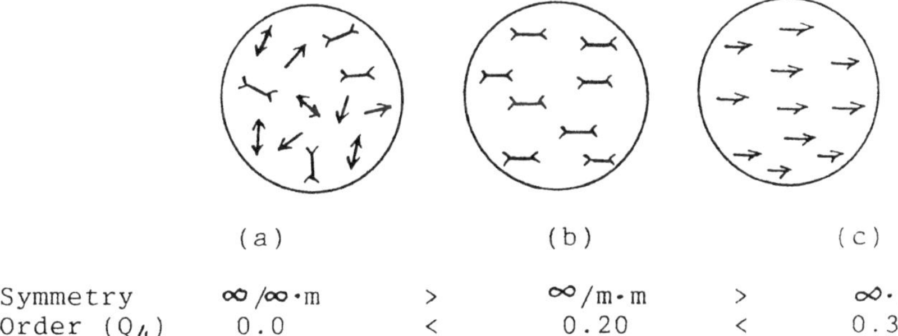

Symmetry	$\infty/\infty\cdot m$	>	$\infty/m\cdot m$	>	$\infty\cdot m$
Order (Q_4)	0.0	<	0.20	<	0.37

Figure 5. HVB in isotropic (a) and anisitropic (b,c) media, the anisotropy being induced by PT (b) or PV (c) influence

It is seen from fig.5 that in any case, when one or another type of non-equilibrium phase transition from isotropic to anisotropic state takes place, the symmetry decreases while order increases:

$$\Delta Q \Leftrightarrow -\Delta \mathrm{Sym} \qquad (7)$$

Such a relation is characteristic for the Landau phase transition of the II order [48], although this is the non-equilibrium phase transition in the case under consideration. It is interesting that in the world of molecules we have observed the opposite relation between the change of order and of symmetry [42].

9. CONCLUSIONS

Of course, calculation of order (disorder) can be a significant method for investigation of the structure of non-crystalline semiconductors on the scales of short-, intermediate-, and long-range order, including defects. Symmetry, which is the basis of amorphography, plays the main role in the process of such calculations. By means of geometrical informational index it succeeded in describing, on the semiquantitative level, the isotropy, anisotropy, and the non-equilibrium transition isotropy -> ->anisotropy. It was shown that the order Q is a function of symmetry, the number of atoms, dimensionality, and the number of structural degrees of freedom; however, it is clear that the real properties of non-crystalline semiconductors depends also on the rate of cooling, concentration of defects and impuruties, etc. The present approach can be extended not only to glassy semiconductors but also to amorphous films; oxide, halide, and other glasses; polymers, etc. The method of obtaining of anisotropic glasses is itself being the method of investigation of their structure. It is possible to predict the new methods of obtaining of anisotropic glassy semiconductors, e.g. under the action of ultrasonic wave in the process of melt quenching, etc., which will be published in a near future.

ACKNOWLEDGEMENTS: This work was supported by the Russian Foundation for Fundamental Research (Grant No. 96-03-33563)

10. REFERENCES

1. Zhdanov, V.G., Kolomiets, B.T., Lyubin, V.M., and Malinovsky, V.K. (1979) Phys. Stat. Sol. (1) 52, 621.
2. Hajto, J. and Janossy, I. (1982) J. Phys. C: Solid State Phys. 15, 6293.
3. Dembovsky, S.A., Chechetkina, E.A. and Kozyukhin, S.A. (1985) JETP Lett. 41, 88.
4. Dembovsky, S.A., Chechetkina, E.A. and Kozyukhin, S.A. (1982) Solid State Comm. 44, 1561.
5. Kozyukhin, S.A. (1986) Thesis, Inst. Gen. & Inorg. Chem., Moscow.
6. Lyubin, V.M. and Tikhomirov V.K. (1991) J. Non-Cryst. Solids 137-138, 993.
7. Lyubin, V., Klebanov, M., Rosenwaks, S. and Volterra, V. (1993) J. Non-Cryst. Solids 164-166, 1165.
8. Tanaka, K. (1989) J. Non-Cryst. Solids 114, 31.
9. Janossy, I., Jakli, A. and Hajto, J. (1984) Solid State Commun. 51, 761.
10. Kastner, M., Adler, D. and Fritzsche, H. (1976) Phys. Rev. Lett. 37, 1504.
11. Fritzsche, H. (1993) J. Non-Cryst. Solids 164-166, 1169.
12. Tikhomirov, V.K. and Elliott, S.R. (1995) Phys. Rev. B51, 5538.
13. Davis, E.A. (1979), in: M.H.Brodsky (ed.) Amorphous Semiconductors. Springer, Berlin.
14. Mott, N.F., Davis, E.A. and Street, R.A. (1975) Philos. Mag. 32, 961.
15. Popov, N.A. (1981) Fiz. Tekh. Polupr. 15, 369; (1982) ibid 16, 344.
16. Dembovsky, S.A. and Chechetkina, E.A. (1990) Glass Formation (Russ.), Nauka, Moscow.
17. Dembovsky, S.A. (1992) Solid State Commun. 83, 761.
18. Dembovsky, S.A. (1989) J. Non-Cryst. Solids 114, 115.
19. Dembovsky, S.A. and Chechetkina, E.A. (1986) J. Non-Cryst. Solids 85, 346.
20. Dembovsky, S.A., Chechetkina, E.A. (1986) Philos. Mag. 53, 367.
21. Dembovsky, S.A. (1992) in: Proc. Int. Congr. Glass-92, Madrid, v.II, p.157.
22. Zhoyludev, I.S. (1976) Symmetry and its Application, (Russ.) Atomizdat, Moscow.
23. Zuybin, A.S. and Dembovsky, S.A. (1994) Solid State Comm. 89, 335.
24. Zuybin, A.S. and Dembovsky, S.A. (1995) in: Proc. Int. Congr. Glass-95, Beijing, v.II, p.402.
25. Zyubin, A.S., Kondakova, O.A. and Dembovsky, S.A. (1996) Fiz. Khim. Stekla 23, No.1.
26. Sulimov, V.B. and Sokolov, V.O. (1996) J. Non-Cryst. Solids (in press).

27. Zyubin, A.S. and Dembovsky, S.A. (1995) in: Proc. Int. Congr. Glass-95, Beijing, v.II, p.408.
28. Zuybin, A.S., Kondakova, O.A. and Dembovsky, S.A. (1996) Fiz. Khim. Stekla (in press).
29. Zyubin, A.S. and Dembovsky, S.A. (1996) Zh. Neorg. Khim. (in press).
30. Zyubin, A.S., Dembovsky, S.A. and Grigor'ev F.V. (1996) Fiz. Khim. Stekla (in press).
31. Hohl, D., Jones, R.O. (1991) Phys. Rev. 1343, 3856.
32. Klinger, M.I. (1988) Phys. Repts 165, 275.
33. Klinger, M.I. Kudryavtsev, , V.G., Ryazanov, M.I. and Taraskin, S.N. (1989) Phys. Rev. B40, 6311.
34. Dembovsky, S.A. (1994) Glass Phys. & Chem. 20, 541.
35. Kitaigorodskii, A.I. (1984), Order and Disorder in the World of Atoms, Nauka, Moscow.
36. Ziman, J.M. (1979), Models of Disorder, Cambridge University Press, London.
37. Boolchang, P., Grothaus, J. and Phillips, J.C. (1983) Solid State Commun. 45, 183.
38. Elliott, S.R. (1991) Nature 354, 445.
39. Wright, A.C., Connell, G.A.N. and Allen, J.W. (1980) J. Non-Cryst. Solids 42, 69.
40. Bonchev, D., Kamensky, D., Kamenska, V. (1976) Bull. Math. Biol. 38, 119.
41. Bonchev, D. (1983) Information Theoretic Indices for Characterization of Chemical Structure, Research Studies Press, Chichester.
42. Dembovsky, S.A. and Koz'min, P.A. (1996) Russ. Chem. Bull, 45, No.8.
43. Koz'min P.A. (1995) Inorg. Mater. 31, 811.
44. Dembovsky, S.A. (1993) Solid State Comm. 87, 179.
45. Nemilov, S.V. (1992) Fiz. Khim. Stekla 18, 3.
46. Lucovsky, G. (1979) in: The Physics of Selenium and Tellurium (eds. E.Gerlach and P.Grosse), Springer, Berlin, p.178.
47. Zuybin, A.S. and Dembovsky, S.A. (1993) Solid State Comm. 87, 175.
48. Toledano, J.-C. and Toledano, P. (1987) The Landau Theory of Phase Transitions, World Scientific, Singapore.

PHOTOINDUCED STRUCTURAL CHANGES IN AMORPHOUS CHALCOGENIDES STUDIED BY RAMAN SPECTROSCOPY

C. RAPTIS, I. P. KOTSALAS and **D. PAPADIMITRIOU**
Department of Physics, National Technical University of Athens ,15780 Athens , Greece

M. VLČEK and **M. FRUMAR**
Department of General and Inorganic Chemistry, University of Pardubice, 532 10 Pardubice, Czech Republic

The use of Raman spectroscopy as probing technique for the study of structure and structural modifications in amorphous materials is briefly discussed, with emphasis being given to amorphous chalcogenides.

Raman scattering results from specific layered and bulk chalcogenide systems are reviewed and discussed in the framework of existing structural concepts. In particular, Ge-S based thin films with a small concentration of either Ga or Sb display reversible structural changes after successive illumination at moderate power densities and thermal annealing treatments. It has been found that ternary compositions having a 3-D amorphous network favour such photoinduced changes. This is explained in terms of the existence of defective bonds in such compositions and the large free volumes (corresponding to 3-D networks) which allow rearrangement or breaking of bonds. For higher power densities of illumination, compositions of the Ge-Sb-S system (in both bulk and layered forms) having large free volumes sustain partial crystallisation of Sb, thus confirming the above conclusion that a key parameter for the occurrence of photoinduced structural changes is the free volume of the material.

1. Introduction

It has been established in several investigations over the past two decades that illumination of amorphous chalcogenides induces, in most cases, substantial changes in their optical and electronic transport properties. These changes are most often accompanied by structural modifications, which can be (depending on the material and/or the extent of treatment) (i) reversible (upon subsequent thermal annealing) involving rearrangement of bonds in the local structure, or (ii) irreversible, usually related with the appearance of different phases (crystallisation, polymerisation, oxidation, etc). The various photoinduced effects in amorphous chalcogenides and the techniques used for their detection has been the subject of recent review articles [1 - 3].

The photostructural changes in amorphous chalcogenides are closely related to optical absorption changes which can be manifested either by photodarkening (PD) or photobleaching (PB) depending whether a red or a blue shift of the absorption edge is observed respectively. The As-based chalcogenides (and particularly the much

A. Andriesh and M. Bertolotti (eds.),
Physics and Applications of Non-Crystalline Semiconductors in Optoelectronics, 291–305.

studied As_2S_3) exhibit PD after illumination, which can be partially reversed upon subsequent thermal annealing [4 - 6]. In contrast, Ge-based binary chalcogenides usually display a PB effect [7 - 12] and this discrepancy has been attributed [3, 8] to the different way the optical band gap varies with composition in the two classes of chalcogenide glasses [7,13]. Whilst in As-based glasses, the optical gap shows a minimum at the stoichiometric composition, the Ge-based glasses display a maximum for this composition [7,13], so that illumination (or annealing) of fresh non-stoichiometric samples produces local stoichiometry, and results in PD of As-based glasses and PB of Ge-based glasses. It has been found [14], though, that the inclusion of a third component in Ge-S glasses alters the dependence of the optical gap on composition.

In general, the photoinduced effects in binary Ge-S based chalcogenides are not very pronounced as in the case of As-chalcogenides and this is most likely owing to the fact that the tetrahedral structural units (and bonds) forming the amorphous network in Ge-based glasses are quite compact (compared to the pyramidal structural units of the As-based glasses) to allow substantial reorientation of bonds. However, photoinduced effects have been observed in binary Ge-S(Se) glasses [7-10], especially in the case of thin film samples [9,10]. It appears that the addition of a third (non - chalcogen) component, such as Ga [11,15,16] or Sb [12,17,18], in binary Ge-S glasses results in enhanced reversible photoinduced changes. Also, Ag-doping of Ge-Se glasses has also increased the photoinduced effects to such an extent, that this glass has been considered [19] as the main constituent for the fabrication of a photoresist mask. Finally , illumination of Ge-Se glasses with laser light causes crystallisation [20,21] which can be reversed [22] providing that the laser power does not exceed a certain threshold . Such photoinduced crystallisation effects have not been observed in binary Ge-S glasses.

In this article both reversible and irreversible structural changes are reported for the ternary Ge-Ga-S and Ge-Sb-S glassy systems. The results are interpreted in terms of the dimensionality and free volume of the glassy network.

2. Raman scattering in amorphous materials

The absence of long range order in amorphous solids implies that the selection rules for Raman scattering by phonons in crystals are lifted, thus allowing simultaneous observation of a variety of scattering processes [23]. In this situation , the optic modes of the crystal become localised modes in the amorphous material within the chemical or intermediate order, while the long wavelength acoustic phonons of the crystal produce the so-called boson band in the glass at low frequencies. The lack of periodicity in glasses results in the broadening of all Raman bands, because the observed Raman spectrum effectively corresponds to scattering from all points of the Brillouin zone of the crystal.

Raman scattering due to vibrations within the local order (high frequency scattering) provides a lot of information about the structural units and bonds, the way these units are linked , as well as the surrounding environment. It becomes evident that Raman spectroscopy constitutes a powerful probing technique for structural characterisation of amorphous materials, unrivalled in most cases even by X-ray and neutron diffraction techniques, as it can detect marginal structural changes.

For the study of photoinduced structural effects in chalcogenide glasses, the high frequency (vibrational) spectra have been measured, and have provided an insight into possible changes of ordering and relative population of bonds, or the appearance of new phases.

3. Materials and treatments

Bulk glasses have been prepared by melting the high purity elements in evacuated fused silica cells (at 950-1000°C for 24 hrs) and quenching the cells at room temperature. Thin films of various thicknesses (0.2-2.0 μm) have been obtained on glass substrates by thermal evaporation under vacuum from bulk glasses of predetermined compositions. Details concerning sample preparation have been given previously [11,12]. The samples were kept under vacuum in the dark to avoid degredation. Two amorphous chalcogenide systems have been studied and their photoinduced structural effects are discussed in this article, namely the ternary Ge-Ga-S glass in thin film form with a stoichiometric Ge:S ratio (GeS_2) and a small inclusion of Ga (≈5%) and the ternary $Ge_xSb_{0.4-x}S_{0.6}$ (x=0.10, 0.15, 0.20, 0.25, 0.30, 0.40) system in both bulk and thin film forms.

The thin films have been illuminated either by a mercury arc lamp with IR cut-off filter for a period of 0.5-2.5 h (for reversible photoinduced changes), or by an intense laser beam (488 nm line of an Ar^+ laser for crystallisation effects). Bulk glasses of the Ge-Sb-S system were, also, illuminated by the same laser beam in order to induce crystallisation.

Annealing of thin films has been performed in air (Ge-Ga-S) or argon atmosphere (Ge-Sb-S) for 1-2 h at temperatures below the glass - transition temperature Tg.

4. Raman set-ups

Two Raman systems were used for the study of chalcogenide glasses depending on the type of photostructural effects to be studied. A double monochromator in conjuction with photomultiplier and photon counting system was used for the observation of Raman spectra of thin films and their photostructural changes. To avoid destruction of the films, a cylindrical lens was used and the laser power density was about 1 W/cm^2. Moreover, *in situ* Raman measurements of thin films in the process of annealing were also performed inside an optical furnace.

A triple monochromator equipped with high spatial resolution (≈2 μm) microscope and CCD camera was used for the study of photoinduced crystallisation. The laser power density was about 10^4 W/cm^2 focused on a spot of about 2 μm diameter via the microscope objective with the same laser beam also being used for the excitation of Raman scattering.

5. Photoinduced structural effects

5.1 REVERSIBLE STRUCTURAL CHANGES IN Ge-S: Ga THIN FILMS.

Interesting photoinduced phenomena were observed when Ga in small amounts was introduced in stoichiometric GeS_2 glass. It was found [24] that a single homogeneous phase can be obtained in this ternary amorphous system for Ga contents up to about 15%.

The study of the optical transmission spectra of amorphous $(GeS_2)_{0.95}Ga_{0.05}$ thin films [11] revealed that reversible photobleaching occurs in such films after suitable illumination and thermal annealing treatments; such transmission spectra are shown in Fig.1 corresponding to three samples of this composition prepared during the same evaporation, but subjected to various treatments. A strong blue shift of the transmission (absorption) edge is observed upon Hg lamp illumination of as-prepared (fresh) films (Fig.1a) A similar blue shift is also observed when the film is annealed

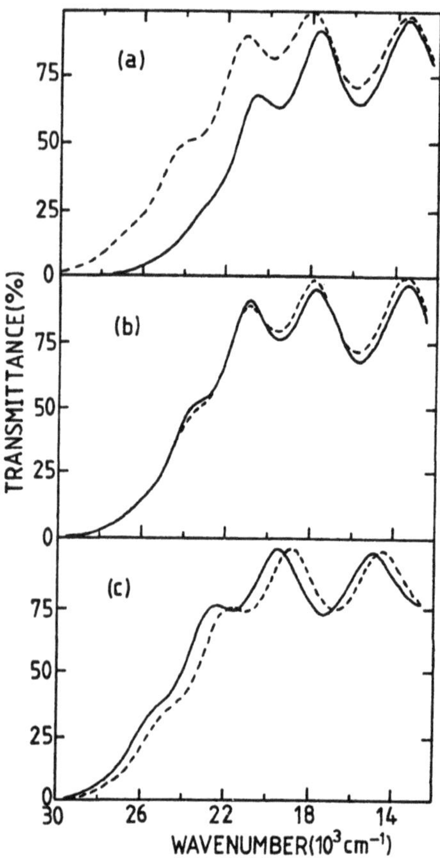

Figure 1 Transmission edge of non-illuminated (solid lines) and illuminated (dashed lines) $Ge_{0.31}Ga_{0.05}S_{0.64}$ films; (a) non-annealed film; (b) annealed to 250°C; and (c) annealed to 320°C [Ref.11]

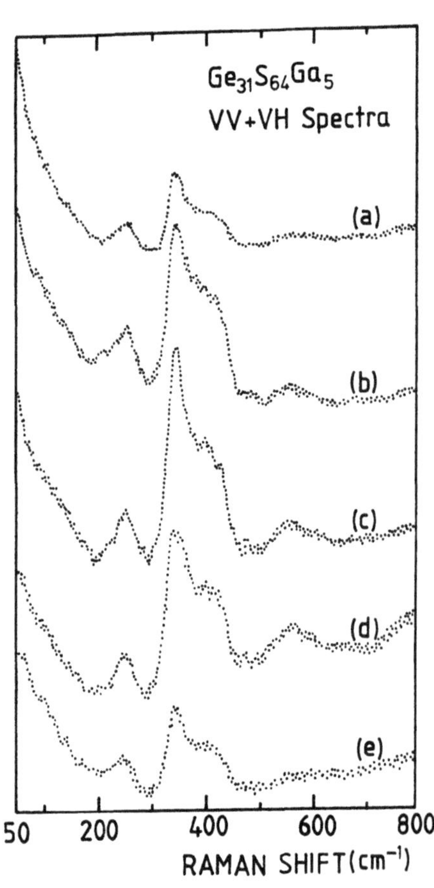

Figure 2 Raman spectra of $Ge_{0.31}Ga_{0.05}S_{0.64}$ films after various stages of treatment; (a) as-prepared film; (b) annealed to 250°C; (c) annealed to 250°C and illuminated; (d) annealed to 320°C; and (e) annealed to 320°C and illuminated [Ref.11]

at temperatures Tan=250 and 320°C, 120° and 50° below Tg respectively (solid curves of Figs 1b and 1c). However , subsequent illumination of annealed films tends to reverse the shift of the transmission edge for Tan about 50° below Tg (Fig.1c).

A close correlation exists between optical transmission and structural changes as they are manifested by the Raman spectra (Fig.2). The latter are dominated by a series of bands in the region 300-450 cm^{-1} corresponding to heteropolar Ge-S (mainly) bond vibrations within the tetrahedral GeS_4 structural units [25]. The band at 250 cm^{-1} is attributed to Ge-Ge and, possibly Ge-Ga bonds, given that the atomic weights of Ge and Ga are very close. In the stoichiometric (compound) composition GeS_2 of the binary system, this band is not expected to be present in the Raman spectra, as almost all Ge and S atoms should be engaged in the heteropolar Ge-S bonds of the tetrahedral amorphous network. This is true for bulk GeS_2 glasses , but not always for thin film samples, because the method of preparation of the latter (deposition of elements from the vapour phase) facilitates the appearance of a much greater number of inhomogeneities (compared to bulk samples) and the formation of homopolar bonds.

The thin film samples used for the measurement of Raman spectra were from the same preparation batch as those used for the optical transmission measurements. Annealing of as-prepared films causes an intensity increase and sharpening of both heteropolar and homopolar vibration bands (Figs 2b and 2d), an indication of structural ordering. Illumination of annealed films induces further increase of order (Figs 2b and 2c) for Tan about 100° below Tg; however , subsequent illumination of films annealed at temperatures Tan approaching Tg, results in an intensity decrease of all bands (Figs 2d and 2e), thus bringing the spectrum to (more or less) its original intensity and shape (comparison of Figs 2a and 2e). This implies that the structure of the glass is restored to that of the untreated film.

The PB effect is quite strong in these films, so that it can be witnessed by direct visual observations of illuminated or annealed films. Furthermore, evidence that the structural changes are related to PB is also obtained from the Raman spectra of Fig.2. After annealing or illumination (Figs 2b, c, d), the broad band at 550 cm^{-1} , due to the glass substrate, becomes substantially stronger, providing a definite indication that the film becomes more transparent in the spectral region around the laser line; however for Tan approaching Tg (Fig. 2e), and subsequent illumination, the band at 550 cm^{-1} almost disappears, which means that the film becomes darker in this spectral region.

5.2 PHOTOINDUCED STRUCTURAL CHANGES IN $Ge_xSb_{0.4-x}S_{0.6}$ THIN FILMS.

A more systematic study of photo-and thermally-induced effects has been carried out in the ternary $Ge_xSb_{0.4-x}S_{0.6}$ glass (x=0, 0.1, 0.15, 0.2, 0.25, 0.3, 0.4). The glassy network of this system consists mainly of GeS_4 tetrahedra [22,25,26] and SbS_3 pyramids [27] which are linked through a common corner or edge. Apart from the heteropolar M-S bonds (M=Ge,Sb), homopolar M-M bonds are expected to exist, as there is M-excess in the system (all the available M atoms cannot be bonded to S), but not homopolar S-S bonds.

This series of compositions was chosen in order to (i) avoid any S-clustering (S-chains or rings which reduce the homogeinity of the sample and (ii) obtain a variety of ratios (k=Nt/Np) between tetrahedral and pyramidal unit populations and different values of the mean coordination number Z which, as will be seen , is a measure of the dimentionality of the glassy network and is related to the free volume of the glass.

In the stoichiometric GeS_2 and Sb_2S_3 compositions of the binary systems, there should ideally be only heteropolar M-S, but not M-M bonds and this is largely true for bulk glasses , but less so for thin films. For the ternary system, one can determine various "stoichiometric" compositions in which there would be only heteropolar M-S bonds, or (for precision) compositions without surplus in either M or S. Such compositions have been prepared, namely bulk glasses and thin films of $Ge_{12.5}Sb_{25}S_{62.5}$ and $Ge_{25}Sb_{10}S_{65}$ having ratios of tetrahedral to pyramidal population k equal to 0.5 and 2.5, and mean coordination numbers Z equal to 2.6 and 2.5 respectively. Measurement of the Raman spectra has produced evidence [28] that homopolar M-M bonds are absent in as-prepared samples of these compositions. In order to investigate the photoinduced effects on the relative homopolar to heteropolar bond population , the compositions studied in this work are slightly off the "stoichiometric" compositions, thus ensuring the presence of homopolar M-M bonds.

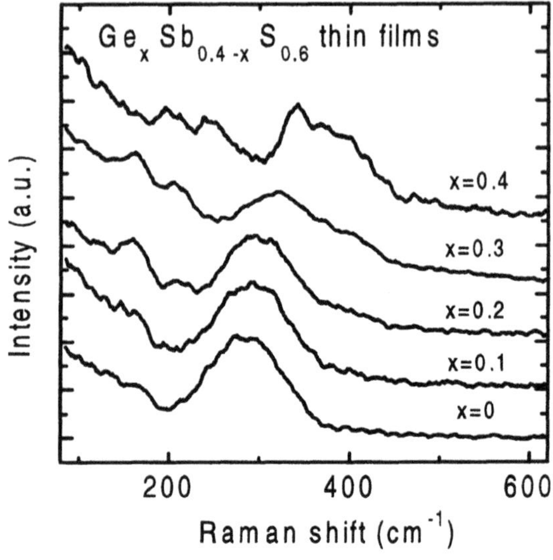

Figure 3 Polarised Raman spectra of $Ge_xSb_{0.4-x}S_{0.6}$ amorphous thin films for various compositions

Fig. 3 shows the polarised (parallel polarisations of incident and scattered light) Raman spectra of as-prepared $Ge_xSb_{0.4-x}S_{0.6}$ thin films for various compositions, including the binary compositions of the series (x=0, 0.1, 0.2, 0.3, 0.4). These spectra were recorded using a low laser power density of 1 W/cm^2 to avoid structural changes. All spectra are dominated by a broad feature which is the superimposition of several bands due to vibrations of heteropolar M-S bonds within the tetrahedra and pyramids. The peak position of this feature depends on the parameter k and lies between 280 (binary $Sb_{0.4}S_{0.6}$) to 340 cm^{-1} (binary $Ge_{0.4}S_{0.6}$). At lower frequencies, the spectra show weaker bands which are attributed to vibrations of homopolar M-M bonds; the frequency and intensity of these bands change with composition. Note that the binary $Sb_{0.4}S_{0.6}$ composition, being stoichiometric, shows only traces of homopolar bands in its spectrum (Fig. 3). Similar Raman spectra have been measured in bulk glasses of the $Ge_xSb_{0.4-x}S_{0.6}$ system.

Since the laser power used for the excitation of Raman spectra has been low, the resulting Raman signal is weak, thus making difficult the detection of either quantitative or qualitative changes of recorded spectra after various treatments. In order for it to become possible to extract some information, the spectra were reduced by subtracting the background, normalising to the Bose-Einstein thermal factor

$n(\omega) = \left[\exp(\hbar\omega / k_B T) - 1\right]^{-1}$ and finally subtracting an appropriate polynomial baseline between the high frequency end of the spectrum (position of zero intensity) and a low frequency limit which corresponds to the first valley of the spectrum (prior to the first peak due to homopolar bonds); the latter procedure serves to eliminate most of the quasielastic and boson scattering tails. Details concerning the procedures for spectra handling and fittings are given in a different publication [28].

The reduced Raman spectra of a freshly prepared $Ge_{0.3}Sb_{0.1}S_{0.6}$ thin film are shown in Fig. 4 after various stages of treatment. The spectrum of this composition consists of two bands at 170 cm^{-1} due to Sb-Sb bonds [27,29] and 215 cm^{-1} attributed to Ge-Ge bonds (the same band is observed in the binary $Ge_{40}S_{60}$ composition, Fig.3), and the broad band of several unresolved lines due to tetrahedral and pyramidal vibrations, showing a peak position at 330 cm^{-1}. Strictly speaking, since this composition is "non-stoichiometric", one cannot determine accurately the parameter k because the fraction of atoms of each metal (Ge and Sb) bonded to S is not known (there is surplus of metal in the composition); however, assuming that the same

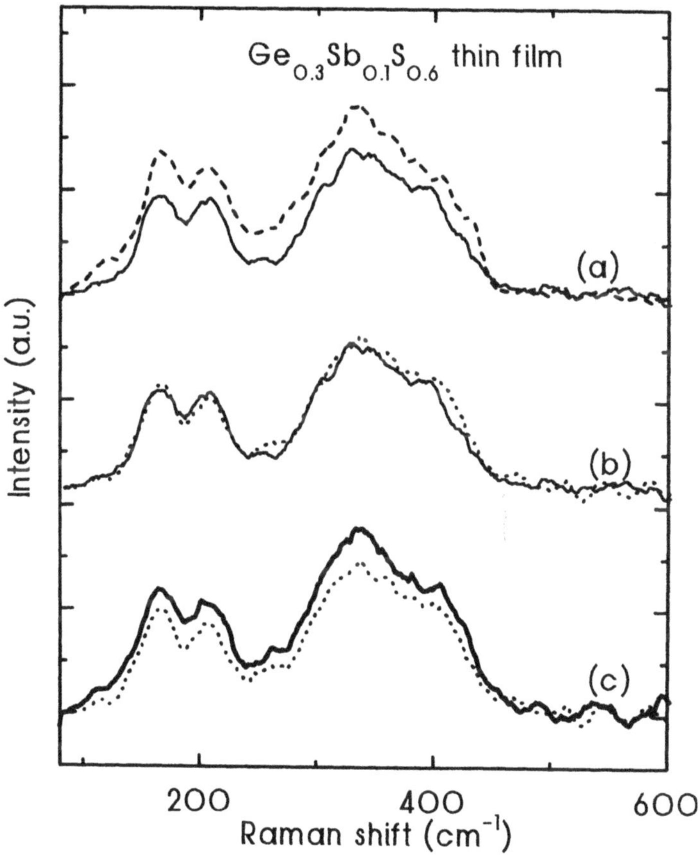

Figure 4. Comparison of reduced polarised Raman spectra of freshly prepared $Ge_{0.3}Sb_{0.1}S_{0.6}$ amorphous thin film; (a) as-prepared (solid line) and illuminated (dashed line); (b) as-prepared (solid line) and annealed to 220°C (dotted line); (c) annealed to 220°C (dotted line) and annealed to 220°C followed by illumination (thick solid line)

proportion of available atoms of each metal is bonded to S, a reduced molecular formula $Ge_{0.24}S_{0.48}$:$Sb_{0.08}S_{0.12}$ (or equivalently $Ge_{0.24}Sb_{0.08}S_{0.60}$) can be obtained for tetrahedra and pyramids, thus leaving a fractional atomic content surplus of 0.06 in Ge and a 0.02 in Sb. In this situation, the tetrahedra population is three times that of pyramids (k=3) and it should be expected that the broad band of the spectrum will be dominated by tetrahedra vibrations; this picture is actually confirmed by the spectra of this composition (Figs 3 and 4). The surplus metal atoms form Ge-Ge and Sb-Sb bonds, giving the two low frequency Raman bands at 215 and 170 cm^{-1} respectively.

Illumination of freshly prepared $Ge_{0.3}Sb_{0.1}S_{0.6}$ thin films induces an overall intensity increase of the spectrum (Fig. 4a), which implies an increase of order of the structural units. The rate of increase, though, is not the same for all bands. For example, the band at 170 cm^{-1} increases at higher rate than the band at 215 cm^{-1}, indicating that the Sb-Sb bond population also increases at the expense of the Sb-S (pyramidal) bond population; this conclusion is supported by a peak position shift of the broad band from 330 (as-prepared film) to 340 cm^{-1} (illuminated film) which implies a population reduction of SbS_3 pyramids.

Compared to illumination, annealing of freshly prepared $Ge_{0.3}Sb_{0.1}S_{0.6}$ thin films results in weaker spectral changes (Fig 4b). Thin film samples of this composition

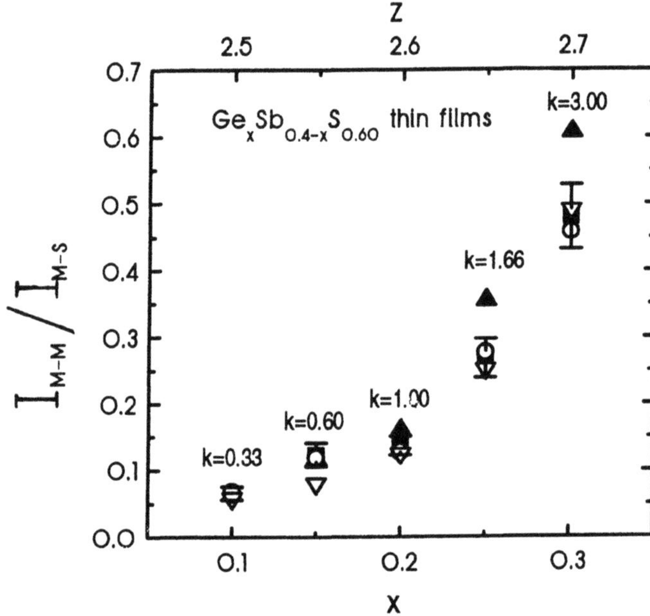

Figure 5. Composition dependence of integrated intensities ratio between homopolar M-M and heteropolar M-S Raman bands of $Ge_xSb_{0.4-x}S_{0.6}$ at various stages of treatment; ■fresh, as-prepared film: ▲ illuminated; O annealed to about 100° below Tg; ▽ annealed and illuminated; Z is the mean coordination (Equation 1) and k the estimated ratio of GeS_4 tetrahedra to SbS_3 pyramids population

were annealed to 220°C, that is, about 100° below Tg. First, the overall intensity of the spectrum remains, more or less, the same, indicating an unchanged order of the structure. There is, however, a relative intensity increase of the 175 cm^{-1} (Sb-Sb) band against the 215 cm^{-1} (Ge-Ge) band, and a shift of the broad band peak by about 10 cm^{-1} towards the higher frequencies. These effects are weak, so definite conclusions cannot be drawn and, for this reason, it is only tentatively suggested that annealing increases the tetrahedra population. Finally, illumination of annealed films results in an overall intensity increase of the spectrum, implying an increase of order of the structure (Fig 4c).

The above mentioned photo-and thermally-induced structural effects are not universal for all thin films of the series $Ge_xSb_{0.4-x}S_{0.6}$. As will be shown, the composition (x) and other related parameters (k, Z) largely determine the extent of these effects. In order to obtain an overall picture of the structural changes over all the compositions of the series, the integrated intensity of the low frequency (homopolar) bands has been compared to the integrated intensity of the broad (heteropolar) band for each composition and after each treatment. The ratio of these intensities, which represents the relative homopolar to heteropolar bond population, is plotted against x and Z and shown in Fig.5. The integrated intensities were determined after fitting the low frequency bands to two or three Gaussian and the broad band to two Gaussians. The parameter k is also shown for each composition, assuming (as before) that the same proportion of available metal atoms of each kind is bonded to S. The parameter Z (mean coordination number) has been determined by the formula

$$Z=4x+3y+2(1-x-y) \qquad (1)$$

which is valid for a ternary system $Ge_xSb_yS_{1-x-y}$ with atomic coordination numbers (nearest neighbours) 4(Ge), 3(Sb) and 2(S).

It has been predicted [30-33] and then verified experimentally [34], that glasses undergo a structural phase transition from a deformable, chain-like network to a rigid, random network at a mean coordination number $Z\approx2.4$; for $Z< 2.4$ the glass is polymeric with only isolated rigid regions, while for $Z>2.4$ most of its regions are rigid. Later, Tanaka [35] studied the compositional dependence of some structural and electronic properties of chalcogenide glasses and discovered striking singularities for $Z\approx2.4$ and, particularly, for $Z\approx2.67$. He disclosed that the atomic volume of chalcogenide glasses has a minimum for Z around the value of 2.4 and a maximum for $Z\approx2.67$. Further, he ascertained that both the optical band gap and the photoinduced absorption edge shift of such glasses show a maximum at $Z\approx2.67$. This coincidence of maxima is significant because it demonstrates that the strength of photoinduced effects is closely related to a critical structure of the glassy network. Tanaka [35] suggested that another structural phase transition occurs at a mean coordination number $Z=2.67$ from a 2-D network ($Z< 2.67$) to a 3-D one ($Z> 2.67$). Recently, Savova et al [14] discovered similar composition dependence of the atomic volume and optical band gap in exactly the same system studied in this work; in agreement with Tanaka's work [35] these quantities show maxima at $Z\approx2.67$ in $Ge_xSb_{0.4-x}S_{0.6}$ thin films.

Returning to the results of this work, it has been found that the extent of photoinduced structural changes is highest for mean coordination values Z in the region of 2.67 and this result is also illustrated in Fig.5 for the ratio I_{M-M}/I_{M-S} of homopolar to heteropolar bonds, which shows the largest variations after illumination or annealing for Z=2.70 (composition $Ge_{0.3}Sb_{0.1}S_{0.6}$). Generally, for Z in the vicinity of the critical value of 2.67, an increase (decrease) of the relative homopolar bond population is observed after illumination (annealing); in fact, the effect of illumination on this relative population tends to reverse the effect of annealing and vice versa. For Z≤2.60 (region of 2-D network) all photoinduced structural effects diminish [28]. The range of Z values for this series of compositions extends between 2.40 ($Sb_{0.4}S_{0.6}$) and 2.8 ($Ge_{0.4}S_{0.6}$); There were, however, no significant photoinduced structural effects observed in the Raman spectra of these binary compositions.

It is interesting to note that the ratio I_{M-M}/I_{M-S} diminishes for small x (or for Z< 2.60). This is because the stoichiometric binary composition $Sb_{0.4}S_{0.6}$ is approached with decreasing x and, therefore, the number of both Ge-Ge and Sb-Sb bonds decreases. In any case, the photoinduced structural effects in either relative (comparison of intensity ratios of different bands) or absolute (comparison of

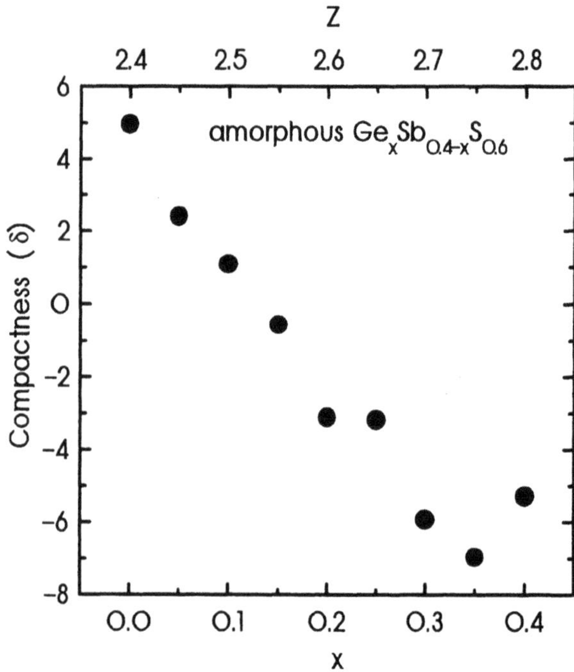

Figure 6. Compactness **δ** (Equation 2) of amorphous $Ge_xSb_{0.4-x}S_{0.6}$ plotted against composition x and mean coordination number Z (Equation 1). Values of **δ** from Ref. 36

intensities of the same band) sense are very weak for compositions close to stoichiometric ones and for Z< 2.60. This conclusion refers also to ternary "stoichiometric" compositions for which the photostructural changes are negligible [28].

The results of this work are in satisfactory agreement with Tanaka's qualitative model [35] of dimentionality change at a critical value of Z. Going further in the data analysis, the strength of the photoinduced structural effects observed in the work is related to the "free volume" of the glasses which reflects the flexibility of the network to sustain changes. A measure of the free volume is the quantity δ known as "compactness" and defined by the expression

$$\delta=\left(\sum V_i - V_{exp}\right)/V_{exp} = \rho\left[\sum A_i\, x_i/\rho_i - \sum A_i\, x_i/\rho\right]/\sum A_i\, x_i \tag{2}$$

where V_i is the volume which would be occupied in elemental form by as many atoms of component i (=Ge,Sb,S) as those existing in the sample under study, V_{exp}. is the experimentally measured volume of the sample, ρ the density of the sample, and A_i, x_i, ρ_i are the atomic weight , the fractional atomic content and the density of component i respectively. Negative values of the quantity δ correspond to large free volumes. The free volumes of several compositions of the system $Ge_xSb_{0.4-x}S_{0.6}$ have been determined previously by Vlcek [36] and Vlcek et al [37] and are shown plotted against composition x and mean coordination number Z in Fig.6. The largest free volumes of the series are witnessed for Z values in the region of 2.70 (Ge-rich compositions), that is, in the region where the strongest photoinduced structural changes occur.

The extent of such structural effects seems to depend on a combination of dimentionality and free volume. Usually 3-D networks display larger free volumes than layered (2-D) and chain (1-D) structures . However, 3-D networks and large free volumes do not always guarantee the strongest photoinduced structural effects. An example supporting this argument is stoichiometric GeS_2 having Z=2.67, δ=-5.36 [36] and a structure dominated by corner sharing tetrahedra (3-D); this compound glass does not display strong photoinduced structural changes. As has been seen, though, the addition of small contents of a third component (Ga or Sb) in this stoichiometric composition and the formation of defective bonds (M-M, Ga-S, Sb-S) loosens, somehow, the compact 3-D structure of GeS_2, thus facilitating the reorientation or break-up of bonds. Similar results were obtained by Tanaka [35], who observed an increase of the photoinduced absorption edge shift when As was added to glassy Ge-S systems. The density of the defective bonds is higher in thin films than in bulk glasses and this explains the stronger photoinduced effects observed in the former.

A final comment in this section should be made about prehistory and ageing effects in amorphous chalcogenides. A characteristic feature of amorphous materials is that they can exist in several metastable states [3,6, 17,38] and transitions between these states can be accomplished by illumination or thermal annealing. It has been found out [12,17,18] that the relative homopolar to heteropolar bond population can increase or decrease after illumination, depending on the metastable state in which the glass exists prior to light exposition. Furthermore, in this work an ageing effect was observed when the Raman spectra of the same films were recorded 12 or 24 months after preparation. Whilst in the freshly prepared films the ratio of homopolar to heteropolar bonds increases after illumination, in the aged films (from the same preparation batch) the situation gradually reverses [28].

5.3 CRYSTALLISATION OF Sb IN AMORPHOUS $Ge_xSb_{0.4-x}S_{0.6}$

Substantial changes in the Raman spectra (Fig. 7) of certain compositions of amorphous $Ge_xSb_{0.4-x}S_{0.6}$ were observed for both thin films and bulk materials when such samples were exposed to high laser power densities of the order of 10^4 W/cm^2 or annealed (for thin films) at temperatures below Tg. Specifically, two new , sharp bands gradually emerged at 110 and 146 cm^{-1}, the intensity of which was stronger for Ge-rich ternary samples; however, such peaks were not observed in the spectra of the binary $Ge_{0.4}S_{0.6}$ glass. The same laser beam was also used to probe the Raman scattering. These sharp bands appeared also during *in situ* observations of Raman scattering from thin films at temperatures 180-230°C using the double monochromator set-up and a laser power density of only 1 W/cm^2 (line focusing by a cylindrical lens).

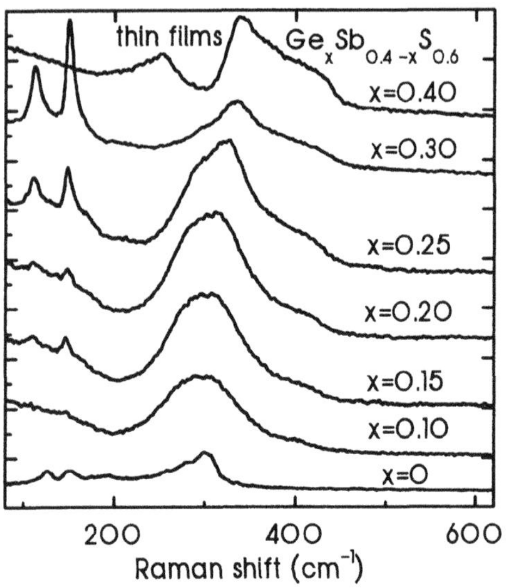

Figure 7. Polarised Raman spectra of $Ge_xSb_{0.4-x}S_{0.6}$ thin films for various compositions after exposure of equal time intervals to the 488 nm Ar^+ laser line at a power density of 10^4 W/cm^2, showing a gradual appearance (for certain compositions) of two sharp bands due to partial crystallisation of Sb

The spectral changes described above are attributed to Sb crystallisation, since it is known [39] that the Raman spectrum of crystalline Sb shows two bands at exactly the same frequency as the sharp bands of the laser exposed or annealed glasses of this work. All spectra of Fig. 7 correspond to identical exposure times by the laser and were recorded by the triple monochromator set-up. At first sight, the results of Fig. 7 appear contradictory, that is, the Sb-rich samples show negligible Sb crystallisation compared to the Ge-rich ones. In order to interpret these results, one has again to take into account the free volume; the rate (threshold) of crystallisation is higher (lower), the larger the free volume of the sample (see also Fig.6). Eventually, all Sb-containing compositions of the system in either bulk or thin film form will sustain crystallisation at various exposure doses which are inversely related to the free volume [40]. It should be noted that high crystallisation rates correspond to mean coordination numbers Z in the region of 2.70 .

Confirmation that the emerging sharp peaks are due to scattering from elemental Sb is obtained by observing the peak position of the broad band of a certain composition at various exposure times (or doses, in general). It has been observed [40] that the peak position shifts towards the higher frequencies with increasing exposure

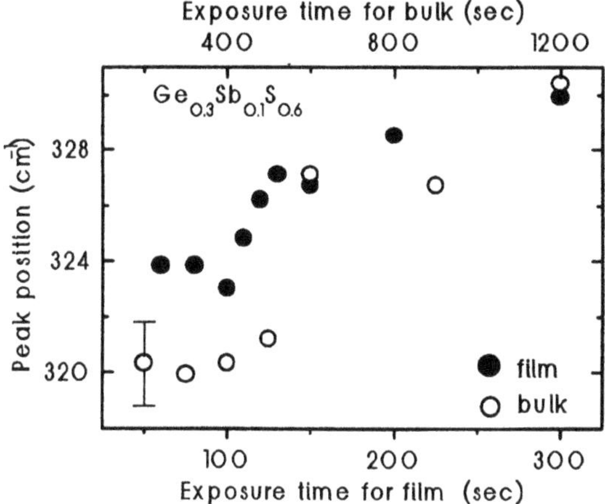

Figure 8. Peak position of the main (broad) Raman band of amorphous thin film (solid circles) and bulk glass (open circles) of the $Ge_{0.3}Sb_{0.1}S_{0.6}$ glass plotted against exposure time; the sudden frequency shift indicates the threshold for Sb crystallisation

time, implying a reduction of the Sb-Sb bond population. The threshold for crystallisation for bulk and thin film samples of composition $Ge_{0.3}Sb_{0.1}S_{0.6}$ is obtained from Fig.8, in which the peak position of the broad band is plotted against exposure time. Note that that the threshold for the thin films is about 4 times lower than that of the bulk sample.

6. Conclusions

It has been shown that Raman spectroscopy remains a unique technique for the structural characterisa-tion of amorphous chalcogenides and the detection of photoindu-ced changes in their structure. The occurrence and intensity of photostructural effects in amorphous chalcogenides depends largely on the existence of defective bonds, which reduce the compactness of the structure. In Ge-S based amorphous systems, a large number of defective bonds can be created by introducing a third (non-chalcogen) component into the system.

In general, 3-D amorphous networks favour photostructural effects. It has been observed that these effects become strongest for ternary compositions, corresponding to mean coordination number Z in the region of, a previously reported [35], critical value of 2.67 for which a structural phase transition from 2-D to 3-D structure has been suggested [35]. Furthermore, it seems that the key parameter for the occurrence of photostructural changes in amorphous chalcogenides is the free volume. Ternary compositioins of the system $Ge_xSb_{0.4-x}S_{0.6}$ having large free volumes have shown strong photostructural effects. Illumination of fresh thin films corresponding to large free volumes causes a reduction of the SbS_3 pyramidal population (with respect to tetrahedra population) and a simultaneous increase of the homopolar Sb-Sb bond population. This result confirms that the addition of a third component in the Ge-S system facilitates the occurrence of photoinduced effects, because the new Sb-S bonds introduced are not as strong as the Ge-S ones, and given that there is space available in the material, all types of bond will rearrange and a proportion of the weaker type will break-up. In general, the possibility of witnessing photostructural effects in amorphous chalcogenides depends on the combination of (3-D) dimensionality, large free volume and existence of an adequate number of defective bonds.

The conditions for the Sb crystallisation in $Ge_xSb_{0.4-x}S_{0.6}$ amorphous thin films are similar to those for reversible photostructural effects, except for the power level of illumination which is 3-4 orders of magnitude higher in the case of crystallisation. Again, the rate of crystallisation is higher, the larger the free volume, and a similar reduction of pyramids with respect to tetrahedra is observed when crystallisation begins.

Acknowledgments

One of the authors (C.R) is grateful Dr. Z.G. Ivanova for the fruitful and constructive collaboration in the past on amorphous Ge-S:Ga thin films. Partial financial support of the project by the Greek Secretariat of Research and Technology is gratefully acknowledged.

References

1. Owen, A.E., Firth, A.P. and Ewen, P.J.S. (1985) *Phil. Mag.* **52**, 347.
2. Tanaka, Ke. (1990) *Rev. Solid State Sci.* **4**, 641.
3. Shimakawa, K., Kolobov, A. and Elliott, S.R. (1995) *Adv. Phys.* **44**, 475.
4. de Neufville, J.P., Moss, S.C. and Ovshinsky, S.R. (1973-1974) *J.Non-Cryst. Solids* **13**, 191.
5. Biegelsen, D.N. and Street, R.A. (1980) *Phys. Rev. Lett.* **44**, 803.
6. Tanaka, Ke. (1980) *J.Non-Cryst. Solids* **35-36**, 1073.
7. Street, R.A., Nemanich, R.J. and Connel, G.A.N. (1978) *Phys. Rev. B* **18**, 6915.
8. Kolobov, A.V., Kolomiets, B.T., Lyubin, V.M., Sebastian, N., Taguirdzhanov, M.A., and Hajto, J. (1982) *Sov. Phys. Solid St.* **24**, 603.
9. Rajagopalan, S., Harshavardhan, K.S., Malhotra, L.K. and Chopra, K.L. (1982) *J.Non-Cryst. Solids* **50**, 29.
10. Tichy, L., Triska, A., Ticha, H. and Frumar, M. (1986) *Phil. Mag.* **54**, 219
11. Raptis, C. and Ivanova, Z.G. (1988) *J.Appl. Phys.* **64**, 2617
12. Vlcek, M., Raptis, C., Wagner, T., Vidourek, A., Frumar, M., Kotsalas, I.P. and Papadimitriou, D. (1995) *J.Non-Cryst. Solids* **192-193**, 669
13. Tichy, L., Triska, A., Frumar, M., Ticha, H. and Klikorka, J. (1982) *J.Non-Cryst. Solids* **50**, 371.
14. Savova, E., Skordeva, E. and Vateva, E. (1994) *J. Phys. Chem. Solids* **55**, 575.
15. Vateva, E., Nikiforova, M., Ivanova, Z.G. and Arsova, D. (1985) *J.Non-Cryst. Solids* **70**, 29.
16. Ivanova, Z.G. and Vasiliev, V.S. (1992) *Phys. Chem. Glasses* **33**, 8.
17. Frumar, M., Ticha, H., Vlcek, M. and Klikorka, J. (1981) *Chech. J. Phys.* B **31**, 441.
18. Frumar, M., Vlcek, M. and Klikorka, J. (1988) *React. Solids* **5**, 341.
19. Yoshikawa, A., Ochi, O., Nagai, H. and Mizushima, Y. (1977) *Appl. Phys. Lett.* **31**, 161.
20. Haro, E., Xu, Z.S., Morhange, J.F. and Balkanski, M. (1985) *Phys. Rev. B* **32**, 969.
21. Sugai, S. (1986) *Phys. Rev. Lett.* **57**, 456.
22. Griffiths, J.E., Espinosa, G.P., Remeika, J.P. and Phillips, J.C. (1982) *Phys. Rev. B* **25**, 1272.
23. Brodsky, M.H. (1975) in *Light Scattering in Solids (Ed. Cardona, M.)*, Springer-Verlag, Berlin, p.205.
24. Ivanova, Z.G. and Vateva, E. (1984) *Thin Solid Films* **120**, 75.
25. Lucovsky, G., Galeener, F.L., Keezer, R.C., Geils, R.H. and Six, A. (1974) *Phys. Rev. B* **10**, 5134.
26. Sugai, S. (1987) *Phys. Rev. B* **35**, 1345.
27. Watanabe, I., Nocuchi, S. and Shimizu, T. (1983) *J.Non-Cryst. Solids* **58**, 35.
28. Kotsalas, I.P., Papadimitriou, D., Raptis, C., Vlcek, M. and Frumar, M., *in preparation*.
29. Lannin, J.S. (1977) *Phys. Rev. B* **13**, 3863.

30. Phillips, J.C. (1979) *J.Non-Cryst. Solids* **34**, 153.
31. Dohler, G.H., Dandoloff, R. and Bilz, H. (1980) *J.Non-Cryst. Solids* **42**, 87.
32. Thorpe, M.F. (1983) *J.Non-Cryst. Solids* **57**, 355.
33. He, H. and Thorpe, M.F. (1985) *Phys. Rev. Lett.* **54**, 2107.
34. Phillips, J.C. (1985) *Phys. Rev. B* **31**, 8157.
35. Tanaka, Ke. (1989) *Phys. Rev. B* **39**, 1270.
36. Vlcek, M. (1986) *PhD Thesis*, University of Pardubice, Chech Republic.
37. Vlcek, M. and Frumar, M., (1987) *J.Non-Cryst. Solids* **97-98**, 1223.
38. Averyanov, V.L., Kolobov, A.V., Kolomiets, B.T. and Lyubin, V.M. (1980) *Phys. Stat. Sol. (a)* **57**, 81.
39. Sharp, R.I. and Warming, E. (1971) *J. Phys. F* **1**, 570.
40. Kotsalas, I.P., Raptis, C., Vlcek, M. and Frumar, M. *in preparation*.

INFRARED REFLECTANCE INVESTIGATION OF THE STRUCTURE OF $xSb_2S_3.(1-x)As_2S_3$ GLASSES

J.A. KAPOUTSIS[a], E.I. KAMITSOS[a], I.P. CULEAC[b] and M.S. IOVU[b]
[a]Theoretical and Physical Chemistry Institute,
National Hellenic Research Foundation,
48 Vass. Constantinou Ave., Athens 116 35, Greece

[b]Center of Optoelectronics of the Institute of Applied Physics,
nr. 1 Academiei Str., Chisinau, MD-2028, Republic of Moldova

1. Introduction

Chalcogenide glasses have attracted much attention over the years in light of their technological applications, including infrared transmitting optical elements, acousto-optic and memory switching devices, and materials useful for image creation and storage [1]. In addition, new chalcogenide glass compositions exhibit superionic conducting properties very promising for electrochemical applications [2].

As_2S_3 is the most extensively studied chalcogenide glass mainly because of its ease of formation, its excellent IR transmission and its resistance to atmospheric conditions and chemicals [3]. Even though As and Sb belong to the same group of the Periodic table, As_2S_3 and Sb_2S_3 do not display the same glass-forming tendency. Glassy Sb_2S_3 is very difficult to form because of the high cooling rates required [4]. However, addition of As_2S_3 to Sb_2S_3 enhances greatly the glass-forming ability of the latter, and thus, glasses in the mixed system Sb_2S_3-As_2S_3 can be formed.

It is generally accepted that the three dimensional network of glassy As_2S_3 is built of trigonal pyramidal units, AsS_3, which are interconnected through As-S-As bridges [5-7]. There is also evidence that the intermediate range order of this glass involves two neighboring pyramids and their shared S-atom, with the correlation length being ~ 7Å [8]. The rearrangement of such coupled pyramids with respect to the neighbors has been used to explain properties such as the reversible photoinduced structural changes [8]. Correspondingly, it has been shown that the basic structural units of glassy Sb_2S_3 are the trigonal pyramids SbS_3 bonded to each other by S atoms [4]. It is of interest to note that the resulting network of glassy Sb_2S_3 exhibits lower degree of local disorder around Sb atoms than that of crystalline Sb_2S_3 [9].

Despite the general agreement on the structure of glasses X_2S_3 (X=As, Sb), the structure of glassy materials in the mixed system Sb_2S_3-As_2S_3 remains controversial. Thus, some authors based on the results of various spectroscopic

A. Andriesh and M. Bertolotti (eds.),
Physics and Applications of Non-Crystalline Semiconductors in Optoelectronics, 307–315.

techniques, such as EXAFS, XANES, IR and ^{121}Sb-Mossbauer, propose the random substitution of As by Sb and the creation of mixed As-S-Sb bridges [10-13]. On the other hand, others have interpreted the results of IR [14, 15], XRD [15], crystallization kinetics [16] and optical gap measurements [17] as suggestive of a glass structure consisting of heterogeneous phases of As_2S_3 and Sb_2S_3 with little or no interactions between them.

It is clear from the above that further work is required to help resolving existing controversies concerning the structure of these mixed chalcogenide glasses. In this paper we present a systematic infrared reflectance study of glass compositions $xSb_2S_3.(1-x)As_2S_3$ in a wide glass forming region, $0 \leq x \leq 0.75$. The purpose of this investigation is twofold; first, to identify the effect of Sb_2S_3 addition on the local glass structure, and second, to explore the possibility of formation of mixed As-S-Sb bridges.

2. Experimental

Glasses were prepared by melting stoichiometric mixtures of glassy As_2S_3 and polycrystalline Sb_2S_3 in evacuated (at 10^{-5} Torr) sealed silica ampoules. Melting was performed at 850 °C for *ca* 8 hrs in a rotating furnace in order to ensure homogeneity. Glasses were then obtained by water-quenching the silica tubes. This technique results in $xAs_2S_3.(1-x)Sb_2S_3$ glasses in a continuous glass forming region, $0 \leq x \leq 0.75$. The color of the obtained glasses varies from red to dark red upon increasing Sb_2S_3 content. The bulk glasses were polished to yield flat samples with good quality surfaces appropriate for infrared measurements. It is noted that glassy Sb_2S_3 could not be prepared in dimensions suitable for infrared reflectance measurements.

Infrared spectra were recorded in the reflectance mode at near normal incidence (11° off-normal) on a Fourier-transform vacuum spectrometer (Bruker 113v), using of a high reflectivity Al mirror as reference. A Hg source, a DTGS detector with polyethylene window and five mylar beam splitters with variable thickness (3.5-50 μ) were used in the far infrared region in order to measure continuous spectra in the range 30-700 cm^{-1}. Each spectrum represents the average of 200 scans with 2 cm^{-1} resolution. The measured reflectivity spectra were analyzed through the Kramers-Kronig inversion technique to obtain the absorption coefficient spectra, as well as the optical and dielectric constants, as described in details elsewhere [18, 19].

3. Results and Discussion

Infrared reflection spectra of $xSb_2S_3.(1-x)As_2S_3$ glasses, with Sb_2S_3 contents spanning the entire glass forming region, are presented in Fig. 1. The presented reflectivity spectra are in good agreement with those reported earlier by Kato et al. in the range $0 \leq x \leq 0.60$ [14]. The spectrum of pure As_2S_3 glass (x=0) is characterized by a strong band at 312 cm^{-1}, a weak feature at 380 cm^{-1} and a weak and broad band at *ca* 100 cm^{-1}. Addition of Sb_2S_3 to As_2S_3 induces systematic spectral changes. Thus, increasing x results in the broadening of the main reflection band and its shifting to lower frequencies (278 cm^{-1} for x=0.75), followed by the progressive weakening of the 380 cm^{-1} feature.

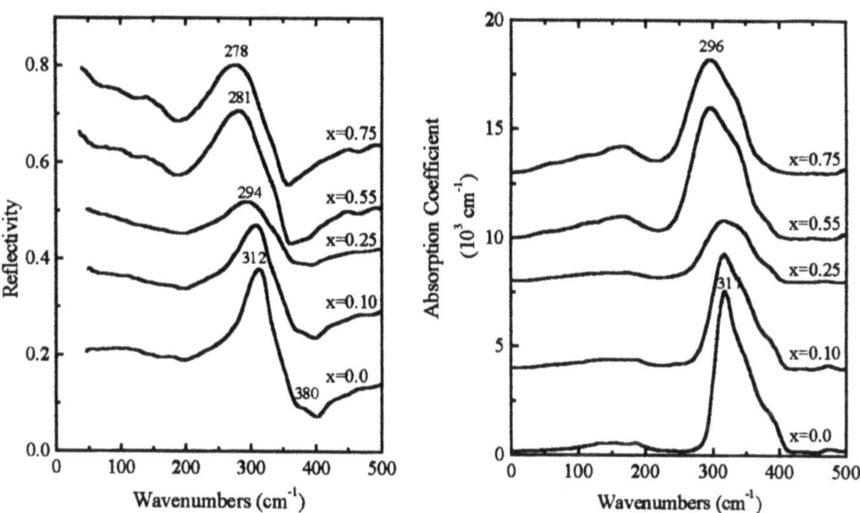

Figure 1(left). Infrared reflection spectra of $xSb_2S_3.(1-x)As_2S_3$ glasses. For x>0, the spectra have been off-set by 0.175, 0.35, 0.35 and 0.5, respectively, to facilitate comparison.

Figure 2 (right). Absorption coefficient spectra of $xSb_2S_3.(1-x)As_2S_3$ glasses. Spectra of glasses $0.1 \leq x \leq 0.75$ have been off-set by 4, 8, 10 and 13 (10^3 cm^{-1}), respectively, to facilitate comparison.

Corresponding changes are observed in the absorption coefficient spectra of these glasses displayed in Fig. 2. The dominant high frequency absorption profile (250-400 cm^{-1}) becomes broader and shifts to lower frequencies upon increasing xSb_2S_3. Also, the absorption envelope below 200 cm^{-1} acquires intensity with x and peaks eventually at *ca* 150 cm^{-1} for x=0.75.

In order to understand the structural origin of these spectral changes with increasing x further analysis of the infrared spectra is required. In particular, we focus attention on understanding the evolution of the complex profile 250-450 cm^{-1}. This is because absorptions due to stretching vibrations of As-S and Sb-S bands are expected in this frequency range. Thus, we have attempted to deconvolute the high frequency envelope, to assign the resulting component bands and to understand their composition dependence. For this purpose we apply in this work a least-squares-fitting program and a deconvolution procedure employed previously to study binary and ternary glasses (18-21). In this approach, we use the minimum number of component bands that gives a reasonable agreement between experimental and calculated spectra. The functional form, the frequencies, bandwidths and intensities of component bands are parameters adjustable by the program. The steps of the spectral analysis followed here are described below.

We have started the spectral analysis by considering first the spectrum of the As_2S_3 glass (x=0). The profile of the 250-400 cm^{-1} absorption envelope of this glass suggests the existence of at least three component bands (Fig. 2). Indeed, a good fit was obtained with three Gaussian bands, as shown in Fig. 3. The three components, designated by 1As, 2As and 3As, have frequencies 317, 348 and 380 cm^{-1}, respectively. The assignment of these bands can be made on the basis of the molecular model, proposed for the interpretation of the vibrational spectra of As_2X_3 (X=S, Se, Te) glasses [5], and found useful to treat the vibrational spectra of binary glasses such as Li_2S-As_2S_3 [22]. This model considers the modes of vibrationally decoupled AsX_3 pyramids and As-X-As water-like bridging bonds [5]. Thus, in terms of the molecular model, the stronger band at 317 cm^{-1} (1As) is assigned to the asymmetric stretching mode, $\nu_3(E)$, of AsS_3 pyramids, while the one at 348 cm^{-1} (2As) to the symmetric stretching mode, ν_1 (A_1), of the AsS_3 pyramidal units. Besides those modes, AsS_3 pyramids exhibit two bending modes, $\nu_4(E)$ and $\nu_2(A_1)$, which are also infrared active [5] and contribute to the weak absorption in the 100-200 cm^{-1} frequency range. The high frequency component at 380 cm^{-1} (3As) is attributed to the asymmetric stretching vibration of As-S-As bridges, ν_{as}(As-S-As). These assignments are consistent with the Raman spectrum of glassy As_2S_3, which shows strong scattering at 343 cm^{-1} (ν_1), a weaker feature at 312 cm^{-1} (ν_3) and a shoulder at 373 cm^{-1} (ν_{as}(As-S-As)) [6].

The infrared spectrum of a glassy Sb_2S_3 film shows the strongest band at 285 cm^{-1} and a weaker feature at 330 cm^{-1} [23]. Bernier et al. [24] reported the corresponding bands of amorphous Sb_2S_3 dispersed in paraffin at 293 cm^{-1} and 332 cm^{-1}. In analogy with As_2S_3, these bands can be attributed to the ν_3 (~290 cm^{-1}) and ν_1 (330 cm^{-1}) modes of SbS_3 pyramidal units. This suggests that consideration of SbS_3 pyramids is sufficient to explain the infrared spectra of glassy Sb_2S_3 in agreement with Ref. [6]. With the above information in mind, and on the basis of the results of deconvolution of the spectrum of As_2S_3 (x=0), we tried to deconvolute the spectra of mixed Sb_2S_3-As_2S_3 glasses (x>0). As input we used the frequencies and bandwidths of bands 1As, 2As, 3As, as determined for x=0, plus two additional bands at *ca* 290 and 330 cm^{-1} to account for the presence of the Sb_2S_3 components. With these five bands we could fit the 250-450 cm^{-1} spectra quite satisfactorily. Using this approach, we found that in all cases the intensity of the 2As band is equal or even higher than that of the 1As band. There is no reason to believe that the symmetric stretching mode, ν_1 (2As), of AsS_3 pyramids, which is the strongest band in the Raman spectrum of the As_2S_3 glass [6], acquires additional intensity in the infrared spectra of the mixed glasses. Thus, we take this result as suggesting the presence of an additional band close to 2As. When the spectra of mixed glasses were fitted with the input presented above and the consideration of a sixth component at *ca* 350 cm^{-1}, then the relative intensities of the resulted 1As and 2As components were similar to that in the spectrum of x=0. Typical results of deconvolution are shown in Figure 3. Following the assignments for glassy Sb_2S_3, bands 1Sb and 2Sb are attributed to ν_3 and ν_1 modes of SbS_3 pyramids in the mixed network. The origin of the component denoted by 3Sb/As in Fig.3 will be discussed below.

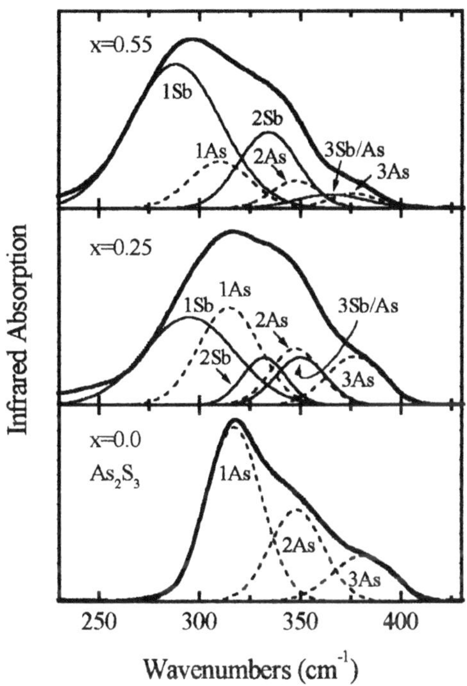

Figure 3. Examples of deconvolution of the higher-frequency envelope 250-450 cm^{-1} (thick lines) of the infrared spectra of $xSb_2S_3.(1\text{-}x)As_2S_3$ glasses. The simulated spectra are shown by dotted lines.

The frequencies of the component bands are plotted in Figure 4 versus mole fraction of Sb_2S_3. Besides the frequency of band 3Sb/As the rest of them do not display a great variation with composition, suggesting that the structure of the basic building units, i.e. pyramids AsS_3 and SbS_3, is retained in the mixed glasses. Nevertheless, closer examination reveals some systematic trends with x which deserve our attention. First, it is of interest to note that the difference ν_1(2As)-ν_3(1As) increases with x, while ν_1(2Sb)-ν_3(1Sb) decreases with x. It was demonstrated by Giehler [6] that the frequency difference ν_1-ν_3 of pyramidal XY_3 units depends mainly on the value of the pyramidal angle, Y-X-Y. In particular, ν_1-ν_3 increases when the angle Y-X-Y decreases. On this basis, the observations made above indicate that the pyramidal angle for both pyramids, AsS_3 and SbS_3, decreases in the mixed glass with respect to the values they have in pure As_2S_3 and Sb_2S_3 glasses. Second, it is observed that the frequency of the 3As band, ν_{as}(As-S-As), decreases with xSb_2S_3. The implication of

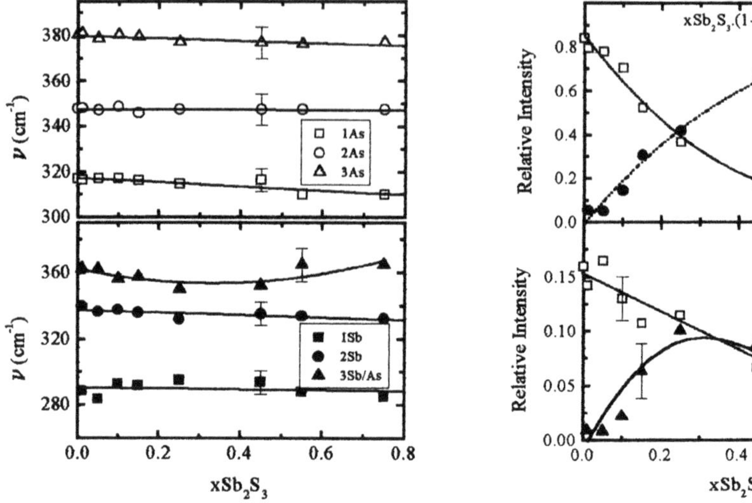

Figure 4 (left). Composition dependence of the frequencies of the component bands resulted from the deconvolution of the 250-450 cm^{-1} absorption envelope (see fig. 3) of xSb_2S_3. $(1-x)As_2S_3$ glasses. Lines are least-squares fitting.
Figure 5 (right). Relative intensities of component bands versus Sb_2S_3 content. Lines are drawn to guide the eye. For details see text.

this trend can be understood in terms of the following formula which gives the frequency of the asymmetric stretching vibration of X-Y-X bridges [25]:

$$\nu^2_{as} = \left(\frac{1}{4\pi^2 c^2}\right)\left[\frac{1}{m_X} + \frac{2}{m_Y}\sin^2\alpha\right]k_r \tag{1}$$

where m_X, m_Y are the masses of atoms X and Y, respectively, k_r is the force constant of bond X-Y, 2α is the angle X-Y-X and c is the speed of light. If the force constant of the bond As-S does not change upon mixing, then the observed trend of ν_{as}(As-S-As) (3As) suggests, in term of Eq. (1), the decrease of the As-S-As angle upon increasing xSb_2S_3. The variation with x of the frequency of the 3Sb/As band is considered below in connection with the composition dependence of its relative intensity.

Figure 5 shows the composition dependence of the relative intensities of the component bands in the 250-450 cm^{-1} envelope. The intensities of bands 1As and 2As, as well as those of 1Sb and 2Sb, have been considered together (Fig. 5(a)), since they both originate from intramolecular vibrations of the XS_3 pyramids (X=As, Sb). The relative intensity of bands attributed to AsS_3 pyramids decreases monotonically with xSb_2S_3, while the relative intensity of bands assigned to SbS_3 pyramids increases, as expected. It is noted though that the intensity of bands 1Sb+2Sb relative to that of 1As+2As is considerably higher than would be expected from the mole fraction of Sb_2S_3 alone. The observed differences can be explained if we consider the changes of dipole moment involved in the vibrations of the X-S bonds. Indeed, it has been shown that the effective charge involved in Sb-S bonding is considerably higher than that involved in As-S bonding [6, 14].

The relative intensity of band 3As, characteristic of As-S-As bridges, decreases monotonically with x (Fig. 5(b)). This is consistent with a progressive destruction of such bridges as the As_2S_3 content in the mixed system decreases. The relative intensity of band 3Sb/As displays a non-linear dependence on x, passing through a maximum value at $x\cong0.35$ (Fig. 5(b)). Since the probability for creating mixed As-S-Sb bridges would be maximized at $x\cong0.5$, the composition dependence of the relative intensity of 3Sb/As band is suggesting its assignment to ν_{as}(As-S-Sb).

According to Eq. (1), ν_{as}(As-S-Sb) would depend on an effective force constant, k_{eff}, and on an effective bond angle, As-S-Sb, both of which could be composition dependent. As shown previously [6] $k_{As-S}>k_{Sb-S}$, therefore k_{eff} is expected to decrease with x. On the other hand, the bridging angle, X-S-X, was estimated to be 99.5° in the As_2S_3 glass and 100.7° in Sb_2S_3 glass [4]. Thus, the effective angle As-S-Sb would increase with x. The combined effect of both factors could lead, according to Eq. (1), to a minimum value of ν_{as}(As-S-Sb) with composition. The composition dependence of the frequency of the 3Sb/As band (Fig. 4) is in accord with the proposed assignment.

The infrared results discussed above are supportive of the models proposing the random substitution of As by Sb, rather than the formation of heterogeneous As_2S_3 and Sb_2S_3 phases connected at the interfaces via mixed As-S-Sb bridges. If the latter proposition were true then one would expect that at least the intramolecular vibrations of XS_3 pyramids, in the two different microphases, would be independent of composition. However, this was not observed in the present work. In particular, the systematic variations with x of especially the ν_1 and ν_3 modes of the pyramids (Fig. 4), is indicating considerable neighboring of the AsS_3 and SbS_3 pyramids.

4. Conclusions

The structure of glasses in the mixed system $xSb_2S_3.(1-x)As_2S_3$ has been investigated in a wide composition range ($0\leq x\leq0.75$) employing infrared reflectance spectroscopy. The high frequency profiles (250-400 cm^{-1}) of the absorption coefficient spectra were deconvoluted into component bands in order to study the composition dependence of the glass structure. The main components of the deconvoluted spectra were understood on the basis of the intramolecular vibrations of trigonal AsS_3 and SbS_3 pyramids. While the structure of such pyramids is basically retained, it was found that the pyramidal S-X-S angle (X=As, Sb) decreases upon mixing. The destruction of As-S-As bridges with increasing Sb_2S_3 content was found to be accompanied by the creation of mixed As-S-Sb bridges with their relative abundance showing maximum value at *ca* x=0.35. It was concluded that the composition dependence of the frequencies and relative intensities of the component bands found in this work is suggestive of a considerable mixing of AsS_3 and SbS_3 pyramids, rather than of the formation of separate As_2S_3 and Sb_2S_3 microphases.

Acknowledgment. *A grant from the NATO Special Fellowships Program supported the stay of I.P.C. at NHRF and made this collaborative work possible.*

5. References

1. Seddon, A.B. (1995) Chalcogenide glasses: a review of their preparation, properties and applications, *J. Non-Crystalline Solids* **184**, 44-50 and references therein.
2. Elliott, S.R. (1991) Chalcogenide glasses, in R.W. Kahn, P. Haasen and E.J. Kramer (eds), *Materials Science and Technology*, VCH, Weinheim, Vol. 9, Glasses and Amorphous Materials, J. Zarzycki (ed.), pp. 375-454.
3. Rawson, H. (1967) *Inorganic Glass-Forming System, Non-Metallic Solids*, Academic Press, London.
4. Cervinka, L., and Hruby, A. (1982) Structure of amorphous and glassy Sb_2S_3 and its connection with the structure of As_2X_3 arsenic-chalcogenide glasses, *J. Non-Crystalline Solids* **48**, 231-264.
5. Lucovsky, G. (1971) Optic modes in amorphous As_2S_3 and As_2Se_3, *Physical Review B* **6**, 1480-1489 and references therein.
6. Giehler, M. (1981) The effect of short-range order on the vibrational spectra of 3:2 coordinated chalcogenide glasses, *Physica Status Solidi (b)* **106**, 193-205.
7. Itoh, S., and Fujiwara, T. (1982) Vibrational properties of As_2S_3 glass, *J. Non-Crystalline Solids* **51**, 175-186.
8. Paesler, M.A., and Pfeiffer, G. (1991) Modeling the structure and photostructural changes in amorphous arsenic sulfide, *J. Non-Crystalline Solids* **137-138**, 967-972.
9. Dalba, G., Fornasini, P., Giunta, G., and Burattini, E. (1989) XRD and EXAFS study of the local structure in some non-crystalline Sb-S compounds, *J. Non-Crystalline Solids* **107**, 261-270.
10. El Idrissi Raghni, M.A., Lippens, P.E., Olivier-Fourcade, J., and Jumas, J. (1995) Local structure of glasses in the As_2S_3-Sb_2S_3 system, *J. Non-Crystalline Solids* **192-193**, 191-194.
11. Durand, J.M., Lippens, P.E., Olivier-Fourcade, J., and Jumas, J.C. (1995) A structural study of glasses in the As_2S_3-Sb_2S_3-Tl_2S system, *J. Non-Crystalline Solids* **192-193**, 364-368.

12. Durand, J.M., Lippens, P.E., Olivier-Fourcade, J., Jumas, J.C., and Womes, M. (1996) Sb LIII-edge XAS study of the ternary system As_2S_3-Sb_2S_3-Tl_2S, *J. Non-Crystalline Solids* **194**, 109-121.
13. Bychkov, E., and Wortmann, G. (1993) ^{121}Sb Mossbauer study of insulating and ion-conducting antimony chalcogenide-based glasses, *J. Non-Crystalline Solids* **159**, 162-172.
14. Kato, M., Onari, S., and Arai, T. (1983) Far infrared and Raman spectra in $(As_2S_3)_{1-x}(Sb_2S_3)_x$ glasses, *Japanese J. Applied Physics* **22**, 1382-1387.
15. Kawamoto, Y., and Tsuchihashi, S. (1969) The properties and structure of glasses in the system As_2S_3-Sb_2S_3, *Yogyo-Kyokai-Shi* **77**, 328-335.
16. White, K., Crane, R.L., and Snide, J.A. (1988) Crystallization kinetics of As-Sb-S glass in bulk and thin film form, *J. Non-Crystalline Solids* **103**, 210-220.
17. Tichy, L., Triska, A., Frumar, M., Ticha, H., and Klikorka, J. (1982) Compositional dependence of the optical gap in $Ge_{1-x}S_x$, $Ge_{40-x}Sb_xS_{60}$ and $(As_2S_3)_x(Sb_2S_3)_{1-x}$ non-crystalline systems, *J. Non-Crystalline Solids* **50**, 371-378.
18. Kamitsos, E.I., Patsis, A.P., Karakassides M.A., and Chryssikos, G.D. (1990) Infrared reflectance spectra of lithium borate glasses, *J. Non-Crystalline Solids* **126**, 52-67.
19. Kamitsos, E.I., Kapoutsis, J.A., Chryssikos, G.D., Taillades, G., Pradel, A., and Ribes, M. (1994) Structure and optical conductivity of silver thiogermanate glasses, *J. Solid State Chemistry* **112**, 255-261.
20. Kamitsos, E.I., Kapoutsis, J.A., Jain, H, and Hsieh, C.H. (1993) Vibrational study of the role of trivalent ions in sodium trisilicate glasses, *J. Non-Crystalline Solids* **171**, 31-45.
21. Kamitsos, E.I., Kapoutsis, J.A., Chryssikos, G.D., Hutchinson, J.M., Pappin, A.J., Ingram, M.D., and Duffy, J.A. (1995) Infrared study of AgI containing superionic glasses, *Physics and. Chemistry of. Glasses* **36**, 142-150.
22. Shastry, M.C.R., Couzi, M., Levasseur, A., and Menetrier, M. (1993) Raman spectroscopic studies of As_2S_3 and Li_2S-As_2S_3 glasses, *Philosophical Magazine B* **68**, 551-560.
23. Droichi, M.S., Vaillant, F., Bustarret, E., and Jousse, D. (1988) Study of localized states in amorphous chalcogenide Sb_2S_3 films, *J. Non-Crystalline Solids* **101**, 151-155.
24. Barnier, S., Guittard, M., Julien, C., and Chilouet, A. (1993) Study of the antimony environment in gallium-antimony-sulphur glasses-Phase diagram and infrared absorption investigations, *Materials Research Bulletin* **28**, 399-405.
25. Herzberg, G. (1945) *Infrared and Raman spectra of polyatomic molecules*, New York, Van Nostrand, Chapter II.

STRUCTURE OF POTASSIUM GERMANATE GLASSES BY VIBRATIONAL SPECTROSCOPY

Y.D. YIANNOPOULOS[a], E.I. KAMITSOS[a] and H. JAIN[b]
[a]Theoretical and Physical Chemistry Institute,
National Hellenic Research Foundation,
48 Vass. Constantinou Ave., Athens 116 35, Greece

[b]Department of Materials Science and Engineering,
Lehigh University, Bethleehem, PA 18015, USA

1. Introduction

The use of germanate glasses in technological applications, such as optical fibers and infrared transmitting windows, has stimulated extensive investigations of their physical properties. The refractive index, density, thermal expansion coefficient, and electrical conductivity of alkali germanates were found to exhibit extrema as a function of alkali oxide content [1-4]. This behaviour is widely known as the "germanate anomaly effect".

To explain this anomalous behaviour, it was proposed that addition of alkali oxide (M_2O) to GeO_2 causes the partial conversion of germanium-oxygen tetrahedra into octahedral units when the alkali content is up to 20-25 mol% M_2O, while higher M_2O concentrations cause the formation of non-bridging oxygen (NBO) containing germanate tetrahedra [5-12]. In a different approach, it was proposed that the formation of non-bridging oxygens is the only modification mechanism of the germanate network [13].

In this work, infrared reflectance and Raman spectroscopies are employed in an attempt to investigate the structural origin of the germanate anomaly effect in the $xK_2O.(1-x)GeO_2$ glass system. In particular, emphasis is placed on understanding the compositional dependence of germanium coordination and the bonding state of oxygen atoms.

2. Experimental

Glass samples in the system $xK_2O.(1-x)GeO_2$ were prepared by melting in Pt crucibles stoichiometric amounts of polycrystalline GcO_2 and K_2CO_3 at ~1150°C for *ca* 0.5-2

A. Andriesh and M. Bertolotti (eds.),
Physics and Applications of Non-Crystalline Semiconductors in Optoelectronics, 317–325.

hrs depending on composition. Splat-quenching the melt resulted in glasses with K_2O contents in the range $0 \leq x \leq 0.55$.

Raman spectra were measured using a Ramanor HG 2S Jobin-Yvon spectrometer and the 488nm line of a Spectra Physics argon ion laser for excitation. The glass sample was placed in a vacuum cell to eliminate hydrolysis effects.

Infrared spectra were recorded in the specular reflectance mode on a vacuum spectrometer (Bruker 113v). The reflectance data in the range 20-4000 cm^{-1} were analysed by the Kramers-Krönig inversion technique. The reported here absorption coefficient spectra $\alpha(\nu)$, were calculated by the expression $\alpha(\nu)=4\pi\nu k(\nu)$, where $k(\nu)$ is the imaginary part of the refractive index and ν is the frequency in cm^{-1} [14].

3. Results and Discussion

3.1. RAMAN SPECTRA

Raman spectra of $xK_2O.(1-x)GeO_2$ glasses are shown in Figures 1 and 2 in the ranges $0 \leq x \leq 0.20$ and $0.24 \leq x \leq 0.55$, respectively. These spectra show considerable agreement with spectra published earlier by Verweij and Buster [8] and Furukawa and White [9].

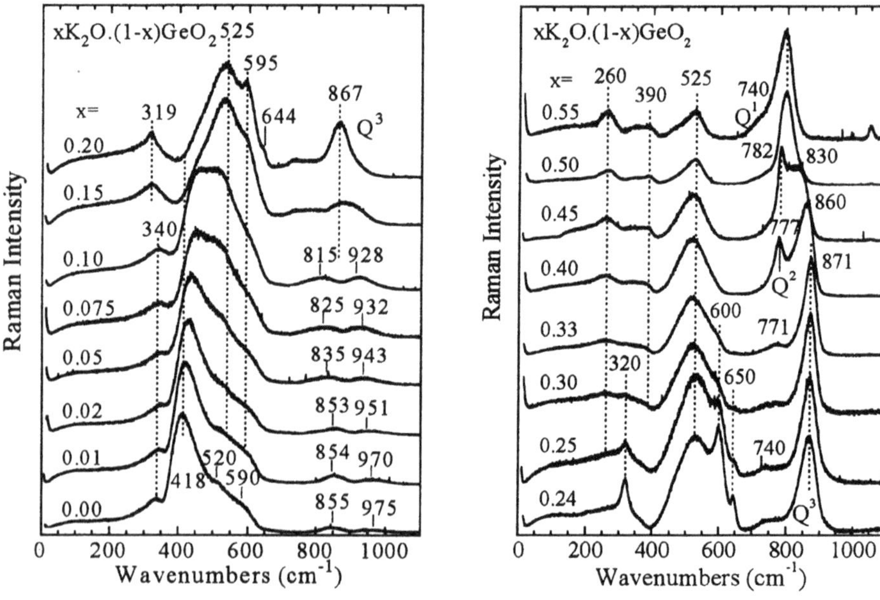

Figure 1. (left) Raman spectra of $xK_2O.(1-x)GeO_2$ glasses ($0 \leq x \leq 0.20$).
Figure 2. (right) Raman spectra of $xK_2O.(1-x)GeO_2$ glasses ($0.24 \leq x \leq 0.55$).

The dominant feature of the GeO_2 glass (x=0) at 418 cm^{-1} is attributed to the symmetric stretching vibration of Ge(4)-O-Ge(4) bridges (ν_s (Ge(4)-O-Ge(4)) in six-membered rings, i.e. rings containing six GeO_4 tetrahedra [15,16]. The number in parenthesis indicates the coordination number of Ge atoms. Upon increasing K_2O content this band decreases in intensity and eventually vanishes at $x \cong 0.25$ indicating the distruction of such tetrahedra containing ring arrangements.

Three bands at *ca* 320, 600 and 650 cm^{-1} develop with increasing x, attain their maximum relative intensity at $x \cong 0.24$ and then diminish for $x > 0.40$. Previous Raman studies of glassy and crystalline alkali germanates have demonstrated that the presence of these bands is related to the formation of interconnected GeO_6^{-2} octahedral units [14,17,18]. Specifically, bands in the range 550-650 cm^{-1} were attributed to ν_s (Ge(6)-O-Ge(6)) [14]. Besides these features, a new band develops at *ca* 525 cm^{-1} with increasing x. As shown elsewhere [14] this band can been assigned to ν_s (Ge(4)-O-Ge(6)), i.e. to bridges connecting a germanium tetrahedron with a germanium octahedron, without excluding some contribution from ν_s (Ge(4)-O-Ge(4)) in three-membered rings of GeO_4 tetrahedra.

For $x \geq 0.10$ a new band develops at 867 cm^{-1} and becomes the dominant Raman feature for x=0.33. This band is assigned to the symmetric stretching vibration of Ge-O^- bonds (O^-=NBO) in germanate tetrahedra containing three bridging and one non-bridging oxygen [14]. This is the so called Q^3 unit, where the notation Q^n indicates a germanium tetrahedron with 4-n non-bridging oxygens per Ge atom. Thus a new type of bridge, Ge(4)-O-Ge(Q^3), is formed with one Ge atom being of Q^3 type. It is expected that the symmetric stretching vibration of the new bridges, ν_s (Ge(4)-O-Ge(Q^3)), will also contribute to the 510-530 cm^{-1} range.

Increasing further the K_2O content (x>0.30) causes additional modifications to the germanate structure as evidenced by the appearance of the new band at 770-780 cm^{-1} and the shoulder at *ca* 740 cm^{-1} (x>0.50). The new features are attributed to the symmetric stretching vibration of Ge-O^- bonds in Q^2 and Q^1 germanate tetrahedra respectively [8,9,14]. Therefore, high alkali oxide contents lead to the progressive depolymerization of the germanate network through formation of non-bridging oxygen atoms.

In an attempt to semiquantify the effect of K_2O on the relative abundance of the various structural units composing the germanate network, we have deconvoluted the complex profile from 350-650 cm^{-1}. A typical example of deconvolution is shown in the inset of Fig. 3 for the x=0.20 glass. The three spectral ranges considered are: 418-450 cm^{-1} (A), characteristic of Ge(4)-O-Ge(4) bridges mostly in six-membered rings; 515-530 cm^{-1} (B) resulting from Ge(4)-O-Ge(4) bridges in three-membered rings and mixed bridges such as Ge(4)-O-Ge(6) and bridges where at least one Ge atom contains NBO's, e.g. Ge(4)-O-Ge(Q^3) and Ge(Q^3)-O-Ge(Q^2); and 600 cm^{-1}+650 cm^{-1} (C) characteristic of *connected* GeO_6 octahedra. The composition dependence of the relative intensity of these selected Raman bands (Fig. 3) show clearly that the relative abundance of GeO_4 tetrahedra in six-membered rings decreases monotonically with x and vanishes for $x \cong 0.40$ (curve A), while the relative population of *connected* GeO_6 octahedra attains its maximum value at ~24mol% K_2O (curve C). This indicates

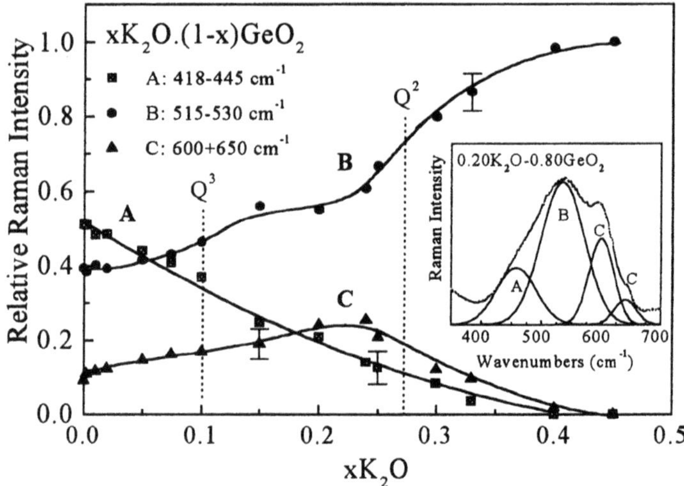

Figure 3. Composition dependence of the relative intensity of deconvoluted bands in the 350-700 cm^{-1} Raman envelope of $xK_2O.(1-x)GeO_2$ glasses. Lines are drawn to guide the eye. The vertical dotted lines indicate the approximate onset for formation of Q^3 and Q^2 germanate tetrahedra. The inset shows an example of deconvolution (x=0.20).

that the $GeO_4 \rightarrow GeO_6$ conversion is the dominant modification mechanism up to $x\approx0.25$, in agreement with the recent studies of K-germanates by XPS and EXAFS [19,20]. The composition dependence of curve B (515-530 cm^{-1}) is more complicated because of the combined contribution of at least four different bridges as discussed above. A more detailed discussion of the Raman spectra will be presented in a future publication.

3.2 INFRARED SPECTRA

3.2.1 *Spectral Assignments*

The infrared absorption spectra of $xK_2O.(1-x)GeO_2$ glasses are presented in Figures 4 and 5 for $0\leq x\leq0.45$. Clearly, addition of K_2O to GeO_2 causes progressive spectral changes in the entire infrared range. However, for the purpose of this work we consider only the high frequency envelope (600-1100 cm^{-1}). For $x\leq0.20$ a monotonic downshift of the high-frequency envelope is observed, while for higher modification levels it splits initially into two components ($0.20\leq x\leq0.33$) and then develops into a broad band ($x\geq0.40$) with a shoulder at *ca* 650 cm^{-1}. The intense band of GeO_2 at 915 cm^{-1} is assigned to the asymmetric stretching vibration of Ge(4)-O-Ge(4) bridges; designated by ν_{as} (Ge(4)-O-Ge(4)) [14-16]. The asymmetry of this band reveals a broad distribution of intertetrahedral Ge-O-Ge angles. For low K_2O contents ($x<0.02$), the

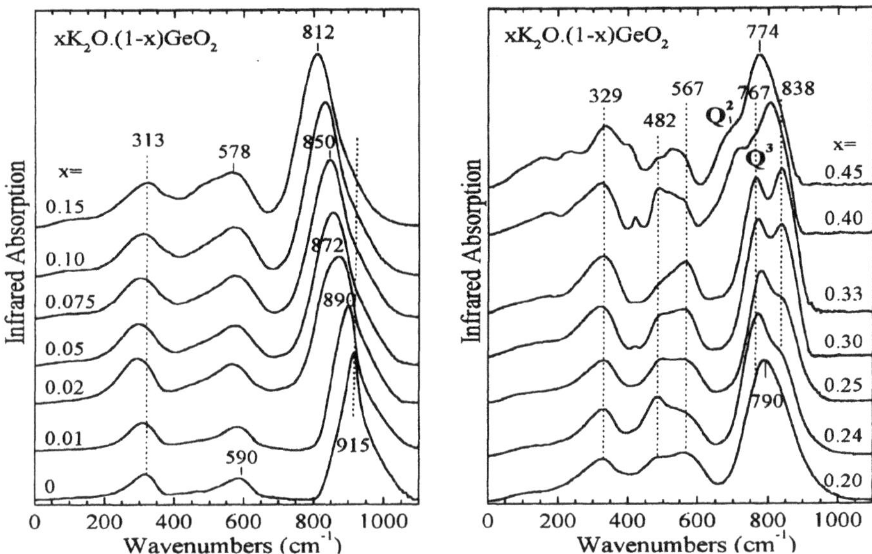

Figure 4 (left). IR spectra of $xK_2O.(1-x)GeO_2$ glasses ($0 \leq x \leq 0.15$).
Figure 5 (right). IR spectra of $xK_2O.(1-x)GeO_2$ glasses ($0.20 \leq x \leq 0.45$).

rapid decrease of ν_{as} (Ge(4)-O-Ge(4)) can be explained in terms of the change in rings statistics in favour of the smaller three membered rings. As shown in the case of rubidium germanate glasses, a decrease of the average Ge(4)-O-Ge(4) angle by ~5° can cause a lowering of ν_{as} (Ge(4)-O-Ge(4)) by as much as 25 cm^{-1} [14].

For higher K_2O contents, the $GeO_4 \rightarrow GeO_6$ transformation mechanism, as well as the formation of NBO containing units take place as shown from the consideration of the Raman spectra. Therefore, the complexity of the induced structural changes makes deconvolution of the 650-1100 cm^{-1} envelope necessary. Typical examples of deconvoluted infrared profiles are shown in Figure 6. For $x>0.01$, the gradual destruction of Ge(4)-O-Ge(4) bridges (intensity decrease of the band centered at ~890 cm^{-1}) is followed by the creation of new types of bridges and therefore the development of new bands. Thus, the formation of octahedral units gives rise to the band at ~850 cm^{-1} (ν_{as} (Ge(4)-O-Ge(6))), and the creation of Q^3 species to the band in the range ~800-770 cm^{-1} (ν_{as} (Ge-O$^-$) [5,14]). In the recent XPS study of potassium germanate glasses NBO's were detected at K_2O content as low as 2mol% [19]. The relative abundance of Q^3 species dominates at $x=0.33$, in agreement with the Raman spectra presented in Fig. 2. The new feature in the range 820-830 cm^{-1} evolving for $x>0.25$ is attributed to the ν_{as} (Ge-O-Ge) where at least one of the germanium atoms contains NBO's (i.e. Q^3, Q^2). The formation of Q^2 units for $x>0.25$ is signaled by the band developing at ~700 cm^{-1}, ν_{as} (Ge-O$^-$).

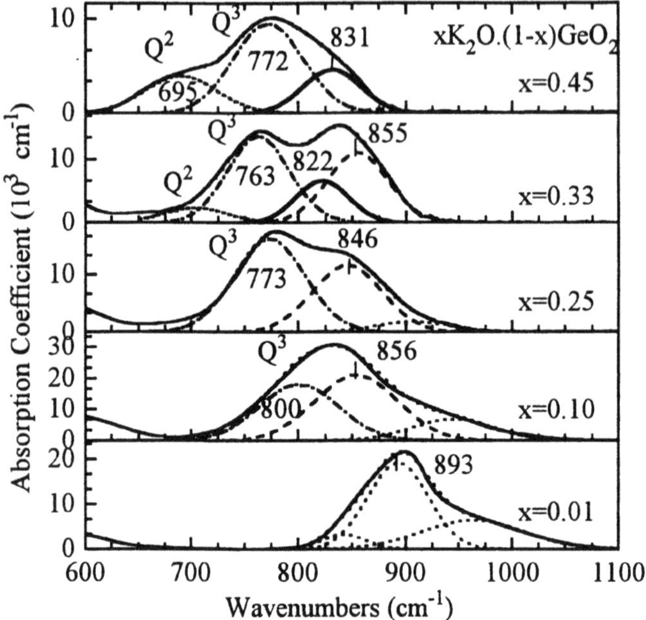

Figure 6. Examples of deconvoluted high frequency envelope of the infrared spectra of potassium germanate glasses. For details see text.

3.2.2 Coordination of Ge atoms in K-germanate glasses

As shown previously [14], the frequency of the asymmetric stretching of Ge-O-Ge bridges perturbed by the presence of GeO_6 octahedral units (ν_{as} (Ge(4)-O-Ge(6)) can be used to estimate the average coordination number of germanium atoms. The Dachille and Roy [21] semiempirical relation between ν_{as} and the coordination number (CN) of the glass forming cation takes the following form for the case of germanate glasses:

$$K=5.6CN/\lambda^2$$

where the constant K was estimated to be K=0.172 and the wavelength $\lambda=(\nu_{as})^{-1}$ is given in μm. The results for the composition dependence of CN (Ge), using for the ν_{as} (Ge(4)-O-Ge(6)) frequencies the values obtained by deconvolution, are shown in Figure 7(a). It is very clear that the CN of Ge atoms changes with K_2O addition to GeO_2 and attains its maximum value (CN≅4.32) at ~25mol% K_2O. This is in good agreement with previously published results by Sakka and Kamiya [7] (CNmax~4.46) and Huang et al. [20] (CNmax~4.2) employing different spectroscopic techniques.

The fraction of Ge atoms in six-fold coordination was estimated from the relation: $N_6=(CN/2)-2$ and is shown in Figure 7(b) *vs* xK_2O. The solid line represents the theoretical curve, $N_6=x/(1-x)$, if no NBO's were formed. N_6 values obtained from

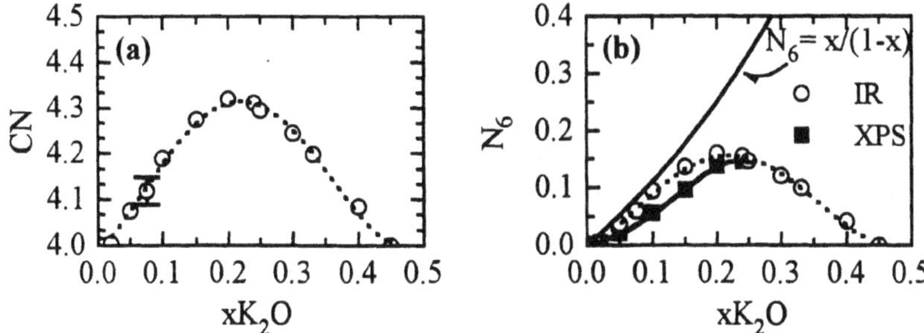

Figure 7. Composition dependence of the coordination of Ge atoms, CN (a) and of the fraction of Ge atoms in six-fold coordination, N_6 (b) in $xK_2O.(1\text{-}x)GeO_2$ glasses. IR and XPS denote N_6 values obtained from the analysis of infrared and XPS data respectively.

XPS spectroscopy [19] are also shown for comparison. It is evident that the results from both spectroscopic techniques are in considerable agreement. Deviation of the experimental N_6 values from the theoretical curve, mainly for $x>0.25$, indicates the increasing rate of NBO formation.

4. Conclusions

Potassium germanate glasses $xK_2O.(1\text{-}x)GeO_2$ were prepared over the entire glass forming region, $0\leq x\leq 0.55$, and studied by infrared reflectance and Raman spectroscopies in order to investigate the composition dependence of the structural modification mechanisms.

In glasses with low K_2O content ($0\leq x<0.02$), six-membered rings rearrange to form smaller rings; probably three-membered ones. This process is manifested by the abrupt downshift of the high frequency envelope of the infrared spectra.

For K_2O contents up to $x=0.25$, the $GeO_4 \rightarrow GeO_6$ conversion mechanism dominates. In the Raman spectra this is manifested by the progressive evolution, and then the disappearance of the bands at *ca* 320, 600 and 650 cm^{-1}, while the infrared analysis showed that the average coordination number of Ge atoms increases and attains its maximum value (~4.32) at $x\cong 0.25$.

The formation of non-bridging oxygens (NBO's) in various Q^n species is the main structural modification process for $x>0.25$. The progressive depolymerization of the germanate network through NBO formation is followed by the development of high frequency bands in the Raman spectra, due to the localized vibration of the $Ge\text{-}O^-$ bonds, and by the increasing deviation of the experimental N_6 values from the theoretical curve.

Acknowledgments. This work was supported by NATO Collaborative Research Grants Program (CRG 931213), NHRF, and USDoE.

5. References

1. Murthy, K. M. and Aguayo, J. (1964) Studies in germanium oxide systems: II, phase equilibria in the system Na_2O-GeO_2, *J. of the American Ceramic Society* **47**, 444-447.
2. Shelby, J.E. (1974) Viscosity and thermal expansion of alkali germanate glasses, *J. of the American Ceramic Society* **57**, 436-439.
3. Mundy, J.N. and Jin, G.L. (1987) Ionic transport in rubidium aluminogermanate glasses, *Solid State Ionics* **24**, 263-272.
4. Sakai, T., Nagoya, T., Takizawa, K. and Sekizawa, T. (1995) Electrical conductivity of K_2O-GeO_2 glasses and germanate anomaly, *Denki Kagaku* **63**, 608-612.
5. Murthy, M.K. and Kirby, E.M. (1964) Infrared spectra of alkali-germanate glasses, *Physics and Chemistry of Glasses* **5**, 144-146.
6. Riebling, E.F. (1972) Infrared study of polymerisation in silver, thallium, and thallium aluminogermanate glasses, *J. of Materials Science* **7**, 40-46.
7. Sakka, S. and Kamiya, K. (1982) Structure of alkali germanate glasses studied by spectroscopic techniques, *J. Non-Crystalline Solids* **49**, 103-116.
8. Verweij, H. and Buster, J.H.J.M. (1979) The structure of lithium, sodium and potassium germanate glasses, studied by Raman scattering, , *J. Non-Crystalline Solids* **34**, 81-99.
9. Furukawa, T. and White, W.B. (1980) Raman spectroscopic investigation of the structure and crystallization of binary alkali germanate glasses, *J. of Materials Science* **15**, 1648-1662.
10. Kamiya,K., Yoko, T., Itoh, Y. and Sakka, S. (1986) X-ray diffraction study of Na_2O-GeO_2 melts, *J. Non-Crystalline Solids* **79**, 285-294.
11. Cox, A.D. and McMillan, P.W. (1981) An EXAFS study of the structure of lithium germanate glasses, *J. Non-Crystalline Solids* **44**, 257-264.
12. Huang, W.C., Jain, H. and Marcus, M.A. (1994) Structural study of Rb and (Rb,Ag) germanate glasses by EXAFS and XPS *J. Non-Crystalline Solids* **180**, 40-50.
13. Henderson, G.S. and Fleet, M.E. (1991) The structure of glasses along the Na_2O-GeO_2 join, *J. Non-Crystalline Solids* **134**, 259-269.
14. Kamitsos, E.I., Yiannopoulos, Y.D., Karakassides, M.A., Chryssikos, G.D. and Jain, H. (1996) Raman and Infrared Structural Investigation of $xRb_2O\cdot(1-x)GeO_2$ Glasses, *J. Physical Chemistry* **100**, 11755-11765.
15. Galeener, F. L. (1979) Band limits and the vibrational spectra of tetrahedral glasses, *Physical Review B* **19**, 4292-4297.
16. Sharma, Shiv K., Matson, D.W., Philpotts, J.A. and Roush, T.L. (1984) Raman study of the structure of glasses along the join SiO_2-GeO_2, *J. Non-Crystalline Solids* **68**, 99-114.
17. Mochida, N., Sakai, K. and Kikuchi, K. (1984) Raman spectroscopic study of the structure of the binary alkali germanate glasses, *Yogyo Kyokai Shi*, **92**, 164-172.

18. Durben, D.J. and Wolf, G.H. (1991) Raman spectroscopic study of the pressure-induced coordination change in GeO_2 glass, *Physical Review B* **43**, 2355-2363.
19. Lu, X., Jain, H. and Huang, W.C. (1996) Structure of potassium and rubidium germanate glasses by x-ray photoelectron spectroscopy, *Physics and Chemistryof Glasses* **37**,201-205.
20. Huang, W.C., Jain, H. and Meitzner, G. (1996) The structure of potassium germanate glasses by EXAFS, *J. Non-Crystalline Solids* **196**, 155-161.
21. Dachille, F. and Roy, R. (1959) The use of infrared absorption and molar refractivities to check coordination, *Zeitschrift Kristallographie* **111**, 462-470.

A REVIEW OF AMORPHOUS CHALCOGENIDES AS MATERIALS FOR INFRARED BULK ACOUSTO-OPTIC DEVICES

A.B. SEDDON and M.J. LAINE
Centre for Glass Research
Department of Engineering Materials
University of Sheffield, Sheffield, S1 3JD, UK

Abstract
The range of chalcogenide glasses with good acousto-optic (AO) properties has recently been extended to cover the 10.6μm operating wavelength. This review puts these recent developments into context. The theoretical background to the selection of materials for AO bulk devices is presented with emphasis on the amorphous chalcogenides. The specific AO properties of sulphide, selenide and selenide-telluride glasses are discussed, with particular reference to: AO figure of merit, refractive index, acoustic velocity, density, acoustic attenuation and optical attenuation.

Introduction

The range of amorphous chalcogenides potentially available for the near-infrared (IR) to far-IR operation of acousto-optic (AO) bulk devices is discussed here from the point of view of the materials' requirements. Such a review is overdue, since the last previous collations of AO infrared materials appear to be those of Sapriel in 1976 [1], and Polyakov *et al.* in 1981 [2]. In addition, it is timely to take stock of IR materials available for bulk devices. This information will be useful for the development of *integrated* AO devices for optical signal processing which is becoming increasingly important, driven by the advent of fibre-optic communication systems and optical sensors [3].

This paper gives as a preface a brief historical outline of the discovery of the AO effect and development of AO devices. The theoretical background which governs the selection of materials for AO devices is presented. Finally the merits of individual chalcogenide materials are discussed. The amorphous chalcogenides are classified, for convenience, for use in the three atmospheric windows as defined by Savage [4]: the near-IR (0 75. to 2.5μm, encompassing the telecom. signal wavelengths of 1.3 and 1.55 μm); mid-IR (3 to 5 μm) and far-IR (7 to 12μm).

A. Andriesh and M. Bertolotti (eds.),
***Physics and Applications of Non-Crystalline Semiconductors in Optoelectronics*, 327–336.**

1. Brief historical outline of the AO effect and AO devices

Acousto-optic devices use acoustic waves to manipulate optical beams. Brillouin [5] first suggested in 1922 that light could be scattered by acoustic waves from thermal excitation in liquids and solids. These effects were demonstrated 10 years later [6,7].

When an acoustic wave propagates through an optically transparent medium periodic regions of strain result *i.e.* regions of compression and rarefaction, which correspond in turn to regions of increased and decreased refractive index relative to the original refractive index of the medium. The magnitude of the index change is proportional to the applied strain *via* a proportionality constant, the Pockles elasto-optic coefficient. For an isotropic material this may be thought of as the effective photoelastic constant, p_{eff}. The periodic index variations can act as a phase-diffraction grating to diffract an optical beam into many orders (see figure 1). These ideas were first developed by Raman and Nath [8], in 1935 to 1936, but real technological development of bulk AO modulator, deflector and tunable filter (AOTF) devices came about after 1960 with the invention of the laser. Thus, using a piezo-electric (pz) transducer as the ultrasonic source, time-varying electrical information can be transferred in the AO cell to light waves in real time. Recently, Rayleigh surface acoustic wave (SAW) devices have been shown to have potential for integrated optical signal processing in which surface optical waveguides increase the AO interaction length and the devices require a relatively low acoustic power [3].

2. Theoretical background to AO materials selection relevant to the development of candidate amorphous chalcogenides

2.1 DIFFRACTION EFFICIENCY

Amorphous chalcogenides ideally are both acoustically and optically isotropic, hence the polarisation of the optical beam is maintained during diffraction and the orientation of the material in the AO cell is not important.

In the low diffraction limit the diffraction efficiency, or ratio of intensities (I) between the first order and zero order beams, of an acousto-optic device is given by [3]:

$$\frac{I_1}{I_0} = \frac{2\pi^2 \; L \; Pa \; M_2}{\lambda_o^2 \; H} \qquad (1)$$

where λ_o is the optical wavelength *in vacuo*; L x H is the rectangular section of the acoustic beam; Pa is the acoustic power in the acousto-optic cell and M_2 is the AO figure of merit of the cell material. M_2 is the most common figure of merit; it is particularly applicable for AO modulator devices, but other figures of merit exist [1].

Two extreme cases of diffraction can be distinguished [3]. Taking a thin phase grating (*i.e.* a narrow sound column) and operating at near zero angles of incidence then Raman-Nath diffraction occurs and the acousto-optic cell gives rise to a fan of diffracted beams about the zero order (figure 1). Motion of the acoustic grating results in a Doppler up-shift in frequency of optical beams which have been diffracted in the same direction as that of acoustic propagation, and a frequency down-shift of optical beams diffracted in the reverse direction. A thick phase grating gives rise to Bragg diffraction in which only first order diffraction is observed.

2.2 AO FIGURE OF MERIT

The acousto-optic figure of merit, here confined to consideration of M_2, is of critical importance for materials choice for use in AO devices. From equation 1, the higher the figure of merit, the lower the acoustic power required to diffract the light and this in turn minimises heating effects. For an isotropic medium with a longitudinally polarised acoustic wave, the value of the figure of merit depends on whether the polarisation of the optical beam is parallel (giving $M_{2//}$) or perpendicular (giving $M_{2\perp}$)to the acoustic plane.

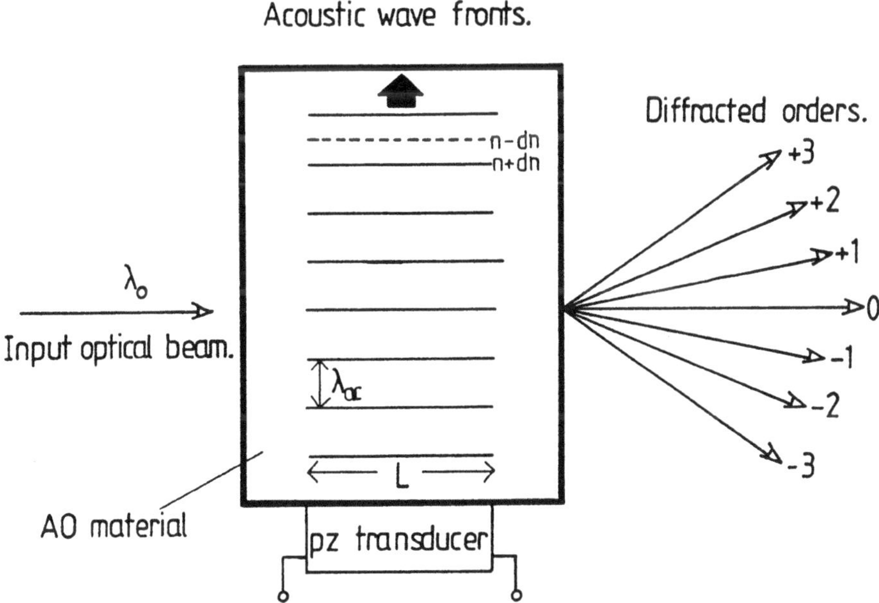

Figure 1. An acousto-optic cell configured as a thin phase grating. Raman-Nath multiple diffraction occurs at near-zero angles of incidence of the optical beam, relative to the plane of the acoustic beam. (Symbols are defined in the text.)

M_2 is given by [9]:

$$M_2 = \frac{n^6 \; p_{eff}^2}{\rho \; v^3} \qquad (2)$$

where n is refractive index; ρ is density and v is acoustic velocity. Unfortunately these are not entirely independent variables as, for a given type of material, a higher index is often incompatible with a lower density. Figure of merit of a particular material tends to decrease as the optical wavelength is increased due to dispersion of the refractive index.

The variables in equation 2 are all readily measurable apart from the photoelastic constant. Figures of merit are, therefore, traditionally measured directly by the Dixon-Cohen comparative technique [10]. A block of the test material is bonded to a known material (usually fused silica) and AO diffraction efficiencies are compared. Use of incident optical beams of differing wavelengths allows materials with a different optical transmission window from silica to be analysed. The Dixon-Cohen method [10] relies on pulsed optical beams and this has led to difficulties for use of the method in the infrared. For instance M_2 of single crystal germanium (for an acoustic wave in the [111] direction and parallel optical polarisation) was originally reported as 815×10^{-15} $s^3 kg^{-1}$ [11] but more recent determinations agree a lower value of 181×10^{-15} $s^3 kg^{-1}$ [12]. Recently, a novel method of measuring M_2 in the infrared has been developed [13 - 16] which involves measurement instead of the total acoustic power contained in the acoustic cavity (see equation 1).

2.3 ACOUSTIC AND OPTICAL ATTENUATION

The diffraction efficiency of a particular glass type depends not only on efficiency of the AO interaction (equations 1 and 2) but also: (a) on the acoustic power sustained in the material, *i.e.* on acoustic attenuation and (b) on optical attenuation. It is interesting that the excellent optical transmission and low acoustic attenuation of fused silica can make it attractive for some applications despite a mediocre figure of merit (table 1). The mechanisms of acoustic and optical loss for chalcogenide glasses are discussed below.

2.3.1 Acoustic Attenuation

According to equation 1, the acoustic power required to diffract an optical signal is proportional to the square of the wavelength of the incident optical beam. Hence it is desirable to minimise acoustic loss when operating in the infrared. For a homogeneous glass free from cracks and inclusions, which does not undergo structural relaxation at the operating temperature (implying a high glass transformation temperature (Tg)), then the dominant mechanism of acoustic attenuation is probably acoustic absorption (α_{ac} / dB cm^{-1}) due to relaxation of the perturbed thermal phonon distribution towards equilibrium, the so-called Akhieser loss. Woodruff and Ehrenrich [17] have developed a model for insulating solids (for $f_{ac}\tau \ll 1$, where f_{ac} is acoustic frequency and τ is thermal phonon relaxation time) which is assumed here to hold for the semiconducting amorphous chalcogenides:

$$\alpha_{ac} = \frac{C_v \; T \; \gamma^2 \; f_{ac}^2 \; \tau}{3\rho v^3} \tag{3}$$

where C_v is heat capacity at constant volume; T is temperature and γ is effective Gruneisen constant. Acoustic loss thus often is observed to increase as the square of acoustic frequency. The acoustic energy lost will be dissipated as heat which could lead to thermal lensing and a distorted optical output from the AO cell. These considerations would imply the need for AO materials of a high acoustic velocity, a high thermal conductivity, a low thermal expansion and low temperature coefficient of refractive index, especially for operation at high acoustic frequencies.

2.3.2 Optical Attenuation

The optical loss mechanisms of chalcogenide glasses may be classified as intrinsic, and extrinsic, absorption and scattering. The glasses are amorphous semiconductors and the optical transmission window is limited at the high energy end by absorption due to the optical bandgap consisting of: (a) a high absorption region; (b) the exponential Urbach region [$\alpha = \exp(h\nu/E_U)$] and (c) the weak absorption tail due to inter-bandgap states [18]. From a study of the optical loss of As_2S_3 chalcogenide glass optical fibres, Kanamori *et al.* [19] have proposed that the weak absorption tail is more important than Rayleigh optical scattering loss in the mid-infrared [19]. The weak absorption tail of some amorphous chalcogenides has been shown to be sensitive to photo-illumination and subsequent annealing [18].

Intrinsic Rayleigh scattering loss [α_{sc}]arises from tiny fluctuations in density, and composition if the glass is multicomponent, which have been frozen in at the glass fictive temperature [α_{sc} = (material constant)/λ_o^{-4}]. Many chalcogenide systems are prone to liquid-liquid phase separation, a potential cause of excess scattering. Extraneous inclusions, including bubbles, will also lead to extrinsic scattering, the magnitude and wavelength dependency of which will depend on the size of the inclusions and the refractive index relative to the medium [20].

At longer wavelengths the intrinsic optical window is limited by the IR multiphonon absorption edge. Heavy atoms, weakly bound absorb at longer wavelengths [$\lambda_o \propto \sqrt{(\mu/K}$), where μ is reduced mass of atoms connected by an atomic bond of force constant K]. However, the main problem for chalcogenide glasses is extrinsic absorption, due to hydride, oxide and hydroxide impurities which are brought in with the chemical precursors [21]. In fact oxide absorption can obscure the intrinsic multiphonon infrared absorption edge.

3. Amorphous chalcogenides as AO materials

3.1 GENERAL OBSERVATIONS

The first requirement of an amorphous chalcogenide for a bulk AO device is that the desired optical wavelength lies within the transmission window of the glass and that glasses of good transparency can be made. For an efficient AO interaction a high figure of merit is required, necessitating a high refractive index and preferably a low acoustic velocity. High refractive index glasses will require matched anti-reflective coatings for optical coupling in the AO device. However, too low an acoustic velocity will limit the switching rate of the device and also raise acoustic attenuation. Being glasses, fine-tuning of the AO properties *via* composition control is possible. Non-toxic glasses, stable against phase separation during preparation, with a high Tg and sufficient chemical and mechanical durability, are desirable. Unfortunately, many of the chalcogenide glasses with useful AO properties are based on arsenic which is toxic.

Refractive index of amorphous chalcogenides increases with polarisability of the chalcogen (and hence also with atomic mass). Thus index rises as tellurium substitutes selenium and as selenium substitutes sulphur [22]. Telluride glasses are poorly stable to devitrification during preparation and hence the highest refractive indices are usually afforded by the selenide-telluride glasses. With increasing atomic mass of the chalcogen then the electronegativity decreases and the optical bandgap thus shifts to lower energies. Increased polarisability of the chalcogen leads to a more weakly bonded network and this, together with the associated larger atomic mass, means that the IR edge is shifted to longer wavelengths. A more weakly bonded material will also be less stable towards phase separation, have a lower glass transformation temperature and be less chemically and mechanically robust. From table 1, it is apparent that acoustic velocity increases with increase in connectivity of the chalcogenide network. Connectivity can be expressed succinctly as the average coordination number of the matrix. Thus, within the bounds of glass stability, maximising the chalcogen (2-fold coordination) and minimising *e.g.* arsenic and germanium (3- and 4-fold coordinated, respectively) will minimise acoustic velocity.

Table 1 lists the main amorphous chalcogenide systems which have been studied for their AO properties. The information in table 1 is supplemented in sections 3.2 to 3.4, below.

3.2 NEAR-INFRARED OPERATION.

Sapriel [1] listed the AO properties of some of the better single crystal materials including αS and αHgS for use in the near-IR and visible regions. From table 1, it can be seen that the amorphous sulphides are similarly transparent in the near-infrared, with limited visible transmission. As_2S_3 is toxic, but exhibits a high M_2 value of 654×10^{-15} s^3kg^{-1} [23]. GaLaS glasses are non-toxic and have excellent transparency in the near-IR region and a high Tg [24]. The relatively low refractive index, and high acoustic

TABLE 1. Acousto-optic data for amorphous chalcogenides (and fused silica as a reference).[# double Tg indicates phase separation.]

Material	fused silica [1]	As_2S_3 [23]	$70Ga_2S_3$. $30La_2S_3$ [24]	Se [25]	Se [25]	As_2Se_3 [26]	$As_{12}Ge_{33}Se_{55}$ [27]	$As_{20}Ge_xSe_{80-x}$ x=0 / x=30 [28]	As.Ge.Se -$Te_{20}Pb_5$ [16]	As.Se.Te -Ge_5Pb_5 [15]	As.Se.Te -Ge_5Pb_{10} [15]
Transparent window/µm	0.2-2.5	0.6-12	0.5-10	0.8-11	0.8-11	0.9-11	1-14 [4]	1-14 [4]	1.35-16.88	1.38-27.8	-
Wavelength /µm	0.633	0.633	0.633	1.15	10.6	1.15	1.06	1.15	10.6	10.6	10.6
$M_2 x 10^{15}$ $/s^3kg^{-1}$	1.51	433	19	1210	1080	779	248	1200 /330	1457	1306	1873
Refractive index	1.46	2.61	2.4	2.497	2.421	2.893	2.70	2.64 /2.64	2.878	3.011	3.205
Light polarisation	perp.	perp.	-	perp.	perp.	perp.	-	parallel	parallel	parallel	-
Density$x10^{-3}$ $/kgm^{-3}$	2.2	3.20	4.1	4.27	4.27	4.64	4.40	4.45 /4.46	5.165	5.336	5.419
Acoustic velocity/ms^{-1}	5960	2600	4375	1830	1830	2250	2518	2000 /2450	2403	2258	2257
Acoustic polarisation	long.	long.	long.	long.	long.	long.	long.	long.	long	long	long.
Acoustic attenuation/ $dBcm^{-1}$at /MHz	3 at 500	-	0.93 at 80	34 at 100	34 at 100	10 at 200	7.1 at 500	0.4 /0.03 at 55	-	-	-
Tg / °C	1180 [ref.33]	205 [ref.33]	561	30	30	178 [ref.34]	368 [ref.34]	89/361 [ref.34]	272	# 174, 255	196

velocity, may account for their low M_2 of 19×10^{-15} s^3kg^{-1}. The high acoustic velocity makes them attractive for high acoustic frequency applications [24].

Selenium and the selenide glasses are potentially sufficiently transparent at telecom. wavelengths, at around 1.3 and 1.55μm, but are opaque in the visible part of the spectrum. Selenium has relatively good AO properties but the exceedingly low Tg of 30°C is unattractive [25]. Fukuda *et al.* [25] showed that a small addition of arsenic to form $As_{10}Se_{90}$ raises Tg to 80°C, while retaining good AO properties. Further addition of arsenic to form As_2Se_3 still leaves a high M_2 of 779×10^{-15} s^3kg^{-1}, at 1.15μm, and Tg is raised to 178°C [26]. Addition of germanium to As-Se systems [27, 28] increases the connectivity of the network hence raising both Tg and acoustic velocity; index is lowered and M_2 values, at 1.15μm, fall from 1200×10^{-15} s^3kg^{-1} for $AsSe_4$, to 330×10^{-15} s^3kg^{-1} for $As_{20}Ge_{30}Se_{50}$ glasses.

In addition to the information in table 1, AO properties of chalcohalide glasses have been reported in the Russian literature. Koperles *et al.* [29] report M_2 (assumed at 0.633 μm), density, acoustic velocity and acoustic attenuation for chalcohalide glasses in the systems: $(As_2S_3)_{1-x}(AsI_3)_x$ for x=0, 0.2, 0.4, and $(AsSI)_{1-x}(GeS_2)_x$ for x= 0.2, 0.4, 0.6, 0.8. Refractive indices were not reported. Glasses with the maximum amount of iodine addition have the highest figures of merit and the lowest acoustic velocity. Thus for $(As_2S_3)_{0.6}(AsI_3)_{0.4}$, M_2 is $585\times10^{-15}s^3kg^{-1}$ and for $(AsSI)_{0.8}(GeS_2)_{0.2}$, M_2 is $575\times 10^{-15}s^3kg^{-1}$. The addition of monovalent iodine decreases connectivity of the matrix and hence acoustic velocity and Tg are lowered. Moreover, the addition of iodine in addition should move the onset of absorption at the optical bandgap to lower energy [30].

Sheloput and Gushkov [31] report that M_2 for $Ge_{20}Sb_5S_{70}I_5$, at 0.633μm, is 120×10^{-15} s^3kg^{-1} and the transmission window is 0.6 to 11μm. $Ge_{18}Sb_{15}S_{52}I_{15}$ has a rather low refractive index ($n_{0.633}$=2.2) leading to a modest M_2 of 75×10^{-15} s^3kg^{-1}, at 0.633μm [31]. Glushkov [32] reports a high M_2 of 1200×10^{-15} s^3kg^{-1} (assumed at 0.633μm) for HgAsS glass.

3.3 MID-INFRARED OPERATION

Most of the chalcogenides in table 1 show sufficient transparency for operation in the 3 to 5 μm region. However, unwanted H-S, H-Se and H-Te vibrational absorption bands occur in the mid-IR [21].

3.4 FAR-INFRARED OPERATION

For a material operating at 10.6μm to have an equivalent AO diffraction efficiency to one operating at 0.633μm, then the figure of merit must be 280 times greater at the longer wavelength because of the variation of diffraction efficiency with λ_o^{-2} (eqn. 1).

The selenide glasses, As-Se and As-Ge-Se, are potentially transparent enough for use at the 10.6μm CO_2 laser wavelength. Recently, Laine and Seddon [15,16] have optimised two chalcogenide glass systems for use in the far-infrared at 10.6μm.

The first series [15] was based on the As-Se-Te system. The base glass with additions of (mole%) 5Ge and 5Pb was demonstrated to have $M_2 = 1306 \times 10^{-15}\ s^3kg^{-1}$. This measured value of M_2 was used to calculate an effective photoelastic coefficient, p_{eff} = 0.328, for this first glass series. Using this value of p_{eff}, the base glass with (mole%) 5Ge and 10Pb, and $n_{10.6} = 3.205$, was calculated to have an outstandingly high figure of merit of $1880 \times 10^{-15}\ s^3kg^{-1}$, at 10.6μm. This figure of merit should increase further at shorter optical wavelengths. Glasses in this series are prone to liquid-liquid phase separation as shown by the double Tg observed *via* thermal analysis (see (a) in table 1), however no evidence could be found of a droplet phase microstructure by means of scanning, or transmission, electron microscopy [15].

The second series was based on the As-Ge-Se system, having significant levels of Ge [16] and hence higher acoustic velocity relative to the first glass series[15]. The glasses of the second series were developed to have better glass stability than those of the first series at the expense of a higher refractive index and figure of merit. The highest refractive index glass ($n_{10.6}$=2.878) was achieved with additions of (mole%) 20Te and 5Pb, giving M_2 equal to $1457 \times 10^{-15}\ s^3kg^{-1}$, at 10.6μm.

4. Conclusions

Recent developments of selenide-telluride based glasses with good acousto-optic properties for use at 10.6μm complete the range of chalcogenide glasses which are now potentially available for operation of bulk acousto-optic devices in the near-, mid- and far-infrared spectral regions. The chalcogenide glasses in general exhibit good acousto-optic figures of merit and acceptable levels of acoustic and optical attenuation.

Acknowledgements
The authors thank Lionel J. Kent of the GEC-Marconi Research Centre, Great Baddow, Chelmsford, Essex, UK, for useful, detailed discussion.

References

1. Sapriel, J. (1979) *Acousto-optics*, John Wiley & Sons, Chichester.
2. Polakov, Yu.A., Makovskaya, Z.G., Dembovskii, S.A., Deryugin, I.A. and Talalaev , M.A. (1981). Translated from *Izvestiya Akademii Nauk SSR, Neorg.* Mater. **17** (7) 1166, Plenum Publishing Corpn. 863-867.
3. Das, P.K. and DeCusatis, C.M. (1991) *Acousto-optic signal processing: fundamentals and applications,* Artech House Inc..
4. Savage, J.A. (1985) *Infrared optical materials and their anti-reflection coatings,*

Adam Hilger Ltd., Bristol.
5. Brillouin, L. (1922) *Annales de Physique* **17**, 88-122.
6. Debye, P. and Sears, F.W. (1932) *Proc. National Academy of Science (US)* **18**, 409-414.
7. Lucas, R. and Biquard, P (1932) *J. Phys. Radium* **3** 464-477.
8. Raman, C.V. and Nath, N.S.N. *Proc. Indian Academy of Science* (1935) **2** 406-420; (1936) **3** 75-84, 119-125, 459-465; **4** 222-242.
9. Smith, T.M. and Korpel, A. (1965) *IEEE J. Quantum Electronics* **QE-1** 283-284.
10. Dixon, R.W. and Cohen, M.G. (1966) *Appl. Phys. Lett.* **8** (8) 205-207.
11. Abrams, R.L. and Pinnow, D.A. (1970) *J. Appl. Phys.* **41** 2765-2768.
12. Fox, A.J. (1985) *Appl. Opt.* **24** (14) 2040-2041.
13. The novel method for measuring the acousto-optic figure of merit was developed by personnel at the GEC-Marconi Research Centre, Great Baddow, Chelmsford, Essex, CM2 8HN, UK.
14. Laine, M.J. (1996) *Preparation of chalcogenide glasses and investigation of their properties for use in acousto-optic devices.* Ph.D Thesis, Centre for Glass Research, Dept of Engineering Materials, University of Sheffield, Sheffield, S1 3JD UK.
15. Laine, M.J. and Seddon, A.B. (1995) *J. Non-Cryst. Solids* **184** 30-35.
16 Seddon, A.B. and Laine, M.J. (June, 1996) *Proc. 10th Int. Symp. on Non-Oxide Glasses,* Corning, NY, US. Editors: Clare A.G., Tick, P.A. and Bartholomew, R. Submitted to *J. Non-Crystalline Solids.*
17. Woodruff, T.O, and Ehrenreich, H. (1961) *Phys. Rev.* **123** (5) 1553-1559.
18. Shimakawa, K., Kolobov, A. and Elliott, S.R. (1995) *Advances in Physics* **44** 475-588.
19. Kanamori, T., Terunuma, Y., Takahashi, S. and Miyashita T. (1984) *J. Lightwave Tech.* **LT-2** (5) 607-612.
20. France, P.W., Drexhage, M.G., Parker J.M., Moore, M.W., Carter S.F. and Wright J.V. (1990) *Fluoride glass optical fibres,* Blackie & Son Ltd., London, chapter 7.
21. Seddon, A.B. (1995) *J. Non-Cryst. Solids* **184** 44-50.
22. Aio, L.G., Efimov, A.M. and Kokorina, V.F. (1978) *J. Non-Cryst. Solids* **27** 299-307.
23. Dixon, R.W. (1967) *J. Appl. Phys.* **38** (13) 5149-5153.
24. Abdulhalim, I., Pannell, C.N., Deol, R.S., Hewak, D.W., Wylangowski, G. and Payne D.N. (1993) *J. Non-Cryst. Solids* **164-166** 1251-1254.
25. Fukuda, S., Shiosaki, T. and Kawabata, A. (1980) *Jap. J. Appl. Phys.* **19** (11) 2075-2083.
26. Ohmachi, Y. and Uchida, N. (1972) *J. Appl. Phys.* **43** (4) 1709-1712.
27. Krause, J.T., Kurkjian, C.R., Pinnow, D.A. and Sigety E.A. (1970) *Appl. Phys. Lett.* **17** (9) 367-368.
28 Adrianova, I.I., Aio, L.G., Asnis, L.N., Kislitskaya, E.A. and Kokorina, V.F. (1976) *Sov. Phys. Acoust.* **22** (3) 250-251.
29. Koperles, B.M (1976) *Akust Zh* **22** (4) 536-539.
30. Seddon, A.B. and Hemingway, M.A. (1994) *Phys. Chem. Glasses* **35** (5) 210-211.
31. Sheloput, D.B. and Glushkov, V.F. (1973) *Izv. Akad. Nauk. USSR Neorg. Mater.* **9** (7) 1149-1152.
32. Glushkov, V.F. (1976) *Izv. Akad. Nauk. USSR Neorg. Mater.* **12** 717.
33. Elliott, S.R. (1990) *Physics of amorphous materials.* Longman Scientific and Technical, Essex, England. 2nd. edition, p52 .
34. Webber, P.J. and Savage, J.A. (1976) *J. Non-Cryst. Solids.* **20** 271-283.

THE POSSIBILITIES TO USE THE INTRINSIC DEFECT'S OPTICAL PROPERTIES FOR OPTOELECTRONICS IN FUSED SILICA

A.R.SILINS
Professor, Secretary General of the Latvian Academy of Sciences, Academy square 1, Riga, LV-1524, Latvia

Abstract

The fused silica is the most spectroscopically investigated non-crystalline material. The main part of the practically used optical fibres are made from pure fused silica or fused silica based materials. Structured emission and excitation spectra for non-bridging oxygen centers were obtained and a model for electronic transitions in this defect center was put forward. Luminescence and absorption bands for twofold coordinated silicon were established and a model for electronic transitions was also put forward. The possibilities to use optical properties of these intrinsic defects for optoelectronics is estimated.

1. Introduction

Among optical glasses the most investigated material is fused silica, in which as the result of long-term experimental work the spectroscopic properties of elementary intrinsic defects, some of its aggregates and main impurity defects are well known[1]. The existence of point defects in glass usually appears as changes of the material spectroscopic (optical) properties in the transparency region. In fused silica non-bridging oxygen atoms, two-fold coordinated silicon atoms and also hydrogen atoms chemically bonded to the non-bridging oxygen atoms belong to this type of point defects. Non-bridging oxygen atoms, threefold and twofold coordinated silicon atoms are intrinsic defects in fused silica and they always exist in this material. In fact, fused silica in ordinary conditions (room temperature) has frozen in intrinsic defect concentrations which reflect the thermal equilibrium defect concentrations at the glass transition temperature. In the stoichiometric glass the main defects are non-bridging oxygen atoms and threefold coordinated silicon atoms [2]. Deviation from stoichiometry lead to an

A. Andriesh and M. Bertolotti (eds.),
Physics and Applications of Non-Crystalline Semiconductors in Optoelectronics, 337–346.

increase of twofold coordinated silicon atoms (oxygen deficit) or an increase of non-bridging oxygen atoms (oxygen excess) [3]. These defects play an important role in the optical wave guides made from fused silica since the high cooling rate in preparation increases the glass transition temperature for fused silica in wave guides. Besides thermally or stoichiometricly induced intrinsic defects, it is possible to generate high concentrations of these defects in fused silica by irradiation [2,4].

For practical use it is more convenient to have only one type of defect centers in the material. Otherwise different types of interactions between defects decrease the stability of the material. In fused silica it is possible by proper choice off stoichiometry or/and irradiation condition to generate mainly non-bridging oxygen atoms [2,5] or twofold coordinated silicon atoms [2,6] which are stable intrinsic defects in this material. This opens possibilities to use fused silica glass with these defect centers for practical needs.

2. Optical properties and concentrations of the intrinsic defects.

All intrinsic defects, discussed above, in fused silica changes the spectroscopic (optical) properties of this material in the transparency region. But threefold coordinated silicon atoms (E'centers) has only one absorption band peaking at 5.7 eV and no luminescence [2]. For this reason such centers are hard to use for optoelectronics and therefore will not be discussed in this review. Here the main attention will be paid to non-bridging oxygen atoms and twofold coordinated silicon atoms, because these defects has several absorption and luminescence bands in the transparency region of fused silica.

2.1. SPECTROSCOPIC PROPERTIES OF NON-BRIDGING OXYGEN CENTERS

There was long history of experimental work until it was found out precisely which spectroscopic properties of fused silica are determined by non-bridging oxygen atoms [2,5,7]. Now it is established that these centers are characterised by red luminescence peaking at 1.9 eV and two absorption bands peaking at 2.0 eV and 4.8 eV in which the red luminescence excites. Red photoluminiscence emission which peaks at 1.9eV has half with around 0.2 eV at 293 K. The half with of weak absorption (excitation) band at 2.0 eV is around 0.2 eV, but the ultraviolet absorption (excitation) band of non-bridging oxygen centers has a nearly Gaussian shape with a maximum at 4.8 eV and a half width around 1.0 eV. The decay of the red luminescence deviates slightly from exponential, with average decay time around 10μs at

293 K. It is estimated from these decay constants that the oscillator strength of the 2.0 eV absorption band is around 10^{-4}. Similar decay curves for red luminescence are observed for excitation in either 4.8 or 2.0 eV excitation bands.

Resonant zero-phonon lines (Fig.1) were observed [5] below 80K

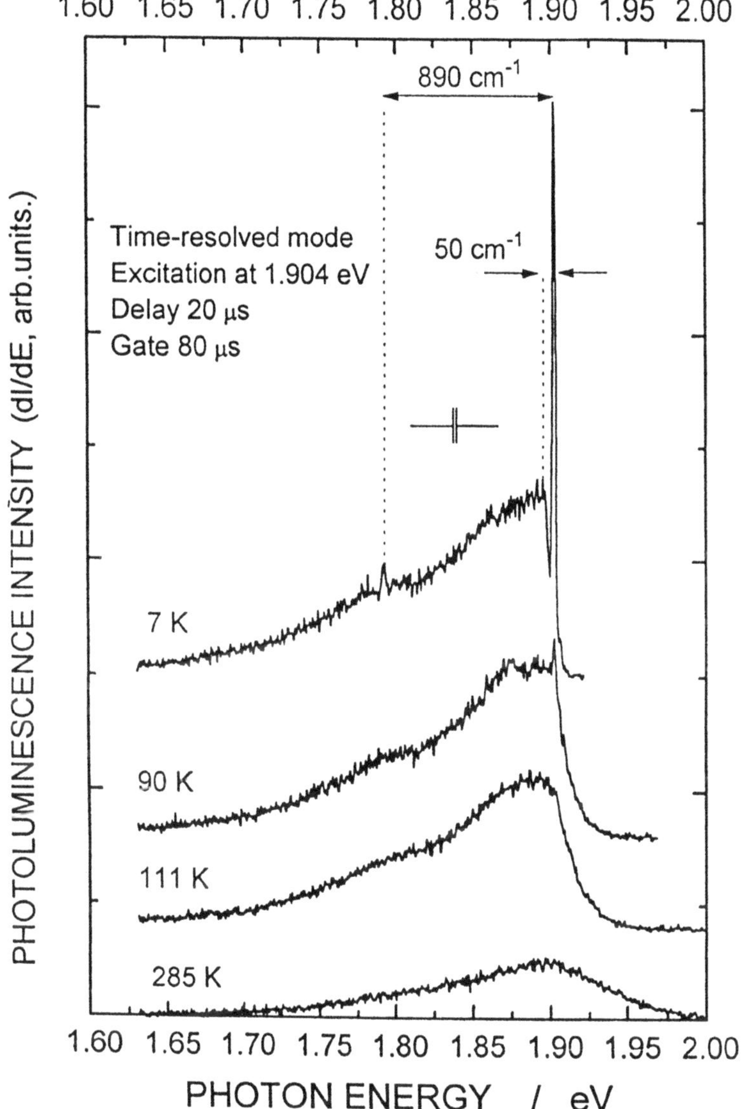

Fig.1. Time-resolved photoluminescence spectra of fused silica with non-bridging oxygen centers [5].

both in luminescence emission and excitation spectra of the non-bridging oxygen centers in the 1.9-2.1 eV region. A vibrational line in emission spectra 890 cm^{-1} below the zero-phonon line energy is attributed to the symmetric stretching vibrations of the silicon-non-bridging oxygen bond in the ground electronic state of the non-bridging oxygen center. A similar line 860 cm^{-1} above the zero-phonon line in the excitation spectra corresponds to the same vibration in the excited state. The intensities of the resonant zero-phonon lines are dependent on the excitation energy and has a nearly Gaussian distribution which describes the concentration distribution of the non-bridging oxygen centers with the respective energies of the excited electronic state.

There are relatively high values of the photoluminiscence degree of polarization in the zero-phonon line regions (+30 to 35)% which indicates that electronic transitions occur between non-degenerate states. Photoluminiscence polarization is observed also upon excitation in the 4.8 eV band, but the degree of polarization is small (around - 1.5%).

2.2. ENERGY LEVEL SCHEME OF NON-BRIDGING OXYGEN CENTERS.

In [5] an energy level scheme (Fig.2) was suggested to explain the optical

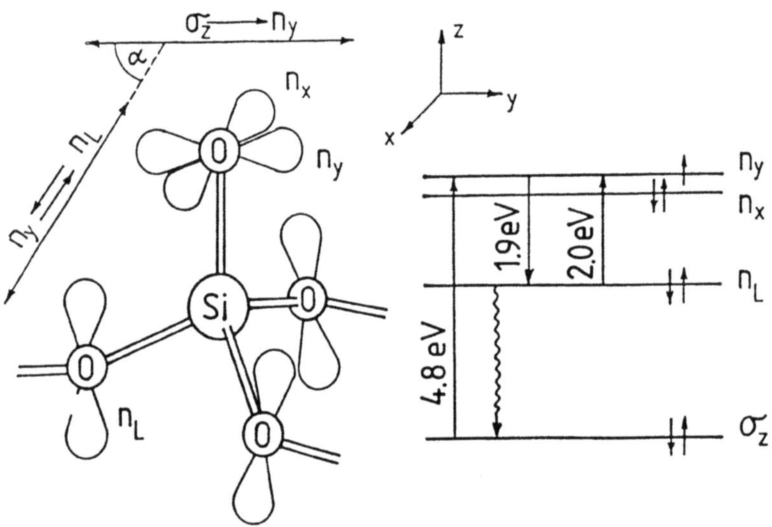

Fig.2. An energy level scheme. suggested to explain optical properties of non-bridging oxygen centers.

properties of the non-bridging oxygen center. According to this scheme the 1.9 eV photoluminiscence and 2.0 eV absorption transition occurs due to the charge transfer transition in this center between the half-filled non-bonding 2p orbital of the non-bridging oxygen atom and the lone-pair 2p orbital of one of the ligand oxygens. The dipole momentum of such transition is oriented approximately along the edge of a SiO_4 `tetrahedron. Since only non-bonding states are involved, the Stokes shift is small and zero-phonon lines can be observed at low temperatures. The relatively long photoluminiscence lifetime is consistent with a charge-transfer transition between weakly overlapping states.

The 4.8 eV absorption is attributed to a transition from the σ-bonding orbital to the 2p non-bonding orbital. The dipole moment of this transition is oriented perpendicular to the direction of the Si-O bond. The hole in the σ-orbital is transferred very fast and non-radiatively to one of the lone-pair 2p orbitals of the ligand oxygens. The charge transfer emission transition is polarized approximately parallel to the 0-0 direction and in this way forms an angle of value near to the "magic angle" (approximately 55°) with the absorption transition and at that angle the polarization dependence crosses zero. This agrees well with the experimentally observed small value (-1.5%) of the degree of polarization in the 4.8 eV band

The splitting between the two non-bonding 2p π orbitals (one half-filled and other full-filled or doubly occupied) of the non-bridging oxygen center varies from site to site due to the glassy disorder. In the excited state of the center, with both non-bonding 2p π orbitals doubly occupied, a finite probability exists for the luminescence transition of hole from lone-pair 2p orbital of the ligand oxygen back to any of doubly occupied 2p π orbitals. Along with the vibrational mixing, this can explain the decrease of the photo-luminescence degree of polarization at lower emission energies.

2.3 NON-BRIDGING OXYGEN CENTER GENERATION.

For practical use it is necessary to have high enough concentrations of active centers in the material. Estimations of the concentrations of frozen in thermal defects in fused silica [3], show, that in the stoichiometric glass the main defects are non-bridging oxygen atoms and threefold coordinated silicon atoms, but concentrations of these defects do not exceed 10^{15} cm^{-3}. By reasonable deviation from stoichiometry, it is possible to increase these concentrations by order of magnitude. Materials with such small concentrations of active centers are inconvenient for practical use.

It is possible to generate high concentrations of non-bridging oxygen centers in "wet" fused silica glass from hydroxyl bonded to the glass network by breaking off O-H chemical bond under vacuum-ultraviolet light irradiation. Isolated hydrogen atoms are also generated in this process. The impurity defects (hydrogen atoms) are stable in the fused silica glass network only at temperatures below 100K. The small activation energy for diffusion of hydrogen atoms allows them to move in the glass network at higher temperatures than 100K and recombine with non-bridging oxygen atoms. Such recombination reactions restore the hydroxyl point defects in the fused silica glass network. At the same time the non-bridging oxygen centers are destroyed.

Until now the only way to generate in fused silica high concentrations of non-bridging oxygen centers which are stable at room temperature is the neutron irradiation, but it must be pointed out that this type of irradiation generates all types of intrinsic defects.

2.4. SPECTROSCOPIC PROPERTIES OF TWOFOLD COORDINATED SILICON ATOMS.

In recent decades, the "B_2"- center, which causes the optical absorption band around 5.0-5.1 eV in fused silica, has been treated in more than 100 papers. There were common agreement that this band is caused by oxygen deficit. By careful spectroscopic study of this center [8] it was possible to attribute absorption in the 5.0 eV region and two luminescence bands with electrically neutral twofold coordinated silicon atoms in fused silica.

Together with further investigations [9] it is established that the optical absorption band at 5.03 eV ("B_2"- band) and two photoluminescence bands, one with the peak between 4.3 and 4.4 eV (high energy band) and other with the peak between 2.6 and 2.7 eV (low energy band) which excites in the B_2 band are caused by twofold coordinated silicon atoms in fused silica. The decay time of the high energy photoluminiscence is less than 10ns, but the decay time of low energy photoluminiscence is around 10 ms at 293K. The intensity of high energy photoluminiscence band increase upon cooling from room temperature to LNT, but the intensity of low energy photoluminiscence band decrease upon such cooling. Luminescence polarization studies showed a positive polarization degree for the high energy emission band and negative polarization degree for the low energy emission band upon excitation in the 5.0 eV absorption band.

In addition to these spectroscopic characteristics later [9] it was established that there is additional peak in the excitation spectra of low energy photoluminescence band around 3.1 eV. The decay time of low energy photoluminescence, when excited in this band is around 10 ms at 293K. The intensity of the low energy photoluminescence band under 3.1 eV excitation increase with cooling.

2.5. ENERGY LEVELS AND OPTICAL TRANSITIONS IN THE TWOFOLD COORDINATED SILICON CENTER.

The described spectral and kinetic features of twofold coordinated silicon center are completely consistent with the usual pattern, provided by orbitally allowed transitions between a singlet ground state (usually denoted as S_0) and the first excited singlet (S_1) and triplet (T_1) states in a single center (Fig.3.). This pattern leads to such features of the center:

a) a strong absorption/excitation band due to the allowed $S_0 \rightarrow S_1$ transition;

b) the existence of two luminescence bands under this ($S_0 \rightarrow S_1$) excitation. The high energy ($S_1 \rightarrow S_0$) emission band is fast with a radiative decay constant less than 100ns. The low energy ($T_1 \rightarrow S_0$) emission band is "slow", since it is caused by the spin forbidden transition;

c) a relatively weak absorption/excitation band, situated at lower energy. It corresponds to $S_0 \rightarrow T_1$ intercombination transition. Its intensity is inversely proportional to decay time of the $T_1 \rightarrow S_0$ emission band and for decay time around 10 ms its oscillator strength is below 10^{-6} and the band is hard to observe.

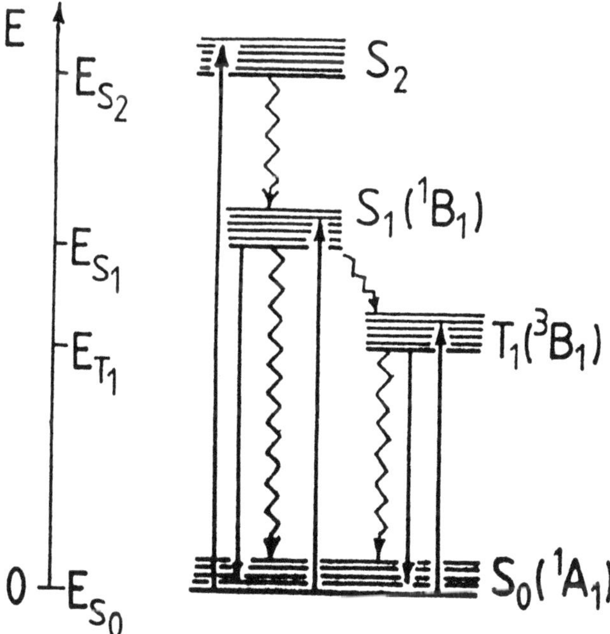

Fig.3. A simplified energy level scheme [9].

2.6. TWOFOLD COORDINATED SILICON CENTER GENERATION.

Twofold coordinated silicon centers appear in oxygen-deficient or irradiated by particles (neutrons, ions) high purity fused silica [9]. By treatment with oxygen these centers are removed from silica glass [10]. As it was established in [3], the increase of the deficit of oxygen in fused silica mainly appears as an increase of the concentration of twofold coordinated silicon centers. There are almost direct proportionality between the deviation from stoichiometry (degree of the oxygen deficit) and the concentration of the twofold coordinated silicon centers. It opens the possibility to generate by thermal treatment of fused silica samples in reducing conditions large concentrations of twofold coordinated silicon centers. The upper level of these concentrations is defined by the beginning of the interaction between twofold coordinated silicon centers and must be find experimentally.

3. Possibilities to use the intrinsic defects optical properties for optoelectronics.

The presence of point defects in glasses is widely used to solve practical tasks . Mainly all glass-based lasers act on electronic transitions between impurity defect (activator) levels. In similar way impurity and sometimes intrinsic defects are used in glass luminofors, scintilators, optical filters and other devices. Large popularity has obtained the refractive grating writing possibilities in the SiO_2 based glass waveguides. This process is connected with the charge state and geometrical structure changes of the existing point defects caused by the light influence.

Until now the intrinsic defects in fused silica are not directly used for optoelectronics. The reason for that is small concentration of these defects in pure stoichiometric fused silica glasses. But recently [11] the ways how to increase these concentrations was analysed.

In this review on the base of the investigated optical properties of two types of intrinsic defects in fused silica the first estimations of the possibilities to use the intrinsic defects optical properties for optoelectronics are done.

3.1. POSSIBILITY FOR PRACTICAL USE OF FUSED SILICA GLASS WITH NON-BRIDGING OXYGEN CENTERS.

From the energy level scheme of non-bridging oxygen centers, discussed above, it is obvious that strong excitation in the 2.0 eV absorption band by which large part of the non-bridging oxygen half-filled non-bonding orbitals are doubly occupied (full-filled) cause the significant decrease of the intensity of the 4.8 eV absorption band because in the excited centers the transition from σ-bonding orbital to the 2p non-bonding orbital is impossible. It means that fused silica glass with non-bridging oxygen centers could be used as an optical element in which by the influence of low energy light quanta (around 2.0 eV) the transparency of the material for high energy (around 4.8 eV) light quanta could be increased. At the same time in the excited centers there appears a possibility for transition of electron from σ-bonding orbital to 2p orbital of one of the ligand oxygens which now is half-filled. It means that by the excitations in the 2.0 eV absorption band the transparency for the middle energy (around 2.8 eV) light quanta could be decreased. Both of these effects might be used to design some kind of optoelectronic or fotonic device.

3.2. POSSIBILITY FOR PRACTICAL USE OF FUSED SILICA GLASS WITH TWOFOLD COORDINATED SILICON CENTERS.

Energy level scheme of twofold coordinated silicon centers, also discussed above, allows to suppose, that strong excitation in the low energy absorption band (around 3.1 eV) by which large part of the centers are transferred to the excited state with lifetime around 10 ms could cause the decrease of the intensity of the 5.0 eV absorption band. It means that fused silica glass with twofold coordinated silicon centers could also be used as an optical element in which by the influence of lower energy (around 3.1 eV) light quanta the transparency of the material for higher energy (around 5.0 eV) light quanta could be increased. But in this case the situation is not so obvious, as in the fused silica with non-bridging oxygen centers.

3.3. COMPARISON BETWEEN FUSED SILICA GLASS WITH NON-BRIDGING OXYGEN AND TWOFOLD COORDINATED SILICON CENTERS.

The analysed optical properties and energy level schemes of both centers allow to conclude that fused silica with these centers could be used as optical element in which by the influence low energy (different for different centers) light quanta the transparency of material for high energy light quanta could be increased. Essential part in this proposal is, that both centers in excited state are not absorbing in the high energy light quanta region. For the fused silica with non-bridging oxygen center it is possible with high probability to approve such situation, because the excitation of these centers is not leading to appearance of electrons in previously empty levels from which the further transitions could occur with possible absorption in the high energy light quanta region. Unfortunately that is not the case for fused silica with twofold coordinated silicon centers. But for other hand it is much easier to prepare material where only twofold coordinated silicon centers are present [3].

4. Acknowledgments.

This work was supported by Latvian Science Council Grant Nr.93/656.

5. References

1. Silins, A.R.(1995) Defects in Glasses, *Radiation Effects and Defects in Solids* **134**, 7-10.
2. Silins, A.R. and Truhkin, A.N. (1985) *Point Defects and Elementary Excitations in Crystalline and Glassy* SiO_2, Zinatne, Riga.
3. Silins, A.R., Lace, L.A.(1992) Influence of stoichiometry on high temperature intrinsic defects in fused silica, *J. of Non-Crystalline Solids* **167**, 229-238.
4. Griscom, D.L., Gingerlich, M.E. and Friebele, E.J. (1993) Radiation-induced defects in glasses: Origin of power-low dependence of concentration on dose, *Phys. Rev.Lett.***71**, 1019-1022.
5. Skuja, L.(1994) The origin of the intrinsic 1,9 eV luminescence band in glassy SiO_2, *J of Non-Crystalline Solids* **179**, 51-69.
6. Skuja, L.(1992) Isoelectronic series of twofold coordinated Si, Ge and Sn atoms in glassy SiO_2: a luminescence study, *J. of Non-Crystalline Solids* **149**, 77-95.
7. Skuja, L., Suzuki, T. and Tanimura, K.(1995) Site-selective laser spectroscopy studies of the intrinsic 1.9 eV luminescence center in glassy SiO_2, *Phys.Rew.B.* **52**, 15208-15216.
8. Skuja, L.(1992) Isoelectronic series of twofold coordinated Si, Ge, and Sn atoms in glassy SiO_2 : a luminescence study, *J. of Non-Crystalline Solids* **149**, 77-95.
9. Skuja, L.(1994) Direct singlet-to-triplet optical absorption and luminescence excitation band of the twofold coordinated silicon center in oxygen-deficient glassy SiO_2, *J. of Non-Crystalline Solids* **167**, 229-238.
10. Imagawa, H., Arai, T., Hosono, H., Imai, H and Arai, K.(1994) Reaction kinetics of oxygen deficit centers with diffusing oxygen molecules in Silica glass, *J. of Non-Crystalline solids* **179**, 70-74.
11. Silins, A.R.(1994) Thermally induced point defects in fused silica, *Glasteehmische Berichte - Glass Sci. Technol.* **67C,** 14-18.

AMORPHOUS SILICON PHOTODETECTORS FOR OXIDISED POROUS SILICON BASED OPTICAL INTERCONNECTIONS.

G. Masini
Dipartimento di Ingegneria Elettronica, Terza Università di Roma
Via della Vasca Navale, 84 - 00146 - Roma, Italy;
A. Ferrari, M. Balucani, S. La Monica, Gabriella Maiello
Dipartimento di Ingegneria Elettronica, Università 'La Sapienza'
Via Eudossiana, 18 - 00184 - Roma, Italy;
V. Bondarenko, A. Dorofeev, V. Filippov, N. Kazuchits
Belorusian State University for Informatics and Electronics
P. Brovki 6, 220600 Minsk, Bielorussia.

Abstract

Optical interconnection of integrated circuits (IC) is a fundamental milestone to be achieved for the next generation of integrated circuit technology: Wafer Scale Integration (WSI). The substitution of the optical signal to the currently used electrical ones in routing the information to the ICs is expected to dramatically cut costs and increase speed.

We have recently introduced a new Oxidised Porous Silicon Optical Waveguide (OPSWG) which shows interesting characteristics for this application. In order to use optical interconnections in WSI technology it is necessary to provide the transducers for transforming the electrical signal 'understood' by the ICs to the optical signal which can be carried on OPSWG.

In this work we focus our attention on the fabrication of amorphous silicon photodetectors to be used as optical/electrical transducers in WSI technology. Hydrogenated Amorphous silicon is well known as a versatile material for the fabrication of photodetectors: in fact it has extremely high absorption coefficient for the light in the visible range thus allowing the fabrication of extremely thin but still efficient detectors. In addition, the low temperature deposition process is fully compatible with the existing C-MOS technology. In the paper some different approaches to the fabrication of such photodetector will be presented ranging from the simple photoresistor to the more complicated junction device. Moreover different coupling geometries between OPSWG and detector are implemented. Characteristics of the fabricated devices are reported together with equivalent electrical models to be used in simulation.

A. Andriesh and M. Bertolotti (eds.),
Physics and Applications of Non-Crystalline Semiconductors in Optoelectronics, 347–359.

1. Introduction

As silicon technology decreases feature sizes allowing the fabrication of lower dimensions and higher speed chips, a further step in silicon based technology is needed: the integration of many chips on a single wafer (Wafer Scale Integration). It is well known that speed to be achieved inside chips is limited by electrical interconnection delay. The use of standard electrical interconnections is even more limiting for routing a clock signal to different parts of a circuit [1].

To overcome this problem the research on suitable interconnections becomes of primary importance. Optical interconnections for routing signals between different ICs have recently received considerable attention as a possible means for removing the shortcomings introduced by standard electrical interconnections, such as high power dissipation, limited bandwidth and crosstalk. [2,3]. Moreover, the fabrication process of these optical links can be very large scale integration compatible.

Optical technologies are becoming increasingly important in areas that were traditionally the domain of electronics. In particular, silica-based optical waveguide fibers are used in long-distance telecommunications and local networks. Recently important developments have been made on smaller scale in optoelectronic integrated circuits in which planar optical waveguides on a semiconductor substrate are employed for optical interconnections between active devices (lasers, detectors, light-emitting diodes, etc) and also in passive devices such as splitters, filters, couplers, etc [4, 5, 6].

The introduction of optical waveguides as interconnects in WSI technology needs the fabrication of transducers to transform the electrical signal flowing inside the ICs to the optical one to be carried on OPSWG. Of course an analogue transducer must be fabricated to perform the optical/electrical transformation of the signal at the OPSWG end, before the input into IC.

1.1 OPTICAL WAVEGUIDES

Standard materials for integrated optical waveguides fabrication are: phosphosilicate glasses, silicon nitride, and different plastics [4, 5, 6]. As regards phosphosilicate glasses, they present an high etching rate, and need additional doping or annealing steps. Silicon nitride deposition results in high mechanical stress; thus only very thin Si_3N_4 films can be used. Finally, it is known that plastics cannot undergo thermal treatments at the temperatures used in microelectronic technology; for this reason their fabrication must be the last step in the electronic system manufactoring process. On the other side, silicon dioxide shows excellent optical and thermo-mechanical characteristics and full compatibility with Si technology. However, confinement of light within a SiO_2 waveguide presents some problems due to the low refractive index of this material. Possible solutions to this problem are the optical confinement by conductive walls or local doping-induced changes in the refractive index. The first solution produces strong confinement accompanied by high losses due to the finite conductivity of cladding; the second one requires an additional doping step (usually with phosphorus, or nitrogen) to the process. Light confinement can be obtained even fabricating a ridge waveguide; this is obtained by growing a thick SiO_2 layer, thermally or by CVD, which presents poor optical properties due to high mechanical stress and presence of cracks. In addition, this process requires further steps to obtain a planar surface which is one among the most stringent needs of nowadays VLSI technology.

In the recent years porous silicon (PS), a material formed by electrochemical partial dissolution of the surface of a silicon wafer, has been successfully used to guide the light. The optical signal confinement was achieved using a PS multilayered structure [7], or by using a Si_3N_4 layer to confine the light [8]. We have presented an original method to confine the visible light by using dense thermally oxidised porous silicon waveguides (OPSWG) [9, 10, 11, 12]. Silicon oxide obtained by thermal annealing of PS shows quite suitable characteristics for WSI technology, such as low losses and strong confinement [13], due to the low mechanical stress of the obtained material. This is principally due to the peculiar feature of PS layer, formed by voids between silicon wires: silicon oxide expands inside pores, and no volume expansion takes place as in usual silicon oxide growth.

Another imporant peculiarity of waveguides obtained by PS thermal oxidation regards the possibility to be arranged both planar to silicon surface, or buried into silicon under electronic components, thus reducing area consumption.

1.2 ELECTRO-OPTICAL TRANSDUCER

As regards electrical/optical transduction, room-temperature visible light emission from reverse biased *p-n* silicon junctions were first reported in [14, 15] and, more recently, in [16]. Unfortunately the efficiency did not exceed 10^{-5} %, not allowing any practical application of the presented devices.

Porous silicon, since Canham first report [17], seems to be an interesting candidate not only as a starting material for dense thermally oxidised PS waveguides, but also to fabricate electrical/optical transducers based on silicon material and operating in the visible range. A number of works were reported in literature in the last years, concerning the fabrication of light emitters based on PS, with different structures. Light emission was observed from Au/PS/Si structures [18] showing a quantum efficiency of 0.01%; from p^+-n^+-n^- PS homojunctions [19] revealing a 0.2% efficiency under pulsed operation; or from an ITO/p-PS:n-Si structure [20] that reached the highest efficiency of 0.1% under continuous operation. All these devices showed degradation phenomena under operation, with a time stability of the initial performances not longer than some hours.

In recent works [21, 22, 23] we reported performances of a light emitting device showing broad band (white in colour to the eye) and extremely stable emission (45 days of continuous operation), fast response (lower than 80 ns under a pulsed driving current), and high external Power Efficiency (PE = 0.01%). This device is based on Schottky junction between aluminium and n-type PS, working in the breakdown region.

It is important to underline the compatibility of porous silicon formation with standard silicon technology, allowing a full integration of Oxidised Porous Silicon interconnections and electrical/optical porous silicon transducers in silicon based WSI.

1.3 OPTICAL/ELECTRICAL TRANSDUCERS

Crystalline silicon *p-n* junctions has been proposed as photodetectors for integrated optoelectronic system [24] These detectors are buried into the crystalline wafer, thus consuming wafer area which can be better used for ICs.

Hydrogenated amorphous silicon has caught our attention as suitable material to transform back the optical signal routed from the Oxidised Porous Silicon Waveguide into electrical signal, at the ICs input. This material is largely used for the fabrication of photodetectors and solar cells because of its extremely high absorption coefficient for the light in the visible range [25]. Because of this a photodetector with high efficiency can be fabricated by using very thin layers, about 1 μm thick. In addition, the low temperature deposition process is fully compatible with the existing C-MOS technology.

In this work a set of photodetectors using a-Si:H as active layer operating for the detection of the optical signal travelling in an OPSWG are presented. Different coupling geometries between OPSWG and detector are implemented. Characteristics of the fabricated devices are reported together with equivalent electrical models to be used in simulation.

2. Light detectors design and fabrication

Design of photodetectors deserves a special attention to the characteristics of the OPSWG. In particular this waveguide is suitable for light guiding in the visible region of the spectrum where amorphous silicon photodetectors show maximum performances.

In this chapter the process used for fabrication of OPSWG is described to the aim of focusing the technological constrains imposed to photodetector design.

2.1 OPSWG FABRICATION

Boron-doped, 0.01 Ωcm resistivity silicon is used as starting material. Porous silicon channels are obtained by selective anodization of silicon wafers using 0.2 μm thick silicon nitride mask. Linewidth of photolitographic mask is 10 μm. Anodization is performed in HF/alcohol solution, under a current density of 20 mA/cm^2. The wafer is then cleaned by standard wet process and rinsed in deionized water. This first steps produces a channel porous silicon waveguide having a width of about 30 μm, and a thickness of about 15 μm. After silicon nitride mask removal the wafer is cleaned by standard wet process and rinsed in deionized water.

In order to obtain waveguiding properties in the visible range, PS channel waveguides must undergo some further technological processes which aim is to create a good quality, dense, silicon oxide core within the channel. This is accomplished by thermal treatment in controlled atmosphere.

Thermal oxidation of the PS waveguides is performed in a diffusion furnace by a three steps process. First the structure of porous material is stabilized [26] and prevented from sintering [27, 28] by a low-temperature oxidation process in dry oxygen at 300°C, during 1 hour. This is a key step to determine optical properties of the final material [29,30]. PS full oxidation is then performed at 900 °C in steam atmosphere. Densification of the created oxide is accomplished by rising temperature up to 1150 °C during 25 minutes, in wet oxygen-nitrogen atmosphere, to improve optical properties of the obtained material [24]. Finally, the temperature is decreased down to 850 °C during 30 minutes before unloading, to reduce thermal stress.

The many steps of the oxidation process together with proper choice of PS parameters provide the formation of oxidised porous silicon waveguides with low mechanical stress and absence of cracks and dislocation.

2.2 LIGHT DETECTORS DESIGN AND FABRICATION

The following issues, grouped by subject, are to be taken into account for photodetector design:

- application needs

specific applications claim for different detector performances. As an example the distribution of the clock signal to hundreds of gates spread on a large surface needs a large fan-out of the clock source. In terms of the design of optical links this corresponds to reduce the 'weight' of the single photodetector on the transmission line, or, in other words the amount of light required to produce a certain signal. Two of the developed photodetectors has been designed having in mind this issue: they make use of the light leakage from the OPSWG surface.

- constrain imposed by OPSWG

OPSWG are electrically insulating: this makes difficult the fabrication of a vertical photodiode-type detector stacked directly on OPSWG. In practice the deposition of a transparent, thin electrode on top of the OPSWG is required. This electrode, while introducing optical losses in light coupling adds a further technological step not always available in usual electronic factory plants. The detectors presented in this work makes use only of intrinsic, not doped, hydrogenated amorphous silicon deposition and aluminium metallization; both these processes are common in electronic factories.

- amorphous silicon advantages/disadvantages

Room temperature Aluminium deposition produces Schottky barriers on intrinsic hydrogenated amorphous silicon [31]. Schottky devices are known to be faster than p-n or p-i-n junction ones [32], thus more suitable for signalling applications. Designer can also take advantage from the quite high (around 3.5 in the visible range) amorphous silicon refractive index. In fact an a-Si:H layer deposited over OPSWG reduces the light confinement within the guide: light escaping from the waveguide is directly coupled into the photodetector active layer.

With the aim of testing the performances of different devices, designed having in mind the issues reported above, we developed the following structures: a Coplanar Photoresistive Detector (CPD), a Pass-through detector (PD), an End-fire Detector (ED).

2.2.1 CPD - Coplanar Photoresistive Detector

Fig. 1(a) reports the schematic view and the optical microscopy picture of a Coplanar Photoresistive Detector. Device operation is based on the change in the conductivity of the amorphous silicon layer grown over the OPSWG induced by the light losses from the waveguide. The two electrical contacts to the device are fabricated by aluminium evaporation followed by photolytographic definition. The gap between the contacts is 100 μm, while contact length is 3 mm.

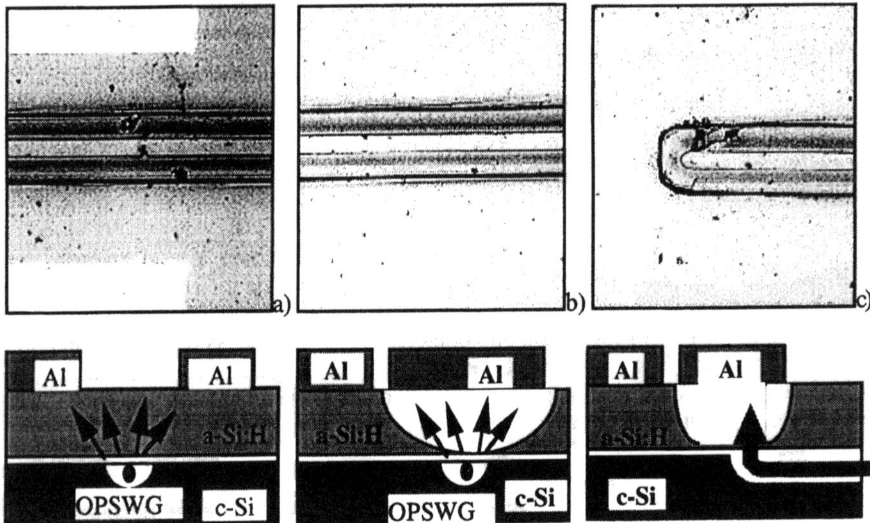

Figure 1. Top view pictures of the CPD(a), PD(b) and ED(c). On the bottom side the corresponding schematic cross sections are reported.

2.2.2 PD - Pass-trough Detector

Pass-through Detector (Fig. 1(b)) is based on the direct illumination of the Schottky junction formed between amorphous silicon and the metal contact; light is 'absorbed' from the waveguide due to the lowering of confinement induced by the high refractive index layer of a-Si:H. The second contact to the device can be either a second Schottky junction formed on a separate pad, or the bulk silicon under the field oxide formed during the oxidation process used for forming the OPSWG. This second approach results in an high-pass behavior of the detector.

2.2.3 ED - End-fire Detector

End-fire detector takes advantage from the ability to introduce a 90 degree bending of the guided light beam in OPSWG technology. The light output is directly coupled to the active layer of an a-Si:H/metal junction, as it is shown in the schematic of Fig. 1(c). The efficiency of this device in terms of signal per device area is expected to be far larger than the others; however this geometry can be used only on the terminal section of the OPSWG.

2.3 AMORPHOUS SILICON DEPOSITION AND ALUMINIUM CONTACTS.

Hydrogenated amorphous silicon film was grown on wafer where OPSWGs were already formed. The deposition was performed in a parallel plate glow discharge reactor operating at a frequency of 13.56 MHz. We used SiH_4 as process gas. The deposition time was set to obtain films 1 μm thick.

Aluminium was finally deposited by evaporation. Pads were defined by standard photolithography.

3. Results

3.1 OPSWG CHARACTERISTICS BEFORE AND AFTER A-SI:H DEPOSITION.

Figure 2 shows SEM micrograph of the OPSWG cross section (a) and a model of the waveguide structure (b). From the micrograph the border between silicon and porous silicon regions is evident at around 15 μm depth. The width of the porous layer is larger with respect to the starting nitride mask, reaching about 30 μm. This is expected from previous experimental results and is due to the low degree of anisotropy (i.e. the ratio between lateral spreading and penetration depth) of the used anodization process [33].

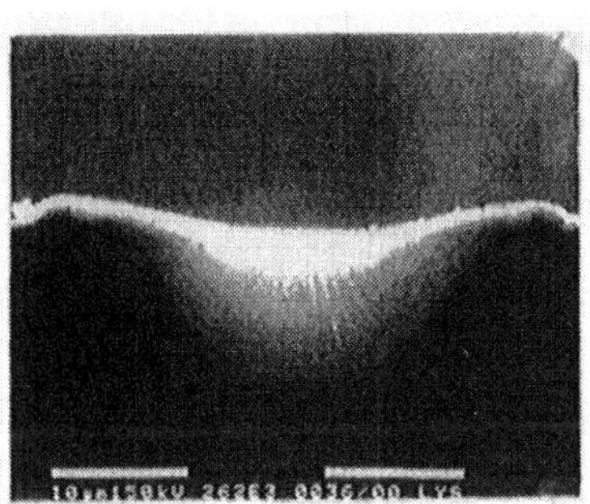

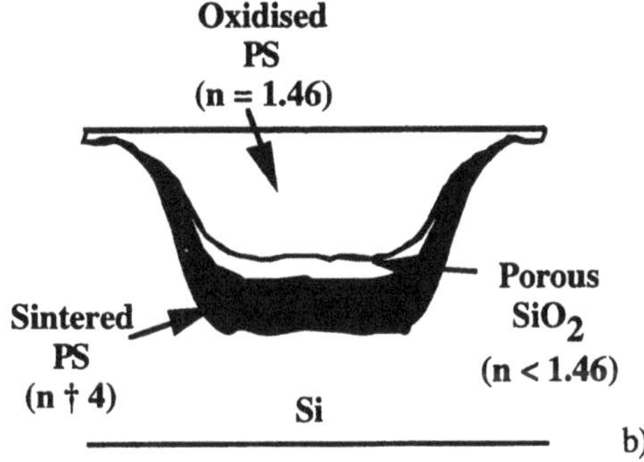

Figure 2. (a) Electron microscopy cross section and (b) shematic model of an OPSWG

Lightguiding in OPSWG is observed in the whole visible range. The output beam due to white light introduced in the OPSWG does not present any appreciable spectral difference compared to the input one.

The near field pattern measured at the end of the waveguide when a He-Ne laser beam (λ = 632.8 nm) is introduced by end-fire coupling, is very well fitted with a gaussian profile with an FWHM of 4-5 μm. This shape accounts well for a low order guided mode.

Out-of-plane scattering losses of the OPSWG has been measured by scanning the surface of the devices using an optical fibre as a probe, under end-fire input coupling [34]. In this experiment the optical fibre mounted on a micrometric stepper motor scans the distribution of the scattered light along the length of the waveguide, perpendicularly to the surface. The probe angle and the distance from the waveguide are maintained constant. Waveguide losses are measured, using this technique, from the slope in log scale of the scattered light level assuming a proportional relationship between guided and scattered light intensities.

In Fig. 3 the intensity of the light scattered along the waveguide length is reported. Three regions are clearly visible from the plot: on the left a low loss region (5-6 dB/cm) where OPSWG is in direct contact to the air, losses are higher (22 dB/cm) on the right side where amorphous silicon layer clads the top side of OPSWG. This difference is due to the high refractive index of the a-Si layer (n=3.5) that decreases the light confinement inside the waveguide.

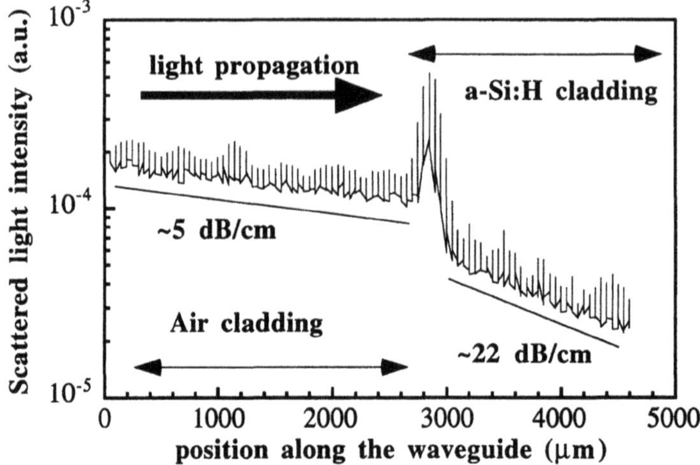

Figure 3 . Intensity of the light scattered from the surface of the OPSWG.

The sharp increase of scattered light intensity recorded at about 2900 μm from the scan origin is due to the existence of a tapered structure at the beginning of the amorphous film originated by border effect of the shadow mask used during a-SI:H deposition. In this region the effect of confinement reduction of the high refractive index film is predominant on its absorption characteristics.

From data reported in Fig. 3 it is possible to calculate the amount of light which is coupled to the amorphous silicon film after x centimeters of propagation. Being I_0 the guided light intensity at x=0 and assuming the light trapping in the a-Si:H film as the main cause of loss, we obtain for the injected light:

$$I \approx I_o\left(1 - 10^{-x}\right)$$

which brings to:

$$I_{abs} = I\left(1 - e^{-\alpha x}\right) \approx I\,0.5$$

for the absorbed light intensity, assuming the absorption coefficient of a-Si:H at the wavelength of 623.8 nm equal to $10^4 cm^{-1}$. Thus about 50% of light loss from OPSWG is coupled in the detector active layer; this value is raised when a high reflective metallic pad is present on the a-Si:H surface as in the PD and ED detectors.

As a matter of fact, the guided light from OPSWG is coupled into the a-Si detector on a certain length, in a distributed way. This is an interesting characteristic to be used for multiple sensing the same waveguide by a number of distributed photodetectors.

3.2 OPTOELECTRONIC CHARACTERISTICS OF THE LIGHT DETECTORS

3.2.1 CPD - Coplanar Photoresistive Detector

Current voltage characteristics of the CPD are reported in Fig. 4 in the dark and at different levels of intensity of the guided light. The expected photoresistive behavior appears only at voltages large enough to breakdown the double backward connected diodes present at the contact pads. At lower voltages, the reverse blocking characteristic of the diode dominates the behavior of the device; in addition, since the light is coupled only in the resistive part of the structure, far from the contacts, no changes in the reverse saturation current of the diode are visible as the light intensity is changed.

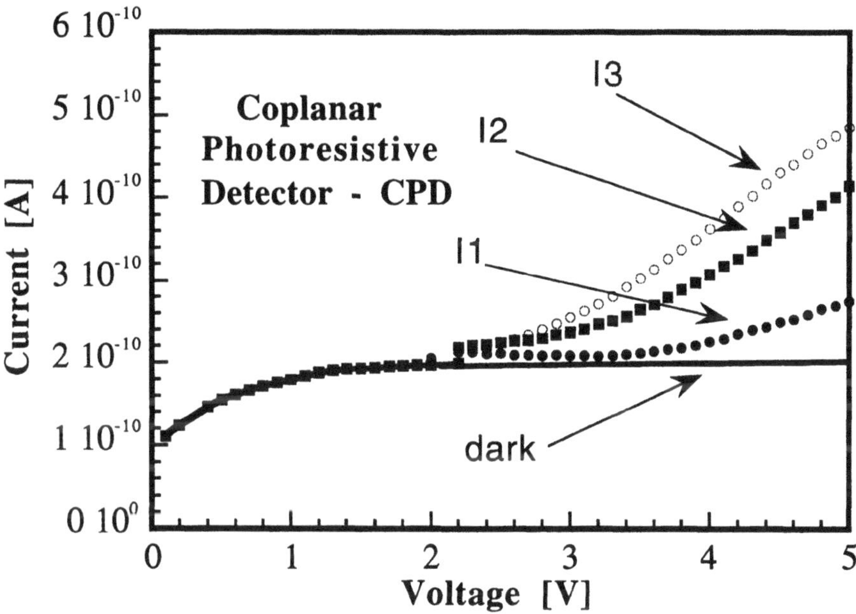

Figure 4. Current-voltage characteristics of CPD in the absence (dark) and for different levels of guided light intensity

3.2.2 PD - Pass-trough Detector and ED - End-fire Detector

Fig. 5 reports the photocurrent measured from PD and ED detectors. Both these devices are based on the illumination of a Schottky junction thus their characteristics are reported on the same graph for comparison. The level of guided light intensity is the same for the two curves. As expected, the collection efficiency of the two devices is greatly enhanced by a reverse bias applied to the junction up to the saturation of photogenerated carriers (unity collection). However, the two devices show almost one order of magnitude difference in the saturated collection efficiency in favour of the ED. This is due to the lower coupling strength of the PD in comparison to the ED.

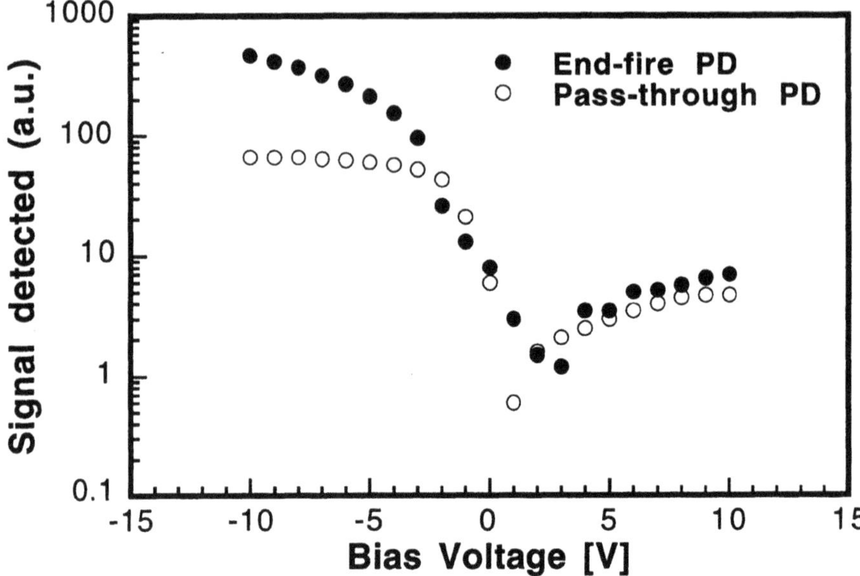

Figure 5. Photocurrent generated by PD and ED devices as a function of the applied voltage.

4. Modelling

Based on the performed measurements, simple electrical model has been developed for the presented devices in order to estimate their high frequency limits. Equivalent model and parameter values are reported in Fig. 6.

CPD is limited mainly by the time constant of the RC network consisting of the photoresistor in parallel to the capacitance of the two facing electrodes. The high value of intrinsic a-Si:H film resistivity results in a quite long time constant (abound 30 μs), independent of the length of the coupling area along OPSWG. This kind of detector, even presenting the interesting feature of being partially transparent and thus allowing a second, stacked, level of detection, is limited to low frequency applications. PD and ED are in principle limited only by the parasitic resistors (not shown in the figure for simplicity) and by the size of the transition capacitance of the Schottky diode. This can be minimized by reducing the dimensions of the metal contact. Unfortunately this corresponds to a decrease in the coupling area for the PD whose result is a reduction of the generated photocurrent. On the contrary ED detector area can be reduced, without losses in signal amplitude, down to the dimension of the OPSWG core, which is of the order of 10 μm diameter. Thus ED seems the most suitable detector for high frequency use. At frequency higher than 100 MHz however other phenomena, than the simple circuit time constants, are to be taken into account: the drift time of carriers over the distance the Schottky barrier depleted region becomes a major limit due to the low mobility of carriers in a-Si:H.

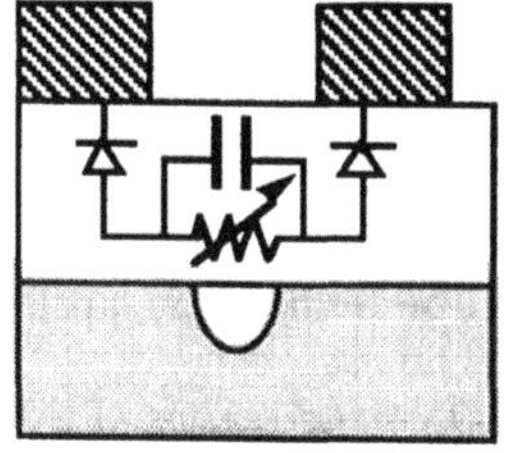
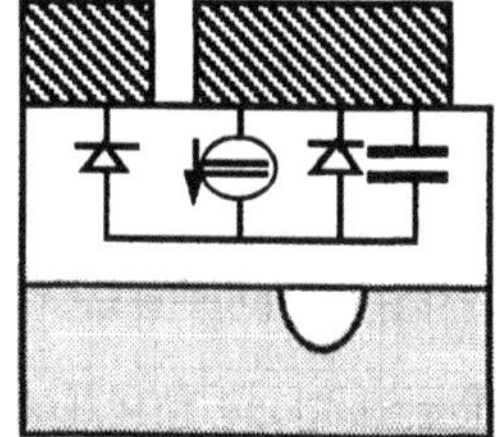
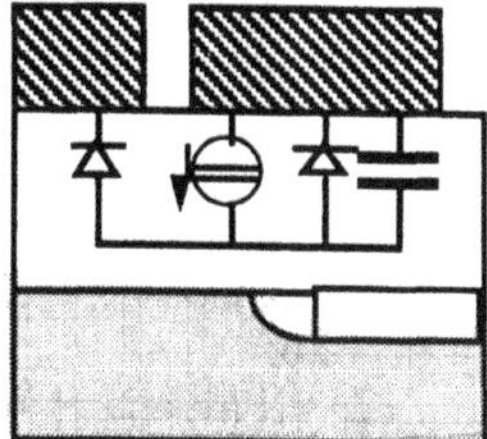

Rdark ≈ 50 MΩ cm
C ≈ 1 pF / cm
$\tau \approx 30\mu s$

C ≈ 1 nF / cm^2

Figure 6. Equivalent electrical model of the CPD, PD and ED devices.

5. Conclusions

In this work a set of optoelectronic systems formed by OPSWG and amorphous silicon photodetector are presented. The low level of losses obtained with OPSWG and the simple technology used for a-Si:H detectors make these systems promising for optoelectronic interconnections in silicon technology.

6. Acknowledgements

This work has been supported by NATO LINKAGE GRANT HTECH.LG 951231. Dr. Giampiero de Cesare is gratefully acknowledged for a-Si:H growth.

References

1. Li, J., Seidel, T.E., Mayer, J.W. (1994) Copper-Based Metalization in ULSI Structures, *MRS Bulletin* **8**, 15.
2. Hornak, L.A., Tewksbury, S.K. (1987) On the feasibility of through-wafer optical interconnects for hybrid Wafer-Scale-Integrated architectures, *IEEE Trans. on Electr. Dev.* **ED -34**, 1557-1563.
3. Soref, A. (1993) Silicon-based optoelectronics *Proc. of the IEEE* **81**, 1687-1706.
4. Boyd, J.T., (1990) *Integrated Optics: Devices and applications*, IEEE, New York.
5. Polman, A., et al. (1991) *J.Appl.Phys.* **70**, 3778.
6. Ayliffe, P., et al. (1994) *Int.J.Optoelectron.* **9**, 179.
7. Loni, A., Canham, L.T., Berger, M. G., Arens-Fischer, R., Munder, H., Luth, H., Arrand, H., and Benson, T.M. (1996) Porous silicon multilayer optical waveguides,*Thin Solid Films* **276**, 143-145.
8. Muller, J., Hilleringmann, U., and Goser, K. (1995) in Shi, L., Spiekman, L.H., and Leitjens, X.J. (eds.), *Proc. 7th Eur. Conf. on Integrated Optics*, Delft University Press, Netherlands, p. 291.
9. Bondarenko, V.P., Varichenko, V.S., Dorofeev, A.M., Kazuchits, N.M., Labunov, V.A., and Stelmakh ,V. F. (1993) *Tech. Phys. Lett.* **19**, 463-465.

10. Bondarenko, V.P., Dorofeev, A.M., and Kazuchits, N.M.(1995) Optical waveguide based on oxidized porous silicon, *Microelectr. Engin.* **28**, 447-450.
11. Bondarenko, V.P., and Dorofeev, A.M. (1995) Why porous silicon for SOI? in J.P. Colinge, V.S. Lysenko and A.N. Nazarov (eds.) *Physical and Technical Problems of SOI Structures and Devices,* NATO ASI Series, Series 3: High Technology **4**, Kluwer Academic Publishers, Dordrecht, The Netherlands, pp.15-26.
12. V.P.Bondarenko, A.M.Dorofeev, and N.M.Kazuchits. Oxidized porous silicon based waveguide for optical interconnections. Proc.7th European Conference on Integrated Optics ECIO'95, April 3-6, 1995, Delft, The Netherlands, edited by L.Shi, L.H.Spiekman, and X.J.M.Leijtens, p.291-294.
13. Maiello, G., La Monica, S., Ferrari, A., Masini, G., Bondarenko, V.P., Dorofeev, A.M., and Kazuchits, N. M. (in press) Light guiding in oxidised porous silicon optical waveguides, *Thin Solid Films.*
14. Newman, R. (1955) Visible light from silicon *p-n* junction, *Phys. Rev.* **100**, 700-709.
15. Cynoweth, A G., and McKay, K.G.(1956) Photon emission from avalanche breakdown in silicon, *Phys. Rev.* **102**, 309-315.
16. Kramer, J., Seitz, P., Steigmer, E.F., Auderset, H., Delley, B., and Baltes, H. (1993) Light-emitting devices in industrial C-MOS technology, *Sensors and Actuators A* , **37-38**, 527-533.
17. Canham, L.T. (1990) Silicon quantum wire array fabrication by electrochemical and chemical dissolution of wafers, *Appl. Phys. Lett.* **57**, 1046-1048.
18. Steiner, P., Kozlowski, F., and Lang, W. (1993) *Appl. Phys. Lett.* **62**, 2700-2702.
19. Linros, J. and Lalic A. (1995) High quantum efficiency for a porous silicon light emitting diode under pulsed operation, *Appl. Phys. Lett.* **66**, 3048-3050.
20. Loni, A., Simmons, A.J., Cox, T.I., Calcott, P.D.J., and Canham, L. T. (1995) *Elect. Lett.* **31**, 1288-1289.
21. Lazarouk, S., Bondarenko, V., Perschukevich, P., La Monica, S., Maiello, G., Ferrari, A. (1995) Visible electroluminescence from Al-porous silicon reverse bias diodes formed on the base of degenerate n-type silicon *Mat. Res. Soc. Symp. Proc.* **358**, 659-664.
22. Lazarouk, S., Jaguiro, P., Katsouba, S., Masini, G., La Monica, S., Maiello, G., Ferrari, A. (1996) Stable electroluminescence from reverse biased *n*-type porous silicon-aluminum Schottky junction device, *Appl. Phys. Lett.* **68**, 2108-2110.
23. Lazarouk, S., Jaguiro, P., Katsouba, S., La Monica, S., Maiello, G., Masini, G., and Ferrari, A. (1996) Visible light from aluminum-porous silicon Schottky junctions *Thin Solid Films* **276**, 168-170.
24. Hilleringmann, U., Goser, K., (1995) Optoelectronic system integration on silicon: waveguides, photodetectors, and VLSI C-MOS circuits on one chip, *IEEE Trans on Electr. Dev.* **42**, 841-845.
25. Sze, S.M., (1981) *Physics of Semiconductor Devices*, Wiley, 828
26. Yon, J. J., Barla, K., Herino, R., and Bomchil, G. (1987) *J. Appl. Phys.* **62**, 1042.
27. Labunov, V., Bondarenko, V., Glinenko, L. , Dorofeev, A., and Tabulina, L. (1986) *Thin Solid Films* 137, 123-126.

28. Dorofeev, A.M., (1986) *Gettering impurities of silicon structures by porous silicon regions,* Ph.D. Thesis, Minsk, Minsk Rdioengineering Institute.
29. Pickering, C., Beale, M.I J., Robbins, D.J., Pearson, P.J., and Greef, R. (1984) Optical studies of the structure of porous silicon films formed in *p*-type degenerate and non-degenerate silicon *J. Phys. C: Solid State Phys.* **17**, 6535-6552.
30. Filippov, V.V., Pershukevich, P.P., Bondarenko, V.P., and Dorofeev, A. M. (1994) *phys. stat. sol. (b)* **184**, 573.
31. Kanicki, J., (1992) *Amorphous & Microcrystalline Semiconductor Devices,* Hartech House, 189.
32. Sze, S.M., (1981) *Physics of Semiconductor Devices,* Wiley, 746.
33. Bondarenko, V.P (1980) Investigation of porous formation process and its application for intercomponent isolation in semiconductor IC's, Ph.D Thesis, Minsk, Minsk Radio-Engineering Institute.
34. Nishihara, H., Haruna, and M., Suhara, T. (1989) *Optical Integrated Circuits,* Mc Graw Hill Book Company, New York.

PHOTOLUMINESCENCE IN HYDROGENATED AMORPHOUS CARBON

M. KOÓS AND I. PÓCSIK
Research Institute for Solid State Physics,
H-1525 Budapest, P.O.Box 49, Hungary

Abstract

Photoluminescence of amorphous carbon has many intriguing properties, a number of which are similar to those of chalcogenide materials and amorphous silicon; however, the underlying processes are very different. In this article we discuss a variety of recent experiments on the photoluminescence properties of hydrogenated amorphous carbon. These experiments include examination of structural effect on the shape and energy distribution of luminescence spectra, the sample's polarization memory with its dependence on emission energy and temperature, moreover the excitation-emission matrix in a wide energy range. Some aspects of photoluminescence decay kinetics and changes in PL properties caused by thermal and light treatment will also be covered. All of our results corroborate that luminescence originates from the excited state of localised π electrons; depending on the extent of π electron localization, they are candidates for non-radiative recombination as well.

1. Introduction

Hydrogenated amorphous carbon (a-C:H) with a wide variation in structure and physical properties can be produced by different deposition techniques and parameters [1]. The source of this peculiar behaviour is the bonding variability of the carbon atom, which can form bonds in three different -- sp^3, sp^2 and sp^1 -- hybridized configurations. Those material, known as diamond-like hydrogenated amorphous carbon (DLHC) [2], which contains a nearly 1:1 ratio of sp^3 sp^2 hybridized carbon atoms and 20-33 at.% hydrogen are used as optical coatings on transparent substrate, or as a wear-resistant protective coating for opaque materials [3]. Another material, with a wider optical band gap than DLHC, -- called soft, polymeric hydrogenated amorphous carbon (PLHC), -- effectively emits light in the visible range at room temperature [4-11]. By virtue of this luminescence and certain other properties of PLHC it can be regarded as a promising material in applications for visible electro-luminescent displays [12, 13]. Light emitting diodes with PLHC as the active material has also been produced [14,15]; however, since these diodes have not demonstrated any great advantages as yet, they are not widely used. Therefore, apart from the pure challenge, there is an interest in understanding the photoluminescence mechanism of a-C:H.

A. Andriesh and M. Bertolotti (eds.),
Physics and Applications of Non-Crystalline Semiconductors in Optoelectronics, 361–378.

The electronic structure of amorphous carbon has peculiar properties because of its special carbon bonds. In the sp^3 hybridized configuration, each carbon atom has four equivalent bonding orbitals arranged tetrahedrally. These orbitals will form strong covalent σ bonds between neighbour carbon atoms. Another possibility is sp^2 hybridisation, which gives rise to three hybrid orbitals lying in the same plane and forming a 120° bond angle between each other. These hybrid orbitals also form strong covalent σ bonds between carbon-carbon or carbon-hydrogen atoms, and the remaining (non-hybridized) p orbital extends above and below the plane of the bonding σ orbitals, perpendicular to it. The degree of overlap between these p orbitals is less than for the σ orbitals, and they form a weaker π bond than the σ bond; this latter feature is mirrored in the binding energy. To break a σ bond, 6.33 eV energy is required whereas 3.98 eV is enough to break a π bond. Electronic states lying deep in the allowed bands of a-C:H are bonding and anti-bonding σ states and the π (bonding) and π^* (anti-bonding) states form the upper valence band and lower conduction band levels respectively. Because level-splitting depends on the extent of the π–bonding region the forbidden gap of amorphous carbon depends on size and shape, olephinic chains or aromatic rings are formed from the π–bonded atoms. As a consequence the electronic properties near the band edges are governed by π states, or by the amount and spatial distribution of these sp^2 hybridized carbon atoms in the amorphous material. The structural model of amorphous carbon developed by applying the Hückel theory is discussed in detail in Refs. [16, 17]. An alternative model for atomic-scale structure and electronic properties has also been proposed [18]. In the present state of knowledge, we must refer to the need of different structural investigations performed on sample series prepared by one and the same method in order to claryfy the real structure of this amorphous material [19].

The observable luminescence effect in a-C:H is of great importance in the study of electronic states near band edges, or closer to the Fermi level. Considerable efforts have been devoted to photoluminescence (PL) measurements [5-11] but the results gained up till now have not made it possible to develop an appropriate model for the luminescence mechanism in amorphous carbon. Confinement of photo-excited carriers into grains of the sp^2 phase was suggested based on the excitation energy dependence of PL spectra, and the observation of strong anti-Stokes emission in a-C:H was used to explain the fast exciton-like decay of PL as well [20]. Analysis of some PL properties parallel with optical absorption and Raman data in two different sample series -- as prepared and thermally annealed -- also led to a similar conclusion: both excitation and emission take place in localised states [21]. More direct evidence of strong localization of electron-hole pairs on the place of their generation was first observed in our laboratory by using linearly polarised light for excitation of PL and analysing the emission spectra in parallel and perpendicular direction to the exciting polarization [22]. Polarization memory near the maximal value, what can be calculated if the same dipole absorbs and emits the light, detected at higher than peak energy of the PL spectra is a good verification of the strong localization of photo-excited carriers [22].

Supposing an energy distribution both of the luminescence centres and of the optical gap values (E_g) of the clusters a good fitting of the low energy side of the PL spectra was successfully calculated in Ref. [23]. The recombination model, -- worked out for amorphous silicon luminescence and known as tail-to-tail luminescence -- was suggested also as being appropriate for a-C:H by some modifications, if we take into account a wider tail state distribution and that the recombination takes place within the cluster of sp^2 sites [24]. Paramagnetic defects are regarded as non-radiative centres in this model.

Let us summarise the main PL features of amorphous carbon observed in different but well defined samples by steady state PL measurements. PL decay properties will also be included despite the sparse results to be found in the literature, dealing with the luminescence decay process in a-C:H.

i). Very broad luminescence band with no fine structure is generally observed [4-9, 22, 23], though two bands with large width and comparable intensity have also been reported [8,21].

ii). PL band position and broadening are unambiguously affected by the structure of a-C:H [6,7,10,21,22, 25], but a clear correlation between the PL peak position and the Tauc gap cannot be established [5, 23]. Mention should be made of the latest results published, indicating increasing PL peak position with E_{04} gap value which means the energy, at which the absorption coefficient reaches the 10^4 cm^{-1} value [28].

iii). Luminescence spectra exhibit significant dependence on excitation energy. Blue shift of the peak position and broadening of emission band with increasing excitation photon energy have been reported [10, 20, 21, 26, 28].

iv). Change in PL intensity with decreasing temperature is very weak and characterised by non-Arrhenius behaviour [6, 20, 22]. A similarly small effect of cooling on the spectral characteristic was found [6,22], but in contrast to this a remarkable change on high energy side of PL spectra was also observed at 80 K [20].

v). Peculiar property of amorphous carbon luminescence is the preservation in a high degree of excitation light polarization in emitted light [22, 27].

vi). Eission energies higher than the excitation i.e. an anti-Stokes photoluminescence tail have been reported at room temperature and at 80 K too [20,23].

vii). Only one laboratory [20] has published experimental results concerning PL decay in amorphous carbon though very fast decay ($\sim 10^{-8}$ s) also has been mentioned in a number of papers [22,27].

viii). Finally attention is drown to a general observation that the photoluminescence effect characterized by properties previously summarized can be detected only in amorphous carbon of higher band gap, i.e. in polymer-like material. With decreasing band gap of amorphous carbon the PL efficiency decreases very rapidly down to the detectable level in material with a gap of ~ 1.5 eV [7,28].

Whatever the model of amorphous carbon PL, there is a need to explain source and mechanism of both radiative and non-radiative recombination. Electrons confined or localized on sp^2 bonded clusters have been suggested by various authors, these electrons recombine radiatively [20,24], but clear proof of strongly localised geminate pair luminescence first has been given in [22]. Energy distribution of luminescence centres [23] and tail states luminescence worked out for a-Si:H with a wider distribution of tail states in a-C:H [24] were considered to describe the different PL properties of carbon. One difficulty with tail state luminescence is that generally in amorphous materials the deeper the states in the gap, the stronger is their observed localization, but the situation is different in amorphous carbon, where the deepest states in the gap are the π states of the largest clusters, and states further from the Fermi level are the π states of smaller clusters representing a good description of the electronic structure. In this case the band tail luminescence would occur between non-geminate electron-hole pairs because at the end of thermalisation geminate pairs diffuse away from each others as a consequence of π state delocalization in a large cluster of sp^2 bonded carbons. This is one reason why care must be taken in applying band tail luminescence for a-C:H. Our view is, that bands of π states of sp^2 sites can be regarded as special sub-bands rather than commonly used tail states in amorphous materials. All of these are because of the very weak correlation between p orbitals which create the π states and sp hybrid orbitals which give σ states.

In the following section our results on PL properties of a-C:H will be presented. At first we deal with the general features of PL spectra and their variation depending on the structure of amorphous carbon: in the as prepared state, after thermal annealing, and prolonged illumination. We also deal in detail with the polarization "memory" of photoluminescence together with its temperature and emission energy dependence. Subsequently excitation energy dependence of light emission studied in a very wide range both of exciting and emitted photons will be discussed. The main peculiarities of carbon luminescence outlined from our experiments are the strong localization of electron-hole pairs recombining radiatively even at the highest excitation energies and the inhomogeneous line-broadening of emission spectra; these provide us with a decisive argument which enables one to understand the real connection between photoluminescence properties and the structure of amorphous carbon, moreover to explain the source and mechanism of photoluminescence of a-C:H.

2. Photoluminescence properties of amorphous carbon

The a-C:H samples studied in our experiments were deposited onto silicon wafer, Corning glass, and fused silica substrates by decomposition of pure methane in a capacitively coupled rf plasma reactor. This coupling accompanied by the appropriate geometrical relations results in negative charging of the powered electrode and the development of negative self-bias voltage ($U_{s.b.}$) -- which also depends on the rf power and the gas pressure. At low self-bias voltage (< 250 V) the structure of amorphous carbon is more polymer-like and exhibits effective luminescence while carbon with diamond-like properties can be deposited only at larger voltages. The substrate is placed on the powered electrode and to prevent its warming up, constant water cooling keeps the temperature at 70 °C. The absolute hydrogen content of carbon films was measured by means of elastic recoil detection (ERD), such measurements are also informative concerning any incorporated impurities. The absolute hydrogen content varied in the range of 45-50 at % and apart from oxygen (1.5 at %) no other impurities were observed. Various experimental methods were used to study the optical and electrical properties and the bonding peculiarities of a-C:H samples; here, however, we are solely concerned with the photoluminescence characteristics.

2.1 PHOTOLUMINESCENCE SPECTRA

Photoluminescence measurements were performed at room temperature on as deposited films, with differing thickness. The PL was excited by $E_{exc.}$ = 2.54 eV photons of an unfocused Ar^{+} laser beam and detected by a photomultiplier (AgCsO photo-cathode) through a grating monochromator using the lock-in technique. All spectra were corrected for the apparatus function.

PL spectra for three a-C:H samples deposited under different conditions are shown in Fig. 1. For comparison of spectral shape, the results are depicted in a normalized form. It should be noted that because of using cutting filters the high energy side is suppressed; however, the real shape of the spectra will be discussed later. Here we should like to emphasize that the energy-distribution of the PL spectrum is very sensitive to the structure of the carbon samples, in this particular case the structure is "characterized" solely by the E_{04} gap, which decreases with increasing sp^2 cordinated carbon content, being 3.54, 3.35 and 2.5 eV as is given in Fig. 1. Scaling with the E_{04} gap from these three spectra cannot be determined, even though the peak energy changes simultaneously with the gap value. All of these spectra are Stokes shifted and can be characterised by structure-less band-shape and with a large half-width.

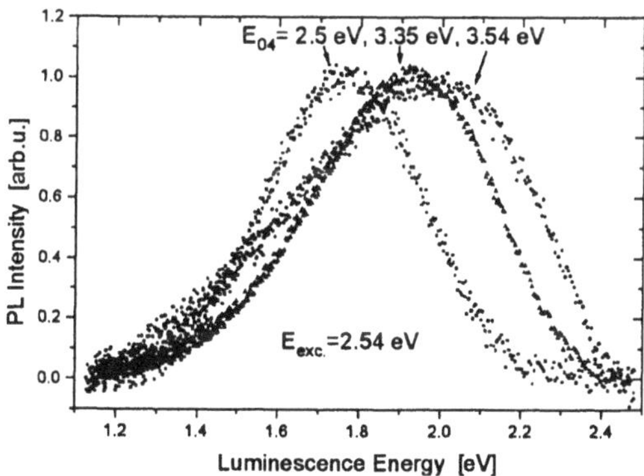

Fig. 1 Room temperature PL spectra of three different a-C:H films excited by 2.54 eV.

The close correlation between the PL band properties and the structure of amorphous carbon can be demonstrated on the same sample if we measure the PL properties of a freshly deposited sample and a relaxed one. For these two states of the same sample, the PL spectra are shown in Fig. 2, where "relaxed" denotes the sample stored at ambient temperature for two months. A red shift of PL position by 40 meV can be determined from normalized spectra, although the PL intensity enhancement on lower energy side of the PL spectra is more pronounced than the change at higher energies. Besides the spectral change it was a general observation that the integral PL intensity increases with storage time.

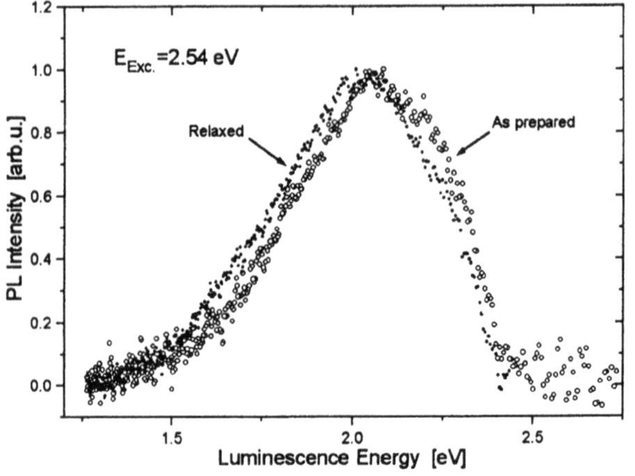

Fig. 2 Room temperature PL spectra for as prepared and relaxed a-C:H samples excited by 2.54 eV.

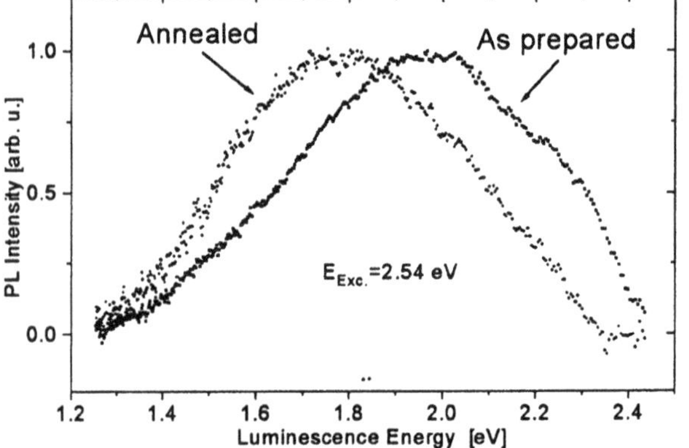

Fig. 3 Room temperature PL spectra excited by 2.54 eV for a-C:H sample before and after annealing.

By annealing the sample the structural relaxation can be speeded up. As was mentioned earlier, our samples contain 45-50 at % hydrogen, which begins to effuse from the amorphous carbon at ~350 °C; this results in a significant change of the structure. In order to avoid effusion of hydrogen the thermal treatment of samples was performed at moderate temperature (200-300 °C) for a longer time. Change in the energy distribution of PL spectra and decrease of PL intensity by one order of magnitude or more have been observed as a consequence of annealing [30]. In Fig. 3, the PL spectra measured before and after annealing are compared in a normalized shape. Red shift of peak position and enhancement on low-energy side of the spectra can be seen in the annealed sample. By means of transmittance measurements the shift of the absorption edge to lower energies was observed -- which is expected from the darkening of the annealed samples [30]. The effect of thermal treatment was also studied by other researchers but at higher temperatures [21]. The results obtained were comparable with ours; PL intensity changes were mentioned, but their PL spectra consists of two bands, which are not comparable to ours. Here we detail the results observed on a-C:H samples deposited under the same conditions, however many samples exhibit a similar tendency of change after heat treatment, apart from the two samples deposited at higher self bias. In the latter case the PL intensity was enhanced after heat treatment, but these carbon layers can be characterised by very low PL efficiency before annealing; which is not expected from scaling with self-bias voltage. Here we deal with the results from the viewpoint of correlation between luminescence spectra and the structure of amorphous carbon, although changes caused by thermal treatment are important as a mean of understanding non-radiative transitions too. This will be discussed later.

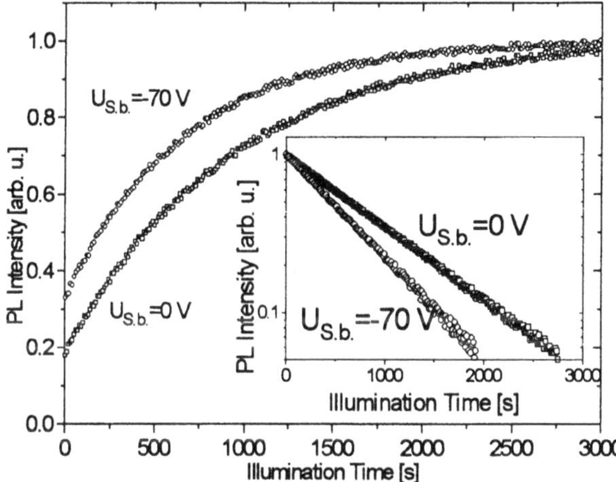

Fig. 4. Time dependence of PL peak intensity under illumination by unfocused laser beam of 2.54 eV photon energy and 9 mW/mm^2 intensity for a-C:H samples deposited at U_{SB} = -70 V (a) and 0 V (b) and annealed at 250 °C for 4 h. Inset shows the exponential character of the process.

Illumination of a sample with the purpose of exciting luminescence or measuring certain other properties can also bring about structural re-arrangement parallel with modification of PL properties [9, 31-33]. In our samples the so called fatigue effect extends to a small percentage (5-10 %) but a more pronounced reverse change i.e. that PL intensity increase under illumination have been detected in heat treated samples [30]. Figure 4 shows the time dependence of PL intensity increase for samples deposited at $U_{S.b.}$ = -70 V and 0 V and annealed at 250 °C for 4 hours. The PL intensity saturates by a rate depending on the structure of a-C:H, after a longer time no further increase can be observed. The experimental points depicted in semi-logarithmic scale (insert of Fig. 4) can very well be fitted by straight lines, suggesting mono-molecular kinetics for increasing PL intensity under illumination. In order to cheque the real kinetics, it is important to measure the excitation intensity dependence of the process. The results are shown in Fig. 5. The time dependence of PL intensity increase is exponential at every excitation intensity, however a non-linear decrease of the time constant with illumination intensity by a rate depending on the structure of a-C:H was found (see Fig. 5). A characteristic feature of all samples studied is that the saturation level of PL intensity reaches the value measured before the heat treatment within the experimental error. As a means of understanding this behaviour, very important observation was the blue shift of PL spectra in the laser treated states. Figure 6 shows PL spectra of as prepared and annealed samples and after saturation by laser illumination. PL spectra of original and laser treated carbon samples are very similar, (disregarding a small difference between the peak positions) giving the impression that the original structural arrangement was restored by laser illumination, which is further confirmed by the transmittance increase or photo-bleaching observed at the end of the laser treatment.

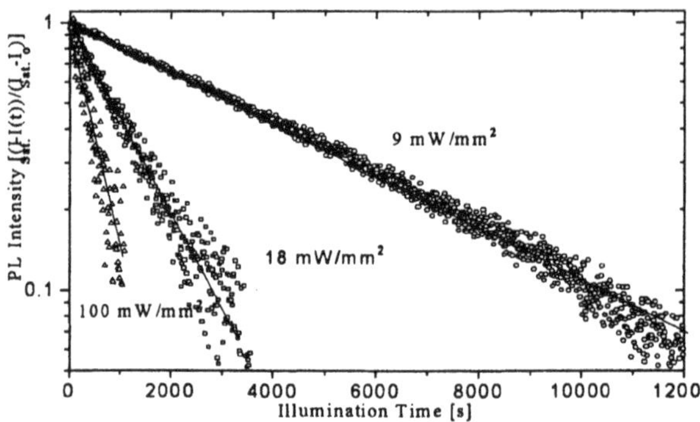

Fig. 5 Excitation power density dependence of PL peak intensity versus illumination time of a-C:H film deposited by - 70 V bias and heat treated at 280°C for 4 h..

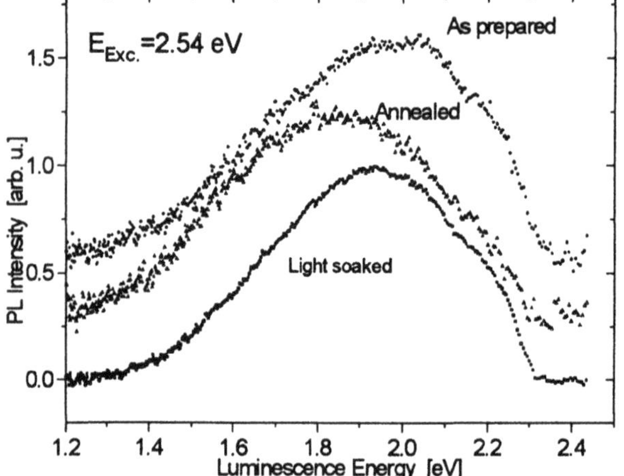

Fig. 6 PL spectra of a-C:H sample deposited at U_{SB} = - 70 V in as prepared state (a), following annealing at 280 °C for 4 h (b), and saturated by laser illumination following annealing (c). The spectra were shifted vertically for the sake of clarity.

A previous line of reasoning was that to display experimental results which prove the correlation between PL properties and the structure of amorphous carbon. This correlation helps if one links both the radiative and non-radiative transitions to some kind of structural configuration. Next we discuss experiments for polarization memory and excitation energy dependence of photoluminescence in order to understand the PL mechanism in amorphous carbon.

2. 2. POLARIZATION MEMORY OF PHOTOLUMINESCENCE

As outlined in the introduction, various authors have suggested that electrons localised or confined on π bonded clusters recombine radiatively, which results in light emission of amorphous carbon. This localization was thought to be a consequence of the amorphous carbon structure, which gives a high barrier for electrons to move between sp^2 bonded clusters embedded into sp^3 bonded matrix. Indeed, carriers recombine radiatively are more localised than migth be expected from this structural argument. Experiments using linearly polarized light to excite luminescence represent a powerful method of verifying localization of photo-excited carriers and offer a good means of studying the extent of localization and its dependence on different parameters. First we introduce our experimental results and then discuss how to relate the polarization memory of PL to the dynamical properties of non-equilibrium carriers.

The experimental set-up in polarization measurements was similar to that used for steady state experiments and the polarization behaviour of PL was analysed by a polariser (Carl Zeiss Bernotar M49) both parallel and perpendicular to the polarization vector of the exciting light. Special care was taken to check the polarization properties of the monochromator (more details about samples and the measuring system can be found in Ref. 22.).

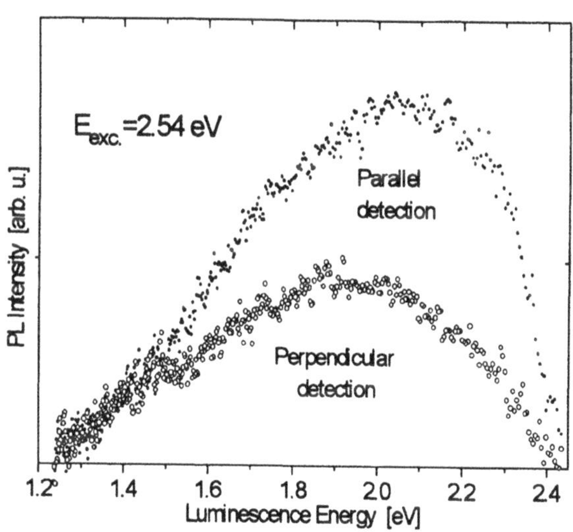

Fig. 7 Polarized PL spectra parallel with and perpendicular to the exciting laser beam polarization at room temperature . E_{ex} = 2.54 eV.

The Stokes shifted PL spectra measured parallel ($PL_{\parallel}$) to the exciting light polarization have a greater intensity through the whole band than PL spectra measured perpendicular to the exciting light polarization ($PL_{\perp}$). Figure 7 displays the polarised luminescence spectra for a-C:H layer deposited at $U_{S,b.}$ = 0 V, at 8 Pa pressure. From Fig. 7 it is obvious that photoluminescence polarization depends on emission energy. For its characterisation we introduce the degree of luminescence polarization (or, as it is also frequently called, the polarization memory of luminescence) defined by the $(I_{\parallel}-I_{\perp})/(I_{\parallel}+I_{\perp})$ ratio, where $I_{\parallel}$ and $I_{\perp}$ respectively mean the luminescence intensity measured parallel with and perpendicular to the exciting light polarization [22]. The emission energy dependence of polarization degree calculated from polarised PL spectra is shown in Fig. 8. At highest PL energy the polarization memory approaches the value of 0.4, and it gradually decreases with decreasing emission energy. Similar behaviour was observed at 80 K too, (see Fig. 9), although the degree of polarization increased by cooling.

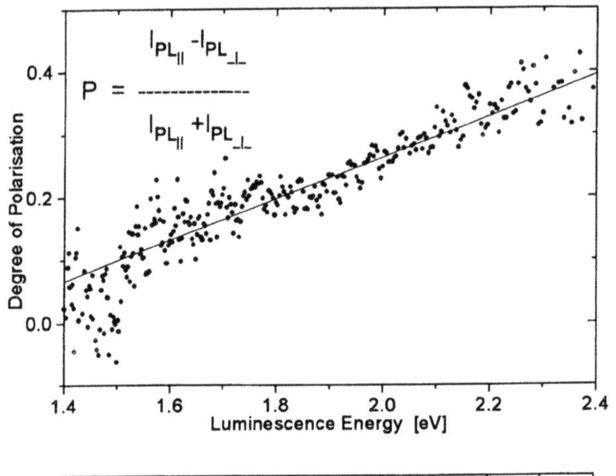

Fig. 8 Emission energy dependence of linear polarization degree calculated from polarized PL spectra at room temperature, $E_{ex} = 2.54$ eV. The line is a guide for the eye.

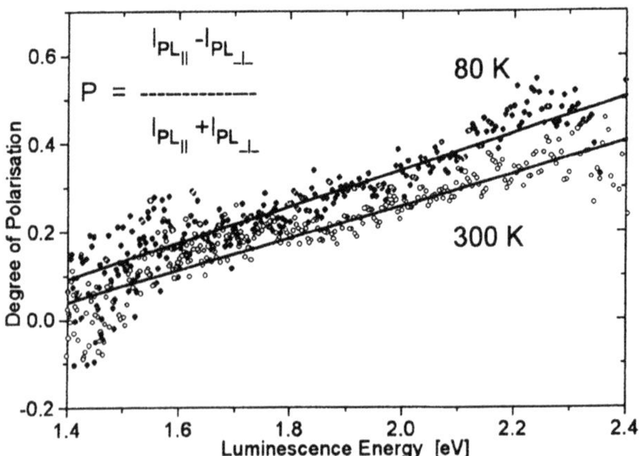

Fig. 9. Emission energy dependence of linear polarization degree at 80 K and at room temperature. Experimental points were calculated from polarized spectra measured at a given temperature. $E_{ex} = 2.54$ eV.

Let us first consider the theoretical value of luminescence polarization. By using electrical dipole for modelling the luminescent centre we can assume that the same dipole absorbs the exciting photon and emits the luminescence photon. The theoretical upper limit for polarization P is 1/2, if linearly polarised light was used for excitation [34]. In that case when luminescence can be described by two oscillators -- one for absorption the other for emission -- which are at an angle of ξ to each other, then the polarization of luminescence propagating perpendicular to the propagation direction of the exciting light will be [35]:

$$P = \frac{2 - 3\sin^2\xi}{4 - \sin^2\xi} \tag{1}$$

Polarization significantly decreases from 1/2 if the absorbing dipole changes its direction before emitting light. The luminescence polarization of amorphous carbon was found to be ~0.4 at room temperature and even larger at 80 K, just below the theoretical limit. This high degree of polarization is thought to be a consequence of strong localization of electron-hole pairs at where they were generated, on the same carbon atom. Decreasing polarization through spectra is caused by increasing the change of dipole direction of light emitting electron-hole pairs in relation to the absorbing dipole direction. This means that electron moves away from its sibling hole, which brings about change of dipole orientation. Bearing in mind the excess energy, which should be dissipated by the electron-hole pair before photon emission increases thorough the PL spectra from the high energy side to the lower one, a substantial decrease in polarization should be expected. Taking into account a change of dipole orientation $\xi = 50°$ in formula (1) we can calculate a drop in the degree of polarization to 0.07. The measured 0.4 value of polarization allows a mean fluctuation of this dipole angle of 23°. In other words we can consider the PL spectra to monitor the localization of photo-generated carriers.

In our opinion the luminescence originates from excited states of localised π electrons. The emission energy dependence of the PL polarisation suggests the existence of at least two type of non-radiative transition of photoexcited carriers in a-C:H. The smallest degree of polarisation on the low energy side of the PL spectra suggests an effective non-radiative relaxation of photogenerated carriers via usual delocalisation through π electron interactions. The high degree of polarisation at larger emission energies proves strong localization, when the previously discussed non-radiative mechanism should be ineffective, and certain other non-radiative channel should be activated. The polarisation value being near to the theoretical upper limit supports that the interaction of electron states should play a minor role in this non-radiative relaxation. Vibrational interactions might play an important role in this process.

What is the structural origin of this strong localisation? In our opinion large potential fluctuations due to the structure of amorphous carbon is one reason for spatial confinement of π electrons. The other reason, not yet taken into account, is the different symmetry of the σ and π bonds. σ bonds are formed by sp hybridized orbitals of sp^2 sites while π bonds are formed by p orbitals which extent above and below the plane of σ bonds perpendicular to it, what gives a zero π electron density in this plane. No overlap between electron wavefunctions of σ and π bonds of sp^2 sites. As a consequence of this localization, electrons will not move away from their sibling hole even when excitation energy is high enough to overcome the potential barrier due to the sp^3 hybridized neighbourhood. This localization offers a good explanation why the degree of polarization is also preserved at higher excitation energy, -- as was very recently observed [27], when electron-hole pairs must dissipate more excess energy because the larger Stokes shift. This localisation explains the light emission at energies larger than band gap itself.

2.3. EXCITATION AND EMISSION IN A BROAD ENERGY RANGE

Excitation energy dependence of spectral shape and peak position [9, 10, 26, 28], and PL efficiency [9, 26, 28] have been studied at different laser wavelengths. In order to obtain more exact dependence the excitation wavelength was changed in the 200 - 500 nm (6.19 - 2.47 eV) range by 5 nm steps and the PL was measured from 200 nm up to 650 nm. The measurements were performed on a Perkin Elmer LS 50 B luminescence spectrometer in which the light source is a xenon flash lamp with 10 μs pulse length, and the PL was detected by a photomultiplier with modified S5 cathode. The measured luminescence matrix was corrected for excitation intensity function. Because of small excitation intensity at different wavelengths a long time integration was needed and special care was taken to analyse spurious light scattering.

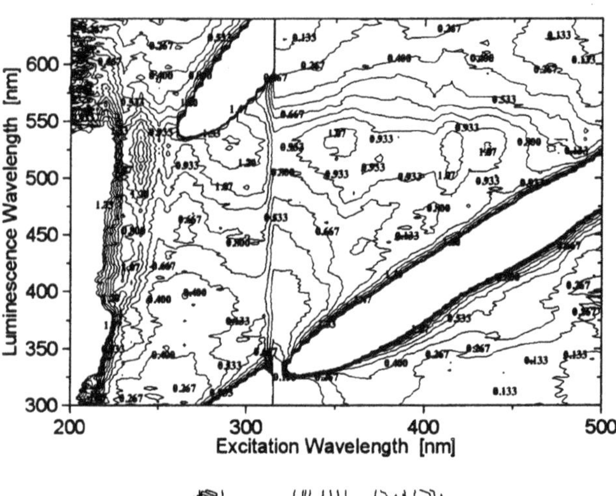

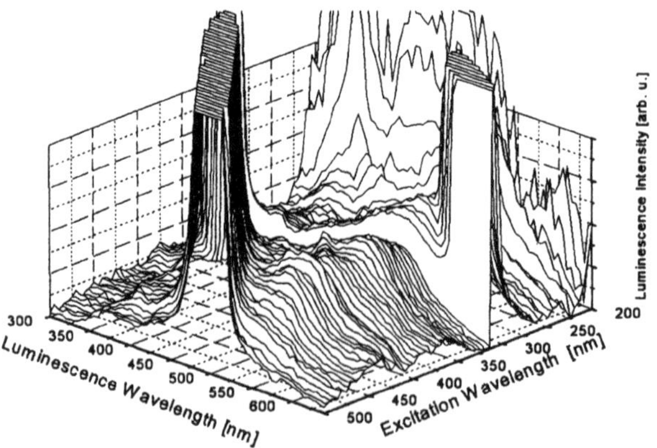

Fig. 10. Excitation energy dependence of PL spectra in contour-plot (upper) and axonometric (lower) representation for a well relaxed a-C:H sample deposited at $U_{s.b.}$ = 0 V by decomposition of methane.

In Fig. 10 we can see the excitation energy dependence (in the range of 6.19 -- 2.47eV) of PL spectra in contour plot and axonometric representations of a well relaxed a-C:H sample prepared by methane decomposition at $U_{S.b.}$ = 0 V. We analyse the results from 250 nm (4.96 eV) excitation because at smaller wavelengths the scattered light disturbs the observation of luminescence -- as can be seen very well on the contour-plot representation. In this excitation range (200-250 nm) further investigations are needed, although the PL band similar to that observed at lower energy excitation can also be effectively excited by 5.64 eV. In order to demonstrate the energy distribution of PL spectra they are shown in Fig. 11 at different excitation energies. The characteristic features of PL spectra excited by different energies are as follows:

i) very broad energy distribution of emitted light (1.9 - 3.8 eV) showing similarly asymmetric shape for different excitation photon energies;

ii) as the excitation energy drops in the PL band anti-Stokes luminescence can be observed;

iii) band narrowing by decreasing excitation energy is smaller than was observed by laser excitation [20,28].

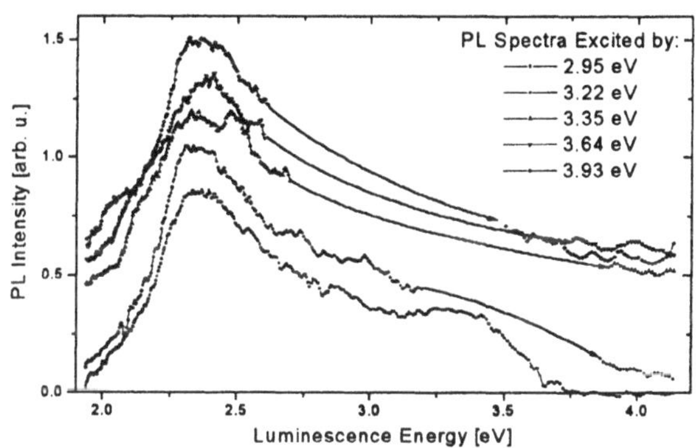

Fig. 11. Photoluminescence spectra at different excitation energies for a well relaxed a-C:H sample deposited at $U_{S.b.}$ = 0 V by decomposition of CH_4. Spectra were shifted vertically for the shake of clarity.

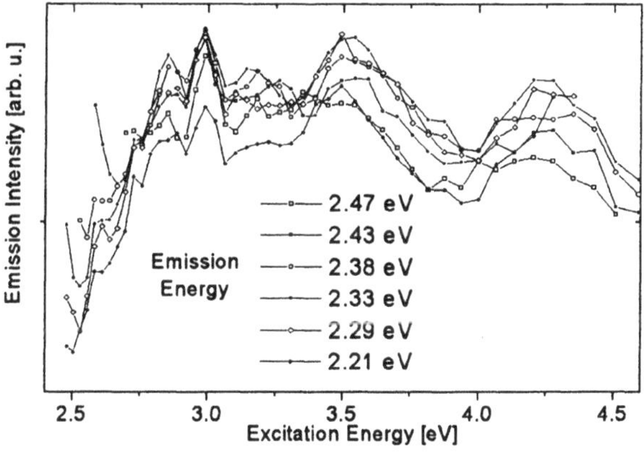

Fig. 12. Excitation spectra measured at luminescence energies near to PL maximum of a well relaxed a-C:H sample deposited at $U_{S.b.}$ = 0 V by decomposition of CH_4.

Excitation spectra for different emission energies can be determined from the excitation-emission matrix. Excitation spectra for emitted energies near to the PL band maximum (Fig. 12) exhibit almost the same efficiency at energies higher than 2.8 eV, and gradually decrease with decreasing excitation energy. From the excitation-luminescence matrix (Fig. 10) we can see that for other PL energies excitation spectra show similar behaviour.

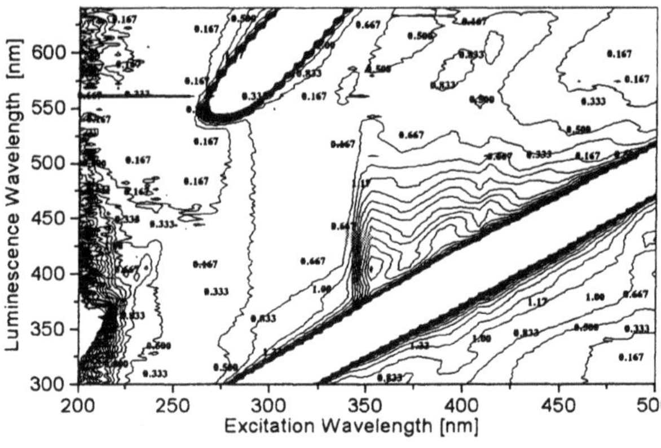

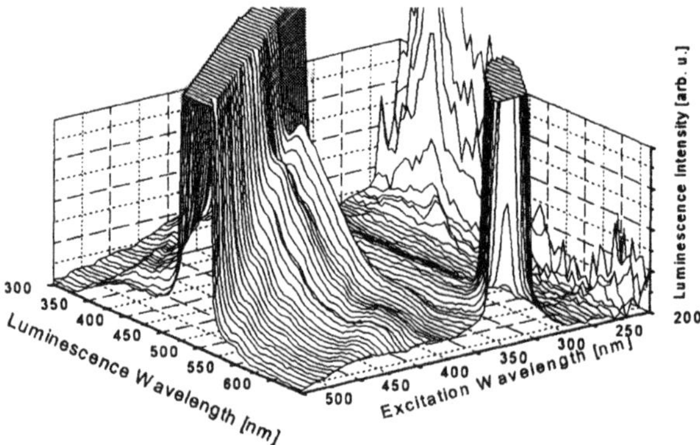

Fig. 13 Excitation energy dependence of PL spectra in contour-plot (upper) and axonometric (lower) representation for freshly deposited a-C:H sample by decomposing benzene (C_6H_6) at $U_{S.b.} = 0$ V.

Figure 13 shows the excitation energy dependence of the PL band measured in a freshly deposited a-C:H sample by decomposing benzene at $U_{S.b.} = 0$V. The emission properties are different from those in the sample prepared from methane.Luminescence can be observed at smaller excitation energies only (<3.65 eV), and the luminescence band peaked at ~3.1 eV at $E_{exc.}$=3.59 eV, which is blue shifted by 0.76 eV compared to the PL spectra of the sample prepared from methane. In Fig. 14 PL spectra at different excitation energies are shown. The characteristic features are as follows:

i) broad energy distribution of emitted light;

ii) a signature of spectral inhomogeneity is represented by the peak position change and narrowing of the PL band as the excitation energy falls in the emission energy range.

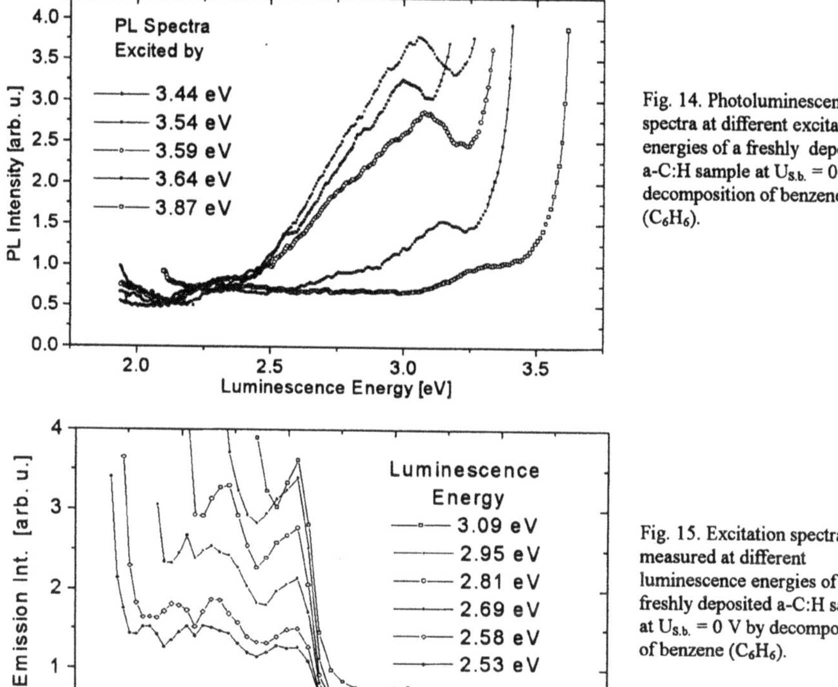

Fig. 14. Photoluminescence spectra at different excitation energies of a freshly deposited a-C:H sample at $U_{s.b.} = 0$ V, by decomposition of benzene (C_6H_6).

Fig. 15. Excitation spectra measured at different luminescence energies of a freshly deposited a-C:H sample at $U_{s.b.} = 0$ V by decomposition of benzene (C_6H_6).

Similar behaviour was observed for the green PL band by laser excitation [20,28]. It should be noted that this green PL band was weak in our sample, and could be excited more effectively by laser excitation. The excitation spectra in a broad range of PL energies are shown in Fig. 15. This PL band can be excited from 3.64 eV only, where the PL intensity increases sharply.

This excitation energy dependence of PL band is in excellent agreement with the luminescence mechanism where the excited states of strongly localised π electrons are responsible for radiative transition. Localization is so strong that electrons do not diffuse away from their sibling hole; in other the way they should recombine non-radiatively. Localisation of carriers due to the structure, and symmetry differences between the σ and π bonds can explain the excitation energy dependence of PL even at higher energy than optical band gap and the PL efficiency of almost independent of excitation energy in a wide range. Localized electrons recombine radiatively, do not correlate with each other, emit luminescence photons from different lowest energy levels which characterize the sp^2 sites, where electrons are localized. A signature of this localization in the PL band is the inhomogeneously broadened luminescence band. The inhomogeneously broadened PL band also explains the non-Arrhenius temperature dependence of PL intensity, while strong localization of carriers causes very weak temperature dependence of PL intensity and spectra as well.

From the excitation energy dependence of PL in a-C:H we have a very important conclusion concerning non-radiative recombination. If PL is excited by 5 eV photon energy, then near 3 eV it must be dissipated non-radiatively in order to emit a photon with 2 eV, and under this dissipation process the electron does not diffuse away from its sibling hole. This proves that very effective non-radiative mechanism works at sp^2 sites to dissipate excess energy. It is for this reason that we consider that electrons do not need to tunnel away from their excitation site to a paramagnetic centre to recombine non-radiatively. The other mechanism for non-radiative transition is delocalization of photoexcited π electrons interacting with other π electrons. Weak luminescence in small gap a-C:H material can be explained by this delocalization. Our results for PL efficiency decrease under thermal treatement corroborate this non-radiative mechanism.

The recombination rate of non-equilibrium carriers in the case of competing radiative and non-radiative recombination processes can be described by

$$\tau^{-1} = \tau_r^{-1} + \tau_{nr}^{-1} = p_r + p_{nr} \qquad (2)$$

where τ denotes decay time, p the probability, and indexes r and nr respectively present radiative and non- radiative recombination. The PL efficiency η given by

$$\eta = \frac{p_r}{p_r + p_{nr}} \qquad (3)$$

shows a strong relationship between efficiency and the probability of both radiative and non-radiative relaxation. In a-C:H, our view is that non-radiative recombination dominates the decay process at room temperature and results in very fast decay. In Fig. 16 PL decay curves are shown excited by laser pulses of 3.68 eV photon energy and measured by 200 ps gate length. This result gives an upper limit of 1.5 ns for the decay time of PL, which is smaller than previously published data [9].

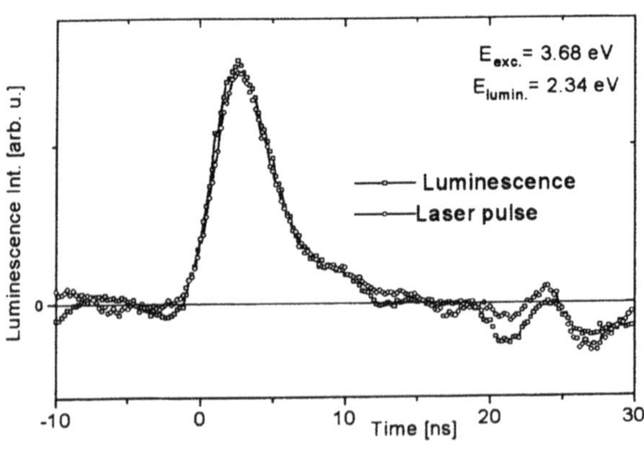

Fig. 16. Photoluminescence decay signal measured at PL peak energy and $E_{exc.}$ =3.68 eV for a-C:H films prepared from methane (a) and excitation pulse (b).

3. Conclusions

A variety of photoluminescence properties of amorphous carbon have been discussed in this paper. The results of our reasoning are as follows:

i). Luminescence originates from excited states of π electrons, which are strongly localised on sp^2 bond sites of amorphous carbon structure. Because of localization electrons do not diffuse away from their sibling hole before light emission.

ii). A high degree of luminescence polarization proves the strong localization of light emitting electron-hole pairs.

iii). Large potential fluctuation, due to the amorphous carbon structure, was suggested as a reason of carrier localization. In our opinion there is another important source of carrier localization: viz. the different symmetry of light emitting π states and the closely located σ states. This localization can explain the effective luminescence at energies higher than the optical gap.

iv). Spectral inhomogeneity is a manifestation of charge carrier localization.

v). Two different non-radiative recombination mechanisms are suggested. On one side, when the interaction between π electrons is strong enough to delocalize π electrons, then they are able to recombine non-radiatively. On the other side radiationless transition can be realised without delocalization of photoexcited carriers, via vibrational interaction with the neighbourhood, similarly to the energy dissipation prior to the radiative transition. This recombination results in a very fast decay process -- as it was proven by the decay curve.

Acknowledgements

This work was supported by the Hungarian Science Foundation under contract number OTKA-T-017371. The authors are grateful for the help of A. Buzádi, J. Erostyák and L. Kozma for excitation measurements, and for their contribution.

REFERENCES

1. Catherine, Y. (1991) Preparation Techniques for Diamond-like Carbon , in R.E. Clausing, L.L. Horton, J.C. Angus and P. Koidl (eds.), *Diamond and Diamond-like Films and Coatings,* by Plenum Press, New York, pp193-229.
2. Matthews, A. and Eskildsen, S.S. (1994) Engeneering application for diamond-like carbon, *Diamond Relat. Mat.* **3,** pp 902-911.
3. Lettington, A.H. (1991) Application of Diamond-like (Hard Carbon) Films, in: R.E. Clausing, L.L. Horton, J.C. Angus and P.Koidl, (eds.) *Diamond and Diamond-like Films and Coatings,* Plenum Press, London, pp. 481-499.
4. Lin, S.H. and Feldman, B.J.(1983) Amorphous hydrogenatad carbon from the plasma deposition of C_2H_2, C_2H_4 or CH_4. *Phil. Mag.B* **47,** 113-116.
5. Watanabe, I., Hasegawa, S., and Kurata, Y. (1982) Photoluminescence of hydrogenated amorphous carbon films, Jap. *J. Appl. Phys.***21,** 856-859.
6. Watanabe, I. and Inoue, M. (1983) Photoluminescence in amorphous C:H films prepared by glow discharge decomposition of CH_4 or C_2H_6, Jap. *J. Appl. Phys.***22,** L176-178.
7. Wagner, J. and Lautenschlager, P. (1986) Hard amorphous carbon studied by ellipsometry and photoluminescence, *J. Appl. Phys.* **59,** 2044-2047.

8. Nonomura, S., Hattori, S.,and Nitta, S. (1987) Photoluminescence of quasi-one dimensonal amorphous Si_{1-x} C_x: H, *Solid State Commun.* **64,** 1261-1264.
9. Vassilyev, V.A. , Volkov, A.S., Musabekov, E., and Terukov, E.I. (1988) Photoluminescence properties of hydrogenated amorphous carbon films, *Pis'ma v JTF.* **18,** 1675-1679.
10. Fang, R., Song, Y., Yang, M., Jiang, W., and Yan, K. (1988) Photoluminescence properties of diamond-like amorphous carbon films, *J. of Luminescence* **40 & 41,** 905-906.
11. González-Hernández, J., Asomoza, R., and Reyes-Mena, A. (1988) Spin defects and recombination in hydrogenated amorphous carbon films, *Solid State Commun.* **97,** 1085-1088.
12. Kim, S.B. and Wager, J.F. (1988) Electroluminescence in diamond-like carbon films, *Appl. Phys. Lett.* **53,** 1880-1881.
13. Yoshimi, M., Shimizu, H., Hattori, K., Okamoto, H., and Hamakawa, Y. (1992) *Optoelectronics* **7,** 69.
14. Hamakawa, Y., Kruangam, D., Deguchi, M., Hattori, Y., Toyama, T., and Okamoto, H. (1988) Highly conductive p-type microcrystalline SiC:H prepared by ECR plasma CVD *Appl. Surf. Sci.* **33&34,** 1276-1284.
15. Hamakawa, Y., Toyama, T., and Okamoto, H. (1989) Blue light emission from a-C:H thin film electroluminescence structure cell, *J. Non-Cryst. Solids* **115,** 180-182.
16. Robertson, J. and O'Reilly, E.P. (1987) Electronic and atomic structure of amorphous carbon, *Phys. Rev.B* **35,** 2946-2957.
17. Robertson, J.(1986) Amorphous carbon, *Adv. Phys.* **35,** 317-374.
18. Frauenheim, Th., Jungnickel, G., Kohler, Th., and Stephan, U. (1995) Structure and electronic properties of amorphous carbon from semimetallic to insulating behaviour, *J. Non-Cryst. Solids* **182,** 186-197.
19. Walters, J. K. and Newport, R.J. (1995) The atomic-scale structure of amorphous hydrogenated carbon, *J. Phys. C: Condens. Matter.* **7,** 1755-69.
20. Chernishov, S.V., Terukov, E.I., Vassilyev, V.A., and Volkov, A.S. (1991) Radiative recombination in a-$Si_{1-x}C_x$:H, *J. Non-Cryst. Solids* **134,** 218-225.
21. Bounouh, Y.,Théye, M.L., Dehbi-Alaoui,A., Matthews, A., Cernogora,J., Fave, J.L., Colliex,C., Gheorghiu, A. and Sénémaud, C. (1993) Influence of π-bonded clusters on the electronic properties of diamond-like carbon films, *Diamond and Related Mater.* **2,** 259-265.
22. Koós, M., Pócsik, I.,and Tóth, L. (1993) Polarization memory of photoluminescence in a-C:H film. *J. Non-Cryst. Solids* **164,** 1151-1154.
23. Demichelis, F., Schreiter, S., and Tagliaferro, A. (1995) Photoluminescence in a-C:H film *Phys. Rev B.* **51,** 2143-2147.
24. Robertson, J.(1996) Recombination and photoluminescence mechanism in hydrogenated amorphous carbon, *Phys. Rev.B* **53,** 16302-16305.
25. Reyes-Mena, A., Asomoza, R., Gonzalez-Hernandez,J., and Chao, S.S. (1989) Influence of the structure on the recombination process in a-C:H films, *J. Non-Cryst. Solids* **114,** 310-312.
26. Xu, S., Hundhausen, M., Ristein, J., Yan, B., and Ley, L. (1993) Influence of substrate bias on the properties of a-C:H films prepared by plasma CVD, *J. Non-Cryst. Solids* **164&166,** 1127-1130.
27. Rusli, Amaratunga, G.A.J., and Robertson, J. (1996) Polarization memeory of photoluminescence in amorphous carbon, *Phys. Rev.B* **53,** 16306-16309.
28. Rusli, Robertson, J.,and Amaratunga,G.A.J. (1996) Photoluminescence behaviour of hydrogeneted amorphous carbon, *J. Appl. Phys.* **80,** 2998-3003.
29. Stief, R. Schafer, J., Ristein, J., Ley, L., and Beyer, W. (1996) Hydrogen bonding analysys in amorphous hydrogeneted carbon by a combination of infrared absorbtion and thermal effusion experiments, *J. Non-Cryst. Solids* **198-200,** 636-640.
30. Koós, M., Pócsik, I., and Tóth, L. (1994) Luminescence efficiency enhancement in laser soaked hydrogeneted amorphous carbon *Appl. Phys. Letter* **65,** 2245-2247.
31. Glesener, J.W., Anthony, J.M., and Cunningham, A. Photoluminescence investigation of a-C:H films, (1993) *Diamond andRelated Mat.* **2,** 670-672.
32. Koropeczki, R.R., Tessler,L.R., Sanjurjo,J.A., and Alvarez, F. (1991) Photoinduced effect in diamond-like hydrogenated amorphous carbon films, *J. Non-Cryst. Solids* **137-138,** 835-838.
33. Schafer,J., Ristein, J., and Ley, L. (1994) Electronic density of states and deep defects of hydrogeneted amorphous carbon (a-C:H) *Diamond and Related Mater.* **3,** 861-864.
34. Stepanov, B.I. and Gribkovskij, V.P.(1971) Theory of Luminescence, Gordon and Breach, London.
35. Levsin, V.L. (1951) Photoluminescence of Liquids and Solids, GITTL, Moskva.

OPTICAL ELEMENTS FOR SENSING AND COMMUNICATION
A Review

T. NECSOIU and C. E. A. GRIGORESCU
Institute of Optoelectronics
76 900 Bucharest, P.O. Box MG-22, Romania

Abstract

This paper reviews some results obtained in the field of amorphous semiconductor devices, used as elements for sensing and communication. The technologies and performances are compared with those involved in the preparation of the corresponding devices based on crystalline materials.

1. Introduction

As a consequence of searching for suitable non-linear materials, the study of optical nonlinearity in amorphous semiconductors has taken into evidence some key features of these materials, which should improve the state - of - the - art in photo- and optoelectronic devices. Thus, sensing and communication technology could find some practical interest in optical switching and signal processing by means of amorphous semiconductor structures.

In comparison with crystalline semiconductor thin films, the amorphous ones exhibit some advantages with respect to making large area and low cost devices. For instance, more than twenty years photovoltaic modules and solar panels based on single-crystalline silicon dominated in most communication and military satellites [1], but because of material cost and complicated processes involved, it is obvious that crystalline silicon technology is expensive enough. Therefore the research has been directed towards some alternative technologies [1, 2], which could bring down the cost. A good evidence would be the need for much thinner amorphous layers, since light is more easily absorbed in a less rigidly ordered atomic structure.

Our work is a review of some results obtained in the field of amorphous semiconductors and their applications in sensing and communication.

The major advantages of (a-Si:H) devices are the wide range of band gap tuning, from 1.5 to 4.5 eV. Optical elements based on either polycrystalline or hydrogenated amorphous silicon, for LCD applications, are compared from the view point of process architecture and device performance.

Since for many applications spatial light modulators (SLM) must strictly meet the requirements of fast frame rates, high contrast ratio and resolution, the results of

A. Andriesh and M. Bertolotti (eds.),
Physics and Applications of Non-Crystalline Semiconductors in Optoelectronics, 379–389.

using N-I-P-I-N a-Si:H structures in bistable optically addressed SLM are underlined in comparison with similar devices based on crystalline AlGaAs/GaAs structures.

A special attention is paid to applications in nuclearmedicine (NM) and related fields. That is why we tried to find a suitable way to open a discussion with respect to X-ray detectors made either of amorphous or crystalline semiconductors.

Even if it could seem surprising, the problem of materials preparation procedures and of their consequence on some material parameters is let at the end of our paper. There is a strong reason in this respect: the technologies used for developing devices involve both thin layers and substrates. This should be kept in mind all the time when discussing about structures, devices, etc. So, will bulk amorphous semiconductor compounds replace the old crystalline ones? Some answers will arise at the end of this paper.

2. Liquid crystal display (LCD) applications

2.1 OPERATION OF LCD

It is to underline that these displays are used in a wide range of applications, for instance in computers and communications.

Active matrix liquid crystal displays (AMLCD) incorporate thin film transistors (TFTs) of either amorphous or polycrystalline silicon, which switch each pixel in LCD.

In brief, the display operation takes place like it follows [2]:

(i) the TFT addresses each pixel element of the LCD;

(ii) the applied data voltage changes the polarisation and the optical transparency of the LC cell;

(iii) the TFT changes the pixel LC capacitance and the change is stored for the whole frame time on the LCD.

A good control of the pixel voltage is needed to obtain a large grey scale range. With the aim of achieving large grey scale and high resolution displays, on-resistance of the TFT and metal data line resistance must be low, and at a time the data and gate voltages must be high. In order to reduce the leakage of charge existing through the TFT when it is switched off it is obvious that the off current of the component should be as low as possible.

2.2 COMPARISON BETWEEN AMORPHOUS SILICON AND POLYCRYSTALLINE SILICON DEVICE PROCESS

First, polycrystalline silicon (poly-Si) thin layers were obtained at 950°C, the only available substrate being the very expensive quartz. The necessity in cheaper substrates led to a lower temperature process, such as PECVD and LPCVD. They develop at a temperature value ranging between 550 and 600°C [2, 3].

Usually, amorphous silicon (a-Si) layers are obtained at 350°C and thus large area glass substrates can be used.

A comparison between the process architectures for poly-Si and A-Si is provided by Thompson [2]. From his work it results that for the poly-Si TFT, obtained by LPCVD at 570°C lower price substrates are available than in case of the devices obtained at 950°C, but the process is not scaled at large area; as regards the a-Si TFT, the process is not only suitable for large area glass substrates, but also enough mature.

The current - voltage characteristics for the TFT obtained from a-Si and poly-Si are presented in the works [2, 4, 5]. It should be stressed on the high difference between the high temperature and respectively the low temperature technologies, that stays in the p - channel performance.

The electron mobility in the a-Si device is about 1 $cm^2/V.sec$, (lower by more than one magnitude than in poly-Si) so a considerably lower drive current appears than in poly-Si TFTs.

The device obtained by high temperature process has higher mobility, lower threshold and sharper subthreshold, due to the lower density of defects.

As it results from many works, for instance [2, 6, 7, 8], for poly-Si TFT the higher drive current improves the performance with integrated driver circuits, but a-Si technology remains more suitable for large area. At a time, the a-Si TFT performance should be improved by applying drive voltages 2 times higher than for poly-Si TFT.

Thus, the users will make a choice in connection with their main goals, taking into account the peculiarities of the involved technologies and, no doubt, the competition between device cost and performance.

The choice of the device design and process structure is a trade-off of the device capabilities, development throughput and possibility to incount large areas.

3. Spatial light modulators (SLM)

Spatial light modulators are components for photonic information processing systems. As active optical devices they modify phase, polarisation and intensity of an optical beam, as functions of space and time.

Since modulation control can be performed either by an electronic or an optical signal, SLM may provide an interface between these two types of signals.

The active pixels in SLM are usually arranged in one - dimensional or two - dimensional arrays. Thus, the device can be operated for free space optics and parallel processing.

Until 1990 liquid crystals, acousto - optic and magneto - optic devices based on garnet films were used to develop SLM arrays. These technologies offer a broad spectral bandwidth but have limited throughput [9].

The modulators based on semiconductor materials provide high throughput, but the wavelength range involved is quite narrow.

For most of their applications SLM must achieve fast frame rates, high contrast ratio and high resolution In individual modulators made of GaAs/AlGaAs, operated in transmission, reflection and Fabri-Perot resonator modes [10-15] the contrast ratio ranges between 2 and 6.

Contrast ratios higher than 100 were obtained in single element devices, in resonantly enhanced reflection mode. To extend this technique to large arrays a very precise control of the semiconductor layers in the device structure is needed.

An interesting SLM has been achieved as self - electro - optic effect device (SEED) [16], which actually is a pin photodiode with multiple quantum wells in the intrinsic region. Very impressive seems the symmetric device.

A symetric SEED [9] working in the reflection mode is an optical bistable device, containing two pin diodes connected in series.

In principle the operation consists in the exciton absorption at the operating wavelength, in the intrinsic quantum well region. If an external electric field is applied, absorption shifts towards longer wavelength; the intrinsic regions in the pin photodiodes switch from the absorption to the transmission mode.

S-SEED can be controlled by an external optical beam too. In this case the state of the device is set by simultaneously shining beams, of different intensities. An intensity ratio of 2:1 is enough for switching.
The devices have been developed as 32×64 arrays, with optical windows designed for beams of 5μm diameter. The components in the array are accessed independently and simultaneously. Thus they offer a potential of massive parallelism.

Read out of the array requires a high power laser source which operates in a narrow wavelength range. Moreover, contrast ratios higher than 10 are unlikely to be obtained for large arrays.

As an alternative solution for optically addressed (OA) SLM, N-I-P-I-N structures made of a-Si:H have been tested [17].

A bit of history is necessary in order to understand why an amorphous material was chosen. Since the OASLM are based on LC layers and LC are associated with a photoconductor, an amorphous silicon photoconductor was proposed to replace the crystalline one [17]. This proposition arose from some advantages of a-Si:H photosensors over the crystalline devices, i.e. good spatial resolution, high uniformity of the deposited thin layer and low temperature technology.

The idea went on with combining a bistable ferroelectric LC (FLC) with an amorphous silicon p-i-n structure. It is known that FLC and other LC need a zero average polarisation, and that is why the photosensor must have, alternatively, positive and negative polarisation.

Thus we reach the requirement for a symmetrical structure, as in the case of S-SEED. The device which satisfies this requirement is a symmetrical n-i-p-i-n structure made of hydrogenated amorphous silicon, that optically addresses an OASLM based on a bistable FLC.

The results on a n-i-p-i-n structure with a high breakdown voltage, low leakage current and fast switching are presented in detail in the paper by Chavrier and co-workers [17]. A remarkable feature of the OASLM based on n-i-p-i-n structures stays in the p - layer, the thickness of which is a key parameter for the device performance.

4. Radiation detectors

During the last period, the performance of medical diagnostic by imaging techniques has noticed an amassing enhancement. It is obvious that the main man-made sources of radiation exposure for the population are ionising radiation methods (Rx) and nuclear medicine (NM) [18]. First of all, their indications are determined and led by the diagnosis performance. In principle, radiology is based on X-ray attenuation by the living human, considering a monoexponential decay of the monochromatic beam [19]. Starting from this point, one should mention that the future of these areas is governed, no doubt, by the diminution of the dosis absorbed by the patients.

If talking about sensors, radiology should be considered as built from two fields [19]:

(a) conventional radiology (CR);
(b) computerised tomography.

Conventional radiology registered a rapid development during the last decade: the X-ray films had been replaced by light intensifiers, read either by TV cameras or CCD systems. Digital imaging makes possible an improved diagnostic performance, as well as data transmission and storage. But for determining film exposition in CR, semiconductor sensors are seldom used instead of the conventional ionisation chambers.

By contrary, in XCT large arrays of detectors are required. Development of XCT techniques has been followed by a high improvement of image quality [20] through optimisation of data collection. Although, this happened at the cost of higher photon fluxes and data sampling, that determine the operation conditions of the detectors.

Actually the detector quality acts directly on the image quality. Thus, when comparing the main features of different X-ray sensors based on semiconductors, it follows that among silicon, mercury iodide and cadmium telluride detectors, the last ones would provide some theoretical advantages upon the others [21-25]. Their disadvantages stay in the economical reasons, because of the extremely difficult technology involved.

An alternative solution is proposed in the work [22], as 2D sensor arrays made of amorphous silicon.

Operation of the sensor in the storage mode requires very low leakage current and long charge retention. Both requirements are satisfied by operating the device at zero bias: under this condition the leakage current is suppressed and at a time the charge collection efficiency remains high enough (approximately 95% [22]) due to the built-in field of the diode.

When taking into account the data presented in [21] and [22] respectively, one would come to the natural conclusion that 2D a-Si arrays are, at least, less expensive. With respect to the energy domain and efficiency, CdTe detectors seems much more interesting.

Another material suitable to replace conventional semiconductor crystals in X-ray detection is a-Se:As [25]. The sensitivity of these receptors depends on the thickness of the amorphous layer and on the applied electric field.

The definition of the sensitivity [25] is given by the following equation:

$$S = \frac{V_0 - V_x}{V_0 R} \qquad (1)$$

where: $(V_0 - V_x)$ is the amount of discharge due to a given amount of radiation;
R is the signal per unit of radiation, expressed in roentgen;
V_0 represents the initial voltage to which the x-ray detector is corona charged;
V_x is the surface voltage resulted after the discharge under exposure to radiation.

The maximum sensitivity is obtained when all the electron - hole pairs, generated under radiation absorption, are involved in the discharge process. The value is specified for given radiation amount and sample thickness. For thin samples the sensitivity is controlled by absorption [25]. Thus the quality of the receptor depends strictly on the material characteristics.

As it is shown in [25], the electron and hole lifetimes depend not only on the impurities, but on the x-ray exposure too. Therefore, after repeated imaging the sensitivity decreases due to the schubweg effects.

Thus the study of the interdependence between the lengths of the carrier schubweg and the sensor thickness leads to the following conclusion: for longer schubwegs (for both types of carriers) in comparison with the thickness, a higher sensitivity should be registered.

Perhaps the future trends in NM and XCT will be directed to replace the conventional crystalline detectors by amorphous structures, wherever it is possible, unless the performance of the devices could be at least comparable.

5. Some considerations on $A^{III}B^{V}$ bulk semiconductor compounds

As we already mentioned in the introductory section, the discussion on bulk semiconductor compounds is let at the end of our review. All the devices we took into consideration the previous sections are actually structures made of thin layers deposited by different technologies on various substrates. If it is question of amorphous semiconductors, the low temperature of the processes involved requires glass substrates, available at lower price.

The main idea that follows consists in making a comparison between the possibilities of obtaining either crystalline or amorphous materials for substrates.

It was not too easily to make a proper choice, but finally we have found that $A^{III}B^{V}$ semiconductor compounds based on GaSb can support our goal.

From the view point of the device preparation, it is clear that the features of the substrate are influencing on the device performance. Here we have to stress on the uniformity of material properties all over the area of the substrate, which acts on the efficiency of the processing steps. Actually it does not matter if the substrate is glassy or crystalline; it is more important how uniformly are its physical and structural properties distributed in the bulk and respectively over its surface. Moreover, it is suitable to remind someone that the technology of preparing the bulk material is responsible for its final characteristically features.

Therefore we will describe the preparation methods and characterisation results for amorphous GaSb and polycrystalline $In_xGa_{1-x}Sb$.

As it is described in the paper by Demishev et al [26], a mixture of amorphous and crystalline GaSb, in bulk form, was obtained by combining the action of high pressure and high temperature, followed by rapid quenching of the melt, on initial GaSb single crystals.

The composition of the mixture as regards the two phases was checked out by X-ray and Raman scattering measurements. It was found that the compositional parameter "x", describing the amorphous phase, is proportional to the heat of crystallisation reduced to the unit mass of the samples.

From resistivity measurements, it resulted that by increasing "x" value important changes are observed in the resistivity dependence of temperature. Moreover, below 300K the resistivity value is time independent, but when increasing the temperature value resistivity becomes time dependent, first in the range 300 - 340K. At higher temperature resistivity decreases. The explanation lies in the different relaxation processes involved for the mixture of two phases.

From the view point of applications, it is sure that random switching is available with respect to the use of this mixture. Although, when looking for substrates, if it is need of bulk Ga compounds, like in the individual SLM device for instance, perhaps homogeneous materials are more desirable than phase mixtures.

It is extremely important that bulk amorphous $A^{III}B^{V}$ semiconductors have been obtained, even mixt with crystalline phase, but the question arises if the procedure itself can lead at least to a lower price if not to uniformity of physical and electrical properties.

That is why we chosen for comparison the results on the direct synthesis and rapid crystallisation of $In_xGa_{1-x}Sb$ (x=0.20) [27- 30].

First of all it is question of Ga based semiconductor ternary compound. Secondly - the process is someway similar to that used for a-GaSb preparation.

It is no need to insist on the details of the technique, since they are given in the papers by Manea and co-workers [27-30]. We just underline that the aim of their work was to obtain highly uniform substrates for opto - electronic devices.

In brief, $In_{0.2}Ga_{0.8}Sb$ polycrystalline ingots were obtained by direct synthesis from the elements, at a maximum temperature of 600°C. The rocking furnace used for preparing the compound was a home - made one, excepting for the EUROTHERM programmer type 125. The rapid crystallisation was achieved by deeping the ampoule, which contented the melt, in liquid nitrogen bath. Homogeneous and nonfragile cylindrical, polycrystalline ingots, 15 mm in diameter and 75 mm in length, were obtained. The composition uniformity along the ingot was checked by Neutron Activation Analysis (NAA). Samples from different regions, with respect to the length of the ingots, were irradiated at neutron flux of 3.4×10^{11} n/cm^2.sec, for five hours and the experiments were repeated within a period of three months. The resulted spectra showed that the composition parameter has a variation of 8% ranging from the top to the rare region of the ingots. A comparison was made between the results of NAA and those obtained by X-ray diffraction on the same samples, and a good agreement was found between NAA and Vegard law approximation.

To confirm the influence of composition uniformity on the optical, electrical and transport properties of the material, the optical absorption coefficient was determined as well as the temperature dependence of the resistivity and carrier mobility in samples taken from different regions of the polycrystalline ingots. The results are remarkable with respect to the uniformity of all these material parameters.

Therefore, taking into account the low temperature process and the very simple but accurate enough technique, we think that $In_xGa_{1-x}Sb$ obtained by direct synthesis and rapid crystallisation could provide suitable substrates not only for crystal based devices, but also for amorphous thin layers deposition. It clearly results that the cost of the product cannot rise too much, but it is no doubt that the performance should be quite satisfactory. Moreover, the technique can be extended to other $A^{III}B^{V}$ ternary compounds.

To close the discussion about crystalline and amorphous semiconductors, with respect to both materials and devices, it is need to emphasise that, from the view point of the intermediate user, i.e. that who develops devices, a careful choice should be made: sometimes the amorphous materials are more easily obtained than the crystals, but they cannot yet meet all the requirements with respect to the device performance. A low cost product might be of high interest only when its performance is comparable with that of the high price product.

6. Final remarks and conclusions

This work was aimed to underline the possibilities to apply noncrystalline semiconductors in optoelectronics. It is to remark that originality stays in the comparison between the capabilities of crystalline and noncrystalline materials with respect to their use in different structures, for sensors and communication.

A special attention was paid to the technology architecture and to the relationship between physical properties of the material and the performance of the corresponding device.

Thus, the conclusion follows that for devices which need materials characterised by high mobility of the charge carrier it is more suitable to use crystalline semiconductors.

With respect to the process architecture for devices development, it is clear that low temperature technologies are more desirable, taking into account the availability of lower cost substrates and scaling to large area.

Deposition of amorphous semiconductor thin layers seems a mature process, but the obtaining of homogeneous, bulk amorphous semiconductor compounds is still a problem.

In the last section we have compared two technologies for preparing gallium compounds, both amorphous and crystalline. It seems that to obtain a polycrystalline Ga based ternary compound is less difficult and less expensive than in the case of an amorphous binary one. The rapid crystallisation process requires only low temperature and direct synthesis.The main feature of polycrystalline $In_{0.2}\ Ga_{0.8}\ Sb$ is the uniformity of its properties. Therefore, we think that when required, it could be easily used as a substrate for amorphous thin layers based devices. Moreover, it is no doubt that a binary compound such as GaSb is even less complicate to be prepared, by direct synthesis and rapid crystalisation, than any ternary semiconductor compound.

Since the problem was treated from the view point of the user, the economical aspect had represented a key point of the discussion. Thus, as we have already temperature and direct sinthesis. The main feature of polycrystalline $In_{0.2}$ $Ga_{0.8}$ Sb in the uniformity of its properties. Therefore, we think that when required, it could by casily used as a substrate for amorphous thin layers based devices. Moreover, it is no doubt that a binary compound such as GaSb is even less complicate to be prepared, by direct sinthesis and rapid crysthalisation, than any ternary semiconductor compound.

Sinec the problem was treated from the view point of the user, the economical aspect had represented a key point of the discussion. Thus, as we have already mentioned in the previous sections, sometimes what is cheep does not always meet perfectly our interest, as it happens in the case of X-ray detectors.

As final conclusions, we would like to stress on some points in our opinion of high importance: it is no doubt that amorphous materials and the corresponding technologies and devices will meet a future extended development, but nobody will stop the development in the field of crystalline semiconductors. That is why, where the device performance will be comparable, amorphous materials will replace crystals, but it seems premature to think that this is a general trend.

7. References

1. Guha, S. (1995) Cost - effective electricity from amorphous silicon alloy, *Photonics Spectra* 7, 111-116
2. Yang, J., Glatffelter, T., Ross, R., Mohr, R., Fournier, J. P. and Guha, S., Crucial parameters and device physics of amorphous silicon alloy tandem solar cells, (1987) *Journal of Non-Crystalline Solids,* **97&98**, 1303-1306
3. Thompson, M. J. (1991) A comparison of amorphous and polycrystalline TFTs for LC displays, **137&138**, 1209-1214
4. Nickel, N., Fuhs, W. and Mell H. (1991) TCS study of n- and p-channel amorphous silicon thin film transistors, *Journal of Non-Crystalline Solids,* **137&138**, 1221-1224
5. Hack, M. and Shur, M. (1987) Analysis of amorphous silicon thin - film transistors, *Journal of Non-Crystalline Solids,* **97&98**, 1291-1294
6. Schropp, R.E.I., Boonstra, A.J. and Klapwijck, T.M. (1987) Reversible dangling - bond generation in amorphous silicon thin-film transistors, *Journal of Non-Crystalline Solids,* **97&98**. 1339-1343
7. Redfield, D. and Bube, R.H. (1991) Defects in amorphous silicon - extrinsic or intrinsic?, *Journal ofNon-Crystalline Solids,* **137&138**, 215-218
8. Hack, M.G., Lewis, A.G. and Shaw, J.G. (1991) *Journal ofNon-Crystalline Solids,* **137&138**,1229-1232
9. Adams, A.C., (1990) Technology Trends: Multiple quantum well spatial light modulators, *Photonics Spectra,* **5**, 191-196
10. Hsu, T.Y., Efron, U., Wu, W.-Y., Schulman, J.N., Haenens, D. and Chang, Y.C. (1988) Multiple quantum well spatial light modulators for optical processing applications, *Optical Engineering,* **27**, 372-376

11. Lee, Y.H., Jewell, J.L., Walker, S.J., Tu, C.W., Harbison, J.P. and Florez, T.L. (1988) Electrodispersive multiple quantum well modulator, *Applied Physics Letters,* **53**, 1684-1688
12. Miller, D.A.B., Chemla, D.S., Damen, T.C., Gossard, A.C., Wiegmann, W., Wood, T.H. and Burrus, C.A. (1984) Novel hybrid optically bistableswitch: the quantum well self - electro - optic effect device, *Applied Physics Letters,* **45,** 13-18
13. Whitehead, M., Rivers, A., Parry, G., Roberts, J.S. and Button,C. (1989) Low - voltage multiple quantum well reflection modulator with on:off ratio >100:1, *Electron. Letters,* **25**, 984-990
14. Barnes, P., Zouganeli, P., Rivers, A.,Whitehead, M., Parry, G.,Woodbridge, K. and Roberts, C. (1989) GaAs/AlGaAs multiple quantum well optical modulator using multilayer reflector stack grown on Si substrate, *Electron. Letters,* **25**, 995-1000
15. Lentine, A.L., Hinton, A.S., Miller, D.A.B., Henry, J.E., Cunningham, J.E. and Chirovsky, L.M.F. (1988) Symmetric self - electro - optic effect device: optical set - reset latch, *Applied Physics Letters,* **52**, 1419-1424
16. Livescu, G., Miller, D.A.B, Henry, J.E., Gossard, A.C. and English, J.H. (1988) Spatial light modulator and optical dynamic memory using a 6×6 array of symmetric self electro - optic effect devices, *Opt. Letters,* **13**, 297-300
17. Chevrier, J.B., Cambron, P., Chittick, R.C. and Equer, B. (1991) Use of n-i-p- i-n a-Si:H structure for bistable optically addressed spatial light modulator, *Journal of Non-Crystalline Solids,* **137&138**, 1325-1328
18. Scheiber, C. and Chambron, J. (1992) CdTe detectors in medicine: a review of current applications and future perspectives, *Nuclear Instruments and Methods in Physics Research,* **A322**, 604-614
19. Cuzin, M., (1992) CdTe in photoconductive applications, *Nuclear Instruments and Methods in Physics Research,* **A322**, 341-351
20. Iwase, Y., Funaki, M., Onozuka, A. and Ohmori, M. (1992) A 90, element CdTe array detector, *Nuclear Instruments and Methods in Physics Research,* **A322**, 628-632
21. Verger, L., Cuzin, M., Gaude, G., Glasser, F., Mathy, F., Rustique, J. and B. Schaub (1992) Electronic properties of chlorine doped cadmium telluride used as high energy photoconductive detector, *Nuclear Instruments and Methods in Physics Research,* **A322**, 357 - 362
22. *** EURORAD Catalog, (1995) 6-7
23. Fujieda, I., Nelson, S., Nylen, P., Street, R. A. and Weisfield, R. L. (1991) Two operation modes of 2D a - Si sensor arrays for radiation imaging, *Journal of Non-Crystalline Solids,* **137&138**, 1321-1324
24. Perez-Mendez, V., Cho, G., Drewery, J., Jing, T., Kaplan, S.N., Qureshi, S., Wildermuth, D., Fujieda, I. and Street, R. A. (1991) Amorphous silicon based radiation detectors, *Journal of Non-Crystalline Solids,* **137&138**, 1291-1296
25. Aiyah, V., Baillie, A., Polischuk, B., Bekirov, A. and Kasap, S. O. (1991) X-ray sensitivity of halogenated a - Se:As photoreceptors for electroradiography, *Journal of Non-Crystalline Solids,* **137&138**, 1329-1332

26. Demishev, S. V., Kosichkin, Yu. V., Lyapin, A. G., Sluchaniko, N. E., Alexandrova, M. M., Larchev, V. I., Popova, S. V., Skrotskaya, G. G. (1987) The bulk amorphous A^3 B^5 semiconductor: technological and physical aspects, *Journal of Non-Crystalline Solids*, **97&98**, 1459-1462
27. Grigorescu, C. E. A., Manea, S. A., Lazarescu, M. F., Logofatu, M. F., Cruceru, M. (1994) Determination of the composition of $In_xGa_{1-x}Sb$ grown by direct synthesis, *Optoelectronica*, **3**, 265-268
28. Munteanu, I., Grigorescu, C.E.A., Manea S.A., Logofatu, M.F., (1995) On some properties of $In_{0.05}$ $Ga_{0.95}$ Sb groun by direct synthesis, *Com.ICCG XI, The Hague, The Netherlands*
29. Manea S.A., Munteanu, I., Logofatu, M.F., Lazarescu, M.F., Grigorescu, C.E.A., (1995) $In_{0.20}$ $Ga_{0.80}$ Sb of highlyuniformcomposition, *IEE-Pro.ISC-CAS-95,233-236*
30. Manea S.A., Munteanu, I., Grigorescu, C.E.A., Logofatu, M.F., Lazarescu, M.F., (1996), On the uniform composition of In_x Sa_{1-x} Sb (x=0.20) bulk crystals for special optoelectronic devices, *Optical Engineering* , **35**, 1356-1359

NEW MATERIALS FOR SOLAR ENERGY CONVERSION

A.V.SIMASHKEVICH, P.V.GAUGASH
Institute of Applied Physics
Academia str. 5, Kishinev, MD2028, Moldova

Abstract: The development of photovoltaic solar energy conversion is considered. A review of the obtaining methods and the photoelectric parameters of the photocells based on different semiconductor materials: silicon, III-V compounds, copper indium deselenide (CIS), cadmium telluride, ternary semiconductor compounds is presented. Future tasks for efficiency increasing and cost reducing of the electrical energy generated by different photocells are discussed.
The results of elaboration and investigation of solar cells based on II-VI semiconductor thin films, semiconductor-insulator-semiconductor structures and tandem solar cells carried out in the Moldavian State University are presented.

The continuous technological development and the population increase up to 8 billion in 2020 result in a ever increasing energy demand. The conventional energy production is based on nonsustainable methods, exhausting our existing natural reserves of oil, gaz, coal, nuclear fuel. It exist more or less pessimistic estimations of these reserves. In conformity with these estimations, concerning the reserves which are now in exploitation, we have oil and gaz no more than for 50 to 100 years.

The conventional energy system causes also the majority of the environmental problems. Only renewable energy systems can meet the growing energy requirements in a sustainable way and without increasing the ecological damage. Therefore the energy system in the future can bee seen as a mix of the conventional energy production systems, which will gradually have to reduce in importance and of the renewable technologies, which will gradually increase in importance. So, the general increase of energy consumption in the next century would be mainly determined by the development of solar energy conversion.

A. Andriesh and M. Bertolotti (eds.),
***Physics and Applications of Non-Crystalline Semiconductors in Optoelectronics*, 391–401.**

Our planet's surface receives at the average 10^{14} KW per hour of a gratuitous solar energy. Conversion of the one thousands parts of this power should allow to provide 10 billions of people with electric energy of 10 KW per ever person that corresponds to a present consumption in the USA. France, for example, receives from the solar radiation an energy which is of 400 times more than the actual energy consumption. The Republic of Moldova is situated at the same latitude of 46°, the annual average sunshine duration is between 2060 and 2300 hours, annually the earth receives from the sun 1100-1400 kW h/m^2. At the same time there are no conventional energy sources, 98% of fuel is imported. The solar energy may be converted into electricity or heat. Let us consider only the photovoltaic (PV) conversion which is a direct conversion of radiation energy into electricity. The PV cells are the most wide-spread installations from simplicity of manufacturing and from an ecological point of view.
The first PV cells were fabricated in 1954 in Bell Telephone Laboratories, the first applications for space came from USA and from the former USSR. The first commercial applications for terrestrial use of PV cells were made in 1966. The oil crisis of 1972 stimulated the research programs on PV all over the world and in 1975 the terrestrial market exceeds the spatial one for 10 times. The total power of produced in 1991 PV cells was equal to 52.1 MW, in 1995 - 81.4 MW. It is interesting to note that the oil company British Petroleum is the main European PV producer in spite of the competition between the conventional and renewable energy sources.
Let us consider the present state and the perspective of application of PV cells and modules based on different solar materials. These materials must be absorbent for photons in the region of the solar radiation and reflective for the "black body" radiation for decreasing of the heating. The dependence of the efficiency on the band gap of different solar materials is represented in the Fig. 1 [1].The maximum efficiency could be expected from materials with the band gap around 1.5 eV. Crystalline silicon is the most used semiconductor material in the manufacture of PV modules today. For crystalline silicon technologies, as well as for other solar materials, ways are sought to improve the performance and reduce the cost of this material. Si PV cells can be made on low-cost wafers, but they generally contain high concentrations of impurities or defects, which can degrade cell performance. The efficiency may be increased by minimizing the reflection at the front surface what can be done by improving of the antireflection coating and by using of a buried pyramidical surface. Some data on the efficiencies for crystalline silicon laboratory cells measured in the calibration laboratory of the Fraunhofer Institute Solar Energy system of Freiburg at standard conditions: irradiation 1000 W/m^2 , T=25° C, light spectrum AM1.5 would be presented below [2]. The maximum efficiency was obtained for cells from South Walles University of Sydney - 23.8% for an area of 4 cm^2 . The increasing of area up to 45.70 cm^2 reduce the efficiency up to 21.3%. An efficiency of 22.5% was obtained in Freiburg for cells with the area of 4 cm^2. A bifacial cell fabricated in Emerthal, Germany with the area of 4 cm^2 has the efficiency of 19.4% for frontal illumination and 16.5% for rear side. The British Petroleum Solar Company fabricates coloured cells with the surface of 143 cm^2 and efficiency of 14.1% for "steel-blue" 12.4% for "gold" and 12.2% for "magenta". In the case of production silicon cells the maximum efficiency was obtained in the USA by the company "Sun Power" - 21.5% for the area of 17.70 cm^2. The efficiency of 19.4% was

obtained for cells with the area of 23.40 cm^2 fabricated by ASE Hellbronn, Germany, for space applications. The company British Petroleum Solar obtained 16.7% for cells with the area of 142.90 cm^2. In the case of multicrystalline silicon the maximum efficiency of 16.8% for cells with the area of 21.20 cm^2 was obtained in Freiburg for laboratory cells. For production cells manufactured by ASE Hellbronn the efficiency is 13.4% (area 100.20 cm^2).

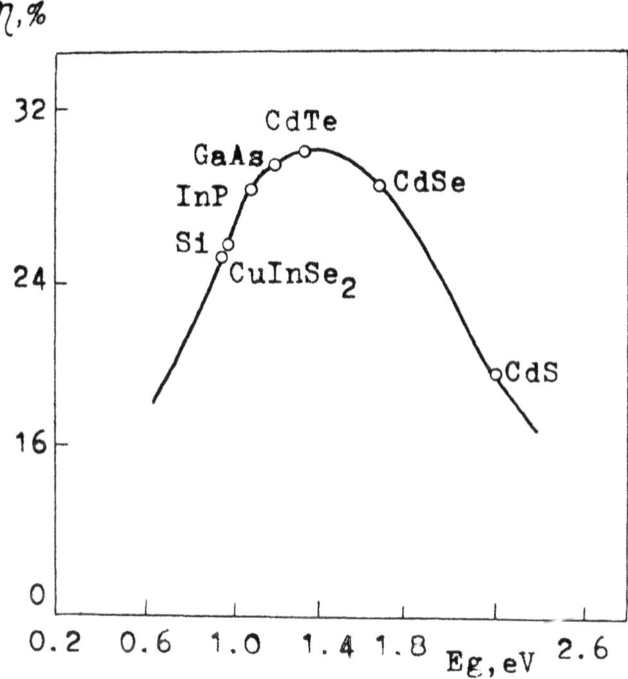

Fig. 1. The dependence of the efficiency on the band gap of different solar materials.

Future tasks for the crystalline silicon cells are:

- the cost reducing by dimension increasing of siliconingots and wafers;
- utilizing of silicon ribbon material;
- optimization of contact systems;
- improvements in cells manufacturing techniques using MISand SIS structures;
- development of new thin-film technologies;
- module design: laser scribing and monolithic interconnection.

The cost of terrestrial GaAs PV devices is currently too high to make these cells a strong contender in the PV market place. But with their high efficiencies, more than 23% under one-sun conditions and more than 30% under concentrated light, GaAs as well as InP cells are stable and resist radiation damage in space. The most effective GaAs PV cells were obtained in The Netherlands (23.9%, area 1cm^2), Germany (23.8%, area 4 cm^2, 20.7% in AM0 conditions) for laboratory cells and 22.9%, area 8 cm^2 (Essex, G.B.) for production cells. In the case of InP the maximum efficiency is 20.3% for area of 4 cm^2.

Future tasks for III-V compounds based PV cells:

- reducing of substrate cost, eliminating of expensive GaAs substrates.
- cost reducing by improvement of crystals and thin films growth methods. More uniform large area epitaxial growth.
- developing of new structures that will improve cell performance, obtaining of tandem cells.

Thin films are a rapidly developing PV technology that has attractive cost benefits. By utilization a small amount of semiconductor material as thin as few microns on a inexpensive substrate, for example glass, it could be achieved a continuous technological process and lower material and manufacturing cost necessary for competitive, large scale power generation. Another advantage of thin film modules is that they can be connected into an electric circuit as a single large piece. The most developed technology in the case of thin film PV cells is for amorphous silicon (αSi). The application of αSi PV modules is limited to the field of micropower (such as watches, calculators) and minipower (such as measuring equipment, automatic features). However a sizable market for amorphous silicon technology was increasing. In 1991 one fourth of worldwide PV production were αSi. In 1994 US manufacturing capacity was more than 20 MW per year, the US industry produces modules with the area up to 1.2 m^2. A key challenge for αSi technology is increasing the performance of production modules in terms of stabilized efficiencies. The main problem is the light-induced instability, an intrinsic effect in this material. The efficiency of stabilized αSi modules has improved significantly. After 1000 hours of light soaking some modules stabilized of greather than 8% efficiency. The efficiency of a module with the area of 4000 cm^2 increasing efficiencies has led to an innovative multijunction structure for αSi devices which contain a stack of two or three PV cells. The light that is unused in top cells can be used effectively in the lower cells. An efficiency of 13.1% for the area of 1 cm^2 was obtained in Hagen University, Germany, in the case of αSi - cSi heterojunction. A p-i-n structure of the same area fabricated by "Sanyo" company has the efficiency of 12.7%.

Future tasks for αSi cells and modules:

- fundamental studies to reduce or avoid the initial light-induced instability;
- development of higher rate deposition processes;
- development of multijunction tandem structures based on αSi and other semiconductor materials.

In contrast with αSi cells, one of the major strenghtes of copper indium diselenide (CIS) technology is that it appears to be extremely stable. Single junction $CuInSe_2$ cells have the potential of achieving 20% efficiency. Now the maximum efficiency is 16.6% for the area of 0.40 cm^2 obtained in National Renewable Energy Laboratory, Golden, USA, 15.4% was obtained in Uppsala, Sweden for a CIS/CdS/ZnO cell with the area of 0.38 cm^2 and 13.9% in the case of 90.6 cm^2 in Stuttgart University.

Future tasks for copper indium diselenide cells:

- investigation of various other alloys such as those incorporating gallium or sulfur to determine whether they increase efficiency;
- improved junction properties to raise voltage and fill-factor;
- elaboration of alternative thin film technologies;
- transfer of cell results to modules.

Cadmium telluride with the band gap of 1.5 eV is one of the most promising solar materials. CdTe cells show about the same durability and potential efficiency at 20% as CIS modules. In 1992 a CdTe thin film solar cell developed by researcher of the University of South Florida achieved 15.8% efficiency for the area of $1 cm^2$ [3]. In the Fig. 2 the increasing of the efficiencies of CIS and CdTe PV cells is presented [4].

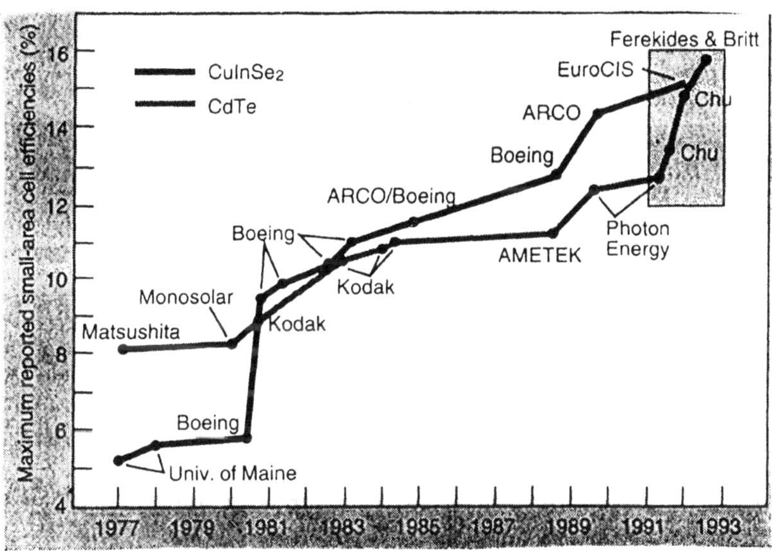

Fig.2. The increasing of the efficiencies of CIS and CdTe PV cells.

Future tasks for CdTe PV cells:

- resolution of contact stability issues;
- improved junction properties to enhance voltage and fill factor;
- transfer of cell results to modules;
- large-area deposition and processing equipment.

We have considered only the most investigated materials but there are some other materials promising for solar energy conversion: wide band gap II-VI semiconductor compounds besides CdTe; ternary compounds with chalcopyrite structure of I-III-VI_2 and II-IV-V_2 type. The first group could be considered as an extension of III-V compounds, the second as an extension of II-VI compounds, therefore ternary

compounds are often used with their analogous in heterojunction pairs [5]. Sensitized fullerene is a new material for solar conversion. On its base it is possible to fabricate photosensitive rectifying heterojunction suitable for application in solar cells. However, PV devices based on fullerene layers obtained by vacuum evaporation give low values of energy conversion efficiency. The improvement of photoelectric properties of fullerene may be .achieved by modification of its layer structure, for example, by creating of the fullerene supercluster of nanoscale dimension [6].
The present situation of PV cells development and production is seen from the Table 1.

Table 1. Development and production of PV cells

PV cells	Present situation			
	Labor.early	Labor stage	Pilot	Series
c Si	■■■■	■■■■	■■■■	■■■■
αSi	■■■■	■■■■	■■■■	■■■■
thin film Si	■■■■	■■		
GaAs	■■■■	■■■■	■■	
CuInSe2	■■■■	■■■■	■■	
CdTe	■■■■	■■		

And now let as consider the most vulnerable point in the solar energy application - the cost of energy generated by PV cells. First, the are situations when the solar energy is the one possible: space, isolated mountain regions, water surface e.t.c. Second,the solar module price in the last years decreases of 5-6 times simultaneously with the rise of the efficiency (Fig. 3).

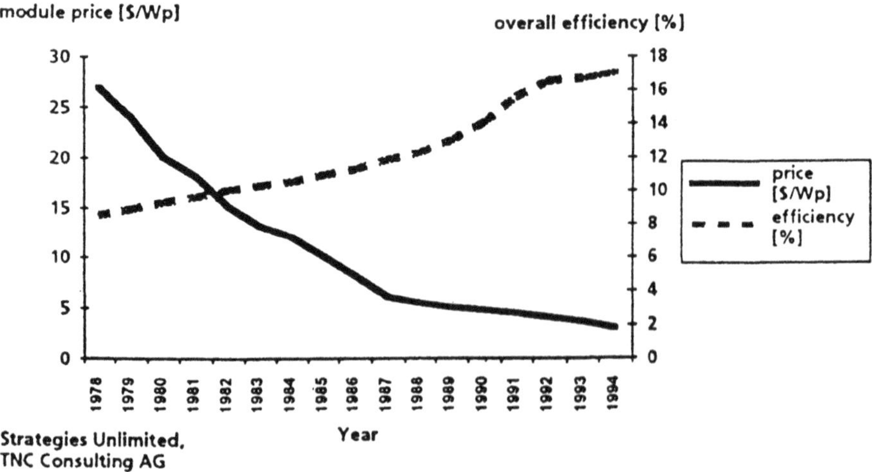

Fig.3. The world development of PV efficiency and price.

The cost of solar cells strongly depends on the quantity of produced modules. Silicon PV modules have been developed during last twenty years as a high quality industrial product. The industrial production worldwide expands every year by 20-25%. The cost for 1992 and the cost estimates for the year 2000, based on the answers by PV manufacturers on a questionnaire are given for crystalline Si in the Table 2 [7].

Table 2. The PV parameters and the cost of PV generated energy

	1992			2000		
	low	average	high	low	average	high
cell efficiency, (%)	12	13	14	14	16	18
module lifetime (years)	20	20	20	30	30	30
cell size (cm)	100	100	100	225	225	225
total cost (ECU/Wp)	1,9	3,0	3,8	1,2	1,8	2,6

The production is almost equally divided over the USA, Japan and Europe. It must be noted that the production of αSi modules is almost constant over the last years due to their use in consumer products which market is in a saturated state. For αSi the average predicted cost for the year 2000 is 1.2 ECU/Wp. In the case of crystalline Si cells 46% of the PV module cost comes from the silicon wafer. This is mainly due to the high cost of purification, of crystallisation and of wafering. The silicon used today is mainly coming from reclaimed material from the microelectronic industry. When the PV market expands further, there will be a need for the production of spatial solar grade silicon. The amount of expensive silicon can also be reduced by improved wafering technique and by deposition of a thin polycrystalline film on a cheep substrate. The larger utilisation of thin film solar cells in an other route of cost reduction. A large research and development effort, a larger manufacturing volume certainly can reduce the cost further.

The perspectives of research and development of PV cells and modules for next years and for tthe period up to 2030 could be considered taking the USA as an example (Table 3) [8].

Table 3. USA PV program achievements and goals

Factor	Mid-term goals 2000	Long-term goals 2010-2030
Module efficiency (%)	10 - 20	15 - 25
Electricity price (c/kWh)	12 - 20	5 - 6
System lifetime (years)	20	30
Installed capacity (MW)	200 - 1000	10000 - 50000

The values of the efficiencies which could be achieved in 2030 are 26% for laboratory cells and 18% for commercial modules in the case of flat-plate crystalline silicon and, respectively, 20% and 15% on the case of flat-plate thin films.

As a conclusion let as consider briefly the results obtained in Moldavian State University. Here the investigations are carried out in three directions:

- obtaining and investigation of semiconductor-insulator-semiconductor (SIS) structures based on silicon, indium phosphide and cadmium telluride;
- obtaining and investigation of InP based PV cells for application in conditions of hard radiation;
- obtaining and investigation of thin-film PV cells based on wide-gap II-VI materials.

Solar cells fabricated on the base of SIS structures are promising for solar energy conversion due to its relatively simple technology and low cost. These structures were obtained by spray deposition of tin dioxide, indium oxide or their mixture, so - called ITO layers, onto different crystal substrates, silicon, InP, CdTe [9,10]. The ITO insulated layer with the thickness around 50 Å is formed during the spraying process. The spectral characteristics of the obtained and investigated solar cells are presented in Fig. 4.

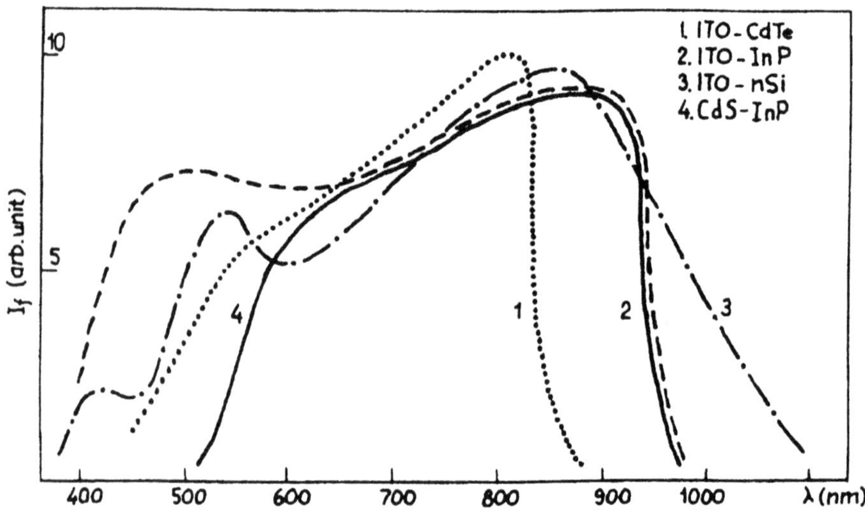

Fig.4. The spectral characteristics of ITO-CdTe (1), ITO-InP (2), ITO/nSi (3), CdS/InP (4) solar cells.

The shape of the characteristics depends on the thickness and on the ambient temperature. The photosensitivity of the investigated samples cover all the visible region of the spectrum. In the long wavelength region the photosensitivity is limited by the absorption edge of the narrow band gap semiconductor. In this figure is presented also the characteristic of CdS/InP heterojunctions, where the CdS layer is a wide gap degenerated semiconductor. This structure is of the same type as well as SIS structures. The photoelectric parameters of SIS solar cells are presented in the Table 4.

Table 4. Photoelectric parameters of SIS solar cells.

Structure	U_{oc} (V)	I_{sc}, (mA/cm^2)	FF	π, (%)	Illumination conditions
ITO-pCdTe	0.43	27.2	0.53	6.3	AM0
ITO-nSi	0.51	26.0	0.65	10.8	AM1
ITO-pInP	0.69	29.6	0.75	11.0	AM0
ITO-p^+-p InP	0.71	30.8	0.73	11.6	AM0
CdS-InP	0.80	29.4	0.75	12.6	AM0
CdS-p^+-p InP	0.83	31.0	0.76	14.0	AM0

It is seen from this table that the photoelectric parameters of InP based structures may be improved by using of p^+-p InP substrates.
The radiation stability is an important parameter when the solar cells are operating in hard radiation conditions. It was shown [11] that the photoelectrical parameters of CdS-InP solar cells after electron and proton irradiation decreases only by 10-15% depending on radiation dose. The results of the investigation of the degradation of photoelectrical parameters of ITO-InP solar cells after they have been irradiated by protons with energy E_p = 20,6 MeV and flux density up to $F_p = 10^{13}$ cm^{-2} and by electrons with E_e = 1 MeV and $F_e < 10^{15}$ cm^{-2} are presented in Fig. 5.

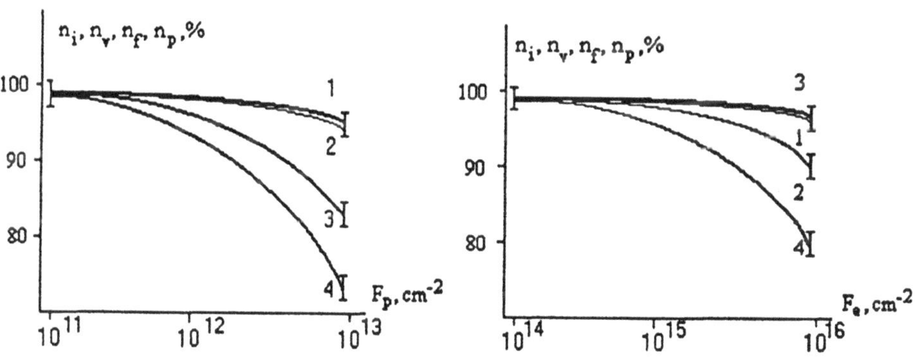

Fig.5. The photoelectric parameters of ITO/InP solar cells after proton irradiation (a) and electron irradiation (b): 1 - I/I_0 2 - FF/FF_0 3 - U/U_0 4 - P/P_0

After irradiation their efficiency decreases by 26% that is more than in the case of InP-CdS heterojunctios but less than for Si and GaAs based cells.

The wide gap materials: ZnSe, ZnTe, CdS, CdSe, CdTe have the band gap corresponding to the photon energies in the visible region of the solar spectrum, are characterized by relatively larg mobility and high photosensitivity. All these compounds are direct gap materials which minimized the requirements for minority carrier diffusions length, what is promising for the fabrication of low cost thin-film solar cells. But for the fabrication of latters the presence of p-n junction is needed, the problem which until recently remained unsolved for the most of above mentioned compounds with expressed monopolar conductivity. In this situation the obtaining of different heterostructures based on II-VI compounds is necessary. Investigations of such heterostructures are described in a series of papers [12-14] and their photoelectrical parameters are given in Table 5.

Table 5. Photoelectrical parameters of solar cells based on II-VI compounds.

Heterojunctions	U_{oc}, V	I_{sc}, mA/cm^2	Efficiency, %	Illuminat conditions
pCdTe/nCdS(thin-film)	0.79	24.0	9.85	AM1
pZnTe/nCdSe	0.80	12.3	7.0	AM1.5
pZnTe/nCdSe(thin-film)	0.84	10.1	6.8	AM1.5
pZnTe/nZnSe	1.20	10.0	9.0	AM1

pCdTe-nCdS heterostructure which is caracterized by close values of electron affinities and as a result there are no conduction band spikes at the heterojunction interface is the most interesting. Besides, a low lattice mismatch promotes the effective separation of light induced nonequilibrium charge carriers. In spite of a considerable lattice mismatch (~ 7%), is at a certain interest the ZnSe/ZnTe heterojunction. Thanks to a $ZnSe_xTe_{1-x}$ solid solutions layer formed at the interface, lattice mismatch diminishes and that leads to a rather high photoelectrical parameters. The three-layer tandem structure ZnSe/ZnTe/CdSe was prepared by successive epitaxial deposition of ZnTe and CdSe layers from vapour phase onto ZnSe single crystalline substrates [15]. The structure was supplied with contacts which allowed to switch on ZnSe/ZnTe and ZnTe/CdSe heterojunctions into separate electrical circuits. At 300 K in AM1.5 conditions the summary U_{oc} was 1.55 V and the efficiency 10.4%. These results were received with non optimized structures without antireflection coating and they could be improved on the expense of diminiching of losses both for current and voltage by selecting the component thickness and doping level.

References:

1. Loferski, J.J. (1956) Theoretical considerations governing the choise of the optimum semiconductor for photovoltaic solar energy conversion *J.Appl.Phys.* **27**, 777-784.
2. The Fraunhofer ISE PV- Charts: Asessment of PV device performance, edition 7 (1996), Freiburg.
3. Britt, J. and Ferekides, C. (1993) Thin film CdS/CdTe solar cell with 15.8% efficiency, *Appl. Phys.. Letters*, **62**, 2851-2852.
4. Photovoltaics: program overview fiscal year 1992, produced by NREL, US Department of Energy, p. 29.
5. Gashin, P.A., Ketrush, P.I., Simashkevich, A.V. (1990) Ternary compounds based heterostructures as perspective structures for solar energetics, in *Proceedings of the 8th Int. Conf. on ternary and multinary compounds*, Kishinev, p. 168-171.

6. Shevaleevskii, O.I., Larina, L.L., Metalnikov, V.M. (1996) Porous super cluster fullerene layers for photovoltaic application, in *Abstracts of 10th Int. Sonnenforum* (EuroSun 96), Freiburg, V, p. 112-113.
7. Van Overstraeten, R.J. (1994) Photovoltaics overview, *Solar energy for sustainable development*, Bucharest, **3**, 7-12.
8. Photovoltaics: Program Plan FY 1991 - FY 1995, prepared by NREL, US Department of Energy, p. 8.
9. Adeeb, N., Kretsu, I.V., Simashkevich, A.V., Sherban, D.A., Sushkevich, K.D. (1987) Spray deposided ITO-CdTe solar cells, *Solar Energy Materials* **15**, 9-19.
10. Gagara, L., Gorcheac, L., Radu, C., Radu, S., Sherban, D., Simashkevich, A. (1996) Photovoltaic converters of solar energy on the base of SIS structures, in *Abstracts of 10th Int. Sonnenforum* (EuroSun 96), Freiburg, V, p. 118-119.
11. Botnariuk, V.M., Gorchak, L.V., Grigorieva, G.M., Kagan, M.B., Kozyreva, T.A., Lyubashevskaya, T.L., Russu, E.V., Simashkevich, A.V. (1990) Radiation degradation of solar cells based on InP/CdS heterojunctions, *Solar Energy Materials*, **20**, 359-365.
12. Gashin, P.A., Simashkevich, A.V. (1973) ZnTe/CdSe heterojunctions II Photoelectric and luminescent properties, *Physica Status Solidi (a)*, **19**, 615-619.
13. Gashin, P.A., Simashkevich, A.V., Sherban, D.A., (1978) Photosensitive epitaxial ZnSe/ZnTe heterojunctions, *Sov. Semicond. Phys.*, **12**, 180-182.
14. Andriesh, A., Simashkevich, A. (1996) Investigations of photovoltaic solar energy conversion on the Republic of Moldova, *Abstracts of 10th Int. Sonnenforum* (EuroSun 96), Freiburg, V, p. 116-117.
15. Gashin, P.A., Dvornic, G.G., Kagan, M.B., Lyubashevskaya, T.L., Simashkevich, A.V., Focsha, A.Ya. (1982) Tandem hetero-photocells based on wide-gap II-VI compounds *Sov. J. Techn. Phys. Lett.* **8**, 930-933.

EXPERIMENTAL INVESTIGATION OF SUBSURFACE STRUCTURE AND SURFACE SYMMETRY OF DISORDERED SEMICONDUCTORS

A.P.Fedtchouk, R.A.Rudenko, E.M.Barnyak, A.A.Fedtchouk
Odessa State University, Physical Department
F.79, Bd.30A, Balkivska Street, Odessa, 270110, Ukraine
phone/fax: 38(0482) 236-427

Abstract

With the use of original laser scanning non-equilibrium carriers life-time monitor we have obtained a various representations of microdefects distribution in subsurface damaged layer (SDL) of crystalline Si wafers. The defects cluster parameters were modified with the use of layer-by-layer etching, equilibrium and non-equilibrium laser annealing and ion implantation procedures. It was discovered for the first time that the SDL microdefects cluster demonstrates quite definitely the multifractal structure, each component of which is connected with one of the subsequent stages of the technological prehistory of the specimen.

We present also the automated complex for the semiconductor surface superlattice (SSS) symmetry type evaluation. The basis of the method is the photo-e.m.f. signal registration as a function of the cell upper transparent electrode rotation angle. As a result of the present work it was demonstrated that the surface charge of the semiconductor is affected mainly by the dipole liquid crystal (LC) layer being in tight contact with the wafer surface micro-curvature (atomic rows potential relief). The proposed method makes it possible to detect the Si(111) surface modification from (2x1) to (7x7) SSS symmetry type caused by the thermal annealing of the wafer.

1. Introduction

The problem of non-intrusive investigation of surface and subsurface damaged layer (SDL) imperfections of semiconductor wafers is the central one that influences very much the maximum possible yield value estimation in microelectronics [1]. The up-to-date methods of electrically active defects' concentration monitoring are either destructive [2] or non-express ones[3] what makes the use of them in real technology almost impossible.

A good deal of various apparata based on the principle of photoelectric monitoring of microdefects' distribution over the disordered region of the wafer were elaborated [4-6], but because of the lack of appropriate quantitative criteria of imperfection degree almost all of them give still non-sufficient help to the highest

A. Andriesh and M. Bertolotti (eds.),
Physics and Applications of Non-Crystalline Semiconductors in Optoelectronics, 403–415.

technological level's approach. The mentioned above need of SDL structure knowledge is oriented mainly to VLSI circuits production and these problems solving can give a significant economical effect in the world-wide competition of the leading plants of microelectronic industry.

There is another vital problem concerning the semiconductor surface superlattice (SSS) symmetry [7] determination as for the regular wafers at the entrance point of technological process and also for it's damage degree as the result of any treatment (such as plasma etching, ion and electron bombardment, physical or chemical vapour deposition, etc.). The precise detection of certain structural defects emerging over the ideal superlattice in the nanometric scale is provided now, using various types of atomic force microscope (AFM) [8]. The only principal difficulty of the AFM's use is the extremely small part of the wafer's surface being displayed in a real-time regime. The mentioned peculiarity also does not make it possible to evaluate a reliable prognosis of superlattice's distortions taking place over the whole wafer. It is quite obvious that the ideal SSS plays the decisive role in epitaxial growth processes which are initialized by lattice matching at the very eve of the surface atomic layer coverings.

So, as we see, one can state that the search of experimental methods of investigation of both geometrical surfaces' and subsurface region's microdefects distribution is still actual as well as elaboration of various apparata for their realization.

2. Experimental set-up and procedure

We have constructed [9] an automated monitoring complex "Photocon" which includes the registration system of condenser photo-e.m.f. (CPEMF) signal generated in the SDL by means of laser excitation of the wafer subsurface region. The "Photocon" is designed in CAMAC-bus standard and is operated with the help of PC IBM.

The apparatus is shown schematically at Fig.1. As the source of stimulating radiation the single-mode He-Ne laser LG-207 (wave length 0.6328 micron) was used. It is operated under the control of the original scheme of intensity stabilization, proposed by [10] and modified by our group. The modulated laser beam is transmitted through the optical fiber light-guiding line into the dielectric potential probe by means of which the CPEMF signal is also discriminated. The chopping frequency was selected to be not less than 1 kHz in order to minimize the role of surface capture portion in the general value of the CPEMF signal.

The CPEMF signal, after appropriate amplification, is being detected and then introduced by analogue-digital converter (ADC) accompanied by interface into PC's port. When needed, the bias voltage is applied across the wafer by means of digital-analogue converter (DAC).

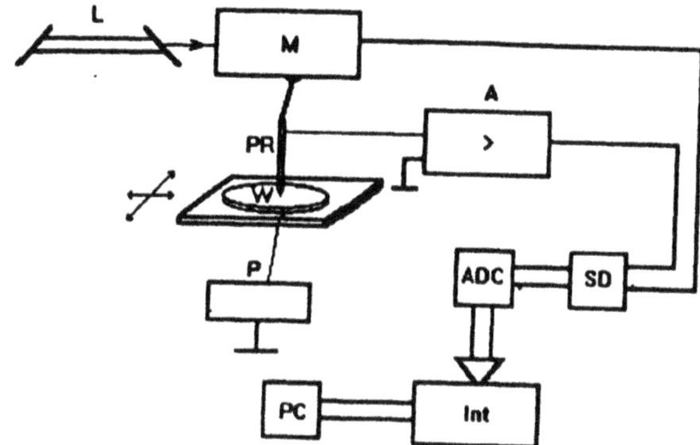

Fig.1. The scheme of "Photocon" monitor. Here we have used the following abbreviations: L - laser source, M - modulator, PR - probe, W- wafer, A - block of amplifiers, SD - synchronous detector, Int - interface, PC - computer, P - positioning apparatus.

The measurements were conducted in a following way. The dielectric potential probe was placed by programmely controlled positioning system onto the point over the wafer surface with known co-ordinates and after this the statistically valuable set of CPEMF signals was taken. After the end of wafer scanning process all the data were being statistically processed. It was shown experimentally that valuable set needed to obtain repeatable topogram or histogram of the wafer under consideration consists of, at least, 10 values of CPEMF signal taken at each point of the wafer surface. Time delay spent for the minimal data set taking does not exceed 0.1 second.

The software of "Photocon" complex is based on Assembler and Pascal languages. Using it allows to introduce the main parameters of the wafer and of the monitoring process itself (e.g. wafer type, it's diameter, scanning step length along both co-ordinates, bias voltage, discretization step on the equilevel signal map, etc.). Data base and original software of the complex both allow to analyze all known types of wafer's SDL modifications which are caused by all the sequential steps of the technological prehistory and process up to the final stage of metallization. We have included special program of SDL microdefects' cluster fractal dimension evaluation which enables to establish the critical value of defect concentration leading to the edge dislocations generation and catastrophic failure of the devices during all the time of use.

For the purpose of SSS symmetry type evaluation we have also elaborated the automated complex "Photosym". Its configuration is shown at Fig.2.

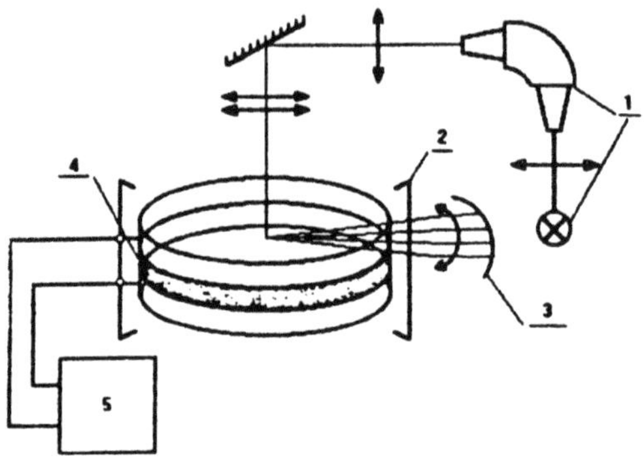

Fig.2. The scheme of the "Photosym" apparata for the determination of SSS symmetry type. 1 - monochromator , 2 - sample holder, 3 - angle's measuring system, 4 - LC twisted layer, 5 - bias voltage source.

This complex is also operated by means of IBM PC. The rotation angle of the upper transparent electrode was controlled by step driving motor and the CPEMF signal was plotted as the function of this parameter. The procedure is based on the determination of vector-director easy orientation axes' map of liquid crystalline (LC) dipole probe at the real semiconductor surface immediately and some time after the removing of the native oxide. The LC dipole probe comprises thin (about 10 micron) layer of planarly oriented molecules of nematic or holesteric LC. This layer is twisted due to the upper transparent electrode rotation. The visco-elastic properties of LC layer enable the rotating mechanical momentum's transfer through the LC layer up to the last monolayer being in contact with the micro-relief of SSS under consideration. This type of co-operative phenomena define the particular value of surface potential drop which value depends significantly upon the penetration depth of LC probe's molecules between the semiconductor surface atoms being organized in SSS. The photo-excitation of the wafer was accomplished using non-coherent light source with photon energy from the intrinsic absorption band of silicon.

The measuring procedure begins with the determination of the working wave length being the most sensitive one to the change of the electrode's rotation angle (see . The working wave length is chosen according to the CPEMF signal angular dependence's maximum contrast vs rotation angle plot. It depends mainly on the dopant's type and concentration as well as on the SDL's cluster non-crystallinity degree. In our case, this wave length was detected to be in the vicinity of 0.96 micron.

3. Experimental results

The results of SDL microdefects cluster monitoring are represented at Fig.3a (for the case of initially uniform wafer) and at Fig.3b (for initially non-uniform wafer of the same parameters taken from the same technological group).

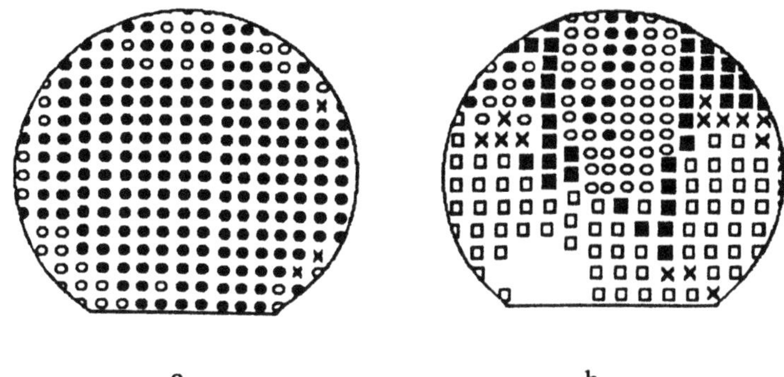

Fig.3. The topograms of CPEMF signal distributions over the wafer surface for the cases of: a) uniform wafer and b) non-uniform wafer

All the results represented below are taken and statistically processed over thousands of wafers during the in-line entrance quality monitoring in the real conditions of semiconductor fab. All the wafers were phosphorous doped and have the resistivity of about 4.5 Ohm.cm.

The ion implantation process ($3 \times 10^{11} cm^{-2}$, 60 keV) leads to serious changes in topogram (see Fig.4).

The lowest implantation dose detectable by means of "Photocon" was established as $10^{10} cm^{-2}$. It is worth noticing that the use of regular measuring methods does not allow to registrate reliably such a low effect.

We have noticed that the sequential stages of isochronous annealing of the same wafer in the temperature range from 748 K to 1373 K modifies the SDL's structure significantly.

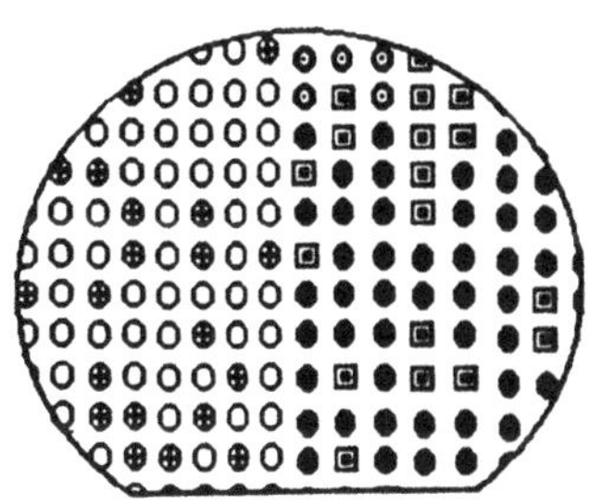

Fig.4. The topogram of CPEMF signal distribution for the case of implanted wafer with half shielded surface

The difference seen at the set of topograms demonstrates the maximum homogeneity of the wafer's SDL obtained as a result of annealing at 1023 K. This result coincides satisfactorily with the known phenomenon of silicon thermodonors annealing [11].

The results of CPEMF signal's angular dependence are represented at Fig.5.

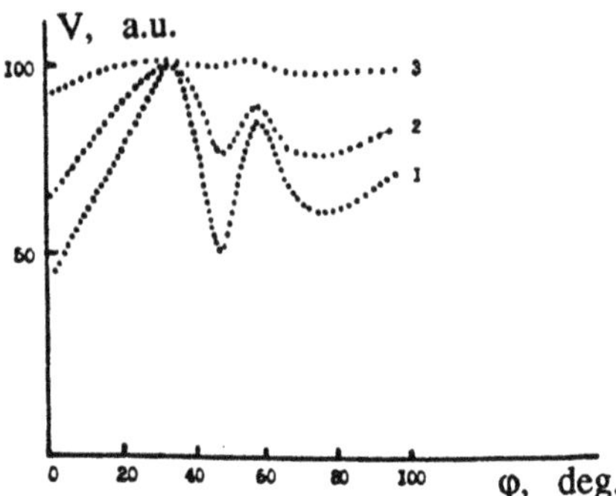

Fig.5. The plot of CPEMF signal's angular dependence for the case of initially untreated wafer and for different times of wafer storage: 1 - newly etched wafer, 2 - after 6 hours storage, 3 - after 24 hours storage in the oxygen containing atmosphere.

It is seen quite well that the native oxide thickness' increase leads to the contrast decrease. This feature of the proposed procedure can be used as a by-path method of detecting the maximum possible time of inter-processing wafer storage. The second and, as to us, more significant feature of the plot of initially untreated wafer was the maxima quantity per quadrant. For the case of Si wafer it was always the same: 2 max/quadrant.

We have discovered also the effect of changing the easy orientation axes density over the quadrant area under the influence of wafer's annealing at 773 K during 1 hour in the vacuum with the pressure of about 0.1 Pa. This effect is represented at Fig.6.

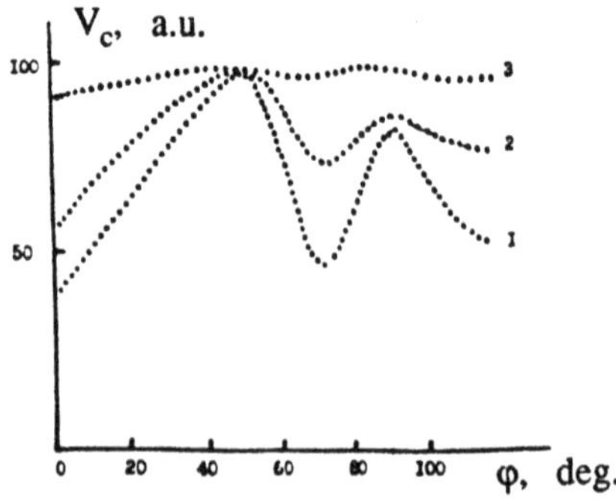

Fig.6. The same plot as at the Fig.5 but taken after annealing the wafer at 773 K.

The nature of this phenomenon is connected, as to our opinion, with the known from the literature [12], Si-SSS symmetry type modification as a result of wafer annealing at almost the same temperature that was demonstrated by quite another experimental technique.

On the basis of SSS symmetry type monitoring with the use of "Photosym" we have experimentally detected the phase transition of chaos-order type in twist-nematic. It is known well enough [13], that the unique experimentally registered up to now form of nematic LC existence is the single-axe ordered state with the ordering only in the long axes ensemble. Phase transition into the double-axed ordered state, including dipole sub-system ordering was predicted theoretically [14], but never has been observed experimentally before. The effect of biaxial ordering was detected from the CPEMF signal vs rotation angle plot of the type, demonstrated at Fig.5 and Fig 6.

As it was mentioned before, the changing of the rotation angle of the upper transparent electrode has led to the rotating movement all the quasi-nematic layers up to the last one situated at the interface of SSS-LC. This effect was accompanied by electric double layer dipole component reorientation [15] and here we have to state that the fixation of this component is due to the free energy minimization of the dipole LC-SSS system, modified by the semiconductor surface energetical profile. And, of course, this profile, in turn, is determined by the SSS symmetry type.

We can consider that the angular dependence itself, registered quite reliably in our experiment and discussed here, is due to the double-axes selforganization of LC layer of the probe which serves as an experimental evidence of the phase transition under consideration.

4. The model and results' discussion

For the purpose of the explanation of the experimental results presented above, we propose the model with the following main suggestions:

1. The registered CPEMF signal value is being controlled under the constant level of the probe potential, by only one of the variety of parameters, the microdefects' concentration which influences mainly the surface recombination velocity and the space charge region width.
2. The cluster of SDL's microdefects is considered to be non-selfshadowed.
3. The local concentration of the microdefects is inversely proportional to the CPEMF signal value.
4. The self-similarity of the fractal cluster of SDL's microdefects gives us the possibility of the linear regions' extrapolation up to the intersection with the horizontal axis and in such a way one can determine the borders of certain type of defects over the surface of the wafer.

For the sake of quantitative approach we use the assumption of one-dimensional description possibility, which is equivalent to the sample uniformity. The one-dimensional boundary problem concerning the spatial distribution of the extrinsic non-equilibrium carriers concentration evaluation being influenced by the non-homogeneous microdefectivity of the wafer's SDL can be written in the form:

$$D_0 \frac{d^2 \Delta p(z)}{dz^2} - \frac{\Delta p(z)}{\tau} + \alpha \Phi_0 e^{-\alpha z} = 0 \quad (1)$$

$$\Delta p(z)|_{z \to \infty} = 0 \quad (2)$$

$$\lim_{\delta \to \infty} D_0 \frac{d\Delta p(z)}{dz}\bigg|_{z=W+\delta} = \lim_{\delta \to \infty} S\,\Delta p(W-\delta) \quad (3)$$

The solution of this boundary problem, according to Chiang and Wagner [16], has the following view:

$$V_c = \frac{kT}{q} \ln\left[\frac{\Phi_0 \alpha L^2 \exp(-\alpha W)}{p_0(1+\alpha L)(D+SL)} + 1\right] \quad (4)$$

where p_0 - the equilibrium concentration of the extrinsic carriers, L - diffusion length of these carriers, D - diffusion coefficient, α - absorption coefficient for the wave length of excitation, Φ_0 - the absorbed photon flux intensity and S - the surface recombination velocity.

Taking into account the experimental fact of the functional linearity of the registered CPEMF signal on the photon flux value, one can rewrite the expression (4) in more appropriate form of:

$$V_c \cong V_b = \frac{kT}{q} \cdot \frac{\Phi_0}{p_0}\left[\frac{\alpha L^2 \exp(-\alpha W)}{(1+\alpha L)(D+SL)}\right] \quad (5)$$

As it was shown before by G.P.Peka [17], the condenser photo-e.m.f. can be presented as the sum:

$$V_c = V_b + V_{sc} + V_d + V_{nh} \quad (6)$$

where V_b - barrier component of CPEMF signal, V_{sc} - surface capture component, V_d - Dember component and V_{nh} - the component caused by the surface non-homogeneities of the sample. Taking into account the real relation between these components, one can see easily that the main part (up to 90 %) of the registered signal is due to the barrier component. This assumption is supported by the fact that while changing the frequency of the excitation modulation in the range 1kHz-1MHz, the CPEMF signal value has changed less than by 10 %.

We have seen that the registered signal exhibited strictly linear dependence on the applied voltage value. That means that additive term in Eq.4 is small in comparison to 1. As we deal, in our case, with the strongly absorbed radiation, it is possible to transform the Eq.5 into more convenient expression:

$$V = \frac{kT}{q}\frac{\Phi_0}{p_0 S}\exp(-\alpha W) \tag{7}$$

because in all practically significant cases we have the fulfilling of the following inequity:

$$S >> \frac{D}{L} \tag{8}$$

In the same book [17] it is demonstrated that the field dependence of the surface recombination velocity can be expressed as follows:

$$S = S_0 \exp[-\frac{q}{kT}(V_{bi} - V_b)] \tag{9}$$

where V_{bi} stands for the built-in subsurface potential of the semiconductor under investigation.

With the scope of the total number of variables minimization, we propose to change the subsurface space charge region's width W by the one-dimensional concentration of electrically active defects $N_{def}^{(1)}$, dispersed in the SDL. It can be done with the use of [18], where the following expression is proposed:

$$W = [\frac{\varepsilon_0 \varepsilon_s}{qN_{def}^{(1)}}\{V_{bi} - V_b - \frac{kT}{q}\}]^{1/2} \tag{10}$$

Inserting the formulas (9) and (10) into Eq.7, we derive the expression, which binds the experimentally measured value of V_b and the averaged linear defect concentration $N_{def}^{(1)}$:

$$N^1{}_{def} = \frac{\alpha\varepsilon_0\varepsilon_s\{V_{bi} - V_b\frac{kT}{q}\}}{q\{\ln(\frac{kTV_b\Phi_0}{qp_0}) - \frac{q}{kT}(V_{bi} - V_b)\}} \quad (11)$$

In the first approximation, we can consider $N_{def}^{(1)}$ to be constant in the borders of the electrode area. And, as a sequence of it, the three-dimensional defect concentration has the form of:

$$N_{def}^{(3)} = \frac{1}{\pi R^2} N_{def}^{(1)} \quad (12)$$

where R - is the electrode's radius.

Besides, we can state also that using the fractal approach to the description of the two-dimensional defects concentration, proposed by us before [19], one can obtain the following formula:

$$N_{def}^{(2)}(r) = \frac{1}{2\pi} \frac{N_0}{r_0^D} \quad (13)$$

where D - the fractal dimension of the SDL microdefects' cluster, r_0 - the minimal length of tetrahedrically bound defects, N_0 - the minimal quantity of defects which composes the tetrahedron and r - radial co-ordinate of the scanning point over the wafer. This value also can be transformed into the three-dimensional expression while considering the defects' concentration to be constant over the whole SDL region accessed by the excitation radiation l_{eff}. In the most rough approach one can treat the l_{eff} as:

$$l_{eff} = \frac{1}{\alpha} \quad (14)$$

This leads to the following expression for the three-dimensional concentration $N_{def}^{(3)}$ of microdefects located in SDL:

$$N_{def}^{(3)} = \frac{\alpha}{2\pi} \cdot \frac{N_0}{r_0^D} \cdot Dr^{-2} \qquad (15)$$

As one considers both in the usual and fractal approaches the same value on $N_{def}^{(3)}$, it is possible to put the equity sign between (11-12) and (15). This makes possible to establish the relationship between the radial dependence $V_b(r)$ and the fractal cluster's dimension D in the form:

$$D\left(\frac{r}{r_0}\right)^D = \frac{\varepsilon_0 \varepsilon_s [V_{bi} - \frac{kT}{q} - V_b(r)] r^2}{N_0 q R^2 \{\ln[\frac{kTV_b(r)}{qp_0}\Phi_0] + \frac{q}{kT}V_b(r) - \frac{q}{kT}V_{bi}\}} \qquad (16))$$

Thus, for the first time, using the fractal properties of the microdefects' cluster of the Si wafers found experimentally over the statistically meaningful quantity of the samples (some thousands units), we have shown the possibility of the cluster's fractal dimension evaluation over the small group of signal values. This makes the fractal approach presented here to be more useful even in the conditions of real technological process of VLSI production.

After evaluation of the fractal dimension D, one can, using Eq.(15), evaluate the three-dimensional defectiveness of the SDL's microdefects cluster $N_{def}^{(3)}$.

5. Conclusions

1. For the first time we have stated experimentally, that the microdefects of the SDL of the silicon wafer are being organized into the fractal cluster.
2. We have discovered that the quantity of the multifractal cluster's components depends mainly on the prehistory of the wafer under investigation.
3. It was demonstrated that the fractal approach gives the possibility to minimize the number of the scanning points and in such a way to make the proposed method of defectiveness monitoring to be really express one.
4. Using the combination of the fractal approach and the surface photo-voltage signal modified by the defects' concentration allows to evaluate the electrically active defects concentration as a function of the radial dependence of CPEMF signal.

5. The surface superlattice symmetry type evaluation gives the possibility of surface atomic ordering early stages' determination what can be considered as the proper addition to the atomic force microscope techniques.

6. All the methods of the subsurface damage and surface symmetry quantitative monitoring are equally applicable to the investigation of photosensitive materials of all types.

6. Acknowledgements

The authors are deeply indebted to ARW Directors, Professors Mario Bertolotti and Andrei Andriesh and to all the audience of Advanced Research Workshop, PANCSO'96 (Chisinau, Moldova) for their interest and fruitful discussion of the presented reports.

This work was supported in part by George Soros Program ISSEP, the grants NN APU 052104 and PSU 062090.

7. References

1. Wang P., Lee F., Chan K.M., Goodner R., Ceton R. (1996) Yield Enhancement in a High-Volume Wafer Fab, *Semiconductor International* **8**, 217-222.
2. Besenbacher F., Norskov J.K. (1993) Oxygen Chemisorption on Metal Surfaces, *Progress in Surface Science* **44**, 5-66.
3. Tepermeister I.,Conner W.T., Alzaben T., Barnard H., Gehlert K., Scipione D. In Situ Monitoring of Product Wafers, *Solid State Technology* **3**, 63-68.
4. Moore A.R., US Patent N 4433288, 21.02.84.
5. Kamienecki E., US Patent N 4544887, 01.10.85.
6. Kamienecki E., US Patent N 4827212, 02.05.89.
7. Fedtchouk A.P., Kornienko Y.K., Shevchenko L.D., Gouzenko N.K., Shapoval V.I., Author Certificate (USSR) N 1597536, 07.10.90.
8. Wilder K., Singh B., Arnold W.H. (1996) Novel In-Line Applications of Atomic Force Microscopy, *Solid State Technology* **5**, 109-116.
9. Fedtchouk A.P., Rudenko R.A., Shevchenko L.D. (1996) The Fractal Approach for the Evaluation of Microdefects in Silicon, *Materials Science & Engineering* **B34**, 164-167.
10. Kositsyn V.E., Timashov A.V. (1987) Stabilization of Laser Radiation Output, *Apparata and Techniques of Experiment (Sov.Phys.)* **1**, 190-192.
11. Dip A. (1996) Fast Thermal Processing: Batch Comes Back, *Solid State Technology* **6**, 113-124.
12. Bolshakov L.A., Veshchounov M.S. (1986) To the Theory of Semiconductor Crystal Surface Reconstruction, *Journal of Experimental and Theoretical Physics (Sov.Phys.)* **90**, 569-580.
13. Blinov L.M., Katz E.I., Sonin A.A. (1987) Physics of the Surface of the Thermotropic Liquid Crystals, *Progress in Physical Sciences (Sov.Phys.)* **152**, 449-477.

14. Kozlov V.A., Petkin N.V., Troufanov N.A. (1988) About the Possibility of Experimental Determination of the Surface Antiferroelectric-Ferroelectric Phase Transition, *Surface (Sov.Phys.)* **6**, 136-138.
15. Alexeyev A.E., Kornienko Y.K., Shevchenko L.D., Fedtchouk A.P. (1990) Phase Transition of Order-Disorder Type in Twist-Nematic Stimulated by Silicon Surface Superlattice, *Letters to the Journal of Technical Physics (Sov.Phys.)* **16**, 77-81.
16. Chang C.L., Wagner S. (1985) The Surface Photovoltage Monitoring of Disordered Semiconductors, *IEEE Transactions* **ED-32**, 1722-1728.
17. Peka G.P. (1984) *Physics of Surface Phenomena on Semiconductors*, Naukova Doumka, Kyiv.
18. Muller R., Cummins T. (1989) *The Elements of Integrated Circuits*, Mir Publishers, Moscow.
19. Fedtchouk A.P., Rudenko R.A. (1996) The Fractal Properties of Microdefects' Cluster of Silicon Wafers, *Ukrainian Physical Journal (in press)*.

MULTIQUANTUM PROCESSES IN SOLIDS. NEW ASPECTS

V.A.KOVARSKY
Institute of Applied Physics
Academy of Sciences of Moldova
Kishinev 2028-MD, Moldova

Introduction

The physics of multiquantum processes in atoms, molecules and solids scored new big success. The review of the results of physics of multiquantum processes as of 1985 one can see in monograph [1].

Let us give a brief summary of the latest advances in this field. First of all, this is a generation of higher harmonics slowly decreasing with the decrease of harmonic number. Experimentally a 130-th harmonic was already generated from Nd-laser interacting with a rare gas falling in the range of soft X-ray radiation [2]. It should be specially emphasized that this radiation is coherent [3]. First experimental results are obtained in the generation of harmonics by surfaces, of solids [4]. The theory of these processes is discussed in a number of works (see e.g. [2,5]). A new trend are multiquantum transitions with the participation of squeezed states (squeezed light, squeezed vibrations). The review on this problem is given in [6]. New is also the trend connected with nonequilibrium vibrations and nonlinear and synergetic effects resulting out of them.

We shell dwell a greater length on some of the above aspects, in particular, those which were developed in Kishinev. Let us start with the analysis of multiquantum processes in amorphous materials and consider also some problems related to crystals.

It is known that it was just in amorphous materials where the high frequency (tera-Hz range) long-lived vibrations were discovered, which make it possible to investigate the influence of nonequilibrium vibrations on the kinetic processes in amorphous materials.

1. Optical and nonradiative transitions in amorphous materials with the participation of nonequilibrium vibrations. Vibration bistability and Mott's law

Let us consider first the so-called frequency effect (the effect of changing vibration frequencies by the electron transition due to the change of electron-vibration value). We shall demonstrate that a well-known Urbach rule for the band edge of

A. Andriesh and M. Bertolotti (eds.),
Physics and Applications of Non-Crystalline Semiconductors in Optoelectronics, 417–429.

optical absorption can be obtained on the basis of the frequency effect and Frank-Condon principle:

$$\alpha_{VIB}(\Omega)=\int_{-\infty}^{+\infty} e^{-q^2/\overline{q^2}}\delta(I_2(q)-I_1(q)-\hbar\Omega)dq \quad (1)$$

Here $I_1(q)$, $I_2(q)$ are the adiabatic potentials of the initial and final state of the electron-vibration system with the vibration frequencies of ω_1 and ω_2, respectively. The $\overline{q^2}$ value determines the temperature-averaged value of the vibration coordinate.

We find:

$$\alpha_{VIB}(\Omega)=\alpha_1 \exp\left\{\frac{\hbar\Omega-\Delta_{CV}}{M\left(\frac{\omega_1^2-\omega_2^2}{2}\right)\overline{q^2}}\right\} \quad (2)$$

$$\hbar\Omega<\Delta_{CV}$$

M is the oscillator mass.

The role of temperature manifests itself primarily in the value of the optical width of the band gap $\Delta_{CV}(T)$. For $\Delta_{CV}(T)$ value the low-frequency vibrational modes are generally of importance. In amorphous materials random electric fields play generally the main role. For example, for As_2S_3 according to [7] a mean intensity square of a random electric field is: $F_0^2 \cong 30\text{x}10^{12}\ V^2/cm^2$.

The computation of the temperature shift of the edge of absorption band due to electron-phonon interaction taking into account Frantz-Keldysh effect brought about by random electric fields can be performed similar to the case of the frequency effect taken into account in the theory of multiphonon optical transitions (see [1]). The expression for the absorption coefficient $\alpha(\Omega)$ with exponential accuracy has the form:

$$\alpha(\Omega)=\mathrm{Re}\, C\int_{-\infty}^{+\infty}\exp(-i\Omega t)I(t)dt \quad (3)$$

$$I(t)=\left\langle\left\langle e^{i\hat{H}_1 t/\hbar}e^{-i\hat{H}_2 t/\hbar}\right\rangle\right\rangle_{F,T}$$

$\hat{H}_1, \hat{H}_2$ are vibrational Hamiltonians taking also into account the influence of the random electric field; $\langle\langle...\rangle\rangle_{F,T}$ means the averaging over the distribution of random electric fields and thermal distribution in the initial electronic state. The expression for I(t) can be reduced for high temperatures ($\bar{n} \cong \frac{kT}{\hbar\omega_1}$) to the form:

$$I(t) = \left\langle \exp\left\{ i\left(\Delta_{CV} - \hbar\Omega\right)\frac{t}{\hbar}\right\} \cdot \frac{1}{\sqrt{1 + iQ\omega_1 t + R\omega_1^2 t^2}} \exp\left\{ i\int_0^t \frac{e^2 F^2 t_1^2}{2\mu\hbar} dt_1 \right\} \right\rangle_F$$

$$Q = 2\bar{n}\left(\omega_1 - \omega_2\right)\frac{1}{\omega_1} \tag{4}$$

$$R = \bar{n}\left(1 - \bar{n}\right)\left(\omega_1 - \omega_2\right)^2 \frac{1}{\omega_1^2}$$

where $\bar{n}$ is a mean filling factor for the oscillator with ω_1 frequency.

In expression (4) for the sake of simplicity only the field electron pulse is taken into account while its thermal pulse is omitted. For the computation of integral over dt in (3) (for large enough F) let us use the method of saddle point analysis, followed by averaging with Gauss distribution function [7,8]. We find:

$$\alpha_{tun}(\Omega) = \alpha_2 \exp\left\{ 3\left(\frac{\sqrt{2\mu}}{3e\hbar F_0}\right)^{2/3} \left(\hbar\Omega - \Delta_{CV}(T)\right) \right\}$$

$$\Delta_{CV}(T) = \Delta_{CV} - 2\hbar\left(\omega_1 - \omega_2\right)\bar{n}\,; \quad \hbar\Omega < \Delta_{CV} \tag{5}$$

Let us consider the case of nonequilibrium vibrations. As it is shown in works of Scholten and his colleagues [9,10] for T=2 K the filling factor $\bar{n}$ for the high frequency mode ω_0 =480 cm^{-1} $\bar{n} \cong 0.2$, which corresponds to the effective vibration temperature T_{vib} =300 K. Here the life time of this vibration is several nanoseconds. If we compare the Raman spectra obtained in [9,10] with the similar ones observed by Cardona and colleagues [11] we can see extra lines shifted further into the red spectrum part than the peak of 480 cm^{-1}. The appearance of these lines is due to the fact that Scholten used laser of higher intensity than that used by Cardona, which according to our opinion, is the manifestation of the frequency effect (the intensity of extra lines increases with the increase of electron concentration in the excited states). In this case the heated-up high-frequency modes can lead to the shift of the absorption band edge even at helium temperatures. The slope of ln $\alpha(\Omega)$ curves now also depends upon the heating-up temperature in the high-frequency modes. Within the range of the intense heating-up one can use formula (2) which follows also from (3) under small fields and high temperatures. The phenomenological results for Urbach rule of the kind cited by Street [12] may also be the consequence of the basic formula (3).

The heating-up effects can be obtained by irradiating the amorphous specimens $\hbar\upsilon > \Delta_{CV}$ with an intensive source. Here it is important to take also into account the linear shifts of the parabola due to the electron-phonon interaction. This computation can be made also by formula (1). Here for the probability of optical absorption of quantum we find:

$$W_{12}(T) \sim \exp\left\{-\frac{A(\Omega)}{T}\right\} \tag{6}$$

The type of constant $A(\Omega)$ is conventional for curves of Gauss type of frequency and we won't need it directly. Within the classical limit of $kT > \hbar\omega_{VIB}$ one can write the following equilibrium equation:

$$k\frac{dT}{dt} = \left(\hbar\Omega - \Delta_{CV}\right) W_{12}(T) - k\frac{\left(T - T_0\right)}{\tau} \tag{7}$$

T_0 is the temperature of thermostat; τ is the life time of the long-lived vibration of ω_{vib} frequency. If $kT<\hbar\omega_{vib}$ then it is possible to use equation (7) where $\bar{n}$ replace for T. The stationary solution of Eq.(7) is found from the condition:

$$\left(\hbar\Omega - \Delta_{CV}\right) W_{12}(T) = k\frac{\left(T - T_0\right)}{\tau} \tag{8}$$

Let $W_{12}(T) = W_0(T)\,\dfrac{I}{I_0}$

Here I is the excitation intensity, I_0 is the dimension parameter, determining $W_0(T)$.

Graphical solution of Eq.(8) determines the vibrational temperature different from the thermostat temperature T_0. We found for T three values of the roots of this transcendental equation $T_1<T_2<T_3$, out of which T_1 and T_3 are stable while T_2 is instable. By changing the system parameters (either T_0 or the excitation intensity I) one can obtain monostable and bistable sections for the hysteretic dependence of the vibrational temperature upon parameters T_0 and I. The bistable sections are in agreement with so-called vibrational bistability which was first obtained for macromolecules by the author and his colleagues in [13]. Similar hysteretic dependences appear also for the rates of optical and nonradiative transitions. Hysteretic dependences can be observed experimentally by successive changing of I from the increase to the following decrease.

Since the process of setting-up of vibrational temperature proceeds rather quickly (a shorter exchange time between the high-frequency vibrations relative to the exchange time with the thermostat), it is possible to use a pulse excitation. The leading pulse front prepares the system on the lower branch of the hysteresis loop while the rear pulse front makes the same on the upper branch of the hysteresis (see Fig.1). Similar hysteresis loops were probably observed in the work of Andriesh, Chumash et al.[14]. However the upper monostable sections were not fixed on them. This problem requires further investigation.
The existence of vibrational bistability is of special interest for the analysis of processes of nonradiative transitions with the excited nonequilibrium vibrations. The well-known Mott's law:

$$\sigma = const \exp\left\{-\frac{B}{T^{1/4}}\right\} \tag{9}$$

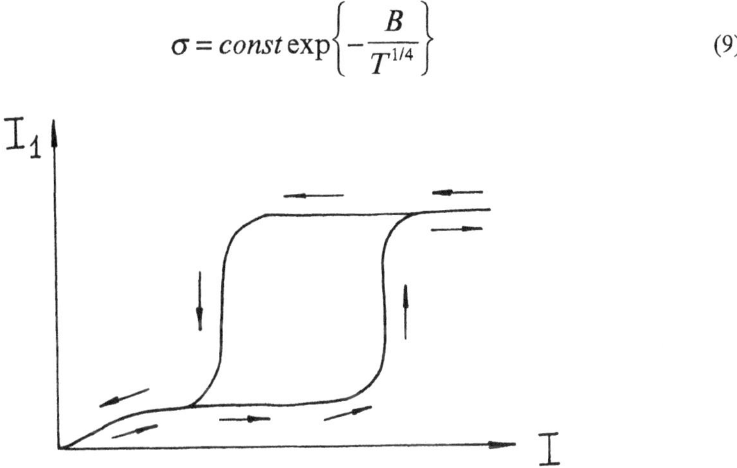

Fig.1 Intensity (I_1) dependence of penetrative light through plate amorphous material via intensity of pumping I.

appears in the assumption about the activation character of the surmount of potential barriers in amorphous materials. The presence of high-frequency vibrational modes facilitates the surmount of the potential barrier. T in Mott's law defines vibrational temperature. The two obtained possible values of the vibrational temperature T_1 and T_3 will lead in accordance with Mott's law to two possible conductivity values of amorphous material, namely, to the low-resistance state. Naturally, the measurement of conductivity must be carried out on the life-time of non-equilibrium vibrations.

2. Squeezed light and squeezed vibrations in solids

The squeezed states of oscillator are determined as the states under which the wave packet composed of a linear combination of oscillatory functions $\varphi_n(x;t)$ has the width of the packet which is larger (super-Poisson case) or less (sub-Poisson case) than that of the packet which is in agreement with the coherent states (Poisson distribution). Below the case will be discussed of the super-Poisson distribution of C_n coefficients in the wave packet $\psi(x;t)$:

$$\psi(x;t)=\sum_n C_n\,\varphi_n(x;t) \tag{10}$$

One can see that by the excitation of the oscillator from the electronic state 1 with a vibrational frequency ω_1 into a higher electronic state 2 with the frequency ω_2 (see Fig.2) under the effect of short pulse F(t) of τ length so that $\tau^{-1} >> \omega_1$, a packet of squeezed states is produced with a wave function of vibrations $\Psi_2(q;t)$:

$$|\psi_2(q;t)|^2=\frac{1}{\sqrt{\pi}\,\sigma(t)}\exp\left\{-\frac{(q-q_0(t))^2}{\sigma^2(t)}\right\}$$

$$q_0(t)=\Delta q\cos\omega_2 t \tag{11}$$

$$\sigma^2(t)=\sigma_0^2\left(\eta^2\cos^2\omega_2 t+\frac{1}{\eta^2}\sin^2\omega_2 t\right)$$

$$\sigma_0^2=\frac{\hbar}{m\omega_2};\qquad \eta=\frac{\omega_1}{\omega_2}$$

where m is the oscillator mass.

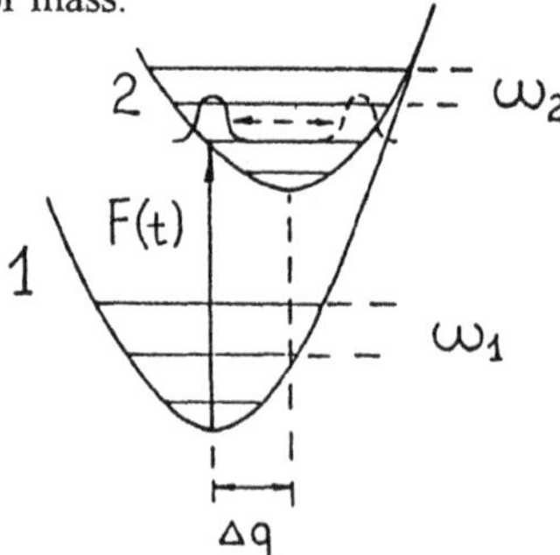

Fig.2 Excitation squeezed states at electronic transition $1\rightarrow 2$ under effect supershort pulse of electromagnetic radiation F(t).

In the author's works [15,16] it is shown that squeezed states for Landau's oscillators can be produced either in crystal or in parabolic quantum wells. The peculiarity of such oscillators consists in the fact that they carry a charge and therefore can emit electromagnetic waves. The analysis has shown that these electromagnetic fields are of nonclassic type and have the dispersion in agreement with the squeezed state. According to the estimate of these works the number of photons generated during the pulse is not less than 10^7 under reasonable excitation intensities. Such squeezed light can be used for the superhigh-frequency modulation of optic signals (for more details see the author's work [17]). As applied to the electron-vibrational systems with long-lived vibrations (some molecules, high-frequency modes of amorphous solids) it can be shown that nonradiative transitions with the participation of squeezed vibrations lead to the so-called nonradiative nonequilibrium transitions [18].

Here the Arrenius transition rate:

$$W_{12} \sim \exp\left\{-\frac{\Delta_{CV}}{kT}\right\} \tag{12}$$

is replaced by the law:

$$W_{12} \sim \exp\left\{-\frac{\Delta_{CV}}{D^2}\right\} \tag{13}$$

Where D^2 is the dispersion corresponding to the super-Poisson distribution $D^2 \sim \sigma^2(t_0)$. For large squeeze coefficients η $D^2 >> kT$ and the rate of nonradiative transitions sharply (exponentially) increases. As applied to amorphous materials the squeezed states for the vibrational high-frequency modes, taking into account the frequency effect, can be prepared by the excitation of the semiconductor with Anderson localization near the valence band in the localized states near the conduction band. In its turn, the jump mechanism of conductivity for electrons is a nonradiative process which can greatly increase by the preparation of squeezed vibrational state. Other manifestations of the squeezed states effect are also possible, e.g. in nonradiative recombination of carriers onto the impurity centres. A significant heating-up of the vibrational subsystem in the area of the light spot (or mask) can lead to the local narrowing of the band gap and the formation of a peculiar dynamic quantum well which does not require matching of the lattice parameters.

3. Generation of higher harmonics on solids

Let us consider the generation of harmonics by the interaction of pulses of intensive laser radiation with a crystal. First we shall derive the basic formulas and then discuss the conditions of their observation.

The value of the mean dipole moment is determined by the formula:

$$\overline{X}(t) = \int \Psi^*(\vec{r};t)\, x \Psi(\vec{r};t)\, d\vec{r} \tag{14}$$

The harmonic intensity at a frequency Ω is determined by the value $|X_\Omega|^2$:

$$X_\Omega = \frac{1}{2\pi}\int e^{-i\Omega t}\,\overline{X}(t)\,dt \tag{15}$$

We use Kane model of a semiconductor governed by the dispersion law:

$$\varepsilon_{\vec{p}} = \varepsilon^c_{\vec{p}} - \varepsilon^V_{\vec{p}} = \sqrt{\Delta^2_{CV} + \frac{\Delta_{CV}}{\mu}\left(\vec{p} + \frac{e\vec{F}}{\omega}\sin\omega t\right)^2}$$

$$\mu^{-1} = m_C^{-1} + m_V^{-1} \tag{16}$$

Here Δ_{CV} is the width of the band gap; m_c, m_v are the effective masses of the electron and hole.

The wave functions of the electron and hole in the field of linearly polarized (along theX-axis) electromagnetic wave $\mathcal{F}$=F cos ωt have the form [19]:

$$\psi^{(c)}_{\vec{k}(t)}(\vec{r}) = e^{i\vec{k}(t)\vec{r}}\exp\left\{-\frac{i}{\hbar}\int_0^t \varepsilon^c_{\vec{k}(t_1)}\,dt_1\right\}U_C(\vec{r})$$

$$\psi^{(V)}_{\vec{k}(t)}(\vec{r}) = e^{i\vec{k}(t)\vec{r}}\exp\left\{-\frac{i}{\hbar}\int_0^t \varepsilon^V_{\vec{k}(t_1)}\,dt_1\right\}U_V(\vec{r})$$

$$\vec{k}(t) = \vec{k} + \frac{eF}{\hbar\omega}\sin\omega t \tag{17}$$

$_c(\vec{r}), U_V(\vec{r})$ are the periodical parts of Bloch function. Since the transverse pulse $\vec{p}_\perp$ ($p_\perp = p_y^2 + p_z^2$) is not perturbed by the electromagnetic wave it is usually assumed equal to zero. Matrix element $X_{12}(t)$ of the interband transition in Kane model has the form [20]:

$$X_{12}(t) = \frac{i\hbar}{2}\sqrt{\frac{\Delta_{CV}}{\mu}}\,\frac{1}{\Delta_{CV} + \frac{\hbar^2}{\mu}\left(k_x + \frac{eF}{\hbar\omega}\sin\omega t\right)^2} \tag{18}$$

We assume that the following condition is fulfilled for the amplitude of the magnetic field:

$$eFX_{12}(0) \ll \Delta_{CV} \tag{19}$$

This makes it possible to restrict to the lowest order of the perturbation theory over the interband transition, but to take accurately into account the effect of the electromagnetic field on the movement of electrons and holes within the bands "c" and "v".

Since the Hamiltonian of the system of zero approximation evidently includes the interaction with the magnetic field, we use the so-called Furry representation [1] and Green function method. In this case we obtain the following expression:

$$\overline{X}(t) = 2I_m\left(\frac{-i}{\hbar}\frac{eF}{\omega\mu}\int_{-\infty}^{+\infty}dt'\sum_{k_x}\theta(t-t')\,X_{12}(t)\,p_{12}(t')\sin\omega t'\exp\left\{-\frac{i}{\hbar}\int_t^{t'_*}\varepsilon_{k_x(t_1)}\,dt_1\right\}\right) \tag{20}$$

Here:

$$p_{12}(t') = \mu\frac{\Delta_{CV}}{\hbar}X_{12}(t')$$

Let us place the preexponential factor $p_{12}(t')$ in Eq.(20) outside the integral sign over dt' at the saddle point t'_* for $k_x=0$:

$$t'_* = \frac{i}{\omega}arcsh\frac{\omega}{\omega_{tun}}$$

$$\omega_{tun} = \frac{eF}{\sqrt{\mu}}\sqrt{\frac{\Delta_{CV}}{\Delta_{CV}^2 - \hbar^2\omega^2}} \tag{21}$$

When taking the integral over $dk_x\left(\sum_{k_x}\ldots = \frac{1}{2\pi}\int\ldots dk_x\right)$ we use the evident form of the interband matrix element $p_{12}(t'_*)$ containing the poles:

$$k_x^{(1)} = ik_0æ^2/2\sqrt{2}; \qquad k_x^{(2)} = -ik_0\sqrt{2}$$

$$\frac{\hbar^2k_o^2}{2\mu} = \Delta_{CV}; \qquad æ^2 = \left(\frac{\hbar\omega}{\Delta_{CV}}\right)^2 \tag{22}$$

Matrix element $X_{12}(t)$ can be reduced due to the application of the conservation law:

$$\hbar\Omega=\varepsilon_{\vec{k}(t)} \tag{23}$$

Choosing the integration contour in the upper half plane, we compute the integral value over dk_x. This allows to proof the applicability of Jordan lemma and the correctness of choice of the saddle point t'_* for k_x=0. The integrals over dt' and dt have no peculiarities and their values can be computed under the parabolic dispersion law. We use the formula:

$$\exp\{iz\sin\varphi\}=\sum_{n=-\infty}^{+\infty}J_n(z)\,e^{in\varphi} \tag{24}$$

where $J_n(z)$ is Bessel function of real argument.

$$X_\Omega=X_o\sqrt{\rho}\left[\sum_{m=-\infty}^{+\infty}\left(J_m\left(\frac{\rho}{2}\right)J_{m-s}\left(\frac{\rho}{2}\right)+J_m\left(\frac{\rho}{2}\right)J_{m-s-1}\left(\frac{\rho}{\varepsilon}\right)\right)\frac{1}{m-s+a}\right]$$

$$a=\frac{1}{2}(q+\rho-1);\qquad \frac{\Omega}{\omega}=2s+1;$$

$$X_0=\frac{\pi}{8}\frac{\Delta_{CV}^{5/2}}{\hbar^2\Omega^2\omega^2\sqrt{\mu}};\qquad q=\frac{\Delta_{CV}}{\hbar\omega};\qquad \rho=\frac{\varepsilon_F}{\hbar\omega} \tag{25}$$

$$\varepsilon_F=\frac{e^2F_o^2}{4\mu\omega^2}$$

Let us use the relationship [21]:

$$\sum_{m=-\infty}^{+\infty}J_m(x)J_{m+l}(x)\frac{1}{m+a}=\frac{\pi}{\sin a\pi}J_{l-a}(x)\,J_a(x) \tag{26}$$

where ***a*** is not an integer.
Finally we find:

$$|X_\Omega|^2=X_0^2\,\rho\frac{\left(J_a\left(\frac{\rho}{2}\right)J_{s-a}\left(\frac{\rho}{2}\right)-J_{a+1}\left(\frac{\rho}{2}\right)J_{s-a}\left(\frac{\rho}{2}\right)\right)^2}{\sin^2 a\pi} \tag{27}$$

For low field intensities $\left|X_{2s+1}\right|^2 \sim \left(F_0^2\right)^{2s+1}$ which is in agreement with the result of the perturbation theory. Fig.3 shows the dependence of the harmonic intensity upon its number for crystal GaAs Δ_{cv}=1.5 eV, $\mu \cong 0.06\ m_0$, $\hbar\omega$= 0.1 eV, F=4x10^6 V/cm.

The plateau is clearly visible in Fig.3 (see also [5]) up to the maximum harmonic $N_{max}=2s_{max}+1 = \dfrac{\Delta_{CV}+\varepsilon_F}{\hbar\omega}$ corresponding to the cut-off effect (ε_F is a ponderomotive energy). The observation of garmonic is possible in the reflected light as it was done in [4], in transmitted light for thin films of 300÷1000 A or in the range between 20th and 30th harmonics falling into the band gap, for voluminous specimens and the quantum of value $\hbar\omega$=0.05 eV.

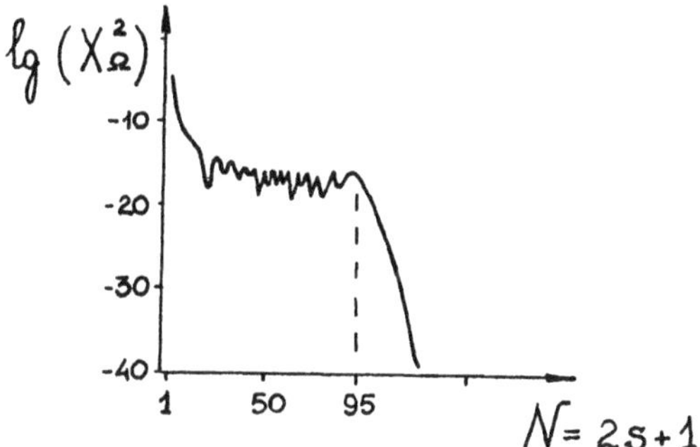

Fig.3. Harmonics intensity dependence on their number for q=15, ρ=40.

Harmonics can also lead to a one-quantum photoeffect out of crystal. Here the photoelectronic energy will, for the most part, repeat the type of distribution over energies shown in Fig.3 since in this area the absorption coefficient of harmonics varies a little. Thus, the generation of harmonics in solids can bring about the production of hard ultraviolet having an advantage over gaseous media due to the compactness of the radiation source of harmonics. In wide-band materials with a small effective mass for fields F=10^7 V/cm and $\hbar\omega$=0.1 eV one can obtain an electric source of soft coherent X-ray radiation.

Acknowledgements

The author expresses his acknowledgement to E.Kanarovsky, the scientific worker of the Institute of Applied Physics of AS, RM, for his helpful remarks.

REFERENCES

1. Kovarsky,V.A., Perelman,N.F.,and Averbukh,I.Sh. (1985) Multiquantum processes, Energoatomizdat, Moscow.
2. Levinstein,M., Balcou,Ph., Ivanov,M.Yu., L'Huiller,A. and Corcum P.B. (1994)

Theory of high-harmonic generation by low-frequency laser field, Phys.Rev. A49, 2117-2130.
3. Balcou,Ph., L'Huillier,A. and Escande,D. (1996) High-order harmonic generation processes in classical and quantum anharmonic oscillators, Phys.Rev. A53, 3456-3468.
4. Von der Linde,D., Engers,T. and Jenke A. (1995) Generation of high-order harmonics from solid surface by intense femtosecond laser pulses, Phys.Rev. A52, R52-R57.
5. Kovarsky,V.A. and Sedletsky,O.A. (in press) High harmonic generation by atoms with hydrogen-like ground state. The account of pole structure. Statistical aspects of the problem, Laser Physics.
6. Bycov,V.P. (1991) Major features of squeezed light, Usp.Fiz.Nauk 161 (10), 145-173.
7. Bonch-Bruevich,V.L., Zviagin,I.P., Kiper,R., Mironov,A.G., Underline,R. and Esser,B. (1981) Electronic theory of unregulated semiconductors, Nauka, Moscow.
8. Dow,J.D. and Redfield,D. (1972) Towards a unified theory of Urbach rule and exponential absorption edge, Phys.Rev. B5, 594-610.
9. Scholten,A.J., Akimov,A.V., Verleg,P.A.W.E., Dijkhuis,J.I. and Meltzer,R.S. (1993) Nonequilibrium phonon dynamics in amorphous silicon, J.Non-Crystalline Solids 164-166, 923-925.
10. Scholten,A.J., Akimov,A.V. and Dijkhuis,J.I. (1993) Nonequilibrium phonons in amorphous silicon studied by pulsed Raman spectroscopy, Phys.Rev. B47, 13910-13913.
11. Bermejo,D. and Cardona,M. (1979) Raman scattering in pure and hydrogenated amorphous germanium and silicon, J. Non-Crystalline Solids 32, 405-419.
12. Street,R.A., Searle,T.M., Austin, I.G. and Sussman,R.S. (1974) Temperature and field dependence of the optical absorption edge in amorphous As_2S_3, J.Phys.C: Solid State Phys. 7, 1582-1593.
13. Perelman,N.F., Kovarsky,V.A. and Averbukh,I.Sh. (1980) Vibrational bistability in nonequilibrium molecular gase on optical excitation, Zh.Eksp.Teor.Fiz. 79 (7), 21-32.
14. Chumash,V., Cojocaru,I.,Bostan,G., Cerbari,P. and Andriesh,A. (1994) Optical hysteresis and nonlinear propagation of laser pulses in chalcogenide glass films, SPIE Proceedings: Properties characteristics of optical glass III, San-Diego, V. 2287.
15. Kovarsky,V.A. (1992) Coherent and squeezed states of the Landau oscillators in solids. Emission nonclassical light, Fiz.Tverd.Tela 34, 3549-3553.
16. Kovarsky,V.A. (1993) Wave packets in quantum wells. Coherent and squeezed light emission, Phys.Stat.Sol.(b) 175, K77-K80.
17. Kovarsky,V.A. (1996) Superhigh frequency modulation of light by Rydberg atoms interacting with the electromagnetic field squeezed vacuum, Phys.Lett. A212, 195-200.
18. Kovarsky,V.A. (1996) Nonradiative transitions in molecules with excited squeezed vibrational states, Zh.Eksp.Teor.Fiz. 110 (10), 1-12.
19. Keldysh,L.V. (1964) Ionization in the field of a strong electromagnetic wave,

Zh.Eksp.Teor.Fiz. 47 (11), 1945-1957.
20. Bychkov,Yu.A. and Dykhne, A.M. (1970) Breakdown in semiconductors in an alternating electric field, Zh.Eksp.Teor.Fiz. 58 (5), 1734-1743.
21. Kovarsky,V.A. and Perelman,N.F. (1973) Peculiarities of dynamic polarizability in the systems with hydrogen-like spectrum in strong electromagnetic field, AS MSSR News: Series of physics-technical and mathematical sciences 1, 37-40.

BOSE-EINSTEIN CONDENSATION OF EXCITONS IN THE PRESENCE OF LASER RADIATION

S. A. MOSKALENKO, V. R. MISKO AND V. GH. PAVLOV
Institute of Applied Physics, Academy of Sciences of Moldova, Academy str. 5, Kishinev, MD-2028, Moldova

Abstract. The exciton absorption and emission bands in the presence of the Bose-Einstein condensation (BEC) of excitons induced by the coherent laser radiation is considered. The laser radiation creates the coherent macroscopic polarization of the crystal, which is described as a virtual BEC of excitons. The energy spectrum of the elementary excitations as well as the probabilities of the quantum transitions depend essentially on the frequency detuning $\tilde{\Delta}$.

In the neighbourhood of the regions where the new waves arise, the peculiarities of the absorption and emission bands appear. They are related with the properties of the nonequilibrium occupation numbers $n_{\vec{q}}^{ex}$ arising in the presence of virtual BEC and of coherent polarization of the crystal. In the vecinity of the instability region the ocupation numbers $n_{\vec{q}}^{ex}$ are finite but anomalous large values. This fact influences on the forms of absorption and emission bands.

1. Introduction

The optical Stark effect in the exciton range of the spectrum was studied in a set of papers [1-5]. The interpretation of the observed phenomenon proposed by Schmitt-Rink, Chemla and Haug [2,3] is based on the idea of the driven Bose-Einstein condensation (BEC) of excitons induced by external coherent laser radiation. In difference from the Keldysh and Kozlov theory [6] of the spontaneous BEC of excitons in the electron-hole description, in the case of the induced BEC the role of the chemical potential is played by the frequency of the laser radiation, whereas its amplitude plays the role of the condensate source. The induced BEC can be real but nonequilibrium

A. Andriesh and M. Bertolotti (eds.),
***Physics and Applications of Non-Crystalline Semiconductors in Optoelectronics,* 431–444.**

[7], when the coherent laser photons excite the resonant excitons with the same wave vector in the exciton band.

The nonresonant laser radiation gives rise to the virtual BEC of excitons [2-5]. Just this variant was carried out experimentally in the papers [1], where the laser photon energy was less than the energy of the lowest exciton level. During the switching on of the ultrashort laser pulse the blue shift of the exciton level was observed.

The level returned to the initial position after the switching off of the laser pulse. The theoretical investigations of this phenomenon [2-5] reveal some peculiarities of the excitons in semiconductors in the presence of laser radiation. For example Schmitt-Rink, Chemla and Hang showed that the virtual electrons and holes fill the band states in momentum space, giving rise to the phase space filling effect. This effect takes place in a similar way as in the case of spontaneous BEC studied by Keldysh and Kozlov [6]. The difference consists in the fact that the density of the virtual electron-hole pairs depends on the intensity of laser radiation. The exciton energy level in the Hartree-Fock-Bogolyubov approximation undergoes a blue shift. But as was shown by Zimmermann [8] this blue shift in the bulk crystals is compensated by the screening effect, i.e by the correlation corrections. As a result in the bulk semiconductors the exciton level remains practically unshifted on the energy scale, whereas the band gap between the valence and conduction bands undergoes the shrinkage when the density of electron-hole pairs increases. These investigations were effectuated in the limit of low density excitons, when the concentration n_{ex} obeys the condition $n_{ex}a_{ex} < 1$. Here a_{ex} is the exciton Bohr radius.

Elesin and Kopaev showed that the density of electron-hole pairs depends ambiguously on the laser intensity as well as on the frequency detuning. The amplitude and frequency hysteresises take place, what is intrinsic to the optical bistability effect [9]. The energy spectrum of the out-of-condensate quasiparticles was discussed in the papers [4-5]. The energy spectrum in the condition of the Optical Stark effect differs essentially from the case of the nonideal Bose gas [10]. These spectra coincide in one particular case, when the frequency detuning $\tilde{\Delta}$ between the shifted exciton level and the laser photon energy equals zero. The new important peculiarity of the energy spectrum is its instability [11], which appears in the case of negative frequency detunings $\tilde{\Delta}$. When the frequency of laser radiation exceeds the bottom of the exciton band, the conversion of two laser photons into two scattered excitons becomes possible. This process is real because the energy and momentum conservation laws can be fulfilled simultaneously. As a result the energy spectrum in some regions of the wave vectors becomes complex. One of two complex-conjugate solution has a positive imaginary part. That means the appearance of the unlimited growing amplitude of the wave, when the time tends to infinite. In such a way at $\tilde{\Delta} < 0$ in the system appear the new growing waves in some regions of the wave vectors. The system can be employed as a generator of the new waves or as an amplificator of the signals propagating through it. These instabilities are named as absolute and convective. Up till now the damping of the initial exciton level was naglected and the occurance of the instabilities was without threshold. If one takes into account the damping of the interacting quasiparticles, one can observe that the instabilities in the system will appear only when the amplitude of the induced Bose condensate exceeds some threshold value.

The instabilities have the threshold for their emergences. This fact was established in the papers [12,13]. Keldysh and Tikhodeev [12] studied the Stokes scattering of the coherent polaritons on the acoustical phonons. They poited out that near the threshold of the induced Mandelstam-Brillouin scattering but below it, in the range of the wave vectors, where the instability can appear, the Green's functions, of the scattering polaritons and phonons as well as the corresponding occupation numbers have the anomalous large values. They have divergency of the type $\frac{1}{\lambda}$ where λ determines the deviation of the concentration of the coherent polaritons n_{pol} from the threshold value n_c: $n_{pol} = n_c(1 - \lambda)$, where $0 < \lambda << 1$. In the paper [13] the induced combinational scattering of polaritons was investigated. It was found a spreaded threshold for this type of induced scattering. But in both cases the thresholds exist and the common phenomena of the induced scattering become available. In the vicinities of the instability regions the anomalous large values of the nonequilibrium occupation numbers of the out of condensate quasiparticles occur. This fact will influence on the forms of the exciton absorption and emission bands and will explain the new pecularities revealed below.

Below the absorption and the gain of a weak light signal propagating through the crystal macroscopically and coherently polarized by the external laser radiation will be studied.

2. The Hamiltonian and the probabilities of the optical quantum transitions

Following the papers [4,5], the Hamiltonian of the excitons interacting between themselves, with the laser radiation and with the photons of the wide-range, weak light source has the form

$$\begin{aligned} H &= \sum_{\vec{p}} E_{ex}(p) a_{\vec{p}}^{+} a_{\vec{p}} + \sum_{\vec{p}} \hbar c p C_{\vec{p}}^{+} C_{\vec{p}} \\ &+ \sum_{\vec{p}} \lambda_p (a_{\vec{p}}^{+} C_{\vec{p}} + a_{\vec{p}} C_{\vec{p}}^{+}) + \lambda_{\vec{k}_0} (a_{\vec{k}_0}^{+} C_{\vec{k}_0} + a_{\vec{k}_0} C_{\vec{k}_0}^{+}) \end{aligned}$$

$$+ \quad \frac{1}{2V} \sum_{\vec{p},\vec{q},\vec{k}} \nu(k) a^+_{\vec{p}} a^+_{\vec{q}} a_{\vec{q}+\vec{k}} a_{\vec{p}-\vec{k}}, \tag{1}$$

where $a^+_{\vec{p}}$, $a_{\vec{p}}$, $C^+_{\vec{p}}$, $C_{\vec{p}}$ are the creation and annihilation operators of excitons and photons correspondingly; $\nu(\vec{k})$ and $\lambda_{\vec{p}}$ are the constants of the exciton-exciton and exciton-photon interactions. It is supposed that the laser radiation has the wave vector $\vec{k}_0$ and the photon frequency $\omega_L = ck_0$. The antiresonant terms of the type $e^{\pm 2i\omega_L t}$ are neglected. The coherent laser radiation is introduced into the Hamiltonian (1) substituting the operators $C^+_{\vec{k}_0}$, $C_{\vec{k}_0}$ by the expressions of type

$$C_{\vec{k}_0} = \sqrt{F_{\vec{k}_0}} e^{-i\omega_L t - i\varphi}, \quad F_{\vec{k}_0} \sim V. \tag{2}$$

The quantum single particle states of the photons with $\vec{p} \neq \vec{k}_0$ describe the electromagnetic field of the vacuum and of the probe signal. The explicit dependence on time of the Hamiltonian (1), arising after the substitution of the expressions (2), can be removed if one goes over into the reference frame rotating with the laser frequency ω_L. This change is achieved with the help of the unitary transformation

$$\hat{V} = e^{-i\omega_L t \hat{N}}; \quad \hat{N} = \sum_{\vec{p}} (a^+_{\vec{p}} a_{\vec{p}} + C^+_{\vec{p}} C_{\vec{p}}) \tag{3}$$

which gives rise to a new Hamiltonian

$$\mathcal{H} = V^+ H V - \hbar\omega_L \hat{N}, \tag{4}$$

This Hamiltonian does not depend on time, but the energies of the quasiparticles are accounted from the laser photon energy $\hbar\omega_L$ as follows

$$\begin{aligned} \mathcal{H} &= \sum_{\vec{p}} \hbar \left[\omega_{ex}(p) - \omega_L\right] a^+_{\vec{p}} a_{\vec{p}} - \sum_{\vec{p}} \hbar \left(cp - \omega_L\right) C^+_{\vec{p}} C_{\vec{p}} \\ &+ \sum_{\vec{p}} \lambda_p \left(C^+_{\vec{p}} a_{\vec{p}} + a^+_{\vec{p}} C_{\vec{p}}\right) + \sqrt{F_{\vec{k}_0}} \left(a^+_{\vec{k}_0} + a_{\vec{k}_0}\right) \\ &+ \frac{1}{2V} \sum_{\vec{p},\vec{q},\vec{k}} \nu(k) a^+_{\vec{p}} a^+_{\vec{q}} a_{\vec{q}+\vec{k}} a_{\vec{p}-\vec{k}}. \end{aligned} \tag{5}$$

The terms linear on the operators $a^+_{\vec{k}_0}$, $a_{\vec{k}_0}$ were excluded after the Bogolyubov displacement operation

$$a_p = \sqrt{N_{\vec{k}_0}} e^{-i\phi} \delta_{\vec{p},\vec{k}_0} + \alpha_{\vec{p}}. \tag{6}$$

The macroscopical number $N_{\vec{k}_0} \sim V$ determines the fillment of the selected mode $\vec{k}_0$. It is related with the macroscopic number $F_{\vec{k}_0}$ of coherent photons of laser radiation by the formula [9]

$$\begin{aligned} n_{k_0} &= \frac{\lambda_{k_0}^2 f_{k_0}}{\tilde{\Delta}^2 + \gamma_{ex}^2}; \; n_{k_0} = \frac{N_{k_0}}{V}; \; f_{k_0} = \frac{F_{k_0}}{V}; \\ \tilde{\Delta} &= \hbar\Big[\omega_{ex}(k_0) - \omega_L\Big] + L_0; \; L_k = \nu(k) n_{k_0}. \end{aligned} \tag{7}$$

It is supposed that between the excitons the repulsion prevails and $L_k > 0$. The damping constant γ_{ex} was introduced phenomenologically. The relation $n_{\vec{k}_0}$ as function of $f_{\vec{k}_0}$ describes the phenomenon of optical bistability in the exciton range of spectrum [9]. Expanding the Hamiltonian (5) on the small operators $\alpha^+_{\vec{k}_0+\vec{k}}$, $\alpha_{\vec{k}_0+\vec{k}}$, with $\vec{k} \neq 0$, one can separate out the additive constant, the quadratic part and the terms of higher orders infinitesimal. Only the quadratic part will be kept below

$$\begin{aligned} \mathcal{H}^{(2)} &= \sum_{\vec{k}} \Big\{\hbar[\omega_{ex}(\vec{k}_0 + \vec{k}) - \omega_L] + L_0 + L_k\Big\} \alpha^+_{\vec{k}_0+\vec{k}} \alpha_{\vec{k}_0+\vec{k}} \\ &+ \frac{1}{2} \sum_{\vec{k}} L_k \left(e^{-2i\phi} \alpha^+_{\vec{k}_0+\vec{k}} \alpha^+_{\vec{k}_0-\vec{k}} + e^{2i\phi} \alpha_{\vec{k}_0+\vec{k}} \alpha_{\vec{k}_0-\vec{k}}\right) \\ &+ \sum_{\vec{k}} \hbar \left(c|\vec{k}_0 + k| - \omega_L\right) C^+_{\vec{k}_0+\vec{k}} C_{\vec{k}_0+\vec{k}} \\ &+ \sum_{\vec{k}} \lambda_{\vec{k}_0+\vec{k}} \left(C^+_{\vec{k}_0+\vec{k}} a_{\vec{k}_0+\vec{k}} + a^+_{\vec{k}_0+\vec{k}} C_{\vec{k}_0+\vec{k}}\right). \end{aligned} \tag{8}$$

The photons of the wide-range light source will be considered as a cause of the quantum transitions. For this reason only the exciton part of the Hamiltonian (8) will be diagonalized. In this case one can put $\phi = 0$ without less of generality. The diagonalization was achieved by the introduction of new operators $\xi^+_{\vec{k}}$, $\xi_{\vec{k}}$, employing the Bogolyubov canonical transformation

$$\xi_{\vec{k}} = \frac{\alpha_{\vec{k}_0+\vec{k}} + A_k \alpha^+_{\vec{k}_0-\vec{k}}}{\sqrt{1 - |A_k|^2}}; \; \alpha_{\vec{k}_0+\vec{k}} = \frac{\xi_{\vec{k}} - A_k \xi^+_{-\vec{k}}}{\sqrt{1 - |A_k|^2}}, \tag{9}$$

The coefficients A_k depend on the energy of the elementary excitations $\mathcal{E}(k)$ in the form

$$A_k = \frac{\tilde{\Delta} + T_k + L_k - \mathcal{E}(k)}{L_k}. \tag{10}$$

Here $\mathcal{E}(k)$ equals

$$\mathcal{E}(k) = \sqrt{\left(\tilde{\Delta} + L_k + T_k\right)^2 - L_k^2} \tag{11}$$

It is the isotropic component of the full energy $E(\vec{k})$ of the elementary excitation

$$E(k) = \mathcal{E}(k) + \hbar \vec{V}_s \vec{k}; \quad \vec{V}_s = \frac{\hbar \vec{k}_0}{m_{ex}}, \tag{12}$$

The full energy $E(\vec{k})$ is an anisotropic value, which depends on the velocity $\vec{V}_s$ of the moving condensate. The velocity $\vec{V}_s$ is determined by the photon wave vector $\vec{k}_0$ and by exciton translational mass m_{ex}. In order to obtain the coefficients $A_{\vec{k}}$, which obey the condition

$$|A_k| \leq 1, \tag{13}$$

in all the momentum space, it is necessary to choose the sign before the square root in formula (11) to be equal with the sign of the expression $(\tilde{\Delta} + T_k + L_k)$ following the rule

$$Sign \mathcal{E}_1(k) = Sign(\tilde{\Delta} + T_k + L_k). \tag{14}$$

This solution will be named $\mathcal{E}_1(k)$ and the corresponding coefficients will be marked as $A_{k,1}$. The energy spectrum determined in such a way remembers in general outline the dispersion law of the initial exciton branch of the spectrum in rotating reference frame $(\hbar\omega_{ex}(\vec{k}_0 + \vec{k}) - \hbar\omega_L + L_0 + L_k)$, which is present in the Hamiltonian (8). This branch of the spectrum will be named quasiexciton branch. Along with $E_1(\vec{k})$ there exists also another branch $E_2(\vec{k})$, which is determined by the value $\mathcal{E}_2(k) = -\mathcal{E}_1(k)$, and has the property

$$E_2(k) = \mathcal{E}_2(k) + \hbar \vec{V}_s \vec{k} = -\mathcal{E}_1(k) + \hbar \vec{V}_s \vec{k} = -E_1(-\vec{k}). \tag{15}$$

In its turn this solution remembers in general outline the initial quasienergy branch $(\hbar\omega_L - \hbar\omega_{ex}(\vec{k}_0 - \vec{k}) - L_0 - L_k)$. The coefficients $A_{k,2}$ are determined by the expression (10), where $\mathcal{E}_2(k)$ substitutes for $\mathcal{E}(k)$. The coefficients $A_{k,2}$ have the properties

$$A_{k,2} \cdot A_{k,1} = 1; \quad |A_{k,2}| \geq 1. \tag{16}$$

In the case of negative values $\tilde{\Delta}$ the values $\mathcal{E}(k)$ are pure imaginary in some region of wave vectors $\vec{k}$. In these regions the coefficients $|A_{k,i}| = 1$, the transformation (9) loses sense, and the generation of new waves takes place at the expense of the external laser radiation. In our case this phenomenon is without any threshold because the attenuation of the exciton level was not taken into account. Though there are two closely situated energy levels, namely, one quasiexciton level and the other quasienergy one nevertheless there is only one set of independent operators ξ_k^+, ξ_k with all possible values of $\vec{k}$. One can choose the operators $\xi_{k,1}^+, \xi_{k,1}$ for their set. They depend on the coefficients $A_{k,1}$ and energy $E_1(\vec{k})$.

The full set of operators $\xi_{\vec{k},1}^+, \xi_{\vec{k},1}$, together with the coefficients $|A_{k,1}| \leq 1$ and energy $E_1(k)$ is sufficient to describe the both type of elementary excitations as well as the quantum transitions in the system. Toward this end it is necessary to take into account both the resonant and antiresonant as regards to the elementary excitations terms and to consider both types of quantum transitions.

Below the index 1 at the operators $\xi_{\vec{k},1}^+, \xi_{\vec{k},1}$ and at the coefficients $A_{k,1}$ will be dropped but it will be kept only at the energy $E_1(\vec{k})$. After the diagonalization of the exciton part of the quadratic Hamiltonian (8) and the introduction of the new operators

$$\eta_{\vec{k}} = C_{\vec{k}_0+\vec{k}},$$

the Hamiltonian (8) will take the form

$$\begin{aligned} \mathcal{H}^{(2)} &= \sum_{\vec{k}} E_1(\vec{k}) \xi_{\vec{k}}^+ \xi_{\vec{k}} + \sum_{\vec{k}} \hbar \left(c|\vec{k}_0 + \vec{k}| - \omega_L \right) \eta_{\vec{k}}^+ \eta_{\vec{k}} \\ &+ \sum_{\vec{k}} \frac{\lambda_{\vec{k}_0+\vec{k}}}{\sqrt{1-|A_k|^2}} \left(\xi_{\vec{k}}^+ \eta_{\vec{k}} + \eta_k^+ \xi_{\vec{k}} - A_k^* \xi_{-\vec{k}} \eta_{\vec{k}} - A_k \xi_{\vec{k}}^+ \eta_{-\vec{k}}^+ \right). \quad (17) \end{aligned}$$

The ground state of the system polarized by the coherent laser radiation is the vacuum state for the annihilation operators $\xi_{\vec{k}}$

$$\xi_{\vec{k}} |0>_{ex} = 0. \quad (18)$$

Though in this state there are no elementary excitations with the energies $E_1(\vec{k})$, nevertheless there are different from zero occupation numbers of the initial exciton states

$$< a_{\vec{k}_0+\vec{k}}^+ a_{\vec{k}_0+\vec{k}} >= n_{\vec{k}_0+\vec{k}}^{ex} = \frac{|A_k|^2}{|1-|A_k|^2|}, \quad (19)$$

$$< a_{\vec{k}_0+\vec{k}} a_{\vec{k}_0+\vec{k}}^+ >= 1 + n_{\vec{k}_0+\vec{k}}^{ex} = \frac{1}{|1-|A_k|^2|}. \quad (20)$$

Thus in the presence of resonant or nonresonant laser radiation the high density exciton gas appears. It consists of real or virtual excitons. The latter differ from the former in the following. The virtual excitons exist only during the action of the laser radiation and disappear simultaneously with its disappearance. Unlike that the real excitons will continue to exist during their life time. In the zero order approximation the states of the polarized crystal and of the probe photon field are independent. The exciton-photon interaction is considered to be weak, serving only as a cause of quantum transitions. The strong interaction between the out-of-condensate excitons and photons in the presence of laser radiation was considered in [4].

3. Exciton absorption and gain bands in the condition of the optical Stark-effect

Following the paper [14] one can consider the quantum transitions under the probe photon action from the ground state of the coherently polarized crystal to the quasiexciton state with the wave vector $\vec{P}$. The initial and final states of the two-component system in the rotating reference frame are the following

$$\begin{aligned} |i> &= |0>_{ex}\, \eta^{+}_{\vec{Q}}|0>_{ph};\;\; E_i = \hbar\left(c|\vec{k}_0+\vec{Q}|-\omega_L\right), \\ |f> &= \xi^{+}_{\vec{P}}|0>_{ex}\,|0>_{ph};\;\; E_f = E_1(\vec{P}) = \mathcal{E}_1(P)+\hbar\vec{v}_s\vec{P}, \end{aligned} \qquad (21)$$

where $|0>_{ph}$ is the ground state of the probe photons. The amplitude of the quantum transition on the exciton-photon interaction Hamiltonian (17) equals to

$$< i|H_{int}|f> = \delta_{\vec{Q},\vec{P}}\frac{\lambda_{\vec{k}_0+\vec{Q}}}{\sqrt{1-|A_Q|^2}}. \qquad (22)$$

The transition probability was averaged on the initial states, which reduce to one state with well definite photon wave vector $\vec{Q}$, and was summarized on the all final states, i.e on the wave vectors $\vec{P}$. After that the following result was obtained.

$$P_{abs}(\vec{Q}) = \frac{2\pi}{\hbar}\cdot\frac{|\lambda_{\vec{k}_0+\vec{Q}}|^2}{|1-|A_Q|^2|}\delta\left(\hbar c|\vec{k}_0+\vec{Q}|-\hbar\omega_L-E_1(\vec{Q})\right). \qquad (23)$$

The probability depends on the wave vector $\vec{Q}$ accounted from the wave vector $\vec{k}_0$. The δ-function can be replaced by the Lorentzian, if one introduces the damping $\Gamma(\vec{Q})$ of the elementary excitation with the energy $E_1(\vec{Q})$ and employes the representation

$$\delta(\chi) = \frac{1}{\pi}\cdot\frac{\Gamma}{\chi^2+\Gamma^2};\;\; \Gamma\to 0. \qquad (24)$$

The full wave vector of the absorbed photon equals $\vec{q} = \vec{k}_0+\vec{Q}$. Its energy in the laboratory reference frame is $\hbar\omega = \hbar cq = \hbar c|\vec{k}_0+\vec{Q}|$. The transition probability as a function on wave vector $\vec{q}$ can be written in the form

$$P_{abs}(\vec{q}) = \frac{2}{\hbar}\cdot\frac{|\lambda_q|^2}{|1-|A_{\vec{q}-\vec{k}_0}|^2|}\cdot\frac{\Gamma(\vec{q}-\vec{k}_0)}{\left(\hbar\omega-\hbar\omega_L-E_1(\vec{q}-\vec{k}_0)\right)^2+\Gamma^2(\vec{q}-\vec{k}_0)};$$
$$\hbar cq = \hbar\omega. \tag{25}$$

The transition probability depends essentially on the orientation between the vectors $\vec{q}$ and $\vec{k}_0$. This dependence is reflected in a more evident way by the coefficients $|A_{\vec{q}-\vec{k}_0}|^2$ and will be discussed below. Side by side with this quantum transition there are also another three ones.

The Stokes process of the light emission together with the concomitant quasiexciton elementary excitation emission has the vacuum initial state and two particle final state as follows

$$\begin{aligned} |i> &= |0>_{ex}|0>_{ph}, \\ |f> &= \xi_P^+|0>_{ex}\,\eta_Q^+|0>_{ph}, \\ E_f &= \hbar|\vec{k}_0+\vec{Q}|-\hbar\omega_L+E_1(P); \quad E_i=0. \end{aligned} \tag{26}$$

This two-quantum transition involves the third term of the interaction Hamiltonian (17). The transition probability summarized on the all possible final states with a given photon wave vector $\vec{Q}$ and with arbitrary wave vector $\vec{P}$ equals to

$$P_{em}(\vec{Q}) = \frac{2\pi|\lambda_{\vec{k}_0+\vec{Q}}|^2}{\hbar}\frac{|A_Q|^2}{|1-|A_Q|^2|}\delta\left(\hbar c|\vec{k}_0+\vec{Q}|-\hbar\omega_L+E_1(-\vec{Q})\right). \tag{27}$$

Here the Lorentzian can be also introduced instead the δ - function. The energy conservation law resulting from this quantum transition looks as

$$\hbar c|\vec{k}_0+\vec{Q}|-\hbar\omega_L+E_1(-\vec{Q})=0. \tag{28}$$

It can be rewritten in the form

$$\hbar c|\vec{k}_0+\vec{Q}|=\hbar\omega_L+E_2(\vec{Q}), \tag{29}$$

that means the emission of the photon due the annihilation of one quasienergy elementary excitation, the energy of which in the laboratory reference frame is $\hbar\omega_L+E_2(\vec{Q})$. If one remembers taht $\hbar\omega_L+E_1(-\vec{Q})$ approximately equals to $\hbar\omega_{ex}(\vec{k}_0-\vec{Q})$, one can represent the equality (28) in the form

$$\hbar c|\vec{k}_0+\vec{Q}|+\hbar\omega_{ex}(\vec{k}_0-\vec{Q})\approx 2\hbar\omega_L. \tag{30}$$

It evidences that in the system takes place the transformation of two laser photons into one photon of the weak light source and in one out of-condensate exciton following the reaction

$$photon(\vec{k}_0) + photon(\vec{k}_0) = photon(\vec{k}_0 + \vec{Q}) + exciton(\vec{k}_0 - \vec{Q}). \quad (31)$$

These considerations show that the only cause of the light emission or of the gain of a weak light signal by the system is the external laser radiation. The difference of two probabilities (25) and (27) gives rise to the probability of a net absorption in one spectral region and of a gain or net emission in other spectral region. It is determined by the formula

$$\begin{aligned} P_{net}(\vec{Q}) &= P_{abs}(\vec{q}) - P_{em}(\vec{Q}) = \frac{2|\lambda_{\vec{k}_0+\vec{Q}}|^2}{\hbar} \\ &\times \left[\frac{1}{|1-|A_{\vec{Q}}|^2|} \frac{\Gamma(Q)}{\left(\hbar c|\vec{k}_0+\vec{Q}| - \hbar\omega_L - E_1(\vec{Q})\right)^2 + \Gamma^2(\vec{Q})} \right. \\ &\left. - \frac{|A_{\vec{Q}}|^2}{|1-|A_{\vec{Q}}|^2|} \frac{\Gamma(-\vec{Q})}{\left(\hbar c|\vec{k}_0+\vec{Q}| - \hbar\omega_L + E_1(-\vec{Q})\right)^2 + \Gamma^2(-\vec{Q})} \right]. \end{aligned} \quad (32)$$

and can be rewritten in dependence on $\hbar cq = \hbar c|\vec{k}_0 + \vec{Q}| = \hbar\omega$,

$$\begin{aligned} P_{net}(\hbar\omega) &= \frac{2|\lambda_{\vec{q}}|^2}{\hbar} \\ &\times \left[\frac{1}{|1-|A_{\vec{q}-\vec{k}_0}|^2|} \frac{\Gamma(\vec{q}-\vec{k}_0)}{\left(\hbar\omega - \hbar\omega_L - E_1(\vec{q}-\vec{k}_0)\right)^2 + \Gamma^2(\vec{q}-\vec{k}_0)} \right. \\ &\left. - \frac{|A_{\vec{q}-\vec{k}_0}|^2}{|1-|A_{\vec{q}-\vec{k}_0}|^2|} \frac{\Gamma(-\vec{k}_0-\vec{q})}{\left(\hbar\omega - \hbar\omega_L + E_1(\vec{k}_0-\vec{q})\right)^2 + \Gamma^2(\vec{k}_0-\vec{q})} \right]. \end{aligned} \quad (33)$$

Here it is reasonable to compare these expressions with the line shapes of the absorption and luminescence exciton bands at $T = 0$ when the spontaneous BEC of excitons was considered [15,18]. As was shown in [15] these line shapes consist from the sharp central peaks at the frequencies close to the energy of the condensed excitons and from the broad wings. The wing of the absorption band is situated on the side of higher energies relative to the central peak. During this quantum transition the elementary excitation is created. The intensity of this wing is determined by the coefficient $u_q^2 = 1 + n_q^{ex}$. The wing of the luminescence band is situated on the side of the lower energies relative to the central peak, because simultaneously with the emission of the light photon is emitted also the elementary excitation.

The intensity of this wing depends on the coefficients $v_q^2 = n_q^{ex}$, which are less than the coefficients u_q^2. Exactly the same properties reveal the expressions (25) and (27) in the case of induced BEC of excitons.

4. The anisotropy of the excitons absorption and emission bands in the case of coherently polarized crystal

In the fig.1 is represented the dependencies of the expresion $|1-|A_{\vec{q}-\vec{k}_0}|^2|^{-1}$ as the function of the argument $|\vec{q}-\vec{k}_0|$. Three different possible orientations were studied: when the probe light propagates parallel to laser radiation, antiparallel or perpendicular to this direction. This factor has the well defined anisotropy which manifests oneself when the coefficients $|A(\vec{q}-\vec{k}_0)|$ are close to unity. At high positive values of $\tilde{\Delta}$ the coefficients $|A_{\vec{q}-\vec{k}_0}|$ for the exciton like energy spectrum are much lower than 1 and the anisotropy of the factor $(1-|A(\vec{q}-\vec{k}_0)|^2)^{-1}$ is very small. At $\tilde{\Delta} = L_0$ the anisotropy is of the order of 2%. At the detuning $\tilde{\Delta} = 0$, the induced BEC of excitons transforms itself into the real but nonequilibrium BEC. In this case the occupation numbers $n_{k_0+k}^{ex}$ at $k \to 0$ diverge. The pronounced difference between the occupation numbers $n_{\vec{q}}^{ex}$

$$n_{\vec{q}}^{ex} = \frac{|A_{\vec{q}-\vec{k}_0}|^2}{1-|A_{\vec{q}-\vec{k}_0}|^2}; \; ; \; \vec{q} = \vec{k}_0 + \vec{k} \tag{34}$$

when $\vec{q}$ tends to $\vec{k}_0$ or when it tends to - $\vec{k}_0$ appears. Because the absorption and emission probabilities depend on the values $(1+n_{\vec{q}}^{ex})$, and $n_{\vec{q}}^{ex}$ correspodingly, for that reason these probabilities become anisotropic too.

At negative values $\tilde{\Delta}$, the instabilities in the system occur. In those regions of the wave vectors, where the energy spectrum is complex the coefficients $|A_k|^2 = 1$ and the corresponding occupation numbers (34) are infinite. In these regions the canonical transformation (9) has no sense. Here the generation of new waves due to the induced exciton combinational scattering takes place. In our case this generation is thresholdless because the damping of the exciton levels was neglected from the very beginning. In the neighbourhood of the instability regions the occupation numbers $n_{\vec{q}}^{ex}$ are finite but anomalous large. The similar behavior reveal the light absorption and emission probabilities. This fact is reflected in fig. 1. It is interesting to outline that the boundaries of the regions with anomalous values of $n_{\vec{q}}^{ex}$ also depend on the geometry of observation. Besides the factor $(1-|A(\vec{q}-\vec{k}_0)|^2)^{-1}$ the probabilities of the net transitions (32) and (33) contain also another factors, which have the Lorentzian forms. The Lorentzians can also reveal different frequency dependences at different

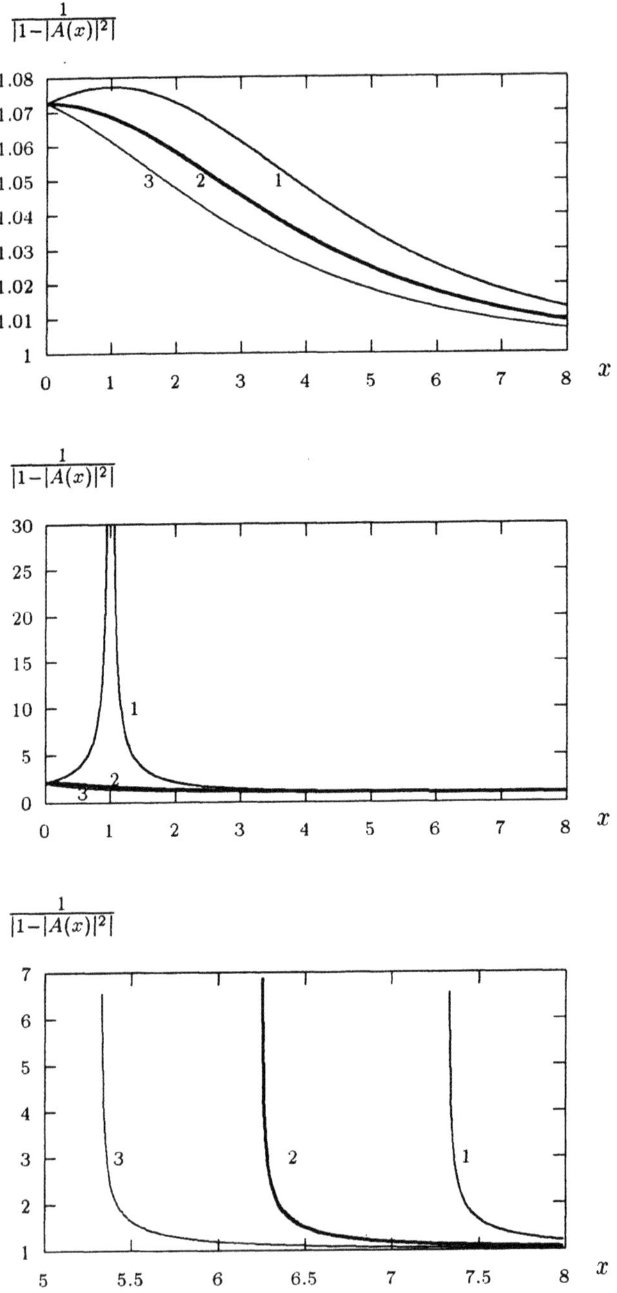

Figure 1. The frequency dependences of the function $(1 - |A(\vec{q} - \vec{k}_0)|^2)^{-1}$ for three geometries of the observation: 1)$\vec{q} \uparrow\uparrow \vec{k}_0$, 2)$\vec{q} \perp \vec{k}_0$, 3)$\vec{q} \uparrow\downarrow \vec{k}_0$. Three different frequency detunings were considered: $a)\tilde{\Delta} = L_0$, $b)\tilde{\Delta} = 0$ and $c)\tilde{\Delta} = -2 \cdot L_0$.

orientations between the vectors $\vec{q}$ and $\vec{k}_0$. But these dependences are much less than that of the first factor.

The fact is that the arguments of the δ-functions and of the Lorentzians represent mainly the small differences of the large values $\hbar\omega$ and $\hbar\omega_L$, each of which is much more than the energy $E_1(\vec{q} - \vec{k}_0)$ of the elementary excitation. In this condition it is difficult to reveal the anisotropy of the energy spectrum.

Nevertheless the presence of the elementary excitation energy $E_1(\vec{q} - \vec{k}_0)$ in the arguments of the expressions (32) and (33) is important. If one neglets these energies $\pm E_1(\vec{Q})$ and supposes $\Gamma(\vec{Q}) = \Gamma(-\vec{Q})$ the formulae (32) and (33) will express the isotropic net absorption. At large damping $\Gamma(\vec{Q}) \geq 2|E_1(\vec{Q})|$ the anisotropy will be not well defined. One can expect a pronounced anisotropy in the opposite case $\Gamma(\vec{Q}) < |E_1(\vec{Q})|$ and at small frequency detunings $\tilde{\Delta}$, when the coefficients $|A_Q|^2$ fall close to unity.

In conclusion one can remember that the anisotropy of the two photon transition from the ground state of the crystal to the ground state of the biexciton in the presence of the laser radiation was studied in the paper [16]. The degenerate exciton levels in the condition of the Optical Stark effect undergoes the splitting and polarization. These questions were studied by Combescot [17]. In our case the exciton level is non degenerate and the anisotropy depends on the propagation direction of the test light signal relative to the wave vector of the laser radiation.

References

1. Mysyrovicz A., Hulin D., Antonetti A. et all (1986) "Dressed excitons" in a multiple-quantum-well structure: evidence for an optical Stark effect with femtosecond responce time, *Physical Review Letters* **56**, 2748-2751
2. Schmitt-Rink S., Chemla D. (1986) Collective excitations and the dynamical Stark effect in a coherently driven exciton system, *Physical Review Letters* **57**, 2752-2755
3. Schmitt-Rink S., Chemla D. and Haug H. (1988) Nonequilibrium theory of the optical Stark effect and spectral hole burning in semiconductors, *Physical Review B* **37**, 941-955
4. Misko V. R., Moskalenko S. A. and Shmiglyuk M. I. (1993) Optical Stark effect in the exciton region inclusion of the polariton effect, *Solid State Physics* **35** (N12), 3213-3221 *(in Russian)*
5. Moskalenko S. A., Misko V. R. (1992) Indused Bose-Einstein condensation of biexcitons and excitons in the field of laser radiation, *Ukrainian Journal of Physics* 37 (N12), (1992), *(in Russian)*
6. Keldysh L. V., Kozlov A. N. (1968) Collective characteristics of excitons in the semiconductors, *Journal of experimental and theoretical physics* **54** (N3), 978-993, *(in Russian)*
7. Moskalenko S. A. (1962) Reversible optical hydrodinamical fenomena in the nonideal exciton gas, *Solid State Physics* **4** (N1), 276-284, *(in Russian)*
8. Zimmermann R.(1976) *Physica Status Solidy B* **76**, 191
9. Elesin V. F., Kopaev Iu. V. (1972) Bose-condensation of excitons in the strong electromagnetic field, *Journal of experimental and theoretical physics* **63** (N4), 1447-

1453, *(in Russian)*

10. Bogolyubov N. N. (1971) *Sobranie nauchnyh trudov v treh tomah*, Kiev: Naukova Dumka, *(in Russian)*
11. Lifshitz E. M. and Pitaevskii L. P. (1979) *Theoretical physics X*, §62-64, Moskwa Nauka, (in Russian)
12. Keldysh L. V. and Tikhodeev S. T. (1986) Nestatzionarnoe rasseianie Mandelishtama-Brilliuana intensivnoi poliarizovannoi volni, *Journal of experimental and theoretical physics* **91** N4(7), 78-85, *(in Russian)*
13. Misko V. R., Moskalenko S. A., Rotaru A. H. and Shvera Iu. M. (1991) Induced Bose-Einstein condensation of polaritons in cristalls of various dimensionalities, *Journal of experimental and theoretical physics* **99** (N4), 1215-1229, *(in Russian)*
14. Bobrysheva A. I., Shmiglyuk M. I., Russu S. S. (1993) Biexciton optical Stark-effect due to dynamical coupling of excitons and biexcitons, *Proc. SPIE* **1807**, 79-88
15. Moskalenko S. A. (1970) *Bose Einstein condensation of excitons and biexcitons*, Kishinev RIO, *(in Russian)*
16. Bobrysheva A. I., Moskalenko S. A. Kam H. N. (1993) Stimulated exciton pairing and polarization of biexcitons in a laser radiation field, *Journal of experimental and theoretical physics* **103** (N1), (1993), *(in Russian)*
17. Combescot M. (1990) Optical Stark effect of the exciton. II. Polarization effects and exciton splitting, *Physical Review B* **41** (N6), 3517-3533
18. Gergel V. A., Kazarinov R. F. and Suris R. A. (1968) O svoystvah Boze-Einshteinovskogo kondensata exitonov nizkoi plotnosti v poluprovodnikah, *in Proceedings of 9-th International Conference on Physics of semiconductors*, vol.1, 472,Moscow (Ed. Nauka, Leningrad, 1969), *(in Russian)*

THERMAL, PLASMA AND PHOTOINDUCED MICROSTRUCTURAL CHANGES ON a-SiC:H FILMS

M.Dinescu, D. Ghica*, N.Chitica, C. Stanciu, G.Dinescu, C.Ghica
Institute of Atomic Physics, P.O.Box MG-16, RO-76 900, Bucharest, Romania;
**Institute of Optoelectronics, P.O.Box MG-22, RO-76 900; Bucharest, Romania;*
A.Ferrari, M. Balucani, G. Maiello, S.La Monica, G.Masini
Dipartimento di Ingegneria Elettronica, Università di Roma "La Sapienza",
Via Eudossiana 18, 00184, Rome-Italy.

1. Introduction

Low cost a-SiC:H films are easily available on different substrates by Plasma Enhanced CVD. The aim of this work is the transformation of these films towards materials with applications in microelectronics (polysilicon, silicon carbide, nitride, oxinitride). Laser, plasma irradiation and thermal treatments in controlled atmosphere are possible candidates to modify the chemical composition and/or the structure of the a-SiC:H films.

2. Experimental

a-SiC:H films were grown on Si(100) wafers by PECVD [1]. Four sets of films, 2 μm thick, were used: a-SiC:H (50% C) and a-SiC:H (30% C), with (M1 and M3) and without (M5 and M2, respectively) hydrogen addition during deposition. Thermal treatments were performed in a N_2 flow furnace varying the temperature with 3°/min slope in the 500-1100°C range. RTA (rapid thermal annealing) were also performed. The processing time was between 4 and 480 minutes.

KrF excimer laser (λ=248 nm, t_{FWHM}=20 ns, E_0=0.1 J/pulse) treatment was performed in a stainless steel vacuum chamber, in Ar atmosphere [2,3]. The fluence on the sample surface was 0.2 J/cm^2, with less than 10% fluctuation in the spot area. Series of up to 200 pulses, with a repetition rate of 1 Hz were applied to the samples.

RF plasma beam treatment (duration 90 min) was performed in a dedicated reactor[4] having 100 W RF emission at 13.56 MHz. The discharge zone is placed outside the reactor and is limited by two electrodes spaced of 1-2 mm (a disk connected to the RF power and a nozzle fitted with the stainless steel wall of the grounded reactor chamber). The sample is placed in the downstream plasma at 100 mm from the nozzle, parallel to it. Two pairs of electrodes were used: graphite (the samples coded as Si 10)and aluminium (samples coded as Si 30).

Before and after the treatment the samples were investigated by different techniques. Optical investigation of the surfaces was made with a LEITZ optical microscope. XRD analysis were conducted with URD6 apparatus with nickel filter, using Cu-Kα line (l = 1.54 Å). The FTIR spectra were recorded with a Mattson 5000 spectrometer, with a resolution set at 0.5 cm^{-1}. XPS analysis were conducted with a VG ESCALAB MK II spectrometer, using the Al-Kα radiation (1486.6 eV) for excitation. The microhardness (MH) was measured with a Vickers Leitz Durimet apparatus. The indentations were performed with 25, 50 and 100 g load.

A. Andriesh and M. Bertolotti (eds.),
Physics and Applications of Non-Crystalline Semiconductors in Optoelectronics, 445–450.

3. Results and discussion

3.1. THERMAL TREATMENT

3.1.1.Optical microscopy investigation

Thermal treatments with low slope temperature increasing (3°C/min) produced mechanical stress and cracks in all the samples. The surfaces divided in separate islands, with average size related to process temperature and duration. Attributing this surface depreciation to the difference in the dilatation coefficients at the interface film ($4.3\text{-}4.7\cdot10^{-6}$ /°C) - Si ($2.6\text{-}3\cdot10^{-6}$ /°C), we used also rapid treatments (RTA). RTA at 900°C for 4 min and 1080°C and 10 min, evidenced smooth surfaces with area up to 10^5 μm^2, only for the 30%C SiC films.

3.1.2.FTIR measurements

The presence of the Si-C bands at 670 and 780 cm^{-1} [5,6] was observed for all the samples. The position of a band is dependent on the electronegativity of the attached group, which tends to rise the frequency towards more stable forms [6]. For a-SiC:H, the looseness of H_2 bonding, simultaneously with an increasing crystallinity may shift the maximum of the absorption to higher energies, due to the electronegativity of SiC with respect to a-SiC:H. Thermal treatments showed a shift of the 780 cm^{-1} peak towards 900 cm^{-1}, as a function of the processing time and temperature (up to 800-900°C) as well as on the C content. A peak intensities increase was also observed.

The samples with 30%C without H_2 presented the clearest modifications, especially when using RTA: the intensity - width ratio of the 900 cm^{-1} peak is double with respect to the same ratio for slow thermal treatments, at the same temperature.

3.1.3.XRD investigations

XRD analysis on 50%C a-SiC films treated by RTA (900°C, 10 min) evidenced the crystallinity of the thin films. A peak centred at 38.4° and attributed to a SiC(102) crystalline phase is clearly evidenced in XRD spectra. Secondary peaks, with lower intensity, centered at 17° and at 64.8° also correspond to the SiC crystalline phase.

3.1.4.Microhardness measurements

We measured 600-800 Kg/mm^2 for the Si substrate, and 200-300 Kg/mm^2 for untreated a-SiC:H films. RTA applied at 900°C and 1080°C for 10 min to 30%C a-SiC films produced the highest value for thermally treated films, 1100 Kg/mm^2.

3.2. LASER TREATMENT

3.2.1.Optical microscopy investigation

The inspection of the irradiated films revealed in all cases a smooth surface. Only rarely some small defects due to impurities overheating can be distinguished.

3.2.2.FTIR measurements

FTIR spectra on 50% carbon content films were measured on sample not irradiated, irradiated with 100 pulses and irradiated with 200 pulses .

The presence of the Si-C bands at 670 and 780 cm^{-1}[4,6] was evidenced in all spectra. After irradiation with 100 or 200 laser pulses, a reduction of the peak width while increasing the number of pulses was observed. For the sample irradiated with 200 laser pulses the reduction of the peak width at 780 cm^{-1} was 10 %. The centre of the same peak shifted toward 800 cm^{-1}. At the same time, the centre of the peak at 670 cm^{-1} shifted towards 700 cm^{-1}. The decrease of the peak from 1100 cm^{-1} showed the reduction of the oxide content. All these facts could be partially explained with the breaking of the bonds and outgassing of hydrogen/oxygen. However, they should be correlated with the strong increase of the bands attributed to Si-CH_3 wagging mode (1240-1340 cm^{-1}) revealing also a transformation towards a film with higher content of such bonds.

The FTIR spectra of the samples with 30% carbon content are quite similar to the previous. The spectra of films deposited in high H_2 partial pressure showed a different general aspect; nevertheless, increasing the pulse number, the Si-C peaks at 670 and 780 cm^{-1} showed an intensity increase and a shift towards 800 cm^{-1}.

3.2.3.XRD investigations

In the diffraction patterns of samples with 50% carbon, a peak centered at 38.3° is present and attributed to a SiC(102) crystalline phase after 200 laser pulses. The appearance of a crystalline phase is as an effect of the laser treatment. For the sample with 30%C, after 200 pulses the 38.3° peak appears, but with a lower intensity and a larger width, revealing in this case the presence of a dominant amorphous phase.

The XRD data related to 30%C samples deposited in a high hydrogen partial pressure are different from the previous. A peak centred at 38.4° and attributed to (102)crystalline phase is present for the unirradiated sample. This indicates the presence of a certain amount of crystalline silicon carbide just in the as-deposited layer. After 200 laser pulses, the intensity of the peak increases (with no change in the FWHM), showing the crystalline SiC increasing amount, due to laser treatment.

3.2.4.XPS analysis

For all the samples the regions of the Si2p, C1s and O1s peaks were investigated. The region Si2p of the XPS spectrum was used in all the cases to estimate the relative amount of SiC, Si and SiO_x, respectively. We considered the recorded signal as the superposition of the signals corresponding to Si atoms bonded in pure Si (denoted by Si-Si), in pure, stoichiometric SiC (denoted by Si-C) and in a silicon oxide SiO_x (denoted by Si-O). We present in Table I the estimated relative amounts of Si-C, Si-Si and Si-O obtained for each sample by fitting the corresponding spectrum using the characteristics of each of these peaks [7-10]. The samples denoted A were irradiated with 100 pulses and the samples denoted B with 200 pulses.

The first series of 100 laser pulses does not modify the amount of Si-C bonds (see M1A in Table I). In the same time, a part of the Si-Si bonds are converted to Si-O. The next series of 100 laser pulses produces an important increase of the amount of Si-C bonds (M1B). Another positive result is the significant decrease of the Si-O relative amount from 62 % to 40 %. This could be the effect of free carbon diffusion from the bulk of the film to the top oxidised layer. Here, carbon atoms could induce, under laser irradiation, the decomposition of the SiOx compound. As a result, the relative amounts of Si-Si and Si-C bonds increase in this layer.

Table I. The relative amount of Si-C, Si-Si, and Si-O bounds (%).

Sample	Si-C	Si-Si	Si-O
M1	26	30	44
M1A	26	12	62
M1B	34	34	32
M2	19	26	55
M2A	14	71	18
M2B	29	38	33
M3	14	60	26
M3A	12	46	42
M3B	5	37	58

Note that XPS apparatus resolves less than 100Å under sample surface. Considering that Si is covered by a some tens of Å thick native oxide layer, the 34 % SiC content detected in the top layer could be considered a large amount.

The sample M2 has a lower amount of Si-C bonds, as compared to the sample M1 (see Table I). This could be due the lower amount of carbon available during deposition. The initial amount of Si-O bonds is also higher. Laser irradiation of the M2 samples induces a very important reduction of the relative amount of Si-O bonds (M2A). The SiO_x compound could be reduced to Si by hydrogen under laser irradiation. It should be noted that in these films a relatively large amount of hydrogen is incorporated during growth. Thus, the amount of Si-Si bonds increases.

The amount of Si-C bonds significantly increases (up to 29 %) after the second series of 100 laser pulses (M2B). The Si-O relative amount is of only 33 %. We suppose that, during the second series of

100 pulses, the Si atoms resulting from the decomposition of SiOx react with C atoms to form SiC. This process could start when the amount of hydrogen was reduced enough (by outgassing because of temperature increasing). When the hydrogen amount decreases, an oxidation process could also take place at the surface, increasing the amount of Si-O bonds.

For samples M3 the "as deposited" film has a low amount of Si-C bonds (only 14 %) and 60 % of the Si atoms are in the Si-Si state. Laser irradiation induces the diffusion of oxygen in the film and the oxidation of silicon. For the first 100 laser pulses, mainly the Si-Si bonds are converted to Si-O bonds (M3A). Thus, after 100 pulses the amount of Si-Si decreases to 26 % and the amount of Si-O increases to 42 %. Only 2 % of Si-C bonds are converted to Si-O. The next 100 laser pulses enhance the oxidation of the layer (M3B). The relative amount of Si-O reaches a value of 51 %, the amounts of Si-Si and Si-C decreasing to 37 % and 5 %, respectively. Thus, the increase of the oxide amount in the top layer of the film is 16 % for each series of 100 pulses. We could assume that the amount of carbon in this layer is too low to avoid or to reduce oxidation.

3.2.5.Microhardness measurements

We measured 600-800 Kg/mm^2 for the Si substrate, and 200-300 Kg/mm^2 for untreated a-SiC:H films.The highest MH value, of 1400 Kg/mm^2, was measured on the M1B sample. This increase can be related to the improvement of the crystallinity of the layer as an effect of the excimer laser irradiation at low incident fluences.

3.3. PLASMA JET TREATMENTS

The emission spectra of the plasma show that treatments take place in presence of many radiative species - OH, NH, γNO, CN. The CN radical is formed by the reaction of nitrogen with carbon atoms sputtered from the electrodes (enhanced by the selfbias of the RF electrode) and/or by the reaction of atomic nitrogen with solid carbon at the electrode surface. This last reaction is exothermic [11] and can be an important channel of transport of carbon from the discharge zone to the treatment zone. Due to all these reactive species, compounds as SiC, Si_3N_4 and CN should be formed during the treatment of the initial deposited layers.Table II presents the codification of all the samples treated by plasma jet.

Table II. Sample codification and treatment conditions.

Sample code	Sample composition	Treatment parameters Gas	Electrodes	Pressure	Temp	Duration
M1-B	C50% noH_2	nitrogen	aluminium	Pd=8torr	200°C	10min
M1-C	C50% noH_2	ammonia	aluminium	Pd=8torr	40-50°C	20min
M1-D	C50% noH_2	nitrogen	aluminium	Pd=8torr	40-50°C	20min
M2	C30% withH_2	nitrogen	graphite	Pd=10torr	40-50°C	90min
M5	C50% withH_2	nitrogen	graphite	Pd=10torr	40-50°C	90min

3.3.1.Optical microscopy investigation

Optical microscopy reveals a smooth, unperturbed surface. The colour of the irradiated areas changed, showing a modification of the structural and/or compositional properties of the film.

3.3.2.FTIR spectra

The FTIR spectra of all targets have some common characteristics as: i) the presence of Si-C specifies bands at 670 cm^{-1} and 780 cm^{-1} [2,3] and ii) a Si-O band in the range of 1000-1100 cm^{-1}. A large peak corresponding to the Si-N bond is present in the spectrum of the M2 sample, at aprox. 3400 cm^{-1}. The peak centred at 2100 cm^{-1} and attributed to the stretching of Si-H of species, as H_2SiC_2 or H_3SiC [2] is slightly shifted and becomes larger than the one of the untreated sample. After deconvolution, contributions of H-SiC_xN_{3-} [12] and C-N [13] can be identified. The band near 1720 cm^{-1} disappears, indicating that the carbonyl-like species are desorbed or reacted. In the FTIR spectra of M5 treated samples, the peak centred at 3400 cm^{-1} does not appear. A contrary behaviour can be observed also to the peak centred at 2100 cm^{-1}, it shifts towards 2000 cm^{-1}. A small peak corresponding to the C-N triple bond is observed at 2200 cm^{-1}. An explanation for the different behaviour could be that M2 samples contain Si in excess as respect to C, this favouring the Si-N reaction. For M5 samples the higher carbon content sustains the appearance of C-Si, C-N reactions, but does not inhibit the Si based reactions.

3.3.3.XRD investigations

The XRD studies confirm the existence of a significant amount of oxide (more precisely a mixture of non-stoichiometric oxides). Indeed, a deconvolving the signal between $2\theta = 32°$ and $2\theta = 34°$, a sharp peak situated at $2\theta = 32.9°$ is found for all treated and untreated samples. It can be attributed to slightly oxidised non-stoichiometric oxides. Small peaks identified to correspond to SiC are also present.

In the XRD spectra recorded from M2 treated sample a small and broad peak at 33.3°, attributed to Si_3N_4, appears [3]. It is slightly shifted (with aprox 0.3°) towards higher values and suggests the presence of very small nitride (or carbonitride) crystallites inside the film. The XRD spectra of M5 treated sample is quite similar with that of M2 sample. Moreover, it evidenced the appearance of a SiC line, the peak of nitride being shifted and diminished [14].

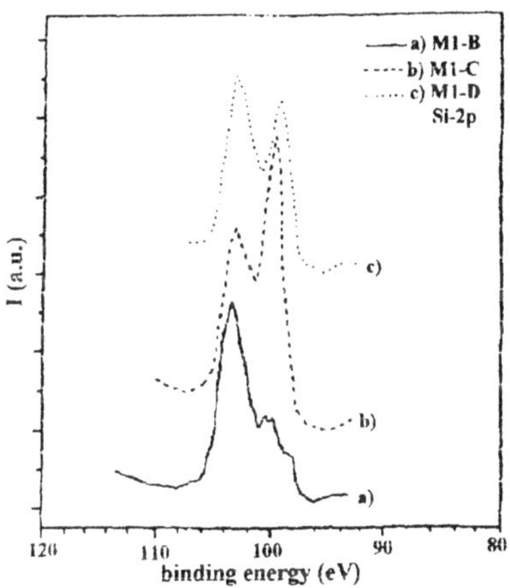

Figure1. XPS spectra of the samples M1B, M1C, M1D -Si2p region.

3.3.4.XPS investigations of a-SiC:H films

XPS spectra have been recorded for samples M1, treated in conditions B, C, D (see Table II). For all samples the regions of Si2p, Si_{KLL}, C1s and N1s were investigated. The region Si2p of the XPS spectrum was used to estimate the relative amount of SiC, Si, SiN and SiO_x. We have considered the signal (Fig.1) as the superposition of the signals corresponding to Si atoms bonded in pure Si (denoted by Si-Si), in stoichiometric Si-C (denoted by Si-C), in silicon nitride (Si-N) and in silicon oxide SiO_x (denoted by Si-O). The identification of peaks has been checked by the Auger parameter as well. As concerning the carbon bonding, the interpretation of the region C1s of the XPS spectrum was done considering the positions of silicon carbide peaks (283.2eV) and those of graphite and adventitious carbon on the surface (284.6 eV) [15]. In samples M1-B and M1-C the position of carbon peak was associated to Si-C bonds but in sample M1-D with C-C bonds. It was possible to identify the C-N bonding type in the sample treated in ammonia (M1-C) where it gives in the spectrum a visible shoulder at the high part of the bonding energy at around 287 eV.

TableIII. The relative amount of Si bonds as estimated from the fit of the Si2p region of the XPS spectra obtained for each sample.

Sample	Si bonds				C bonds	
	Si-Si(%)	Si-O(%)	Si-C(%)	Si-N(%)	C-Si,C-C(%)	C-N(%)
Reference	30	44	26	0	100	0
M1-B	6	68	26	0	100	0
M1-C	0	51	49	0	82	18
M1-D	37	55	0	8	100	0

As concerning the nitrogen bonds (N1s core level) the XPS spectra have shown the absence of the corresponding peak in the probe treated as M1-B and its presence in the samples treated as M1-C and M1-D. We present in Table III the relative amounts of bonds obtained for each sample by fitting the corresponding spectrum. The plasma beam treatment seems to induce the apparition of new types of bonds in the a-SiC material. The C-N bond is favoured to appear in case when the treatment is performed in ammonia plasma. The conversion of Si-Si bonds to Si-C and Si-O bonds is to be noted as well. For the samples treated in nitrogen plasma the apparition of Si-N bonds is favoured, being accompanied by the disappearance of Si-C bonds.

3.3.5. Microhardness measurements

All a-SiC:H treated samples exhibit an MH increasing. The highest MH value, of about 1850 Kg/mm^2, was measured on the sample with higher initial carbon content (50%) treated as described. The Si substrate MH was 600-800 Kg/mm^2, while the unirradiated deposited films were in the 400-800 Kg/mm^2 range. This increase can be related both to the compositional changes and to the improvement of the crystallinity character of the layer as an effect of the rf plasma treatment.

4. Conclusions

We succeeded in transforming the composition and the structure of as deposited a-SiC:H / Si layers by several treatments, arriving to complex composite thin films containing new Si-C-N-Si bonds. Treatments by RTA have been prooved to be appropriate for films modifying without destroying them. Laser treatments result in an evolution towards a better crystallinity. In the case of plasma beam treatments, for the same experimental parameters the evolution of treated samples containing less carbon, towards a compound rich in SiN, can be evidenced. On the samples having higher carbon content a upper layer rich in Si-C and sometimes even in C-N bonds was identified. In all cases, the obtained films exhibit an increase of microhardness values and good adherence to the substrate.

5. References

1. G. De Cesare et al. Diamond and Related Materials, **2** (1993) 773
2. G. De Cesare et al. Surface and Coating Technology, **80**, (1996), 237
3. G. De Cesare et al. Applied Surface Science, **106**, (1996), 193
4. M. Dinescu, O. Maris, G. Musa, Contrib. Plasma Phys. **31**, (1991), 49
5. N. T. Tran, Springer Proceedings in Physics, Vol. **34**, Amorphous and Crystalline Silicon Carbide. Editors G.L. Harris, C.Y.W. Yang, Springer Verlag Berlin Heidelberg (1989), 134
6. K. Yoshii et al. Thin Solid Films, **199**, (1991), 85
7. Practical Surface Analysis (2-nd edition), Vol **1**. Auger and X-ray Photoelectron Spectroscopy (Eds. D. Briggs and M. P. Seah), 1990. John Wiley & Sons Ltd.
8. C. Maillot, H. Roulet, G. Dufour, J. Vac. Sci. Techn., **B2**, (1984), 316
9. F. Rocher et al. Adv. Phys, **35**, 3, (1984), 237
10. R. Alfonsetti, et al. Thin Solid Films, **213**, (1992), 158
11. S. Veprek, J. Weidman, F. Glatz, J. Vac. Sci. Techn., **A13**, 6, (1995), 2914
12. G. Ramis et al.J. Am. Soc., **72**, 9, (1989), 1692
13. H. X. Han, B. J. Feldman. Solid State Commun. 65, 9, 921 (1988)
14. C. Gomez-Aleixandre et al. J. Mater. Res., 7, 10, (1992), 2864
15. A.S Byrne et al. in Springer Proceedings in Physics, Vol. **43**, Eds. M.M. Rahman, G.Y.-W Yang, G.L. Harris, Springer Verlag Berlin, Heidelberg, 1989, 80

SOME ELECTROPHYSICAL PROPERTIES OF C_{60} THIN FILMS

A.ANDRIESH[a], M.IOVU[a], E.KHANCHEVSKAIA[a], K.TURTA[a], I.GERU[b], D.SPOYALA[b]
[a]*Academy of Sciences of Moldova, Chishinau, 2028 Moldova*
[b]*State University of Moldova, Chisinau, Moldova*

Recent success in the production and separation of new forms of carbon, fullerenes such as C_{60} and C_{70} has stimulated a variety of experimental and theoretical studies due to their interesting properties and possibilities of applications including optoelectronics.

The fullerene C_{60} thin films of 1.4 μm thickness used in our experiments were prepared by the thermal vacuum evaporation on the glass substrate covered with the SnO_2 semitransparent electrode. The top Al electrode was deposited for electrical and photoelectrical measurements.

The current-voltage (I-V) characteristics are non-symmetrical, the magnitude of the current in the case of positive polarity on the bottom electrode is greater than for the opposite polarity. The I-V characteristics show a power low behavior $I \propto U^n$, where n=1 . . . 2.5.

The spectral distribution of photoconductivity of SnO_2-C_{60}-Al sandwich structures depends on polarity and magnitude of the applied field. The maximum of photoconductivity corresponds to $\sigma_{ph}^{max} \approx 1.8$ eV.

The capacitance-voltage (C-V) characteristics were measured in the dynamic regime with a slowly varying triangular voltage pulse with the frequency from 10^{-1} to 10^{-3} Hz. The C-V characteristics depend on the frequency of the applied field and on the illumination. An increase of capacitance and the displacement of the C-V characteristics under illumination was observed.

The optical absorption spectra of these films in the 500-50 000 cm^{-1} spectral range were investigated. The IR spectra contain 4 intensive bands which were interpreted in terms of theoretical group analysis of Y_h point symmetry group of C_{60} cluster. Thin bond between C_{60} molecules in fullerite was taken in to account. In the UV-VIS range the rich structure of spectrum with the low frequency Urbach energy (320 meV and 250 meV for 500 A and 1500 A thicknesses, respectively) indicate to high degree of structural disorder which is typical for amorphous semiconductors.

A. Andriesh and M. Bertolotti (eds.),
Physics and Applications of Non-Crystalline Semiconductors in Optoelectronics, 451.

DEPOSITION OF $CuInSe_2$ THIN FILMS BY PULSED ELECTRON BEAM ABLATION

A.M. ANDRIESH, S.A. MALKOV, M.S. IOVU
Institute of Applied Physics, Academy of Sciences of Moldova,
Academiei str. 5, Chisinau, 2028, Moldova.

Technology of chalcopyrite compounds thin films and their alloys attracts much attention because of promising applications as solar cell material, light-emitting diodes and non-linear optic materials. One of the most known is the $A^1B^3C^6_2$ compound - $CuInSe_2$ (CIS) and its analogs due to usage in high efficiency polycrystalline thin film heterojunction solar cells of high efficiency.

This report presents the results of application of the pulsed electron beam ablation - the new method for deposition of the CIS thin films. It is successfully employed for depositing epitaxial thin films of high temperature multinary superconductors. This method is similar to the pulsed UV-laser ablation but the capital costs are many times lower. The main advantages of this method are the low deviation in composition from target to film, the large area deposition, the absence of dangerous gases, the simple handling and low cost.

The CIS deposition was implemented by a pulsed high current and magnetically self pinched electron beam produced in a low pressure channel spark camera at the following operation conditions: high voltage - 15-20 kV, pulse duration - 100 ns, repetition rate - 1-5 Hz, power density $< 5\ 10^8$ W/cm^2, beam diameter - 1.5 mm, argon pressure - 1-3 Pa, substrate temperature - 20- 500°C. The target was performed by cutting the ingot prepared by fusing components in evacuated sealed quartz ampoule. The common soda-lime glasses were used as substrates.

It was experimentally established basing on the quantity of impulses, the film thickness and the depth of target crater that the beam penetration depth into target and deposition rate reached 0.5-1.5 mkm and 0.5-20 A per impulse, respectively, at various beam parameters. The used characterization techniques have indicated on the reproducible transfer of properties of the target to the substrate under determined growth conditions and beam parameters. The CIS films were found to be p-type and to have the resistivity in the range of 100-1000 (Om cm). The observed surface microstructure of prepared films consisted of close arranged grains with diameters of 0.03 - 0.5 mkm in dependence on substrate temperature. The drastic edge in absorption spectra near the band gap of 1 eV for direct allowed transitions and the addition absorption due to the direct forbidden transitions with $E_g = 1.2$ eV were observed.

We believe that the properties of prepared thin films are suitable for application of them in heterojunction solar cells but the pulsed electron beam ablation method can be also applied for deposition of other semiconductor amorphous and polycrystalline thin films.

A. Andriesh and M. Bertolotti (eds.),
Physics and Applications of Non-Crystalline Semiconductors in Optoelectronics, 452.

OPTICAL PROPERTIES OF Ge-As-Se GLASSES AND FILMS ON THEIR BASIS

V. BAZAKUTSA, S. GAPOCHENKO, V. BELOZERTSEVA,
V.MUSSIL, Ye. LEMESHEVSKAYA, and A. RYABCHUN
Kharkov State Politechnical University, 310002 Kharkov, Ukraine

We have determined the optical parameters (refraction index n, absorption coefficient α of bulk and thin film samples of $Ge_xAs_ySe_{1-x-y}$ ($O \leq x \leq 0,3$, $0,1 \leq y \leq 0,4$) chalcogenide glasses (CG)

The chemical composition of samples was controlled by laser energy-mass analyze.

This report represents the results of measurement of changes in optical properties photoinduced in $Ge_xAs_ySe_{1-x-y}$ thin films prepared by flash evaporation.

Both spectral and expositive dependencies of photoinduced changes in optical parameters of structures were evaluated at wavelength λ=0,442 μm and 0,633 μm.

The variation of n is accompanied by shift of absorption edge transparency range. We have obtained the maximum value of $\Delta n \approx 0,15$. Variation of glass chemical composition, tech- nological parameters of thin film deposition, optical irradiation conditions allow to control the characteristics of photoinduced processes in thin films. The optical properties and photoinduced effects in two-layer systems "CG-silver" have been studied. The optical density, sensitivity and contrast of these systems have been measured. These parameters are essentially dependent on the composition of Ag-$Ge_xAs_ySe_{1-x-y}$ film structures.

The obtained results are analysed in dependence on the mean coordination number <m> of covalent bonds per atom ($\langle m \rangle = 2+2x+y$ for $Ge_xAs_ySe_{1-x-y}$ - like glasses). This parameter is an important quantitative characteristic, which determines physical properties of disordered materials. In was shown that dependencies of optical parameters CG and two-layer systems on their basis on <m> correlate with similar dependencies of elastic properties of bulk glasses.

Note that interest in the optical properties of non-crystalline $Ge_xAs_ySe_{1-x-y}$ alloys have been stimulated by some perspective for their use in optoelectronics and integrated optics. Due to photoinduced effects in optical parameters the amorphous thin films have possible applications as holographic media. The obtained data demonstrated the potential of Ag-$Ge_xAs_ySe_{1-x-y}$ for production of diffraction gratings with diffraction efficiency ~20 %.

A. Andriesh and M. Bertolotti (eds.),
Physics and Applications of Non-Crystalline Semiconductors in Optoelectronics, 453.

OPTICAL REGISTRATION MEDIA WITH NEAR IR SENSIBILITY FOR XEROGRAPHY

A. BUZDUGAN
Center of Metrology, Automation & Scientific Research Work, Academy of Sciences, 3/2, Academiei str., MD-2028, Chisinau, Moldova

In this paper the results on investigation of photoelectric properties in xerographic regime of the multistructure photoreceptors are presented.

The technological requirements for manufacturing of a new registration media based on heterostructures from binary and ternary vitreous compounds with improved performances have been developed. The xerographic parameters of the investigated photoreceptors are presented in Tabl.

Tabl.

The main xerographic parameters of the investigated photoreceptors :

Registration Media Formula	Δ_λ, nm	S, $(lx\ s)^{-1}$	DRSP, %/ min
n-InP-$(As_2S_3)_{0,55}(Sb_2S_3)_{0,45}$ -As_2S_3/As_2Se_3	400...1200	0,85	16- 22
p-InP-$(As_2S_3)_{0,55}(Sb_2S_3)_{0,45}$-$As_2S_3/As_2Se_3$	400...1200	0,92	16- 21
p-InP-$(As_2S_3)_{0,95}(Sb_2S_3)_{0,05}$ -As_2S_3	400...1200	0,25	10
$TlSbSe_2$-$(As_2S_3)_x(Sb_2S_3)_{1-x}$ - As_2S_3	400 ...1300	1,5	23
In_2S_3-$(As_2S_3)_{0,5}(Sb_2S_3)_{0,5}$ - As_2S_3	400 ... 800	0,78	14
In_2Se_3-$(As_2S_3)_{0,5}(Sb_2S_3)_{0,5}$ - As_2S_3	400 ...1300	0,88	12

Δ_λ - Spectral Sensitivity Range; S - Integral Sensitivity;
DRSP - Dark Relaxation of the Surface Potential

A. Andriesh and M. Bertolotti (eds.),
Physics and Applications of Non-Crystalline Semiconductors in Optoelectronics, 454.

NONLINEAR PROPAGATION OF STRONG LASER PULSES IN NON-CRYSTALLINE SEMICONDUCTOR FILMS

V. CHUMASH[@], I. COJOCARU[@], G. PARA[@],
E. FAZIO[&], F. MICHELOTTI[&] and M. BERTOLOTTI[&]
[@] *Center of Optoelectronics, Academy of Sciences of Moldova, Academiei str. 1, Chisinau, 2028, Moldova*
Università degli Studi di Roma "La Sapienza", Dipartimento di Energetica, GNEQP of CNR and INFM, Via A. Scarpa 16, 00161 Roma, Italy

The interest to the study of the nonlinear propagation of laser radiation in non-crystalline semiconductors (NS) is determined not only by the new fundamental physical mechanisms present in these materials, but also by a wide spectrum of possible applications in high speed optoelectronic devices and photonic switching. This report illustrates the present state of knowledge and some problems of the nonlinear interaction of micro-, nano-, pico-, and femtosecond laser pulses with NS (chalcogenide glasses: As_2S_3, As_2Se_3, AsSe, $GeSe_2$, $As_{22}S_{33}Ge_{45}$, and a-Si:H; 0.2 - 5 μm thick).

When the input light pulse intensity (at interband excitation of samples, $h\nu \geq E_g$) increases over some threshold value (I_t), the transmission of the NS films decreases nonlinearly. The peculiarities of nonlinear light transmission include a nonlinear increase of the light absorption coefficient; a nonlinear refractive index change, whose magnitude depends strongly on the excitation intensity; the threshold character of the nonlinear transmission; the change of the time profile of the laser pulses, leading to a hysteresis transmission dependence; the characteristic time constants of the sample photoinduced darkening, which does not exceed the laser pulse duration; the independence of the nonlinear transmission character on the light beam diameter, on the film thickness, and on the substrate material; the weak dependence from the laser pulse wavelength; the successive step character of restoration of light transmission initial state; the full restoration of light transmission initial state in a microsecond time interval; the temperature behavior of the parameters of the nonlinear interaction (the invariance of the light intensity threshold values and of the transient time into the nonlinear light absorption state); and the correlation with the photostructural transformations.

The physical mechanisms, which can contribute to the nonlinear light absorption in NS under pulsed excitation, are critically reviewed. It is shown that these characteristics cannot be explained satisfactorily within the limits of the existing physical models. A new mechanism of nonlinear light absorption in NS, taking into account the interaction with equilibrium and nonequilibrium phonons and localized vibrational modes is considered for the explanation of the experimental results. The nonlinear change of the light absorption coefficient may be reached in the NS at much lower irradiation intensity than in the crystalline ones. This is possible when interband transitions with the participation of nonequilibrium phonons and nonequilibrium localized vibrational excitations play an important role.

A. Andriesh and M. Bertolotti (eds.),
Physics and Applications of Non-Crystalline Semiconductors in Optoelectronics, 455.

THE PECULIARITIES OF PHOTOINDUCED ABSORPTION IN CHALCOGENIDE GLASS FIBRES

I. P. CULEAC
Centre of Optoelectronics of the Institute of Applied Physics,
No 1, Academiei str., Chisinau, Republic of Moldova

Photoinduced absorption (PA) provides a rich source of information about the localized states in the gap of chalcogenide glasses. Employment of fibre samples instead of thin film or bulk samples offers the possibility to increase the sensitiivity of PA measurements. Respectively one can study photoinduced absorption characteristics at very low excitation light intensities, to extend the PA spectral and temperature range measurements.

For excitation light intencities at least up to 10^{-2} W/cm^2 the PA in chalcogenide glass fibres exhibits a reversible behavior. After cessation of bandgap light the samples restore their initial optical transmittance. The restoration time varies from several seconds to several hours depending on the temperature, probing light photon energy, glass composition, etc. The spectral dependency of PA coefficient exhibits an exponential tendency for all investigated compounds (i.e. As_2S_3, As-S-Se, As-Ge-Se, etc.). The dependence of PA coefficient versus the intensity of exciting light has a power-law behavior, at least in the investigated range of excitation light intensity (10^{-6} - 10^{-2} W/cm^2). PA kinetics measurements were carried out in a time interval of 10^{-2} - 10^4 s after switching on/off the exciting light. The character of PA kinetics changes drastically with measurements conditions, such as temperature, the power of the exciting light, the probing light photon energy, etc.

At low temperatures, irradiation of fibre samples with subgap light leads to well-known photobleaching effect. It was found that illumination of chalcogenide glasses with subgap light at 77 K can lead not only to bleaching of previously induced absorption but it can produce an enhancement of induced absorption too. In this case PA photobleaching kinetics exhibits a non-monotonous character. It can be supposed that the potobleaching effect is a result of two opposite processes, photoinduced bleaching and photoinduced absorption.

Experimental results on PA in chalcogenide glass fibres can be interpreted in terms of the multiple trapping model with excess carriers redistribution on localized gap states with an exponential distribution. Analysis of experimental results in terms of multiple trapping model can be used for evaluation of the parameters of localized states distribution. Computer simulation of mathematical solutions obtained from PA rate equations shows a good corelation of the model with experimental results.

A. Andriesh and M. Bertolotti (eds.),
Physics and Applications of Non-Crystalline Semiconductors in Optoelectronics, 456.

MASS-SPECTROMETRIC INVESTIGATION OF THE PHOTOSTRUCTURAL CHANGES IN CHALCOGENIDE GLASSES

V.T.DOLGHIER
Center of Optoelectronics, Academy of Sciences of Moldova
1, Academiei Str., Chisinau, MD-2028, Moldova.

In this report we suggest that a study of the composition of condensed molecules by means of mass spectrometry may yield complementary information helpful in building structural models of chalcogenide classes and explaining the mechanism of the photostructural changes in chalcogenide glasses.

For an initial study we have chosen the systems As-S and As-Se prepared by different techniques (bulk, fiber and thin film) before and after laser illumination. The mass spectra were registered at an interval of vaporization temperature from 393 to 553 K and on ionizing electron energy from 5 to 70 eV.

Data of mass spectrometric analysis of $As_2S_3(Se_3)$ prepared by different techniques before illumination shows the visible difference in ratio of $As_mS_n(Se_n)$ molecular units registered at low vaporization temperature (393 K for As_2S_3 and 433 K for As_2Se_3) and low ionizing electron energy (20 eV). After laser illumination we registered some new kind of molecular units as As_m, S_n and Se_n. The observed changes in the mass spectra may be explained taking into account some re-arrangement under illumination of the constituent atoms in the short-range order.

For thin films just after the evaporation such molecular units as As_4, $As_4S_3(Se_3)$; $As_4S_4(Se_4)$; $As_2S_3(Se_3)$; $As_2S_4(Se_4)$; $S_8(Se_8)$ may be bonded by Van-der-Waals forces and the observed photoinduced changes in the infrared absorption spectra are interpreted by rearrangement of the bonding configuration. Thus the illumination can change the densities of chemical bonds in the glassy $As_2S_3(Se_3)$.

As a result of this research we suggest that there is a correlation of structure and photostructural changes in chalcogenide glasses $As_2S_3(Se_3)$ with the structural molecular units As_m; $As_mS_n(Se_n)$; $S_n(Se_n)$.

A. Andriesh and M. Bertolotti (eds.),
Physics and Applications of Non-Crystalline Semiconductors in Optoelectronics, 457.

COOPERATIVE TWO-PHOTON EMISSION IN THE MICROCAVITY

N.A. ENAKI, M.A. MACOVEI
Institute of Applied Physics, Academy of Sciences of Moldova,
Academiei str., 5, Kishinev MD-2028, Moldova

The one photon cooperative emission in the microcavity is very well studied in the literature [1- 4]. In such processes it the amplification and inhibition of spontaneous emission in function of the geometrical dimensions of microcavity are possible . When the energy distances between the quantized modes of electromagnetic field in the microcavity coincide with the transition energy between the exited and ground levels of atoms, the cooperative one-photon emission is amplified. In opposite case , when the spontaneous emission is inhibited by the absence of the cavity mode for the transition energy of atoms, the cooperative emission rate essentially decrease.

In case of two-photon cascade cooperative emission of excited impurity or excitons in semiconductors microcavity with dimensions of order of emission wavelength the inhibition of emission rate between the excited and intermediate state of cascade three level system is possible. In this situation the two-photon dipole forbidden transitions between the upper and ground states of three level system are possible. Such a dipole forbidden transitions occur through the intermediate level that is out of resonance with microcavity modes. The problem is very interesting from physical points of view, because we can obtain the powerful pulse of two-photon highly correlated light.

1. Weisskopf V., and Wigner E., (1930) Z. Phys., **63**, 54.
2. Parker J. and Stroud C. R., (1987) Phys. Rev. A, **35**, 4226.
3. Agarwal G. S., Longe W. and Walther H., (1993) Phys. Rev. A, **48**, 4555.
4. Lewenštein M., Mossberg T. W. and Glonber R. J., (1987) Phys. Rev. Lett. **59**, 775.

A. Andriesh and M. Bertolotti (eds.),
Physics and Applications of Non-Crystalline Semiconductors in Optoelectronics, 458.

A LAYERED-INHOMOGENEOUS MODEL OF THE STRUCTURE OF VITREOUS GeS_2 - BASED VACUUM CONDENSATES

V. GERASIMOV and V. MITSA
Uzhgorod State University, Department of Solid State Electronics
32 Voloshyn str., 294000 Uzhgorod, Ukraine

In the report the results of studying the depth dependence of Raman spectra, measuring the refractive index profile and SIMS-profile of a film based on GeS_2 have been summed up. Based on these data a model of the layered-inhomogeneous structure of films (Fig.1) is proposed. The peculiarity of the model is presence of a wide transition film-substrate region (up to 300 A). Its composition differs from stoichiometric GeS_2 and may be shown in the form of Ge_xS_{1-x}, where x>33 at. %.

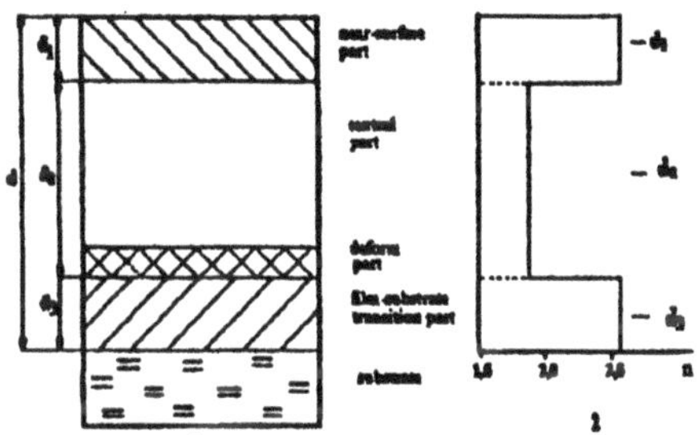

Fig. 1 Model of non-uniform layer structure for a-GeS_2-type films (1) and model non-uniform of refractive index at λ_{ex}=6300 A for GeS_2 (2).

An average refractive index of the transition region is n=2,6 and in Ge_xS_{1-x} system it correspond to $Ge_{50}S_{50}$. In accordance with Raman spectra analysis of GeS_2 films internal strains appear in the film at the initial stage of its growth which with increasing of the film thickness relax in respect to it.

At film-vacuum interface a near-surface layer is revealed which is rich with Ge in comparison with the central part of the film.

By many-angle ellipsometry method it is revealed that while irradiating a layered-inhomogeneous film by a laser radiation the most marked photoindiced changes in the refractive index occur in the central part of the film.

The influence of the layered-inhomogeneous structure of a highly refractive layer based on GeS_2 upon spectral characteristics of two-component interference filters of vacuum deposition has been discussed.

A. Andriesh and M. Bertolotti (eds.),
Physics and Applications of Non-Crystalline Semiconductors in Optoelectronics, 459.

ELECTRICAL PROPERTIES OF p-Si/α-C:H HETEROJUNCTIONS

M. KOOS, I. POCSIK and M. IOVU*
Research Institute for Solid State Physics
H-1525 Budapest, P.P.Box 49, Hungary
**Center of Optoelectronics,*
Nr.1 Academiei Str., Chisinau MD-2028, Moldova

Hydrogenated amorphous carbon (α-C:H) is a very promising material for different electronic devices, IR-optics and for passivation coatings. The α-C:H thin films (L=0.8÷4.2 μm) for electrical and photoelectrical measurements were deposited on p-type silicon wafer by RF glow discharge decomposition of methane (CH_4) under different self-bias voltages $U_{s.b.}$= -100 ÷ -600 V. The top Au, Al and Sb semitransparent electrodes of 4 mm in diameter on the surface of α-C:H thin films were deposited by thermal vacuum evaporation.

The forward I-V characteristics are characterized by an exponentially increases of the current up to 1.0 V. Then between 1.0 and ~2.0 V the dependence of the current versus applied voltage show a power-low dependence ($I \sim V^n$, were n takes the values between 1.65 and 6.6) and that behavior has to be due to space-charge-limited-current (SCLC) flows in the amorphous carbon films. At higher forward voltage the I-V characteristics have a subliniary dependence which is caused by the consecutive resistance of the α-C:H film. That resistance have been estimated for different α-C:H thin films of the heterojunction obtained at different self-bias voltage $U_{s.b.}$ and takes the values between $4 \cdot 10^6$ and $5 \cdot 10^7$ Ohm·cm. The p-Si/α-C:H heterojunction exhibit the light sensitivity at the forward bias in the region more than ~ 2 V, which is to be due to the photoconductivity of the bulk amorphous carbon. The reverse I-V characteristics of the p-Si/α-C:H heterojunction in the dark show a linear dependence for different metallic top electrodes (Au, Al, Sb) to the α-C:H film.

The obtained p-Si/α-C:H heterojunctions at different self-bias deposition $U_{s.b.}$ exhibit a photosensitivity at the reverse bias. That photocurrent is due to the generation of the electron-hole pairs at the interface of the p-Si/α-C:H heterojunction and its separation by the applied field.

The maximum in the photoresponse around λ=0.9 μm may be associated with the absorption of the incident photons in the p-Si substrate. The observed shoulder at λ=0.6 μm may be due by the absorption in the band tails of the α-C:H thin film. The difference between the two maxima corresponds to the value of the barrier height φ_b=0.66 eV determined from the C-V characteristics. At the same time the amplitude of the photocurrent in the maxima λ=0.9 μm depends very strongly upon the reverse applied voltage (from 2.5 μA for U=0.1 V up to 8000 μA for U=8.0 V). The same dependence of the photocurrent upon the applied voltage ($I_{ph} \sim U^2$) was obtained from the reverse illuminated I-V characteristics.

A. Andriesh and M. Bertolotti (eds.),
Physics and Applications of Non-Crystalline Semiconductors in Optoelectronics, 460.

ELECTROMOTIVE FORCE IN THE STRUCTURES METAL-GRADIENT FILMS <As_2S_3-Al(Bi)>-METAL

I. MIGOLINETS, I. SHOVAK, I. POPOVICH, A. LADA,
V. PINZENIK and S. MIKULANINETS
Uzhgorod State University, Pidhirna 46, Uzhgorod, Ukraine

It drew the interest to investigate the behavior of the gradient structure, which is created as a combination of glass-like semiconductor and metal, under the action of electric fields. For the creation of the gradient structure such elements as well known glass-like As_2S_3 and Al and Bi were used. The choice of Al and Bi is explained by value of their normal electrode potentials, relative to electrode.

Films of the changeable structure <As_2S_3-Al> were obtained by the method of joint evaporation. Simultaneously from one direct-channel, tantalum evaporator the simple temperature evaporation Al(Bi) was carried out, as well as discrete evaporation As_2S_3. These processes were carried out using small particles of glass. The process was carried out the increasing in time amount of discrete evaporated particles of As_2S_3, that supplied uninterrupted increase of steam flow of As_2S_3 on the substrate, relatively to static flow of steams of Al. To guarantee the conditions of discrete As_2S_3 evaporation in such non-stationary regime the evaporator was warmed to 1700-2000 K. To obtain of the results from evaporation to evaporation concerning the change of the profile of Al(Bi) in the films, the calculation of the parameters of technologic process were carried out.

In normal conditions without the applying of electric fields the gradient structure is the source of electromotive force (EMF), being static. The polarity of EMF of the gradient structure As_2S_3-Al(Bi) is defined by the included metal, and its value depends upon the profile of the concentration of the included materials, temperature and environment. In the even distribution of the metals in As_2S_3 the EMF is 4 to 5 levels smaller than EMF of the gradient structure. Decreasing the pressure the EMF also decreases, and than remains static. Heating the structure in vacuo till 373 K does not change to a great extent the EMF.

In the region of the electric fields $E \sim 10^{-4}$ - 10^{-5} V/cm the volt-ampere characteristics (VACh) of the structure also depend upon the profile of the concentration of the included metals and have linear and sub-linear behavior as well. Linearity as a rule is observed in the even distribution of the included Al(Bi) in As_2S_3, but for the gradient structure the VACh become sub-linear and non-symmetric, and are described by the sublinear dependence ($I \sim U^n$). The index of the dependence n changes from 1.5 to 3. When the temperature lowers the index of the dependence goes to 1 .

A. Andriesh and M. Bertolotti (eds.),
Physics and Applications of Non-Crystalline Semiconductors in Optoelectronics, 461.

THERMOSTIMULATED LUMINESCENCE OF AMORPHOUS OXIDE FILMS OF ALUMINIUM

V.V. MIKHO, L.N. VILANSKAYA, Y.V. ROBUL and M.K.NAFTULOVICH
Odessa State University, 2 Dvorianskaya str., Odessa, 270100, Ukraine

The influence of preliminary X-rays exciting on thermostrimulated luminescence (TSL) of oxide films of aluminium obtained by electrochemical method have been studied. Those films had been excited by ultraviolet (UV) with $\lambda<300$ nm then TSL spectrum was observed. The measurements carried out in temperature range from room temperature to 650 K. The rate of heating was 0.5 K/s. By heating of the samples some maxima on TSL curves were observed. It was noticed that the samples which had emitted their lightsum before, might store it again after UV-exciting in atmosphere contained water vapour. For making clear X-ray influence on TSL the samples of oxide films of aluminium were excited X-rays. As a result by X-rays exciting of the samples that had not been excited by UV before the lightsum storage did not take place. So the samples, excited by UV emitted the lightsum after X-rays exciting. The intensity of TSL maxima reduces with increasing of X-ray influence duration. The X- ray influence on low-temperature maxima is grater than on high-temperature ones. It is important that the process of storing of the lightsum and its emission may be repeated many times with the same result.

The depending of intensity of TSL maxima on X-ray exciting duration may be explained in the following way. It has been shown before that appearance of TSL maxima is due to the centers, formed by OH^- ions, arisen as a result of water vapour molecules dissociation on catalitically active centers of micropore surface of oxide films of aluminium. These centres may play role of traps filled of electrons. The intensity of TSL maximum is proportional to the area under TSL curve. This area is proportional to the number of the centres, responsible for TSL. X-rays influence, accompanied by emission of the lightsum leads to liberation of electrons from the centres, formed by OH- ions and desorbtion of neutral ion-radicals from the oxide film's surface. The liberation of electron from those centres may be the result of ionization by collision of those centres by the electrons, accepted the additional energy from X-ray. The liberation of electrons must be accompanied by reducing of oxide films resistance. This reducing of resistance was noticed in some experiments. This fact supports the proposed mechanism of X-ray influence on TSL of oxide films of aluminium. Thus the possibility of recording of information caused by X-ray influence was determined. The oxide films may be used for information recording many times.

A. Andriesh and M. Bertolotti (eds.),
Physics and Applications of Non-Crystalline Semiconductors in Optoelectronics, 462.

LOW FREQUENCY RAMAN SPECTRA, SIZES OF FRACTALS AND LONGITUDINAL ELASTIC MODULES IN WIDE GAP CHALCOGENIDE GLASSY SEMICONDUCTORS

V. MITSA
Uzhgorod State University, Department of Solid State Electronics
32 Voloshyn str., 294000 Uzhgorod, Ukraine

In the present report the results of processing of low-frequency (LF) Raman spectra in Alexandr-Orbach model have been given as the coordination dependence of size of fractal clusters (ξ) and longitudinal elastic modulus (C_l) in ternary glasses while changing the composition along As_2S_3-GeS_2, Ge-As_2S_3, As_2S_3-Ge_2, As-GeS_2, As-Ge_2S_3 sections. The same has been done for binary As-S and Ge-S glasses.

The analysis of dimensionality dependence of LF Raman spectra at $w>w_b$, where w_b is the spectral position of LF maximum, points to the availability of linear ordering of fractal clusters in all the glasses.

In As-S system there is a peculiarity at average coordination number r=2.2 and minimum at r=2.4 on the plot of fractal sizes versus r. C_l versus r gradually increases. The peculiarity at r=2.2 and maximum at r=2.4 is observed on it.

Along As_2S_3-GeS_2 section $\xi(r)$ increases and in one-dimensional (chain-like) limit the length of chain (L) reaches L=22A at r=2.66. The minimum C_l at r=2.66 correspondents to the maximum L.

Along As_2S_3-Ge_2S_3 section up to r=2.65 increase in ξ=f(r) is observed, at r>2.65 a sharp decrease in the fractal dimensions approaching the minimum ξ at r=2.8 (Ge_2S_3) is observed. At r>2.65 the growth in C_l is observed.

The introduction of Ge into As_2S_3 causes a little growth in C_l up to r=2.47, and it is weakly reflected on the magnitude of fractal dimensions up to r=2.42. With further growth in r the sizes of fractals increase and at r=2.47 the crystallization with β-As_4S_4 separation is noticed. The process of crystallization is accompanied by the decrease in C_l up to r=2.65. At r>2.65 the growth in C_l is observed.

The introduction of As into GeS_2 causes a little growth in C_l up to r=2.67. At larger r>C_l a certain decrease and then, at 2.68 the growth in C_l is observed. The peculiarity at r>2.67 of ξ=f(r) dependence decrease in value, compared with the sizes of fractals which are typical for binary GeS_2.

The introduction of As into GeS_2 causes the decrease in C_l and growth in ξ. In Sen-Thorpe model the changes in Ge-S-Ge angle while introducing As into GeS_2 and Ge_2S_3 are calculated.

Thus, the peculiarites at r=2.2; 2.4; 2.66; 2.8; on the plot C_l=C_l(r) are well defined.

A. Andriesh and M. Bertolotti (eds.),
Physics and Applications of Non-Crystalline Semiconductors in Optoelectronics, 463.

OPTICAL BISTABILITY IN NONLINEAR INTERACTION OF SHORT LIGHT PULSES WITH THIN-FILM RESONANT STRUCTURE

V. MUSINSCHI and M. CARAMAN
State University of Moldova, Department of Physics
60 Mateevich str., Chisinau, 2009 Moldova

Nonlinear optical systems based on $A^{III}B^{VI}$ layered crystals have now an attracting interest not only because they offer useful candidates for studying non-equilibrium systems with rich spectrum of temporal and spatial behaviour, but because they provide a promising basis for future signal elements for the possible applications in digital optical signal element processing schemes.

To achieve optical bistability requires both an optical nonlinearity and some form of feedback mechanism. The optical nonlinearity can be provided by a change on optical absorption or refractive index with increasing pulse power. Band gap resonant excitation allows creation of electrons via band tail states and excitonic levels. These carriers in turn alter the absorption and refractive index of the material.

Results of the experimental investigation and the theoretical modelling of the nonlinear absorption and propagation of short pulses in GaSe crystals and thin amorphous ZnTe Films are presented. Bulk single crystalline GaSe and evaporated thin film of ZnTe have been used for optical bistability. The structures consisted of the Fabry-Perot etalon of various thickness of semiconductors platelets (GaSe) or ZnTe films. Natural crystal facets served as the resonator mirrors with the reflectance $R \approx 0{,}25$. By means of two-beam experiments, the transmission properties of highly excited GaSe crystals of different thicknesses (from 2 μm to 1.5 μm) have been investigated in the fundamental absorption and free exciton region. Induced transmission was observed in the whole considered region for lower pump intensities.

As a result of the nonlinear absorption a change of time profile of the light pulses was recorded at the output.

The results of numerical solution of nonlinear equation, describing the propagation of pulses, and experimental transmission of the light pulses ($\Delta t \approx 10^{-6}$ s) are discussed.

The nonlinear absorption in GaSe single crystals stems from Coulomb interaction of nonequilibrium electrons and holes at high intensities of pump beam. It is assumed that the nonlinearity of medium Fabry-Perot cavities (GaSe crystals or ZnTe films) is due to the saturation of absorption.

A. Andriesh and M. Bertolotti (eds.),
Physics and Applications of Non-Crystalline Semiconductors in Optoelectronics, 464.

RECORDING EQUIPMENT BASED ON PHOTOTHERMOPLASTIC MEDIUM FOR SPACE AND AIRBORN APPLICATIONS

L.M. PANASYUK, V.K. ROTARU, I.V. CHAPURIN and
O.Ya. KORSHAK
State University of Moldova, Department of Physics
60 Mateevich str., Kishinev MD-2009, Moldova

Photothermoplastic medium (PTPM) is a non-silver two-layer semiconductor - thermoplastic structure for optical data recording, which is radiative resistant. It allows to record photographic, holographic and other kinds of optical data without any wet chemical development that giving the maximal effect for space and airborn usage.

Optical data of photothermoplastic recording are characterized by simplicity of developing and erasing processes; unlike the xerography it does not need any special developing material; unlike vidicons it does not need any complex electronic equipment. To record data on PTPM only electric power (≅ 60 W) is necessary.

The property of radiation resistance makes this systems particularly useful for remote sensing of the different objects from Space. The main PTPM's disadvantage when comparing with *AgHal* films is relatively low photographic sensitivity values $S \cong 7.0 \div 10.0\ lx^{-1}\ s^{-1}$, but it can be compensated by using of high-speed lens. We suggest a PTPM which is photosensitive in different spectral ranges, stable to radiation light-mark, being able to record halftones and increasing the resolution of lens - PTPM system at the frequencies close to limiting characteristics of the lenses.

Such a medium can work in a circular scheme providing the multiple write - erasure mode of recording. The image obtained on PTPM can be transmitted to the observation station instantly or with a necessary delay. During image recording process a PTPM is placed into the camera and the heating element is switched on. When applying the voltage to the coroning electrode with simultaneous exposure corresponding to the photosensitivity of a PTPM, the creation of visual image takes place in the form of phase relief of the layer surface being deformed. A duration of the exposure is determined both by the slit width of the thermostatting unit and the diaphragm size of the apparatus optical train (in the case of slit recording technique). The equipment foresees image erasing mode to provide repeatable recording. A number of repeating record-erasure cycles may be up to 30 depending on the kind of PTPM. The resolution of PTPM is a function of a thermoplastic layer thickness and varies from 400 to 1800 mm^{-1}. The apparatus optical system with focal length F=99.3 mm and angle of field-of-view $2\omega=16°$ and aperture ratio A=1 : 1.2 can be used. When slit recording method is used the record quality depends on flexible substrate noise. With the OKS-100-1 lens and relative aperture 1:2 the resolution up to 200 mm^{-1} is achieved, visual resolving power being about 960 mm^{-1}. The reserve on the PTPM film cassette is about $40 \div 60$ m as depend on the substrate thickness.

A. Andriesh and M. Bertolotti (eds.),
Physics and Applications of Non-Crystalline Semiconductors in Optoelectronics, 465.

PHOTOTHERMOPLASTIC MEDIA FOR NON-DESTRUCTIVE CONTROL SYSTEMS

L.M. PANASYUK, O.Ya.KORSHAK, V.K. ROTARU and
I.V. CHAPURIN
State University of Moldova, Department of Physics
60 Mateevich str., Kishinev MD-2009, Moldova

Steady rise demands at quality and safety of the turning out industrial predetermines a search of new and improvement the present non-destructive testing methods, which permits to avoid waste of time and pecuniary expenditures. In according with this, the physical testing methods which don't leads at destructive of the finished articles and first of all holographic and radiation acquiring a big importance. A special place in the mentioned methods has registration media which in some cases must secure the record of article defects in real scale of time with a great resolution. The traditional *AgHal* medium hasn't such operativeness.

The results of two-layer photothermoplastic media (PTPM) applications in holographic, interferometric, speckle-photographic, and radiation systems of the non-destructive testing has been shown in this communication at the successive and other methods of the simultaneous photothermoplastic recording.

As the investigation objects were utilized the articles of responsible destination from metal to ceramics (pine-lines, aerohoneycomb constructions, vibration systems, turbine blades); optical elements, printed circuit boards; integral microcircuits; different plants and biological objects. The PTPM were used for holographical and speckle interferogram recording. The shift in the plane of the diffusion-reflected object and its slope were studied as a whole, according to the double-exposure method of *Berch* and *Tockarsky*. The diffusion-reflected objects was lighted by an expanded He-Ne laser ray. Resolution power, while measuring objects plane shifts constitutes, is ~ 10 μm, but at the object slope - 10". Also we saw, that it is possible to execute on the PTPM not less 5 ÷ 10 exposures which are sufficient for the most practical supplements. When doing investigation of the oscillating and continious object shift them it is necessary to record the speckle interferogram in one exposure. The angular distance between them is determined by the speed of the test-object shift.

Using of vitreous chalcogenide semiconductors with different compounds in the PTPM structure provides easy control of spectral range of PTPM sensitivity from the X-ray to near IR (up to 900 nm) range. There were received shadow X-ray pictures of the different objects, resolution power of which are ~ 100 mm^{-1}. The contrast coefficient is $\gamma \cong 0.6$, and the photographic latitude is $L \cong 0.8$. The sensitivity value of PTPM in the X-range is ~ 300 R^{-1}. As a conclusion, the possibility of equipment created for non-destructive interferometric and radiation control with PTPM working in real time has been shown.

A. Andriesh and M. Bertolotti (eds.),
Physics and Applications of Non-Crystalline Semiconductors in Optoelectronics, 466.

INVESTIGATION OF INTEGRATED-OPTICS DEVICES ON THE BASIS OF THIN FILMS OF CHALCOGENIDE GLASSY SEMICONDUCTORS

A. POPESCU
Institute of Applied Physics, Center of Optoelectronics
Chisinau, Moldova

Thin-film waveguides on previously purified glass substrates with attenuation 2-3 dB/cm were obtained by the thermal evaporation method in vacuum 10^{-6} Torr. In multilayered structures As_xS_{1-x} planar waveguides with attenuation less than 0,5 dB/cm were obtained. This experimental fact indicates a considerable contribution of the surface scattering in a monolayered waveguide. Using the waveguide spectroscopy method we have found significant anisotropy of the refractive index ($\sim 10^{-2}$) which may be decreased by posterior annealing.

Many authors paid their attention to creation of channel waveguides in thin films (irradiation by electron beam, by argon laser), however in this case light scattering increases due to photostructural transformations. Together with other authors (M.Bertolotti et al.) we have obtained channel waveguides by the method of thermal laser annealing by CO_2 laser radiation. We investigated in detail kinetics of the waveguide light modulator in As_2S_3 films which for the first time was found by K.Tanaka. More rapid reverse switching in As_2Se_3 compounds with the addition of tin up to 3% was found. On the basis of the waveguide ray diffraction on the plane holographic grating the spectral demultiplexer of two wavelengths (630 nm and 1150 nm) was created.

New possibilities of usage reveal non-linear and photoelastic properties of these materials. In contrast to the works of V.Ciumas et al. we are interested in Kerr-like non-linear phenomena in the field of optical transparency. A cycle of works by the mass-spectroscopy method on revealing basic molecular forms and chemical bonds in As-S-Se compounds was carried out in order to understand physical nature of photostructural transformations.

A. Andriesh and M. Bertolotti (eds.),
Physics and Applications of Non-Crystalline Semiconductors in Optoelectronics, 467.

REDUCED TEMPERATURE GROWTH AND CHARACTERIZATION OF InP /SrF_2 / InP(100) HETEROSTRUCTURE

S.L. PYSHKIN[a], V.P. GREKOV[a], J.P. LORENZO[b], S.V. NOVIKOV[c] and K.S. PYSHKIN[c]

[a]*Institute of Applied Physics, Chisinau, 2028 Moldova;*
[b]*USAF Rome Laboratory, RL/EROC, MA, USA;*
[c]*A.F.Ioffe Physico-Technical Institute, St.Petersburg, Russia.*

Tu et al. [1] reported the growth of double heterostructures consisting of epitaxial InP/CaF_2/InP and InP/Ba_xSr_{1-x}/InP by MBE. Due to a big mismatch in lattice parameters for the first case (-7.2%), inhomogeinity of the fluoride composition along the substrate for the second case and, as a consequence, unsatisfactory quality of the dielectric and semiconductor films these results can not be evaluated as final and suitable for optoelectronics. Crystalline insulator films of SrF_2 were grown by MBE at ultra-high vacuum (10^{-8} Pa) and reduced temperature (≤ 350°C) conditions under *in situ* RHEED control onto substrates prepared with a week (formation of protective layer of phosphorus) pre-treatment from individually packed atomically clean InP(100) wafers produced by Japan Energy Corporation.

It has been shown that a small (1.18%) mismatch between lattice parameters of InP(100) and SrF_2 does not prevent from a good interface formation of SrF_2/ InP(100) structure and well-ordered single crystal insulator films grow from 50 nm of thickness. 500 nm of thickness crystalline SrF_2 films on InP(100) were used as substrates for the next reduced temperature epitaxial growth of InP by two different methods.

The first method was MBE growth from two Knudsen-type ovens filled with bulk InP crystals and having different temperatures that to choose an optimal In:P ratio in vapour flow. The second one was laser vacuum epitaxy (LVE, [2]) using a stoichiometric plasma, evaporated by a pulse laser from InP bulk crystal at P overpressure created by additional Knudsen-type oven with bulk InP. A few tenth of micron InP(100) crystalline overlayers have been grown under RHEED control. It has been demonstrated that perfect InP films can be obtained at the temperature of SrF_2 / InP(100) substrate equal to or even less than 400°C by the second method, while in the case of MBE growth from two Knudsen-type ovens reduced temperature epitaxy until yet gives unsatisfactory results. Growth mechanisms for InP/SrF_2 /InP(100) heterostructure as well as its electrical properties have been discussed.

1. C.W.Tu, S.R.Forrest, and W.D.Johnston, Jr., Appl. Phys. Lett. **43** (1983) 569
2. S.L.Pyshkin, S.Fedoseev, S.Lagomarsino. and C.Giannini, Appl. Surf. Sci. **56-58** (1992) 39

A. Andriesh and M. Bertolotti (eds.),
Physics and Applications of Non-Crystalline Semiconductors in Optoelectronics, 468.

FRESNEL ZONE PLATES ON CHALCOGENIDE GLASS WITHOUT CHROMATICAL ABERRATIONS

R. RADVAN[a], R. SAVASTRU[a], D. GHICA[a], V. BIVOL[b],
A. PRISACARI[b], and G. TRIDUH[b]
[a]*Institute of Optoelectronics SA Bucharest - Magurele, Romania*
[b]*Center of Optoelectronics Chisinau, Rep. Moldova*

This paper proposes two systems of Fresnel Zone Plates (FZP) which can focus two parallel monochromatic and coherent beams (with different wavelenghts), eliminating the wellknown chromatical aberrations of the classical construction. FZP is one of the most important means of forming images, especially, in the spectral region of IR, x-rays, gamma rays. When a plane wave upon a FZP whose transmission is T(r) [1], the transmitted light is focused to a point. Having in view a technological "simplification", and a mathematical determination of the radii which describe the pattern, it is necessary to find the radii values for T=0.5. The same FZP functions like a lens with focal distance f for two beams with different wavelengths (λ_1, λ_2) only if $\lambda_1 f$ - $\lambda_2 f$, for every k. Of course, this equality can not be satisfied and the element has an important chromatical aberration.

The principal disadvantage - high chromatical aberration-could be essentially reduced, even canceled, using a structure made from two patterns of FZP drawing in the same plane, on the semicircle surfaces. Each element contains a succession of strips calculated for the same focal distance f. The supports for both sections of FZP are the interferential filters. This choice was justified by the filter's special property concerning the intensity of the transmitted wave. Another compact configuration, with minimal assembly problems is described in this paper. Two FZP-s are realized on the different faces of the same substrate (i.e. a plane-parallel plate from common optical glass). A classical FZP contains opaque and transparent zones. The new construction proposes the pattern realization from optical filterable materials properly chosen, in concordance with the functional wavelenghts. The incident plane waves have λ_1 and λ_2 wavelengths.

The pattern R1, on the first surface, is made from a filter material for λ_1, and transparent for λ_2. It has the refractive index n_1, and the layer thickness is e_1. For the second pattern (R2), there is a similar notation. This pattern will filter the wave with λ_2 and will transmit the wave with λ_1. The system's design will generate the first pattern for f_1, and the second pattern f_2. This system will focus both radiations in the same point. The difference between the focal distances is necessary for the compensation of the path through optical glass (through plane-parallel plate).

This paper presents the main technologies used for different diffraction elements manufacturing, especially for micrometer dimensions or less. The advantages

A. Andriesh and M. Bertolotti (eds.),
Physics and Applications of Non-Crystalline Semiconductors in Optoelectronics, 469-470.

and the specific problems of the electron beam lithohraphy, the laser beam photolithography and the interferential recording are underlined. The paper includes information about a system of zone plates without chromatic aberrations and its technology. A complete analysis of the emergence laser beam generated some interesting conclusions.

The following applications are only for examples that solve critical problems from laser field. Generally, laser applications in biomedical field used wavelenghts out of visible range (λ = 1064 nm, 904 nm, 820 nm etc.). For this reason, almost all devices have two laser sources, one for therapy or surgery and other for aiming. In these situations, mechanical solutions are complex and request high accuracy for the all dimensions and surfaces. The paper shows a schematical arrangement of the system available for a laser biostimulator. Both beams are focused in the same point. A similar system as suggested for laser optics in an endoscopical equipment. The modern ophthalmology uses laser's advantages in many affections. For example, the YAG:Nd lasers are used for iridectomy or capsulotomy.

The FZP's have many applications in industry and research. The lithographic technologies use UV beams that can be marked by a visible spot.

1. Teruhiro Shiono, Kentaro Setsune, (1988) Elliptical Micro-Fresnel Lenses Fabricated by Electron-Beam Writing Technique, Electronics and Communications in Japan, Part. 2, Vol.71, No. 5.

BIPOLAR TRANSPORT AND BIPOLAR PHOTOCONDUCTIVITY IN AMORPHOUS FILMS CONTAINING ARSENIC CHALCOGENIDE VITREOUS SEMICONDUCTORS

Sh.Sh. SARSEMBINOV, O.Yu. PRIKHODKO, M.G. MALTEKBASOV, S.Ya. MAKSIMOVA, A.P. RYAGUZOV, M.H.WASFY[&]
Kazakh State National University Al-Farabi, Almaty, 480121 Kazakhstan,
[&]Suez Canal University, Egypt

It is well known, that in bulk samples of arsenic containing chalcogenide vitreous semiconductors and in amorphous thin films of these materials produced by different methods of thermal evaporation monopolar transport and monopolar photoconductivity take place. This is attributed to hole transport phenomenon, due to significant prevalence of hole drift mobility over that of electron mobility.

In contrast to this hole monopolar phenomenon, we have for the first time observed bipolar transport and bipolar photoconductivity, i.e. transport of both holes and electrons with almost the same average drift mobility, in As_2Se_3 thin films prepared by radio frequency sputtering (RF). Similar properties has been found in other arsenic containing chalcogenide vitreous semiconductor thin films with stoichiometrical compositions prepared by RF.

It is for the first time also to observe a monopolar electron transport phenomena and a monopolar electron photoconductivity in RF-films of As-Se system containing excess of arsenic.

The results are interpreted by considering the presence of differences in the electronic structure of the films obtained by different methods. These differences cause differences in the depth and concentration of localized states in the electron energy spectrum bounded to charged structure defects of both Se and As atoms.

A. Andriesh and M. Bertolotti (eds.),
Physics and Applications of Non-Crystalline Semiconductors in Optoelectronics, 471.

ELECTRON-INDUCED ANISOTROPY IN CHALCOGENIDE GLASSES: STATE OF THE PROBLEM, PHYSICAL FEATURES AND MICROSTRUCTURAL MECHANISM

O.I. SHPOTYUK[1,2], V.O. BALITSKA[1] and M.M. VAKIV[1]
[1]*Lviv Scientific Research Institute of Materials*
Stryjska St. 202, lvov, UA-290031, Ukraine
[2]*Physics Institute of Pedagogical University of Czestochowa*
Al.Armii Krajowej 13/15, Pl-42201, Czestochowa, Poland

Effect of electron-induced anisotropy (EIA) was observed in the samples of glassy As_2S_3 (samples in the form of cube with length of rib near 8 mm) irradiated by accelerated electrons beam with 2,8 MeV energy and more than 10^{15} cm^{-2} fluences in the perpendicular plane to the probe light. The polarization of the probe light was chosen in correspondence with electrons beam direction (perpendicular and parallel to the latter). All investigations were carried out one day after electron irradiation using "Specord-40" spectrophotometer in the range of 200-900 nm. The effects produced by inhomogeneity of scalar electron-induced darkening were excluded from consideration due to the procedure of 100 % line balance putting at the 180° turning of the sample around the electron beam direction.

The magnitude of EIA was described by $\Delta\tau=\tau_{\perp}-\tau_{11}$ parameter, i.e. transmission coefficients difference for the probe light with perpendicular and parallel orientations of polarization plate (relatively to the direction of electrons beam). We discover that the effect of EIA decays completely at the 300 K during 10-15 days, while the photoinduced anisotropy decays only partially at this temperature. It may be concluded that EIA mechanism in chalcogenide glasses is related to electron- induced formation of new oriented defects. Such defects are more probably undercoodinated atoms of glass matrix, created due to chemical bonds breakings induced by accelerated electrons. These defects with a lower coordination form new quasistable structural state similar to the one formed during condensation of thermally evaporated chalcogenide glasses on a substrate. Using the model of random structural network, we find that defects with a lower coordination are (As_2^+, S_1^-), (As_2^+, As_2^-) and (S_1^+, S_1^-). The charge of defect center is designed by superscript and the number of nearest atoms-by subscript. Annihilation of these defect pairs requires that some activation energy barrier be overcome. It follows that such defects are quite stable at low temperature and unstable at high temperature (more than room temperature). The processes of defects annihilation has a long-time component with character duration of ten days. Hence, the EIA effect decays not only at the thermal treatment of the investigated samples (T>300 K), but also at the prolonged samples preservation without additional heating (T=300 K) during 10-15 days. Described defects annihilate when the broken chemical bonds are restored due to thermal annealing.

A. Andriesh and M. Bertolotti (eds.),
Physics and Applications of Non-Crystalline Semiconductors in Optoelectronics, 472.

SUB-BAND-GAP ABSORPTION IN As2Se3 FILMS FROM PHOTOCAPACITANCE SPECTROSCOPY

S.D. SHUTOV and I.A. VASILIEV
Institute of Applied Physics, Center of Optoelectronics
5 Academiei Str., MD 2028 Kishinev, Moldova

The barrier photocapacitance spectra were measured in Al/a-As_2Se_3 thin (1.2 μm) film diodes in the photon energy range from 0.8 to 1.8 eV at the temperature 297...322 K. With these data the values of optical absorption coefficient α were determined from 10^{-3} to 10^{4} cm^{-1} using the initial rate of photocapacitance relaxation [1]. The measurements were performed by quasi-static C-V method under continuous optical excitation with sub-band-gap light. It was found that the spectral dependence of the absorption coefficient could be divided into three regions. In the region I within 1.5 to 1.8 eV the absorption coefficient is described by a typical for the Urbach edge exponential function with the slope of 20 eV^{-1}. The photon energy 1.76 eV corresponding to the value of $\alpha=10^4$ cm^{-1} coincides with the optical gap of amorphous As_2Se_3 at 300 K. As the temperature rised the strong parallel shift of the Urbach edge to lower energy occurred. Below the band edge in the region II between 1.1 and 1.3 eV the tail of weak absorption was observed. The origin of this tail is often associated with impurities or charged defects. The model calculation, which incorporates the density of conducting states and a Gaussian-form distribution of the gap states yields the value of peak energy 1.36 eV and a standard deviation 0.095 eV. Finally in the spectral region III from 0.8 to 1.1 eV a broad shoulder (up to the region II) with flat dependence on photon energy and extremely low (10^{-2} cm^{-1}) absorption coefficient was revealed. In this region photoinduced optical absorption of about 10^{-1} cm^{-1} stimulated by illumination with photon energy 1.8 eV during 30 min has been observed at 300 K. If one assumes that the upper limit for the optical cross section of an induced state is 10^{-17} cm^2 then the estimated lower limit of 10^{16} cm^{-3} is obtained for the density of states in the mid-gap from the observed induced absorption coefficient. This value is in good agreement with the estimation of photoinduced spin concentration in As_2Se_3 [2].

[1] N.M.Johnson, D.K.Biegelsen. Phys.Rev.,B31,4066 (1985)
[2] S.G.Bishop, V.Strom and P.C.Taylor. Phys.Rev.Lett., 34, 1346 (1975)

'rtolotti (eds.),
ıs of Non-Crystalline Semiconductors in Optoelectronics, 473.
ıic Publishers.

ROOM-TEMPERATURE PHOTOLUMINESCENCE OF AMORPHOUS HYDROGENATED SILICON CARBIDE DOPED WITH ERBIUM

E.I. TERUKOV, V.Kh. KUDOYAROVA, A.N. KUZNETSOV and W.FUHS*,
Ioffe Physico-Technical Institute, 194021, St.Peterburg, Polytechnicheskaya 26, Russia
**Hahn-Meitner Institut, Rudower Chaussee 5, D-12489, Berlin, Germany*

Recently the luminescent properties of erbium-doped crystalline silicon (c-Si:Er) have attracted much attention. The reason for this interest originates from the idea to fabricate LED's which are integrable into silicon electronic devices and emit at a wavelength of 1,537 μm where the absorption of silica glass optical fiber is the lowest. The photoluminescence (PL) of c-Si:Er is strongly quenched with increasing temperature.

It has been shown that the intensity of the Er emission strongly depends on the band gap energy of the host semiconductor, mainly for the room temperature emission.

In this report, the results obtained on erbium-doped amorphous hydrogenated silicon (a-Si:H:Er) are extended to erbium-doped amorphous hydrogenated silicon carbide (a-$Si_{1-x}C_x$x:H:Er) to increase the band gap energy limit.

It is shown that a-$Si_{1-x}C_x$:H:Er exhibit strong room-temperature PL at 1,54 μm, which is assigned to the internal 4f-shell transition in Er ions.

Films of a-$Si_{1-x}C_x$:H:Er have been prepared by cosputtering of graphite and Er targets applying the magnetron-assisted silane-decomposition (MASD) technique with mixtures of Ar and SiH_4 used as sputtering gas. The composition of films (x) and the presence of Er in the films have been monitored by Rutherford back scattering (RBS), (x) was varied in the range 0 - 0,29. The concentration of incorporated Er-ions was 6 x 10^{19} cm^{-3}. The content of hydrogen was estimated by IR spectroscopy at 9-12 at. %. The optical band gap was determined by CPM as an energy where the absorption coefficient is 10^3 cm^{-1} varied from 1,59 (x=0) to 1,82 eV (x=0,29).

A detailed comparison between the temperature dependences of PL in a-$Si_{1-x}C_x$:H:Er, a-Si:H:Er and Er-implanted c-Si:Er is presented. It was shown that the onset of temperature quenching of PL in the case of a-$Si_{1-x}C_x$:H:Er is observed at higher temperatures than for a-Si:H:Er and c-Si:Er and the quenching less pronounced. The weak temperature dependence of PL in a-$Si_{1-x}C_x$:H:Er is discussed in the terms of the model previously proposed by us for a-Si:H:Er. In this model the mechanism of electronic excitation of Er ions is based on defect-related Auger excitation.

This work was supported in part by the Arizona University.

A. Andriesh and M. Bertolotti (eds.),
Physics and Applications of Non-Crystalline Semiconductors in Optoelectronics, 474.

ANALYTIC INDEX

GPSR Compliance
The European Union's (EU) General Product Safety Regulation (GPSR) is a set of rules that requires consumer products to be safe and our obligations to ensure this.

If you have any concerns about our products, you can contact us on

ProductSafety@springernature.com

In case Publisher is established outside the EU, the EU authorized representative is:

Springer Nature Customer Service Center GmbH
Europaplatz 3
69115 Heidelberg, Germany

www.ingramcontent.com/pod-product-compliance
Ingram Content Group UK Ltd.
Pitfield, Milton Keynes, MK11 3LW, UK
UKHW021900190726
13853UKWH00003B/1357